IET RADAR, SONAR AND NAVIGATION SERIES 23

Tracking Filter Engineering

Other volumes in this series:

Volume 1	**Optimised radar processors** A. Farina (Editor)
Volume 3	**Weibull radar clutter** M. Sekine and Y. Mao
Volume 7	**Ultra-wideband radar measurements: analysis and processing** L. Yu. Astanin and A.A. Kostylev
Volume 8	**Aviation weather surveillance systems: advanced radar and surface sensors for flight safety and air traffic management** P.R. Mahapatra
Volume 10	**Radar techniques using array antennas** W. Wirth
Volume 11	**Air and spaceborne radar systems: an introduction** P. Lacomme (Editor)
Volume 13	**Introduction to RF stealth** D. Lynch
Volume 14	**Applications of space-time adaptive processing** R. Klemm (Editor)
Volume 15	**Ground penetrating radar, 2nd edition** D. Daniels
Volume 16	**Target detection by marine radar** J. Briggs
Volume 17	**Strapdown inertial navigation technology, 2nd edition** D. Titterton and J. Weston
Volume 18	**Introduction to radar target recognition** P. Tait
Volume 19	**Radar imaging and holography** A. Pasmurov and S. Zinovjev
Volume 20	**Sea clutter: scattering, the K distribution and radar performance** K. Ward, R. Tough and S. Watts
Volume 21	**Principles of space-time adaptive processing, 3rd edition** R. Klemm
Volume 22	**Waveform design and diversity for advanced radar systems** F. Gini, A. De Maio, L.K. Patton
Volume 101	**Introduction to airborne radar, 2nd edition** G.W. Stimson
Volume 102	**Low-angle radar land clutter** B. Billingsley

Tracking Filter Engineering

The Gauss–Newton and Polynomial Filters

Norman Morrison

The Institution of Engineering and Technology

Published by The Institution of Engineering and Technology, London, United Kingdom

The Institution of Engineering and Technology is registered as a Charity in England & Wales (no. 211014) and Scotland (no. SC038698).

First published 2013

The Institution of Engineering and Technology
Michael Faraday House
Six Hills Way, Stevenage
Herts, SG1 2AY, United Kingdom

www.theiet.org

British Library Cataloguing in Publication Data
A catalogue record for this product is available from the British Library

ISBN 978-1-84919-554-6 (hardback)
ISBN 978-1-84919-555-3 (PDF)

Typeset in India by MPS Limited
Printed in the UK by CPI Antony Rowe, Chippenham, Wiltshire

For Adriénne Juliet

without whose presence in my life
this book would never have been written.

הִנֵּה אֲנִי מֵבִיא בָכֶם רוּחַ וִחְיִיתֶם

וְנָתַתִּי עֲלֵיכֶם גִּידִים וְהַעֲלֵתִי עֲלֵיכֶם בָּשָׂר

וְקָרַמְתִּי עֲלֵיכֶם עוֹר וְנָתַתִּי בָכֶם רוּחַ

וִחְיִיתֶם

Behold, I will cause breath to enter into you, and ye shall live.

And I will lay sinews upon you, and I will bring up flesh upon you,

and cover you with skin, and put breath in you,

and ye shall live.

Ezekiel XXXVII

Contents

Preface

Why would anyone write a book about Gauss–Newton tracking filters? There are many reasons. Here are just a few.

◊ The Kalman filter has become an extremely complex topic with a bewildering number of mutations. By contrast, Gauss–Newton filters are easy to understand and easy to implement. What we are offering here is a simpler, more practical approach to tracking filter engineering that will appeal to both experienced practitioners and to newcomers.

◊ The Gauss–Newton filters work, and they work extremely well. We and our colleagues have established that – by using them in a variety of operational hardware as well as in simulations.

◊ Once launched, Kalman filters[1] have only one type of memory – expanding – something that is impossible to change.

By contrast, the Gauss–Newton filters can have fixed length, variable length or expanding length memories and any of these can be changed at will during filter operation.

The Gauss–Newton filter models can also be changed during filter operation – if need be from one cycle to the next.

This ability to change their memory lengths and filter models opens up powerful new ways by which to track manoeuvring targets, ways that are difficult to implement with the Kalman filters.

◊ The Gauss–Newton filters can be used when the observations are stage-wise correlated, something that cannot be done with the Kalman filters, and while it is true that observations in filter engineering are often stage-wise uncorrelated, that is not always the case.

[1] The words 'Kalman filter' mean different things to different people. When we use them here we are referring specifically to the mainstream version appearing in References 3, 6, 22, 23, 24, 25, 30, 31, 48, 50, 53, 54, 55, 56, 57, 173 and 174. The equations of the mainstream version appear in Appendix 11.1.

◊ Perhaps most important of all, the Gauss–Newton filters are not vulnerable to the instability problems that plague the extended Kalman filters – in which, unpredictably, the estimation errors become inconsistent with the filter's covariance matrix.

The Gauss–Newton filters therefore do not require what is known in Kalman parlance as 'tuning' – searching for and including a $\boldsymbol{Q}$ matrix that one hopes will prevent something that is both unpredictable and almost inexplicable, and in the process degrading the accuracy of the Kalman estimate.

This in turn means that in practice the Gauss–Newton filters are more accurate, and that they are easier to implement, require shorter project time, are extremely robust and, best of all, unconditionally stable.

And what can one say about execution time?

Using Gauss–Newton on a 1958-era machine, the extraction of a satellite's orbital parameters from a set of radar observations would have added about one hour of processing time *after* the observations had been obtained.

However, on today's readily available desk-top or lap-top machines, that hour has shrunk to a few hundred milliseconds, and as computer performance and other technologies such as FPGA continue to advance, it is safe to say that in the not too distant future it will have shrunk to tens of milliseconds or even less.

◄►

As envisaged here, tracking filters are computer programs that extract estimates of the states of dynamic processes in near real-time from sequences of discrete real observations.

Tracking filter engineering is the collection of techniques, knowledge and activities that we employ when we create and operate such filters.

The following is a partial list of fields in which tracking filters are used:

Air-traffic control, artificial intelligence, astronomy, atmospheric re-entry, ballistic missile defence, chemical engineering, civil engineering, control engineering, econometrics, electrical engineering, GPS and WAAS, industrial engineering, inertial navigation, mechanical engineering, missile engineering, neural networks, physics, pilotless-aircraft engineering, radar and tracking, robotics, satellite engineering, space navigation, statistics, telecommunications, telescope engineering and wind-power engineering.

This book will be of interest to three groups of people:

◊ Practitioners working in the aforementioned or similar fields.

◊ Graduate-level newcomers wishing to learn about Gauss–Newton and polynomial filters, and how they can be used in filter engineering.

◊ University lecturers who might wish to include material on Gauss–Newton and polynomial filters in graduate-level courses on tracking filter engineering.

◄►

Tracking filter engineering came into existence around 1958 at the start of the Satellite Era.[2]

Engineers and scientists had long been fitting curves to data by least squares, but this was different. Now there were huge tracking radars making observations on artificial satellites, from which estimates of orbital parameters – known as Keplerians – were to be extracted, and the filtering algorithms for doing so were far more sophisticated than simple least squares.

Astronomers had been estimating Keplerians of objects in orbit since the time of Gauss, using what was known as the Gauss–Newton algorithm. However, such calculations were done by hand and it took many months to obtain results, whereas with artificial satellites the need was for the observations to be processed by computers and for the estimates to be available in near real-time.

Running Gauss–Newton on a 1958-era machine would have taken roughly an hour to extract a satellite's Keplerians from a set of radar observations *after* the latter had been obtained. Such a delay was clearly too far removed from real-time, and so Gauss–Newton could not be used.

In its place two new filters were devised – both related to Gauss–Newton but computationally different. The first of these was published by Swerling in 1958[3] and the second by Kalman[4] and Bucy[5] in 1960/61, both of which could extract Keplerians in near real-time on the existing machines.[6]

Starting from the extraction of Keplerians, use of the Kalman filter spread rapidly to the many other fields listed above where today it occupies the dominant place in

[2] Sputnik-1 was placed in orbit by the Soviet Union on October 4th, 1957. See Reference 40.

[3] Peter Swerling (American mathematician and radar theoretician), 1929 – 2000. See References 7, 8, 9, 91.

[4] Rudolf Emil Kalman (Hungarian-born American scientist), b. 1930. See References 1, 2, 92.

[5] Richard S. Bucy (American mathematician), b. 1935. See Reference 93.

[6] See Chapter 11 for what we mean by 'Swerling filter' and 'Kalman filter'.

filter engineering. The Kalman filter has also come to occupy the dominant place in filter-engineering curricula throughout the academic world.

◀▶

But while all of this was taking place, something else was also happening.

Since 1958, computer technology has made advances that are difficult to comprehend, and the end is still nowhere in sight. Consider for example the following item that appeared on the Intel corporate website:

> 'Intel remains at the forefront of Moore's Law. Our 22 nanometer technology-based Intel microprocessors will enable never-before-seen levels of performance, capability, and energy-efficiency in a range of computing devices.'

And so, while it may have been appropriate to reject Gauss–Newton in 1958, it is no longer appropriate to do so today. In this book we accordingly do something different.

Many books have been written about the Kalman filter – the author is aware of at least fifteen. However, to our knowledge nobody has yet written a book about the Gauss–Newton filters and their remarkable fit to certain important areas of filter engineering.

This then is our attempt to do so, and to present those filters in a readable and self-contained way.

◀▶

The Gauss–Newton filters possess attributes that make them particularly well-suited for use in tracking filter engineering, among them the following:

◊ Ideal for tracking both manoeuvring and non-manoeuvring targets.

◊ Can be used when the observations are stage-wise correlated.

◊ Are not vulnerable to the instability phenomena that plague both the Kalman and Swerling extended filters, and hence require no tuning.

◊ Require no initialization in the all-linear case and very little in the three nonlinear cases.

◊ Possess total flexibility with regard to

  ◊ Memories that can be configured as fixed length, variable length or expanding length – and if need be, reconfigured cycle by cycle.

  ◊ Filter models that can also be reconfigured, if need be, from cycle by cycle.

◊ Offer immediate access to the residuals which can then be used to run goodness-of-fit tests.[7]

◊ Unconditionally stable, meaning that they are always error/covariance-matrix consistent.[8]

◊ Cramér–Rao consistent, meaning that their accuracy is the best that it can possibly be.

◊ Can employ nonlinear models of almost any complexity.

◊ Nonrecursive, and hence devoid of internal dynamics.

◊ Easy to implement.

◊ Extremely robust.

The techniques proposed in this book have all been tested in simulations, and many of them have been used in operational hardware built by industrial organizations with which the author has been associated.

The engineers who created the filters for those applications have made it known how easy it was to do that, and how robust were the resulting filters.

◀▶

Regarding Gauss–Newton execution time on today's machines:

We have carried out simulator-based timing studies in which the Gauss–Newton filters were used to track a mix of manoeuvring targets such as high-performance military jets, cruise missiles and helicopters. We found that readily available modern computers – either desktop or laptop – can readily handle the load imposed by these filters while tracking many such targets, including all of the overhead calculations, namely track formation, track deletion and plot-to-track allocation using the auction algorithm.

Indeed, depending on the level of radar clutter, our studies have suggested that concurrent tracking of a mix of possibly as many as fifty manoeuvring targets of this kind can be accomplished on such computers using these filters, and as processing power continues to increase, the conservative figure of fifty will continue to rise.

[7] Goodness-of-fit tests are used by the Master Control Algorithms (MCAs) to control a Gauss filter's memory length and filter model when tracking manoeuvring targets, in such a way that its performance is always optimal. The MCAs are discussed in Chapter 10.

[8] Error/covariance-matrix (ECM) consistency is discussed in Chapter 1. When a Kalman or Swerling filter becomes unstable then it is ECM consistency that is absent, a condition which is fatal.

And we would add the following: The execution time of the overhead calculations in the above studies far outweighed that of the Gauss–Newton filters, and the overhead calculations would also have to be performed if Kalman filters were used. Thus under similar circumstances, the number of targets that could be tracked using Kalman filters would have been almost the same.

◄►

Our discussion of filter engineering centres on two concepts mentioned in the aforementioned list:

◊ Error/covariance-matrix (ECM) consistency, meaning that a filter's covariance matrix is an accurate representation of the actual covariance matrix of the estimation errors, something that we examine in depth and whose absence in a filter is fatal.[9]

◊ Cramér–Rao (CR) consistency, meaning that the filter produces estimates of the highest possible accuracy.

In this regard we discuss six tests:

◊ Three to determine if a filter is ECM consistent.

◊ Three to determine if it is CR consistent.

A number of other concepts are also discussed, among them the following:

◊ Tracking of manoeuvring targets using what we call the Master Control Algorithms (MCAs) that dynamically control a Gauss–Newton filter's memory length and filter model in such a way that its performance is always optimal.

◊ A goodness-of-fit test which determines if a Gauss–Newton filter's estimate is an unbiased fit to the observations.

◊ A real-time test which monitors the observation errors.

◊ Prefiltering, in which the polynomial filters form a high-data-rate front end for the main tracking filters such as Gauss–Newton, Kalman or Swerling, which then extract the desired estimates at their low data rates.

[9] ECM consistency of a filter is by no means guaranteed, and as is well known and mentioned above, both the Kalman and Swerling filters can spontaneously lose their ECM consistency – known in Kalman parlance as 'becoming unstable' – prevention of which is accomplished by the use of what is known as a $\boldsymbol{Q}$ matrix for 'tuning' the filter, but which also degrades their accuracy. We examine all of this in Chapter 11.

We include fourteen computer programs that were created as research tools over a period of many years, and which serve to demonstrate almost every aspect of our discussion:

- ◊ In many of them we have implemented one of the Gauss–Newton filters.
- ◊ In almost all of them we have implemented one or more of the eight tests mentioned earlier.
- ◊ In two of them we have implemented prefiltering using the polynomial filters, and in which the main filters are Gauss–Newton, Kalman and Swerling – all three of which process *precisely* the same observations.

  These two programs

  - ◊ Demonstrate how prefiltering works.
  - ◊ Enable you to witness instability in both the Kalman and Swerling filters[10] and the fact that Gauss–Newton, operating on exactly the same data, is always stable.
  - ◊ Make it possible to time the executions of the three main filters.
  - ◊ Enable you to compare their performance in a way that is 'apples-to-apples'.

◄►

The polynomial filters discussed in this book were first devised when we worked at Bell Labs Ballistic Missile Defense in Whippany New Jersey from 1964 to 1968. They have been written up twice before – once in our first textbook (Reference 6) and again in Reference 24.

However, over the succeeding years we have learned a great deal more about them and continue to do so[11] and in this book we provide a complete discussion which includes much of that new material.

◄►

We have tried to keep the book as generic as possible, i.e. applicable to multiple fields. However, in some places we were forced to place it in a specific context, and the field that we selected was that of radar and tracking.

[10] The Swerling filter becomes unstable whenever the Kalman filter does, and in all regards their performances are essentially identical.
[11] See References 112, 118 and 202.

Some of our discussion is thus slanted towards target motion sensing using two types of radar – those with constantly rotating antennas (track-while-scan or TWS radars) and those with steerable antennas (tracking radars).

In this regard we beg forgiveness from those readers who use filters in other fields. However, we assume very little knowledge of radar and tracking and instead place our primary emphasis on filtering techniques, and so we hope that the book can be read and used by practitioners and students who are interested in using tracking filters in almost any field.

◂▸

The material in the book has been taught on six occasions on three continents to a mix of practicing filter engineers, graduate students and university lecturers.

It is well suited for a complete single-semester graduate course of 24 lectures or else for partial incorporation into other graduate courses, and contains more than a sufficiency of examinable material that such courses require. The fourteen computer programs and the end-of-chapter problems and projects will enable students to envisage and apply the concepts that are discussed in the chapters.

We have made every effort to keep the book readable and friendly. We can only hope that we have succeeded.

Norman Morrison

Cape Town, October 2012

Viewing the track-while-scan video clips

In order to see, first hand, how well suited these filters are to tracking manoeuvring targets, we suggest at this point that you turn on your computer and retrieve the down-loadable material. (See downloading instructions at the start of Chapter 1.)

Once that is done, please read the following document so that you understand fully what you will be viewing:

Video_Clips\TWS\Documents\Readme.pdf

Then please take a few minutes to run the track-while-scan video clips that are contained in the folder

Video_Clips\TWS\Flights.

This will enable you to see how effectively the Gauss–Newton filters under MCA control are able to perform when tracking manoeuvring targets.

Words containing the letter *Z*

We lived in the USA for thirty years and were under the impression that the use of the letter *z* in words like *maximize* and *minimize* was a distinctly American thing, and that on our side of the Atlantic one would spell them as *maximise* and *minimise*.

Imagine our surprise when we consulted the Shorter Oxford Dictionary and found that it also uses the letter *z* in many such cases. Throughout the book we have attempted to follow the usage in that dictionary.

Acknowledgements

Many people made this book possible.

Dr. Richard Lord, formerly of the University of Cape Town and now of the Karroo Array Telescope (KAT), stepped forward to bring his remarkable creativity and outstanding programming skills to bear on our tracking problems. I will always be grateful to him.

Johan Kannemeyer, an extremely inventive software engineer at Reutech Radar Systems, programmed and tested many of the algorithms. He also made many suggestions, all of which were adopted. It was Johan who came up with the ratio matrix that is discussed in Chapter 7.

Professors Linda Haines and Tim Dunne of the University of Cape Town's Department of Statistical Sciences kept their doors open for me and made available whatever time it took to discuss my problems. They also read parts of the manuscript, found errors and made suggestions, all of which were adopted. Without their help this book would not have been written.

Professors Daya Reddy and Ronnie Becker of the University of Cape Town's Department of Mathematics and Applied Mathematics read parts of the manuscript, found errors and offered many excellent suggestions.

Professor Mike Inggs of the University of Cape Town's Department of Electrical Engineering posed a problem to me that ultimately led to the writing of this book. He also read parts of the manuscript and made suggestions. It was Mike who connected me with the IET and with the University of Cape Town's Open Access Database.

My colleagues at the University of Cape Town, Professor Christopher Gilmour, Kenneth Hughes, Alan Reynhoud, Neill Robertson and Anneliese Schauerte checked the proofs of many of the propositions, and Peter Bruyns made time for discussion and clarification.

Professor Pieter Willem van der Walt, formerly Dean of Engineering at the University of Stellenbosch and now Technology Executive of Reutech Radar Systems, read an early version of the manuscript and set me on the right path.

Pieter-Jan Wolfaardt, Technology Executive of Reutech Radar Systems, constantly challenged me with his brilliant mind and his fresh ideas. It was Pieter-Jan who brought me into their radar project, and I will always be beholden to him for his support and encouragement.

Ivan Gibbons, Chief Engineer of Denel Aerospace Systems, and Pieter Reyneke, a Denel software engineer, quickly caught on to the merits of Gauss–Newton. Ivan encouraged me to keep on going, and Pieter gave of his time and programming skills to derive results in Symbolic Matlab.

Drs. Yossi Etzion and Michael Maidanik of IAI also quickly caught on to the merits of Gauss–Newton. Yossi made a number of suggestions, all of which were adopted, and Michael's enthusiastic understanding of the advantages of Gauss–Newton served as a constant reason for me to keep on going, against the odds.

Jean-Paul de Conçeaçao, a graduate student at the University of Cape Town, asked a question that caused me to rewrite the entire book. In his Master's thesis, Jean-Paul successfully implemented a Gauss–Newton filter in an FPGA.

Roaldje Nadjiasngar, also a graduate student at the University of Cape Town, read the manuscript, programmed a number of the algorithms, found errors that we fixed and then verified that the algorithms were correct. Roaldje has since made use of Gauss–Newton in his doctoral thesis with great success.

Professor Stephen Hodgart of the University of Surrey revived my interest in tracking filters during our walks on Table Mountain, after I had been away from them for many years. Thank you, Stephen.

Professor Dov Hazony, my doctoral thesis adviser at Case Western Reserve University in Cleveland, Ohio, taught me, almost fifty years ago, what the word 'complete' means. I can only hope that I have succeeded.

Professor Dick (Ricsike) Varga, one of my lecturers at Case Western Reserve University, gave me a love for linear algebra that I have never lost. He and I are still discussing it, even today.

The late Professor Richard Duffin of the University of Pittsburgh was an enduring inspiration to so many us. While writing this book I thought often of him, and also of the late Professor Brian Hahn of the University of Cape Town who died under tragic circumstances, much too long before his time. I learnt a great deal from both Dick and Brian. They were good teachers and good friends, and I miss them both.

I was privileged, almost half a century ago, to work with many brilliant people at Bell Labs in New Jersey who first introduced me to tracking filters, among them Drs. Marvin Epstein, Paul Buxbaum, Alphonse Claus and Jack Riordan. While writing this book, memories of those friends and what they taught me were constantly in my mind.

My very special thanks to Professor Hugh Griffiths of University College London, IET Series Editor, and to Nicki Dennis and Paul Deards, both IET Commissioning Editors. Hugh and Nicki gave me the encouragement I needed at a most critical time, and Paul helped with the shortening and reorganization of the typescript. My thanks also to Helen Langley, IET Production Controller, for her endless patience, and to the many others at the IET who participated in the publication of this book.

◆

A number of people found errors and omissions in the manuscript of this book and I have made every effort to correct them and to ensure that it is error-free. However, I assume full responsibility for any and all errors, omissions and incorrect statements that might still remain.

The Greek alphabet

A	α	*alpha*	I	ι	*iota*	P	ρ	*rho*
B	β	*beta*	K	κ	*kappa*	Σ	σ	*sigma*
Γ	γ	*gamma*	Λ	λ	*lambda*	T	τ	*tau*
Δ	δ	*delta*	M	μ	*mu*	Y	υ	*upsilon*
E	ε	*epsilon*	N	ν	*nu*	Φ	φ	*phi*
Z	ζ	*zeta*	Ξ	ξ	*xi*	X	χ	*chi*
H	η	*eta*	O	o	*omicron*	Ψ	ψ	*psi*
Θ	θ	*theta*	Π	π	*pi*	Ω	ω	*omega*

Why this book?

For many years the Kalman filter has been the accepted approach to tracking filter engineering.

The purpose of this book is to present an alternate approach, one based on the Gauss–Newton filters which we believe offer certain important advantages. We also provide a full discussion of the polynomial filters.

- ◊ In Chapters 1 to 7 we cover the necessary background.
- ◊ In Chapters 8 and 9 we discuss the details of the Gauss–Newton filters.
- ◊ Chapter 10 is devoted to the Master Control Algorithms (MCAs) that are used to control the Gauss–Newton filters when manoeuvring targets are being tracked.
- ◊ In Chapter 11 we provide a brief discussion on the Kalman and Swerling filters.
- ◊ Chapters 12 and 13 are devoted to the polynomial filters.

Welcome, then, to tracking filter engineering based on the Gauss–Newton and polynomial filters.

Organisation

As shown below, the material comes in two parts – *the printed part* and the *supplementary material.*

Printed part	***Supplementary material***
Preface	Computer programs
Contents	Miscellaneous
Chapters 1 to 13 and their appendices	Problems and solutions
References	Projects
Index	Video clips

The printed part is the book that you are now reading. It is completely self-contained and is published by the Institution of Engineering and Technology (IET).

The supplementary material is for those readers who wish to explore the subject in greater depth.

Please note that the IET is not responsible for any of the supplementary material, and that the author has the sole responsibility for its creation and maintenance.

The supplementary material is housed in the University of Cape Town's *Open Access Database* and can be retrieved from the following (case-sensitive) URL:

http://bit.ly/MorrisonSSP

The material can also be retrieved from the following (also case-sensitive) URL:

http://goo.gl/DD82m

Kindly send the author an email at the following address

norman.morrison@uct.ac.za

so that he can notify you when a new release of the supplementary material becomes available. Please include the words *Tracking Filter Engineering* in the title of your email.

◂▸

The supplementary material consists of the following:

◊ ***Computer programs:*** Fourteen computer programs, all of which either verify or demonstrate something important that is discussed in the book. Included are both the executables and the source code listings. Documentation for running each of the programs has been included in the form of a short *Quickstart* and also, in many cases, a more detailed *Readme*.

◊ ***Miscellaneous:*** Contains a number of files, executables and documents that are referred to as needed.

◊ ***Problems and solutions:*** Contains sets of problems for each of the 13 chapters. These enable you to test your understanding of the concepts discussed there, and some of the problems also include further theoretical material. Complete solutions to the problems have been included.

◊ ***Projects:*** Contains projects for most of the 13 chapters. Some of these involve writing short computer programs and others involve reading the material that has been provided. All of them are intended to deepen your understanding of the material contained in the printed chapters.

◊ ***Video clips:*** Contains two sets of video clips, one set with, and the other set without MCA control. (The latter is discussed in Chapter 10.) The set with MCA control shows the output of a Gauss–Newton filter that is being used to track a variety of manoeuvring targets such as fixed-wing aircraft, helicopters and cruise missiles. In the preface we suggested that you view the track-while-scan video clips (these are the ones with MCA control) before doing any further reading. If you missed that, then perhaps now would be a good time to do so.

  However, first please read the document ***Video_Clips\TWS\Documents\Readme.pdf*** so that you understand fully what you will be viewing. Then run the video clips in the folder ***Video_Clips\TWS\Flights***.

Part 1

Background

Chapter 1

Readme_First

In this first chapter we do the following:

◊ Define the scope of the book – what it covers and what it doesn't.

◊ State the book's objectives and what we believe it can do for you.

◊ Define much of the notation and symbols that we use.

◊ Explain the meanings of three concepts that will be with us throughout the book:

  ◊ Error/covariance-matrix consistency

  ◊ Cramér–Rao consistency

  ◊ Memory, as it relates to filter engineering

There are times when we are forced to defer further consideration of an item until a later stage, when more theoretical background has been covered. There are other times when we cite a result that is first derived in a later chapter.

We do our best to keep such out-of-sequence discussion to a minimum, but that is not always possible. Hopefully, on a second reading it will constitute less of a problem than on the first.

1.1 The reader's educational background

Every effort has been made to keep this book completely self-contained. However, we do assume reasonable familiarity with the material taught in the following five undergraduate courses, particularly the first:

Linear Algebra,   Ordinary Differential Equations,

Elementary Probability and Statistics,   Vector Calculus,   Numerical Methods

A considerable amount of the discussion involves concepts from Linear Algebra. If you feel that you need to brush up in this area then we suggest that you consult books similar to References 27 or 28, or at a somewhat more advanced level, Reference 16.

1.2 Motivation for writing this book

To understand what motivated us to write this book it is necessary that we recall a few historical facts.

Data-smoothing's big bang took place in 1801, when Carl Friedrich Gauss combined his newly invented Gaussian probability density function with another of his inventions, the method of least squares. This took place during the year that he spent determining the orbit of Ceres – the first of the minor planets to be discovered – and it was perhaps the first time that the science of probability had been fully applied to the combination of observations.[1]

Gauss invented the method of least squares in the final decade of the 1700s, and he used it regularly from 1801 onwards in astronomical calculations, although it was Legendre who first publicized it in an appendix to an astronomical memoir in 1805. However, Legendre gave no justification for the method, whereas Gauss gave two, one in 1809 and another in 1820.[2] (We discuss both in Chapter 9.)

When working on the orbit of Ceres, Gauss also made use of the differential calculus invented by Isaac Newton, thereby creating the *Gauss–Newton algorithm*, but the version of the algorithm that he created was not complete.

In 1935, A.C. Aitken drew it all together, and it was through Aitken's contributions that the Gauss–Newton algorithm attained its full present-day form.[3]

In 1958, Peter Swerling, whose name is well-known in the radar field, started from the Gauss–Newton algorithm and arrived at the Swerling filter (Chapter 11).[4] That was shortly after Sputnik-1 had been placed in orbit by the Soviet Union, and American engineers and scientists were hurriedly preparing to put their own first satellite into orbit.

Computers in those days had clock rates of somewhere around 1 MHz, and random access memory – in the form of hand-threaded metal-oxide core planes – was extremely slow and inordinately expensive (approximately one US dollar for one byte of RAM). Using Gauss–Newton to extract a set of orbital parameters on such

[1] See Reference 5.
[2] See References 4 and 17.
[3] See Reference 10.
[4] See Reference 7 to 9.

machines required something of the order of an hour of computation time *after* the observations had been obtained, thereby ruling out its use in satellite engineering.

The newly invented Swerling filter, on the other hand, gave estimates of the same orbital parameters almost from the beginning of the tracking interval, initially with accuracy that was poor, but which improved steadily until it equalled that of Gauss–Newton by the time the interval had ended – and the further one-hour delay was then not added.

It was about two years after Swerling's ground-breaking contribution that R.S. Kalman and R.E. Bucy published their algorithm – now known as the Kalman filter – whose structure was not unlike that of Swerling's.[5]

Kalman and Bucy started from a different place than Swerling, and used a different approach.[6] However, their algorithm can be derived from Swerling's, which in turn means also from Gauss–Newton, and so the three are closely related. We discuss all of this in Chapter 11.

After its initial use in satellite engineering, the succeeding years saw the spread of the Kalman filter to many other branches of engineering, and tracking filter engineering centred on the Kalman filter also became a field of study and research in the academic world.

◂▸

But while all of this was taking place, something else was happening.

Since 1958 computers have gone through the most spectacular of changes so that their throughput today dwarfs anything that was available back then, and we are not talking here about supercomputers but about today's readily available desktop and laptop machines. Moreover, the end is nowhere yet in sight, and we can expect the throughput of the next wave of computers similar to dwarf that of today's machines.

The Swerling and Kalman filters were unquestionably what were needed as replacements for Gauss–Newton in 1958 when it could not be used in satellite tracking. However, times have changed since then:

◊ When running Gauss–Newton filters on today's machines, the one hour that it took in 1958 has shrunk to seconds or to hundreds of milliseconds, and in the not-too-distant future it will inevitably shrink to tens of milliseconds or even less.

[5] See References 1 and 2. For an easier to read discussion see Chapter 15 of Reference 6.

[6] See Reference 6, p. 610.

Add to that the following:

◊ Gauss–Newton filters possess a number of properties that make them particularly well-suited to tracking filter engineering.

◊ Gauss–Newton filters are not vulnerable to the instability phenomenon that plagues both the Kalman and Swerling filters.

It was a combination of these three factors that motivated us to write this book.

Regarding the polynomial filters, when you finish reading Chapters 12 and 13 you will almost certainly agree that no book on filter engineering can be complete without them.

Note: See Projects 1.1 to 1.3 that relate to the increase in computer speed since 1958.

1.3 Filter names

◊ The name *Gauss–Aitken filters* will be used for the all-linear implementations, and in Chapter 8 we will see that there are two of them.

◊ The name *Gauss–Newton filters* will be used for the implementations that are employed when nonlinearities are present, and in Chapter 9 we will see that there are six of them.

◊ We will refer to the Gauss–Aitken and Gauss–Newton filters collectively as the *Gauss filters*.

◊ The name *polynomial filters* will be used for the sets of compact algorithms that are discussed in Chapters 12 and 13.

1.4 Scope of the book

The techniques discussed in this book are perhaps not applicable to all situations in which filter engineering might be called for, although they do cover most of the cases that occur in practice. We now consider where they do, and do not, apply.

In what lies ahead we will speak about two topics – *dynamic models* and *observation equations*. Setting aside observation equations until Chapter 3 (which is devoted solely to that subject) we consider dynamic models.

As they relate to filter engineering, dynamic models occur in two places:

1. The dynamic models that we believe adequately describe the physical processes that we are observing. These are mathematical constructs in the form of ordinary differential equations (ODEs, or more simply DEs) that have been formulated by physicists or applied mathematicians, and are external to any smoothing filters that we may create. In any given situation such models are seldom unique, and in the end their selection comes down to a matter of choice.

2. The dynamic models on which tracking filters are based. These are mathematical constructs – also in the form of DEs – which are internal to the filters that we create.

We will refer to the first kind as *external models*.

In tracking filter engineering, models of the second kind – the internal kind – are often called process models or plant models, but this has led to confusion. Throughout this book we will always refer to them as *filter* models.

Although it is necessary that the filter model should emulate the external model, the two are totally distinct entities, and they need not be the same.

◊ The external model is usually implemented when we are running a simulation, in order to generate observations that will be processed by the filter.

◊ The filter model, on the other hand, forms an essential part of the filter and is implemented within it.

Dynamic models will be explored further in Chapter 2.

◀▶

We differentiate between two types of physical processes that involve filter engineering.

◊ The first is distinguished by the property that either a deterministic or a stochastic forcing function, or both, is an essential part of the DE of the external model.

◊ The second is distinguished by the property that there is no such forcing function present in the DE of the external model – neither deterministic nor stochastic.

The following is an example of the second type of external model in which there is no forcing function.

Example 1.1: Given the following:

◊ Newton's inverse square law of universal gravitation.

◊ His second law, that force equals rate of change of momentum.

◊ That the earth can be represented by a point mass.

◊ That the earth's gravitational attraction is the only force that must be considered.

Then the DE that models the motion of a small object in orbit about the earth is as follows:

$$D^2\boldsymbol{r} = -(\mu/r^3)\boldsymbol{r} \tag{1.5.1}$$

The quantities in (1.5.1) have the following meanings:

◊ D^2 signifies the second derivative with respect to time, namely d^2/dt^2.

◊ $\mu = GM$, where G is the universal gravitational constant and M is the mass of the earth.

◊ $\boldsymbol{r}$ is the vector from the centre of the earth to the object.

◊ r is the magnitude of $\boldsymbol{r}$.

We observe that there is no forcing function present in the DE – neither deterministic nor stochastic – and that had a forcing function been present, then t would have appeared explicitly, as in, for example

$$D^2\boldsymbol{r} = -(\mu/r^3)\boldsymbol{r} + f(t) \tag{1.5.2}$$

or even possibly as

$$D^2\boldsymbol{r} = -\left(\mu(t)/r^3\right)\boldsymbol{r} \tag{1.5.3}$$

■■

In presenting the concepts discussed in this book we will always assume

◊ That the external model is a known deterministic differential equation.

◊ That there is no forcing function present in the external model – neither deterministic nor stochastic.

Extensive research has been, and continues to be conducted on filter engineering related to systems modelled by DEs that are driven by either stochastic or deterministic forcing functions. However, we will regard that subject as being outside the scope of our discussion, and so the only filter models that we consider will be deterministic DEs without forcing functions of any sort.

◀▶

Given that we will only be considering models without forcing functions, the reader may well wonder how we propose to deal with targets that can manoeuvre by changing their heading and/or their speed, and so the presence of a forcing function is implied.

Manoeuvring targets will be fully examined in Chapter 10, where we discuss what we call the Master Control Algorithms (MCA-1 and MCA-2).

1.5 The simulator

There are two environments in which tracking filters operate – in the field and in a simulator.

1.5.1 In the field

The *field* is our general term for the exterior location in which a tracking filter will be deployed. We think of two stages of that deployment:

◊ *Initial testing in the field*: During this stage it might be possible to place instrumentation on the objects or targets that are being observed – such objects being classified as *cooperative*. Valuable data can be obtained under these circumstances that can be used for further debugging of the filter.

◊ *Final operation in the field*: During this stage the objects under observation may or may not be cooperative, and it is here that the filter will be expected to function fully in the manner for which it was intended.

1.5.2 In a simulator

In filter engineering there can be no substitute for extensive testing, and the best way to do that is to create and run a computer program called a *simulator*.

Operation of a filter in a simulator should always precede operation in the field. In Figure 1.1 we show the basic elements of such a program, whose purpose is to assist with the design, testing and debugging of the filter.

◊ *Target generator* creates a data stream that represents a target with its respective characteristics. This is where the external model is usually implemented.

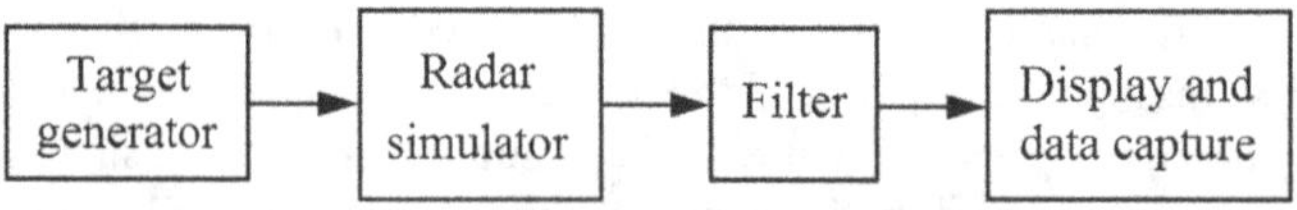

Figure 1.1 Basic elements of a simulator

We pause to comment that in modern tracking systems there are often multiple targets involved. However, the multi-target tracking problem[7] is a complex one that is beyond the scope of this book, and so we will limit ourselves throughout to the tracking of a *single* target.

The outputs of this first block constitute what are called the *exact values* or the *true values* against which the outputs of the filter will be compared.

◊ *Radar simulator* is a mathematical model of the essential aspects of the observation instrument with which the filter is intended to work. The radar simulator accepts the target data-stream as its input and modifies it to appear as though it were being observed by the observation instrument that will be used in the field. The radar simulator includes a random-number generator that creates observation errors that correspond as closely as possible to those of the actual observation instrument.

◊ *Filter* is the computer implementation of the filter that is being built and tested. This is where the filter model referred to earlier is implemented.

◊ *Display and data capture* displays results in real time on a computer screen, and also captures data from one or many runs for later analysis.

Simulators operate in two basic modes.

1.5.2.1 Single shot

In this mode a single stream of target data is created that simulates an object/target. The radar simulator operates on this data and its output is then processed by the filter. The filter's output is displayed on a computer screen, and possibly also saved for further analysis. The simulator program is run only once, hence the name *single shot*.

It can often be extremely convenient if programming is done in such a way that a given target data-stream can be repeated and emerge from the radar simulator with precisely the same errors, and then be processed again by the filter in single-shot mode. This typically involves the ability to store and re-use what is called the *noise*

[7] See for example References 30, 31, 50 and 100.

seed of the random-number generator that is used to create the radar simulator's errors.

In this way various options can be implemented in the filter, and their effectiveness compared against each other.

The following eight computer programs in the supplementary material are all single-shot simulators:

02_Bias_Errors ***03_Gauss_Aitken*** ***05_Pol_Filter_EMP***

06_Pol_Filter_ FMP ***08_Radar_Cartesian***

11_Gauss_Newton_A ***13_Gauss_Newton_C*** ***14_Orbit_Track***

Each of them contains a subroutine called *Makedata* that combines the functions of the target generator and the radar simulator appearing in Figure 1.1. The programs also contain one or more tracking filters, as well as both graphic and numeric displays.

None of them captures data for later analysis, and none of them has noise-seed re-use capability. In other simulators that we have constructed both data-capture and noise-seed re-use have been included.

1.5.2.2 Monte-Carlo

In this mode the simulator functions as follows:

◊ The target generator issues repeated runs of precisely the same data stream.

◊ These are passed through the radar simulator, where observation errors are added for each run. The observation errors differ randomly from run to run, but are drawn from the same PDFs.

◊ The filter processes each run, and results are accumulated from which values such as averages, variances and covariance matrices are derived and saved for further analysis.

The central assumption on which Monte-Carlo is based is that all three subsystems of the simulator – the target generator, the radar and the filter – perform *identically* for each run and only the observation errors being added in the radar simulator are allowed to vary randomly from run to run.

The following computer programs are simulators with Monte-Carlo capability:

04_Monte_Carlo ***09_Random_Vectors*** ***12_Gauss_Newton_B***

1.6 Notation

We make a start at defining the notation, some of which is shown in Figure 1.2.

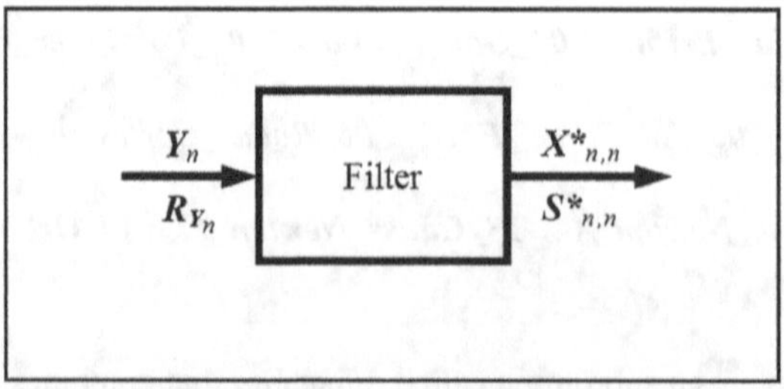

Figure 1.2 Notation

The reader who is familiar with Kalman filtering will observe that we are using double-subscript notation that differs from that used in some books on the Kalman filter. The choice was made out of necessity, and without it we would not have been able to write this book.

- ◊ All mathematical items have been set in italics.
- ◊ Vectors and matrices are always shown in bold type, and scalars are shown in normal (not-bold) type.
- ◊ The subscript n signifies most recent or now, $n - 1$ signifies previous and $n + 1$ signifies *next.*
- ◊ t is always time, and t_n signifies the most recent time instant, or *now*. When used as subscripts, n and t_n mean the same thing, as in $\boldsymbol{X}_n$ and $\boldsymbol{X}_{t_n}$. The first is a bit more compact, but on occasions we must use the second.
- ◊ $\boldsymbol{Y}$ is a vector of observations that is the input to a filter. $\boldsymbol{Y}_n$ is the most recent observation vector or the one that was made *now*. $\boldsymbol{Y}$ and y will always be used as the filter-input symbols.
- ◊ $\boldsymbol{R}_{\boldsymbol{Y}}$ is the covariance matrix of $\boldsymbol{Y}$. The symbol $\boldsymbol{R}$ will always designate an input covariance matrix and r an input variance. The subscripted $\boldsymbol{Y}$ in $\boldsymbol{R}_{\boldsymbol{Y}}$ shows that $\boldsymbol{Y}$ is the vector for which $\boldsymbol{R}_{\boldsymbol{Y}}$ is the covariance matrix.

◊ We will also use the *sans-serif* symbols Y and R_{Y} for the observation vector and its covariance matrix, and in Chapter 3 it will become clear what the differences are between these and the symbols $\boldsymbol{Y}$ and $\boldsymbol{R}$ that come *with* serifs.

◊ The symbol $\boldsymbol{X}$ without a star is always used for what we call the *true state vector*. For example, $\boldsymbol{X}_n$ and $\boldsymbol{X}_{t_n}$ mean the same thing, or $\boldsymbol{X}_{n+1}$ and $\boldsymbol{X}_{t_{n+1}}$ also mean the same thing.

◊ A star * signifies that the quantity is produced by the filter:

- ◊ $\boldsymbol{X}^*_{n,n}$ (or equivalently $\boldsymbol{X}^*_{t_n,t_n}$) is the filter's estimate of the true state vector.

- ◊ $\boldsymbol{S}^*_{n,n}$ (or equivalently $\boldsymbol{S}^*_{t_n,t_n}$) is the covariance matrix produced by the filter. This is the *supposed* covariance matrix of the errors in $\boldsymbol{X}^*_{n,n}$.

◊ The letters $\boldsymbol{S}^*$ and $\boldsymbol{S}$ will always be associated with filter-output covariance matrices. These are either produced by the filter, such as $\boldsymbol{S}^*$, or are related in some way to the quantities produced by the filter such as $\boldsymbol{S}$.

In Figure 1.3, we summarize the notation discussed thus far.

	True state vector	Filter inputs	Filter outputs
Vector	$\boldsymbol{X}_n$	Y_n	$\boldsymbol{X}^*_{n,n}$
Covariance matrix		$\mathsf{R}_{\mathsf{Y}_n}$	$\boldsymbol{S}^*_{n,n}$

Figure 1.3 Notation pertaining to vectors and covariance

Definition 1.1: The two subscripts appearing in $\boldsymbol{X}^*_{n,n}$ have the following meanings. The first signifies that the **validity instant** of $\boldsymbol{X}^*$ is t_n. The second signifies that $\boldsymbol{X}^*$ is an estimate that was **based on** or **conditioned on** observations that were made up to, and including, time t_n. ■■

As examples of our notation:

◊ $\boldsymbol{X}^*_{n,n}$ or $\boldsymbol{X}^*_{t_n,t_n}$ are called *current estimates*, and are estimates of what the true state vector $\boldsymbol{X}_n$ or $\boldsymbol{X}_{t_n}$ is at time t_n (the validity instant) based on observations made up to time t_n. The corresponding filter covariance matrices are written as $\boldsymbol{S}^*_{n,n}$ or $\boldsymbol{S}^*_{t_n,t_n}$.

◊ $\boldsymbol{X}^*_{n+1,n}$ or $\boldsymbol{X}^*_{t_n+1,t_n}$ are called *1-step predictions* and are estimates of what $\boldsymbol{X}_n$ or $\boldsymbol{X}_{t_n}$ will be at time t_{n+1} based on observations made up to time t_n. The corresponding filter covariance matrices are written as $\boldsymbol{S}^*_{n+1,n}$ or $\boldsymbol{S}^*_{t_{n+1},t_n}$.

◊ $\boldsymbol{X}^*_{t_n+\zeta,tn}$ is an estimate of what $\boldsymbol{X}_n$ or $\boldsymbol{X}_{t_n}$ will be at time $t_n+\zeta$ based on observations made up to time t_n. The corresponding filter covariance matrix is written as $\boldsymbol{S}^*_{t_n+\zeta,t_n}$. If ζ is positive then $\boldsymbol{X}^*_{t_n+\zeta,t_n}$ is a ζ-second *pre*diction and if ζ is negative then $\boldsymbol{X}^*_{t_n+\zeta,t_n}$ is called a ζ-second *retro*diction.

◊ From time to time, in order to simplify the notation we show vectors like $\boldsymbol{Y}$, $\boldsymbol{X}$ and $\boldsymbol{X}^*$ and matrices like $\boldsymbol{R}$ and $\boldsymbol{S}^*$ without any time subscripts. However, whenever we do that it is always to be understood that they are in fact $\boldsymbol{Y}_n$, $\boldsymbol{X}_n$, $\boldsymbol{X}^*_{n,n}$, $\boldsymbol{R}_n$ and $\boldsymbol{S}^*_{n,n}$.

◊ In all definitions throughout this book, the item being defined will be shown in **bold type**, and the symbol ■■ will be used to signify the end of a definition, an example or an important development in the discussion.

◊ The superscript symbol ^{T} will be used to designate the transpose of a matrix or a vector, as in $\boldsymbol{M}^T$ or $(x, y, z)^T$. The prime symbol $'$ is often used in textbooks to designate a transpose. We will use the prime symbol on occasions, but never to designate a transpose.

◊ The non-bold italic D always means the derivative with respect to time, that is d/dt.

◊ Similarly, D^2 means d^2/dt^2 and so on. Letters with dots above them, like $\dot{x}$ and $\ddot{x}$ mean Dx and D^2x. If the derivative is with respect to another variable, for example ζ, then we show it as $d/d\zeta$.

◊ The words *differential equation* will always be abbreviated as DE and *probability density function* will be abbreviated as PDF.

◊ The expression $diag(x_1, x_2, x_3, \ldots)$ means a *diagonal matrix* whose elements on the diagonal are $x_1, x_2, x_3, \ldots$.

◊ The bold non-italic Greek letter $\boldsymbol{\Phi}$ will always be used to designate a *transition matrix*. It will usually be followed by its one or more arguments in brackets. As examples:

  ◊ $\boldsymbol{\Phi}(\zeta)$, $\boldsymbol{\Phi}(t_n+\zeta, t_n; \bar{\boldsymbol{X}})$

  ◊ $\boldsymbol{\Phi}(n, n-1)$ and $\boldsymbol{\Phi}(t_n, t_{n-1})$, which mean the same thing

  Transition matrices will be fully explored in Chapter 2.

◊ The following are called *observation equations* and we will see them frequently in what lies ahead:

$$\boldsymbol{Y}_n = \boldsymbol{M}_n\boldsymbol{X}_n + \boldsymbol{N}_n \tag{1.6.1}$$

and

$$\mathsf{\mathbf{Y}}_n = \mathsf{\mathbf{T}}_n\boldsymbol{X}_n + \mathsf{\mathbf{N}}_n \tag{1.6.2}$$

◊ The letter m will be used for the dimension of vectors like $\boldsymbol{X}$ and $\boldsymbol{X}^*$, where m is typically 3, 6 or 9. On rare occasions m could be of the order of 36 or 48. Values like these are small and we will refer to vectors with such dimensions as *short-vectors*. $\boldsymbol{Y}_n, \boldsymbol{X}_n$ and $\boldsymbol{N}_n$ in (1.6.1) are short-vectors.

◊ The letter k will be used for the dimension of vectors like $\mathsf{\mathbf{Y}}_n$ and $\mathsf{\mathbf{N}}_n$ in (1.6.2), where k is typically in the hundreds and could even be in the thousands. These dimensions are large, and we will refer to such vectors as *long-vectors*.

◊ We will follow the standard practice for specifying the sizes of matrices, namely, if a matrix is $m \times k$ then m is the number of its rows and k is the number of its columns.

◊ In (1.6.2) $\boldsymbol{X}_n$ is of dimension m and $\mathsf{\mathbf{Y}}_n$ is of dimension k, where $k > m$. The matrix $\mathsf{\mathbf{T}}_n$ is therefore $k \times m$ and has more rows than columns. We will refer to such a matrix as a *high-matrix*.

◊ In Figure 1.4, we show a pictorial representation of (1.6.2).

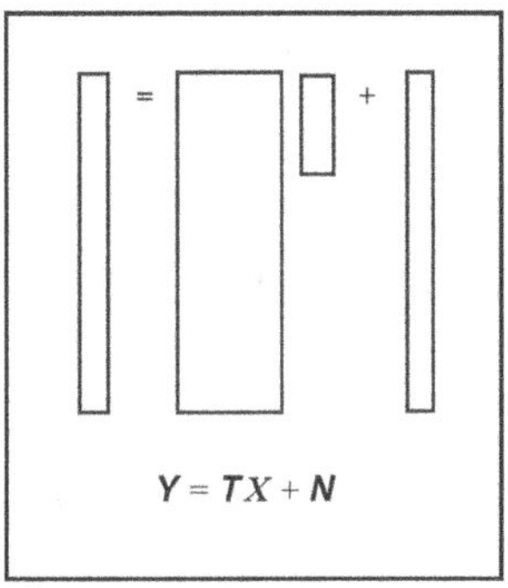

Figure 1.4 Two long-vectors, one short-vector and a high-matrix

◊ The symbols $\boldsymbol{N}$ and $\mathsf{\mathbf{N}}$ in (1.6.1) and (1.6.2) are not *En's* – they are instead upper-case *Nu's* from the Greek alphabet. If we had used upper-case *En's* then their vector elements would have been lower-case *en's*, namely n, and the letter n has already been used to refer to the most recent, as in t_n.

The elements of the vectors $\boldsymbol{N}$ and $\mathsf{\mathbf{N}}$ are therefore always the Greek lower-case *nu*, namely ν. Thus

$$\boldsymbol{N} = (\nu_1, \nu_2, \ldots \nu_m)^T \quad \text{and} \quad \mathsf{\mathbf{N}} = (\nu_1, \nu_2, \ldots \nu_k)^T \tag{1.6.3}$$

Unfortunately, the Greek symbol *nu* (ν) is the same as the English symbol *vee* (v), which is the usual symbol for *speed*, but it will be clear from the context which of the two we have in mind.

◊ A vector is always a column vector, unless we specifically state that it is a row vector. In (1.6.3) we show N and $\mathbf{N}$ as the transposes of row vectors, which means that they are in fact column vectors. We will do this wherever possible in order to save trees.

◊ In addition to the observation equations in (1.6.1) and (1.6.2), we will also encounter what we call *filter equations*. As examples

$$X^*_{n,n} = W_n Y_n \quad \text{and} \quad X^*_{n+1,n} = W'_n Y_n \tag{1.6.4}$$

Given that X^* is of dimension m and $\mathbf{Y}$ is of dimension k, where $k > m$, it follows that the rectangular matrices W and W' in (1.6.4) must be $m \times k$, and so they have more columns that rows. We will refer to such matrices as *wide-matrices*.

In Figure 1.5 we show a representation of the first filter equation in (1.6.4).

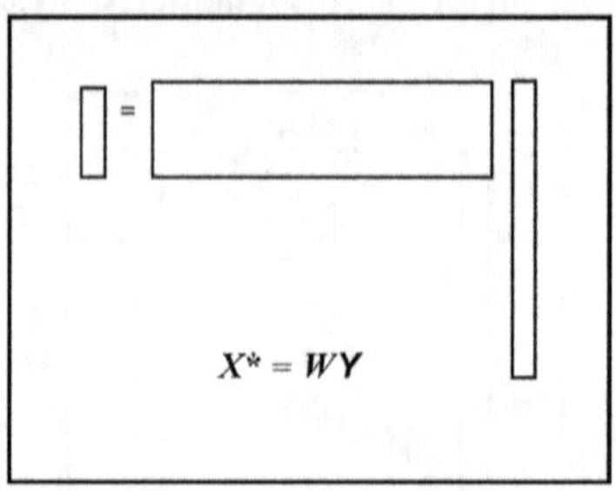

Figure 1.5 One short-vector, one wide-matrix and a long-vector

◊ The symbols $\bar{X}$ and $\bar{x}$ mean the same as $\bar{X}$ and $\bar{x}$. The latter were not present in the fonts that came with Word 2003, and so they had to be created by cut and paste using a graphics editor. This is a time-consuming process and so we instead used $\bar{X}$ and $\bar{x}$ in the problems, solutions and projects in place of $\bar{X}$ and $\bar{x}$.

1.6.1 Accuracy

Accuracy is a concept that plays a key role in filter engineering. Although it is not really a part of notation, it needs to be specified and this is as good a place as any.

In this book we will only speak about *relative* accuracy.

Definition 1.2: Let a and b be two estimates of the same quantity and let the errors in them have standard deviations σ_a and σ_b, where $\sigma_a \geq \sigma_b$. Let the ratios of the sigma values be

$$R_{a/b} \equiv \sigma_a/\sigma_b \quad \text{and} \quad R_{b/a} \equiv \sigma_b/\sigma_a \tag{1.6.5}$$

Then the following are three equivalent statements:

◊ b is $R_{a/b}$ times more accurate than a.

◊ The errors in a are $R_{a/b}$ times larger than those in b.

◊ The errors in b are $R_{b/a}$ as large as those in a.

■■

Example 1.2: Let $\sigma_a = 10\ m$ and $\sigma_b = 2\ m$. Then $\sigma_a > \sigma_b$, $R_{a/b} = 5$ and $R_{b/a} = 1/5$

◊ b is *5* times more accurate than a.

◊ The errors in a are *5* times larger than those in b.

◊ The errors in b are $1/5$ as large as those in a.

Let $\sigma_a = \sigma_b$. Then $R_{a/b} = 1$ and so

◊ b has the same accuracy as a.

◊ The errors in a are the same size as those in b.

◊ The errors in b are the same size as those in a.

■■

1.7 Two important words

There are two words that we will use frequently:

1. ***Actual***

When something is *actual* then it exists in fact.

◊ A simple test to determine whether or not a tree, for example, is *actual* would be to run into it at full speed. The answer will become immediately apparent.

◊ The fact that we cannot see, taste, feel, hear or smell an object has no bearing on whether or not it is actual (e.g. try running into the tree in the dead of night when you cannot see it).

◊ Something that is actual is out there, regardless of whether or not we are aware of it.

◊ Mathematical constructs such as covariance matrices can be actual (or not).

2. ***Supposed (sup-póse-ed)***

If something is *supposed* then it is held as a belief or an opinion.

◊ The existence of the above tree can be supposed, which means that we are of the opinion that it is there. It may or may not be actual.

◊ One covariance matrix can supposedly be a good representation of a second one that is actual.

Another word for *supposed* is *alleged*. Newspapers frequently use the word *alleged* in statements like 'It is alleged that …' when they want to stress that something is supposed, and may or may not be actual.

1.8 The filtering procedure

Covariance matrices will be discussed in detail in Chapters 4 and 5. If you are already familiar with them, then what now follows will not pose any problems. However, if they are new to you, then we suggest that you continue reading anyway, keeping in mind that the material will make much more sense when you read it again after having read those later chapters.

In Figure 1.6 below we show the basic elements of the filtering procedure.

(1) The observed process has true state vector $\boldsymbol{X}$. In the field $\boldsymbol{X}$ is only conceptual, and so the values of its elements are unknown to us. However, when running a simulation $\boldsymbol{X}$ is *actual* and its values are known. The primary objective of filter engineering is to provide an estimate of $\boldsymbol{X}$.

(2) The observation instrument (e.g. radar) observes the process during what is called the *observation window* or the *observation interval.*

(3) The observation vector $\boldsymbol{Y}$ is produced by the observation instrument.

(4) The vector $\boldsymbol{Y}$ contains errors that have an actual covariance matrix that is shown as $\boldsymbol{R}_{Y,\,actual}$. Both the errors in $\boldsymbol{Y}$ and their covariance matrix $\boldsymbol{R}_{Y,\,actual}$ are unknown to us, unless we are running a simulation.

(5) The covariance matrix $\boldsymbol{R}_{Y,\,engrs}$ is the supposed covariance matrix of the errors in $\boldsymbol{Y}$. This matrix comes to us from the engineers who built the observation instrument, and is supposedly a good representation of the actual covariance matrix $\boldsymbol{R}_{Y,\,actual}$ in (4).

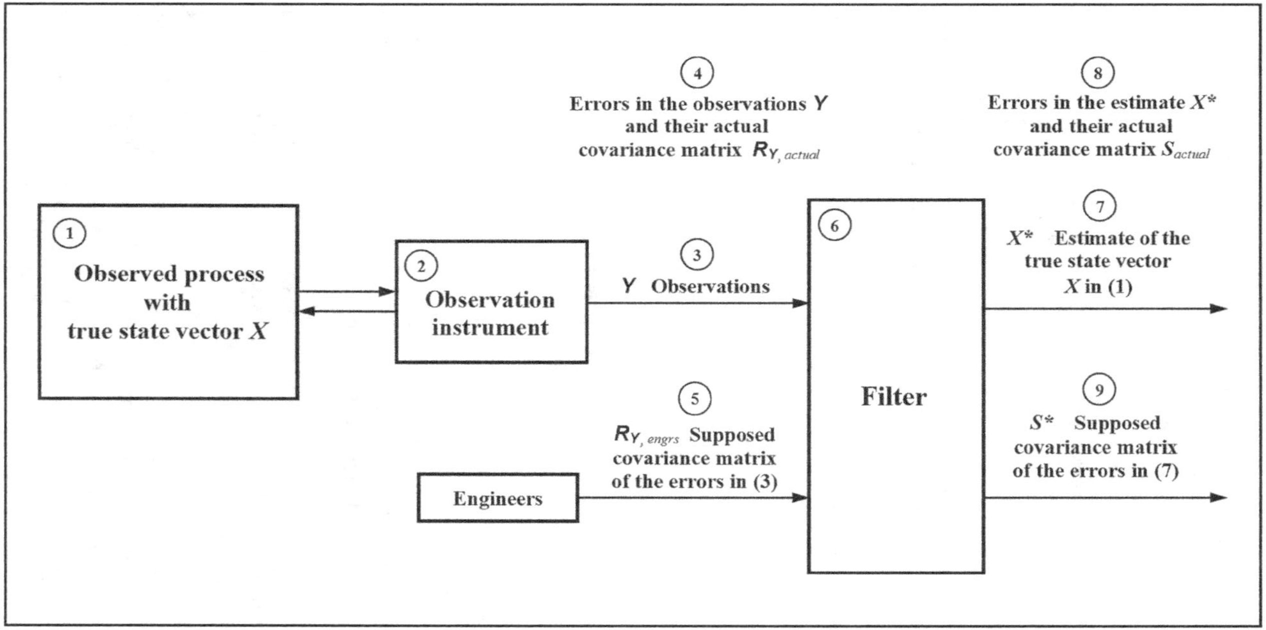

Figure 1.6 Basic elements of the filtering procedure

(6) The filter inputs are $\boldsymbol{Y}$ in (3) and $\boldsymbol{R}_{Y,\,engrs}$ in (5), and the outputs are $\boldsymbol{X}^*$ in (7) and $\boldsymbol{S}^*$ in (9). $\boldsymbol{X}^*$ is the required estimate of the true state vector $\boldsymbol{X}$, and contains errors that have an actual covariance matrix $\boldsymbol{S}_{actual}$ in (8). The errors in $\boldsymbol{X}^*$ and their actual covariance matrix are unknown to us in the field, but if we are running a simulation then the errors can be evaluated, and their actual covariance matrix can be estimated by a Monte-Carlo.

(9) The covariance matrix $\boldsymbol{S}^*$ is always produced as part of any filtering algorithm. The matrix $\boldsymbol{S}^*$ is supposedly a good representation of the actual covariance matrix $\boldsymbol{S}_{actual}$.

Definition 1.3: The covariance matrix $\boldsymbol{S}^*$ in (9) is called the **filter covariance matrix**. The actual covariance matrix of the estimation errors $\boldsymbol{S}_{actual}$ in (8) (in the figure) is called the **actual estimation–error covariance matrix.** ■■

The expressions *filter covariance matrix* and *actual estimation-error covariance matrix* may sound a bit cumbersome, but we will have to use them in order to avoid confusion.

◄►

We pause to make a few comments regarding the actual estimation-error covariance matrix $\boldsymbol{S}_{actual}$.

When running a simulation we are always able to compute the errors in a filter's estimate vector by means of the equation

$$\boldsymbol{N}^*_{n,\,n} = \boldsymbol{X}^*_{n,\,n} - \boldsymbol{X}_n \tag{1.8.1}$$

in which $\boldsymbol{X}^*_{n,\,n}$ is the estimate being produced by the filter at time t_n and $\boldsymbol{X}_n$ is the true state vector of the observed process at that time, both of which are known. $\boldsymbol{N}^*_{n,\,n}$ is then the known vector of *actual estimation errors* contained in $\boldsymbol{X}^*_{n,\,n}$.

However, in the field it is by definition not possible to implement an equation like (1.8.1) because that requires that $\boldsymbol{X}_n$ be known, and if that were the case then we would not need a filter in order to estimate it.

Note: During testing and debugging of the filter in the field, and assuming that the target or targets are fully cooperative, it might be possible to include on-board instrumentation such as inertial platforms, GPS and similar devices, which provide us with knowledge about $\boldsymbol{X}_n$. However, we cannot always count on this during actual field operation, and so it is in this sense that we say that $\boldsymbol{X}_n$ is unknown in the field.

Assume that we are running a simulation, and that we have obtained one error vector $\boldsymbol{N}^*_{n,\,n}$ using (1.8.1). Then that vector is a single draw from a population of such vectors, all of which share a common covariance matrix, and it is *that* covariance matrix to which we are referring when we speak about the actual estimation-error covariance matrix, which at time t_n would be written as $\boldsymbol{S}_{actual,n,n}$.

In Chapter 5 we will give a more precise definition of $\boldsymbol{S}_{actual}$ and we will also discuss how we can use a Monte-Carlo to estimate it.

◀▶

In Figure 1.7 below we provide a first look at the Gauss–Aitken filter. (In Chapter 8 it will become clear why it is referred to it as Version 1.)

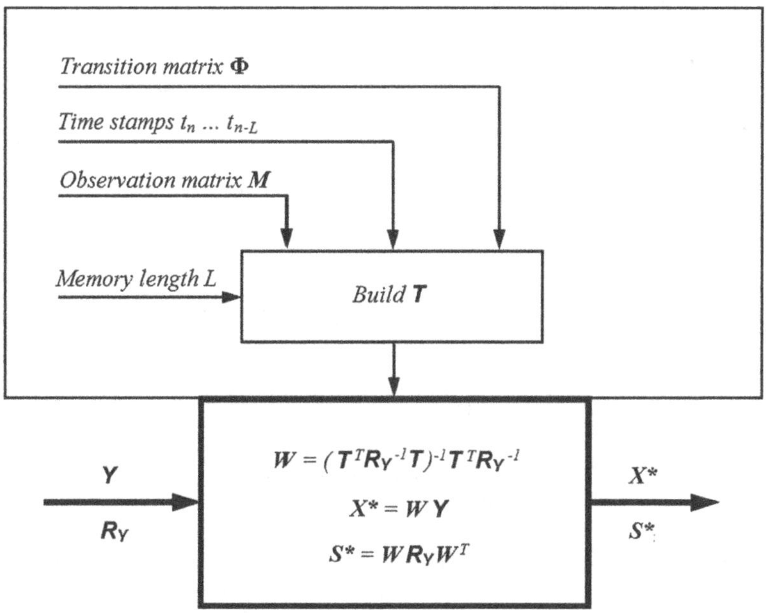

Figure 1.7 The Gauss–Aitken filter (version 1)

◊ The vector $\boldsymbol{Y}$ corresponds to item (3) in Figure 1.6.

◊ For the matrix $\boldsymbol{R_Y}$ we use $\boldsymbol{R}_{\boldsymbol{Y},engrs}$ that was numbered (5) in Figure 1.6.

◊ The vector $\boldsymbol{X}^*$ corresponds to item (7) in Figure 1.6, and is the estimate of the true state vector $\boldsymbol{X}$ in (1) in Figure 1.6.

◊ The matrix $\boldsymbol{S}^*$ is the filter covariance matrix, and corresponds to item (9) in Figure 1.6. $\boldsymbol{S}^*$ is supposedly a good estimate of $\boldsymbol{S}_{actual}$, the actual estimation-error covariance matrix in item (8).

1.9 Error/covariance-matrix consistency

In what lies ahead we will often speak about two types of consistency – error/covariance-matrix consistency and Cramér–Rao consistency – both of which are central to filter engineering, and will be with us throughout the entire book. In this section we examine the first, and in the next section we examine the second.

In addition to putting out an estimate vector $\boldsymbol{X}^*$, every properly constructed filter also puts out a filter covariance matrix $\boldsymbol{S}^*$, which is supposedly the covariance matrix of $\boldsymbol{X}^*$.

Error/covariance-matrix consistency deals with the issue of whether or not $\boldsymbol{S}^*$ is acceptable as the covariance matrix of $\boldsymbol{X}^*$.

Definition 1.4: Referring to Figure 1.6, if the filter covariance matrix $\boldsymbol{S}^*$ in (9) is a good representation of the actual estimation-error covariance matrix $\boldsymbol{S}_{actual}$ in (8), then we say that the filter is **error/covariance-matrix consistent**.

On the other hand, if the matrix $\boldsymbol{S}^*$ is not a good representation of the actual estimation-error covariance matrix $\boldsymbol{S}_{actual}$, then we say that the filter is **error/covariance-matrix inconsistent**. ■■

In Chapter 7 we will specify precisely what we mean by 'a good representation'.

From now on we will use the term ECM consistency as the abbreviation for error/covariance-matrix consistency.

ECM consistency is by no means guaranteed, and in Section 5.3 we will consider some of the reasons why it could be absent.

In Chapter 7 we will discuss three tests that can be used to determine, with good reliability, if a filter is ECM consistent. Those tests can only be used in a simulation and are impossible to run in the field, but we hope that what we learn from the simulation will continue to apply in the field.

ECM consistency is of fundamental importance for a number of reasons:

◊ A filter puts out an estimate vector $\boldsymbol{X}^*$ together with its filter covariance matrix $\boldsymbol{S}^*$. If the filter is not ECM consistent – meaning that $\boldsymbol{S}^*$ is not a good representation of $\boldsymbol{S}_{actual}$ – then better not put $\boldsymbol{S}^*$ out at all. But if we did that, then the filter would be giving us estimates $\boldsymbol{X}^*$ with no bounds on their accuracy, which makes those estimates of dubious value, and so better not put them out as well.

◊ When operating in the field we cannot know the value of the true state vector $\boldsymbol{X}$ and so we also cannot know how large or small the filter's estimation errors are.

We must therefore turn to its filter covariance matrix $\boldsymbol{S}^*$, which is supposedly a good representation of $\boldsymbol{S}_{actual}$. Critical judgments might be made based on $\boldsymbol{S}^*$, and if ECM consistency is absent, then those judgments could be in serious error.

◊ There are times when the estimates from a number of filters are to be combined, in order to arrive at an improved composite estimate. Such a procedure – known as *data fusion* and discussed briefly in Chapter 8 – will require knowledge of both the estimates $\boldsymbol{X}^*$ from each filter, as well as their filter covariance matrices $\boldsymbol{S}^*$. If one or more of the latter are in error (i.e. those filters are ECM inconsistent), then their use in the combination procedure will cause a degradation of accuracy in the composite result.

We will accordingly regard ECM inconsistency in a filter as a condition that is *fatal*, and to be avoided at all costs.

1.10 Cramér–Rao consistency[8]

We have examined briefly the first type of consistency, namely ECM consistency. We now examine the second, namely Cramér–Rao consistency.

Cramér–Rao consistency deals with the issue of accuracy. When a filter is Cramér–Rao consistent, then the accuracy of its estimate is the best that it can possibly be for the given set of observations from which that estimate was derived, and when not Cramér–Rao consistent, then its accuracy is less than what it can possibly be.

1.10.1 Univariate Cramér–Rao consistency[9]

We wish to estimate the mean of a population and we accordingly make a number of observations. In Cramér–Rao theory it is always assumed that

◊ The estimate will be obtained by a *linear* combination[10] of the observations.

◊ The estimate produced by the linear combination is *unbiased*, that is it is zero-mean.

The acronym for such an estimate is *LU* (*linear unbiased*).

The question is then asked: What is the *best* linear unbiased (*BLU*) estimate that can possibly be obtained from a given set of observations, where best means smallest variance?

[8] Cramér–Rao theory was developed independently by H. Cramér and C.R. Rao, both of whom published their results in 1946. See References 51 and 52.
[9] See References 11 or 12.
[10] We elaborate on the concept of linearity in Chapter 2.

The answer is provided by univariate Cramér–Rao theory, and is as follows:

Let a set of independent observations be made on a population whose mean is P. Let P be estimated from those observations using many different *LU* algorithms, the *j*th one producing the estimate P_j together with its variance σ_j^2.

Then there exists a lower bound on all of those variances called σ_{CR}^2, such that

$$\sigma_j^2 \geq \sigma_{CR}^2 \qquad j = 1, 2, \ldots \tag{1.10.1}$$

The variance σ_{CR}^2 in (1.10.1) is called the Cramér–Rao lower bound, and the best estimate of P that can be obtained from the given set of observations is the estimate whose variance is σ_{CR}^2.

Cramér–Rao theory provides us with a way of finding σ_{CR}^2. (See Problems 1.1 to 1.5.)

Definition 1.5: If the variance of the estimate of P obtained from an *LU* estimator is equal to σ_{CR}^2, then we say that the estimator is **Cramér–Rao consistent**.

On the other hand, if the variance of the estimate of P obtained from an *LU* estimator is greater than σ_{CR}^2, then we say that the estimator is **Cramér–Rao inconsistent.** ■■

It is important to keep in mind that the value of σ_{CR}^2 in (1.10.1) is not in any sense absolute but depends on the set of observations that is used, and if a different set is used then σ_{CR}^2 will almost certainly change.

1.10.2 Multivariate Cramér–Rao consistency[11]

We have obtained a specific vector of observations $\boldsymbol{Y}$ from which we wish to derive an estimate of the true state vector $\boldsymbol{X}$ – an estimate that will be called $\boldsymbol{X}^*$ – and since the dimension of the latter two vectors is almost always greater than 1, we are involved in a process of multivariate estimation. The following is a brief statement of what multivariate Cramér–Rao theory provides.

We are restricted to estimators with the following two properties:

1. They are all *linear transformations* of the observation vector of the form

$$\boldsymbol{X}^*_j = \boldsymbol{W}_j \boldsymbol{Y} \tag{1.10.2}$$

[11] See Reference 13.

in which X^*_j is the jth estimate of X, both assumed to be m-vectors, W_j is the jth $m \times k$ filter matrix and Y is the observation vector of dimension k.

2. The estimates that they produce are all *unbiased*, that is

$$\mathcal{E}\{X^*_j\} = X \tag{1.10.3}$$

in which $\mathcal{E}$ signifies the expectation operator (to be discussed in Chapter 4).

We will use the acronym *LU* (*linear unbiased*) for estimates X^*_j that satisfy (1.10.2) and (1.10.3).

Given that $S_{actual,j}$ is the actual estimation-error covariance matrix of the jth *LU* estimate X^*_j, and that $[S_{actual,j}]_{i,i}$ is its ith diagonal element, then Cramér–Rao theory proves the following:

There exists an $m \times m$ covariance matrix S_{CR} whose diagonal elements and those of every $S_{actual,j}$ satisfy

$$[S_{actual,j}]_{i,i} \geq [S_{CR}]_{i,i} \qquad 1 \leq i \leq m \qquad j = 1, 2, \ldots \tag{1.10.4}$$

By analogy to the scalar case above, S_{CR} is called the Cramér–Rao lower-bound covariance matrix, or more simply the Cramér–Rao covariance matrix, and Cramér–Rao theory provides us with a way in which to find it. (We will see an example of how that is done in Chapter 9.)

Simply put:

◊ The best (i.e. most accurate) *LU* multivariate estimate that can be obtained from a given set of observations is one which has an S_{actual} whose diagonal elements are all equal to those of the Cramér–Rao covariance matrix.

◊ No *LU* estimate can have an S_{actual}, any of whose diagonal elements are smaller than the corresponding elements of the Cramér–Rao covariance matrix.

As with the univariate case, S_{CR} in (1.10.4) is not an absolute, but depends on the observation vector Y from which the *LU* estimates X^*_j are derived. If a different observation vector Y is used, then S_{CR} will change.

Definition 1.6: If at time t_n a filter produces an *LU* estimate of X_n whose $S_{actual,n,n}$ has diagonal elements that are all equal to the diagonal elements of $S_{CR,n}$, namely if

$$[S_{actual,n,n}]_{i,i} = [S_{CR,n}]_{i,i} \qquad 1 \leq i \leq m \tag{1.10.5}$$

then the filter is said to be **Cramér–Rao consistent** at time t_n, and the estimate is the most accurate possible.

On the other hand, if at time t_n a filter produces an *LU* estimate whose $\boldsymbol{S}_{actual,n,n}$ has one or more diagonal elements that are larger than the corresponding elements of $\boldsymbol{S}_{CR,n}$, then the filter is said to be **Cramér–Rao inconsistent** at time t_n, and the estimate is then not the most accurate possible. ■■

From now on we will use the term CR consistency as the abbreviation for Cramér–Rao consistency.

In Chapter 9 we will elaborate on three tests that can be used to determine if CR consistency is present or absent. As with the three tests for ECM consistency mentioned earlier, these can also only be used in a simulation and are impossible to run in the field, and so again we must hope that what we learn from the simulation will continue to apply in the field.

Just as ECM consistency is of fundamental importance, so also is CR consistency.

When an *LU* filter gives us an estimate, we would like it to be of the best possible accuracy.

◊ If the filter is CR consistent, then the variances of the errors in its estimate vector *are* the smallest possible, and so that estimate is as accurate as can be expected.

◊ If the filter is not CR consistent, then the variances of the errors in some or all of the elements of its estimate vector are greater than the smallest possible, and so that estimate is less accurate than an estimate which *is* CR consistent.

Absent other factors, we thus always choose a CR-consistent filter over one that is CR inconsistent. However, CR inconsistency is not a fatal condition, as is the case with ECM inconsistency, and we will see in Chapter 8 that there are situations in which we are prepared to sacrifice CR consistency in order to achieve an objective, but *only* if ECM consistency is maintained.

1.11 Considering ECM and CR consistency together

We have examined ECM and CR consistency separately, and in Figure 1.6 we referred to the following two matrices:

◊ The actual estimation-error covariance matrix $\boldsymbol{S}_{actual}$.

◊ The filter covariance matrix $\boldsymbol{S}^*$.

We have now introduced a third matrix:

◊ The Cramér–Rao covariance matrix $\boldsymbol{S}_{CR}$.

In Figure 1.8 we show how the first of these is related to the second by ECM consistency and to the third by CR consistency.

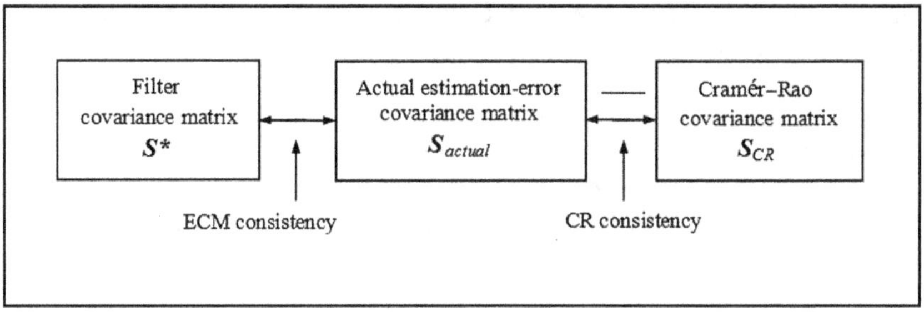

Figure 1.8 ECM and CR consistency

Note: Statisticians do not usually include $\boldsymbol{S}_{actual}$ in their thinking, and simply assume that $\boldsymbol{S}^*$ *is* the filter's actual covariance matrix. As a result, for them there are only two covariance matrices – $\boldsymbol{S}^*$ and $\boldsymbol{S}_{CR}$ – and they regard a filter as being CR consistent if the diagonal elements of these two matrices are equal. However, we will see in Chapter 11 that it is possible for a certain well-known filter to spontaneously enter a condition in which $\boldsymbol{S}^*$ is *not* a good representation of its actual covariance matrix. This means that in filter engineering we are forced think in terms of *three* matrices, and to differentiate between the filter covariance matrix $\boldsymbol{S}^*$ and the actual estimation-error covariance matrix $\boldsymbol{S}_{actual}$.

◀▶

ECM and CR consistency are independent of each other and so at a given time all four possibilities shown in Figure 1.9 can occur.

ECM consistent CR consistent	ECM inconsistent CR consistent
ECM consistent CR inconsistent	ECM inconsistent CR inconsistent

Figure 1.9 The four ECM/CR possibilities

We naturally attempt to build filters that are both ECM and CR consistent. However, as stated earlier, CR inconsistency is not fatal whereas ECM inconsistency *is* fatal.

1.12 Kalman/Swerling instability

We have mentioned the word *instability* in relation to the Kalman filter both in this chapter as well as in the preface and the same word will be used again in the same context in later chapters.

Having examined the concepts of ECM and CR consistency, we are now in a position to provide a brief introduction to what we are referring to when we speak about *Kalman instability*. (In Chapter 11, which is devoted to the Kalman and Swerling filters, we discuss the matter in greater detail.)

We are by no means the only ones to have observed the phenomenon, and the literature is replete with references to Kalman instability.[12]

In Figure 1.10 we display values obtained from one of the computer programs – this one entitled ***13_Gauss_Newton_C*** – in which instability in both the Kalman and Swerling filters is clearly present. (It never occurs singly in either of them, and we have never seen it in any of the Gauss filters.)

Four filters are implemented in this program, a polynomial prefilter, a Kalman filter, a Swerling filter and a Gauss–Newton filter.

High data-rate samples of a sinusoidal waveform are fed to the polynomial prefilter from which, once every *1, 200* cycles, a value is extracted and sent to the Kalman and Swerling filters where it is processed upon receipt. The same values are also placed in storage and are later processed as a batch by the Gauss–Newton filter.

The filter model implemented in the three main filters is the DE for sinusoidal motion whose frequency is unknown (see (2.1.36)). The objective is for all three filters to estimate that unknown frequency.

The filter model is nonlinear, and so the Kalman/Swerling filters are implemented in one of their *extended forms* and the Gauss–Newton filter is implemented in its appropriate form.

Note: These topics – *prefiltering, nonlinear models, Kalman and Swerling extended forms* and *Gauss–Newton filter* – will all be discussed in later chapters.

◊ All runs used the same sequence of observations that differed from run to run only in their errors, all of the latter being drawn from the same PDF which in this case was Gaussian, and all three filters – Kalman, Swerling and Gauss–Newton – processed *exactly* the same values on each run.

[12] See for example References 3, 170, 171.

```
                    Kalman Filter (Cycle #68)

             Error              Sigma          Error/Sigma

  x      -1.2699769e-2     1.1798091e-3     -10.764258
x-dot    -1.3465111e-6     6.4989266e-8     -20.718977
omega     1.0411155e-7     3.8223478e-9      27.237593

           Chi-squared = 1.7265719180911e+03 (11.35)

                    Swerling Filter (Cycle #68)

             Error              Sigma          Error/Sigma

  x      -1.2699769e-2     1.1798091e-3     -10.764258
x-dot    -1.3465111e-6     6.4989266e-8     -20.718977
omega     1.0411155e-7     3.8223478e-9      27.237593

           Chi-squared = 1.7265719180982e+03 (11.35)

                          Gauss-Newton

             Error              Sigma          Error/Sigma

  x       1.5556049e-3     .00191947          .81043459
x-dot     6.5501149e-10    1.6519109e-8      3.9651744e-2
omega     5.1388082e-9     4.7932243e-9      1.0720984

                 Chi-squared = 1.529577 (11.35)
```

Figure 1.10 Kalman/Swerling instability

◊ Neither the Kalman nor the Swerling filter contained what is known as a $\boldsymbol{Q}$ matrix.

◊ Two of the three tests that we discuss in Chapter 7 for detecting ECM inconsistency were implemented, namely the *3-sigma test* and the *Chi-squared test*.

At some random instant after a run was started, the ***D*** key was pressed, calling for a display of the current status of each of the three filters. (It was then that the values in storage were processed by the Gauss–Newton filter.) If all three filters were ECM consistent, then the program was restarted and run again with the same sequence of observations that differed from the other runs only in their errors.

The values in Figure 1.10 came from what was perhaps the 30th run. In the first *29* runs both the Kalman and Swerling filters were ECM consistent when the ***D*** key was pressed, and then, unpredictably, on about the 30th run, the key was pressed and Kalman/Swerling values like these appeared.

We pause to examine the numbers.

The first thing we notice is that the Kalman/Swerling values appear to be identical. However, that is only because we did not have sufficient space to display more of the digits, and it was for this reason that we extended the displays of their Chi-squared values, which show that, although not exactly the same, there is a remarkable *11 digit agreement* between their outputs. We will say more about this in Chapter 11.

Next we observe the following:

A. In absolute magnitude the Kalman/Swerling error/sigma ratios are all far in excess of *3*, and while such values are possible, for Gaussian errors they are *highly* improbable. With a high degree of certainty, in absolute magnitude they should all be less than *3*.

B. The Kalman/Swerling Chi-squared values are approximately *1726*, a figure which is also possible but again, *highly* improbable. We have here *3* degrees of freedom, and so with *99%* confidence their Chi-squared values should both be less than *11.35*.

C. The Gauss–Newton error/sigma ratios are all comfortably less than *3*, and its Chi-squared value is less than *11.35*.

The correctness of the computer code for the Kalman/Swerling filters is beyond doubt, as is the way in which they were initialized, because the code and the initialization gave values that were ECM consistent for each of the first *29* runs.

Given the values shown in the figure for the 30th run, we can therefore be almost certain that the Kalman/Swerling filter covariance matrices are not good representations of their actual estimation-error covariance matrices, and so they are both is a state of ECM inconsistency.

When they behave in this way, the Kalman/Swerling filters are said to be *unstable*.

Despite many hundreds of runs of this and another of the programs (in which the Kalman and Swerling filters on occasions also become unstable) we have never yet seen a run in which a Gauss filter has become unstable, and so instability is something to which the Gauss–Newton filters appear to be immune.

Note 1: The Gauss–Newton filters can on occasions fail to converge, but that is altogether a different problem, for the following reasons: (1) If this takes place, then we become aware of it, both in simulations as well as in the field. (2) We can then apply one of the convergence-enhancing techniques that are mentioned in Section 9.7.

Note 2: We have seen instances in which Gauss–Newton filters with extremely long memory lengths (in the hundreds of thousands) develop Chi-squared values that are somewhat over the *99%* threshold, but never ones in which their Chi-squared values exceed 10^4 or even more, which is what one sees when the Kalman/Swerling filters become unstable.

Instructions for running this program and witnessing Kalman/Swerling instability are given in the file entitled

Programs\13_Gauss_Newton_C\Documents\Quickstart.pdf

◀▶

The above results came from a simulation, but in the field we would be entirely unaware that the Kalman/Swerling filters were ECM inconsistent, thereby making the problem particularly insidious. That is because, without values of the true state vector (available only when running a simulation), we would be unable to compute either their error/sigma ratios or their Chi-squared values.

To enable you to see this in a more direct way, we have done the following: On each of the data displays in the programs ***13_Gauss_Newton_C*** and ***14_Orbit_Track*** you will find an option called ***T (Toggle)***. Pressing ***T*** shows the displays as they would be seen in the field, from which you would be unable to tell whether or not any of the three filters is in a state of instability.

◀▶

The way in which Kalman/Swerling instability is typically prevented from occurring is to include in them what is called a ***Q*** matrix (discussed in Chapter 11), and if an appropriate such matrix is included, then stability is assured. However, the following must then be borne in mind:

- ◊ The occurrence of Kalman/Swerling instability is unpredictable, and so one is forced to include a ***Q*** matrix on *every* occasion when running one of the extended Kalman or Swerling filters.
- ◊ There is no serious theory to guide one for the selection of ***Q*** matrices, and so one is forced to use largely trial and error, hoping that what one finally selects will be appropriate for all occasions, and that the 'medicine' is neither too strong nor too weak. This is particularly problematic in the field where we cannot even know if stability is present or absent.
- ◊ Including a ***Q*** matrix in the Kalman/Swerling filters comes at a significant price, because now they become CR inconsistent – their estimates are less accurate than what they would have been had no such matrix been included (see Chapter 11) and therefore less accurate than the estimates produced by a Gauss–Newton filter processing the same data, and sometimes significantly so (see Appendix 11.4).

This means that whenever a Kalman or a Swerling filter *with* a $\boldsymbol{Q}$ matrix is selected for an application, the person doing so should be aware that there is an alternate filter – Gauss–Newton – that provides estimates with smaller sigma values, making them more accurate.

And if a Kalman or a Swerling filter *without* a $\boldsymbol{Q}$ matrix is selected for the application, then the possibility exists of its becoming unstable, resulting in ECM inconsistency which is fatal.

1.13 Filter memory

The concept of *memory*, as it relates to filters, is a subject that is central to filter engineering, and we now examine what it means.

Referring to Figure 1.11, we assume that observations are being made on a physical process.

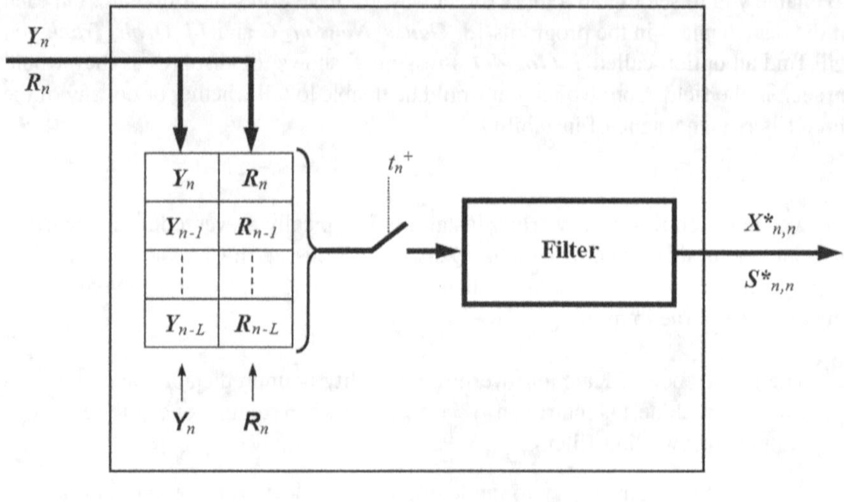

Figure 1.11 A filter and its input stack

The most recent observation is called $\boldsymbol{Y}_n$, which could be either a scalar or a short-vector. The covariance matrix of $\boldsymbol{Y}_n$ is called $\boldsymbol{R}_n$.

The quantities ($\boldsymbol{Y}_n$, $\boldsymbol{R}_n$) together with their predecessors ($\boldsymbol{Y}_{n-1}$, $\boldsymbol{R}_{n-1}$), ($\boldsymbol{Y}_{n-2}$, $\boldsymbol{R}_{n-2}$) etc., are stored in a stack. Just after time t_n the switch is closed, the filter is cycled, and we obtain the estimate vector $\boldsymbol{X}^*_{n,n}$ and its filter covariance matrix $\boldsymbol{S}^*_{n,n}$.

Definition 1.7: The contents of the stack constitute what is in the filter's **memory**.

■■

The stack can be operated in three modes, and the filter's memory-type is correspondingly given one of three names:

◊ ***Mode 1:*** The stack has fixed depth. We then say that the filter has a *fixed-length memory* (also called *fixed memory*). The Gauss filters operate naturally with fixed memory.

◊ ***Mode 2:*** The stack has a variable depth. We then say that the filter has a *variable-length memory*. The Gauss filters also operate naturally with variable-length memory.

◊ ***Mode 3:*** The stack has an expanding depth. We then say that the filter has an *expanding-length memory* (also called *expanding memory*). The Gauss filters also operate naturally with an expanding memory.

A further attribute known as *fading* can be superimposed onto each of these modes.

The Kalman and Swerling filters operate naturally in expanding-memory mode, and fading can be added if required.

It is only with a considerable amount of filter engineering that they might be configured to operate in fixed-memory or variable-memory modes.

Further discussion of the three types of memory has been placed in Appendix 1.1. In Project 8.5 we consider how fading is implemented.

1.14 The eight performance monitors

When constructing complex systems such as filters – and despite our best efforts, they invariably do become quite complex – it is always reassuring if we can somehow satisfy ourselves that they are operating properly. With that objective in mind, we will discuss eight tests for monitoring filter performance.

The first three were alluded to in Section 1.9 when we considered ECM inconsistency, and have the following names:

◊ The matrix-to-matrix ECM-consistency test.

◊ The *3*-sigma ECM-consistency test.

◊ The Chi-squared ECM-consistency test.

These are discussed in Chapter 7, where we will see that they can only be used when running a simulation.

The next three – called the CR-consistency tests – will be discussed in Chapter 9 when we discuss CR consistency in further detail. Like the ECM consistency tests, they have the following names:

◊ The matrix-to-matrix CR-consistency test.

◊ The *3*-sigma CR-consistency test.

◊ The Chi-squared CR-consistency test.

The CR-consistency tests can also only be run in a simulation.

The seventh test – called the goodness-of-fit (GOF) test – will be discussed in Chapter 10, where it will constitute the central element of what we call the MCAs. The GOF test can be used both in the field as well as in simulations, either as part of the MCAs or by itself, making it particularly powerful.

The final test – called the sigma monitoring (SM) test – will be discussed in Chapter 12, where we will see that it can be used to estimate the *sigma* of the observations being processed by a polynomial filter. Like the GOF test, the SM test can also be used both in the field as well as in a simulation, making it particularly powerful as well.

1.15 What can this book do for you?

The following four items are covered in this book:

◊ Two sets of filters, for use in low to medium data-rate environments, that are the 'heavy artillery' of what we offer:

  ◊ The Gauss–Aitken filters for use when both the filter model and the observation equations are linear. (The meanings of these words will be discussed in Chapters 2 and 3.)

  ◊ The Gauss–Newton filters for use when nonlinearities are present in either the filter model or the observation equations.

◊ Two sets of highly compact filters for use primarily in medium to high data-rate environments, but can also be used at low data rates in a variety of applications, namely

◊ The expanding memory polynomial (EMP) filters.

◊ The fading memory polynomial (FMP) filters.

This book will introduce you to these four items and will show you how they can be used – either alone or in combination with each other – to create practical filters for a wide variety of situations.

We also touch briefly on the Kalman and Swerling filters in Chapter 11, but only for completeness.

Figure 1.12 shows one example of what the Gauss filters can do. The display comes from a multiple target tracker that employs the auction algorithm[13] for plot-to-track allocation, and three Gauss–Aitken filters under MCA control, for data smoothing.

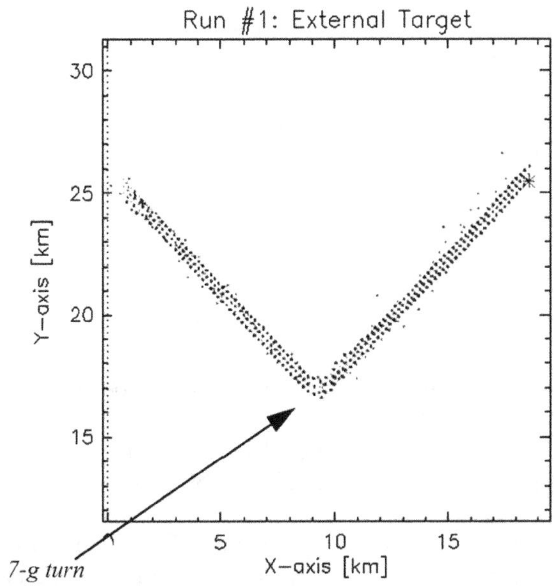

Figure 1.12 Gauss–Aitken filters under MCA control, tracking three manoeuvring targets

[13] See References 106–110.

◊ The targets are three simulated high-performance jet aircraft, flying parallel tracks.

◊ The radar is track-while-scan (TWS) whose observations are range, azimuth and elevation with a (*1-Hz*) data rate. A few of the radar observations can be seen spattered in the figure.

◊ The outputs of the filters are shown in the figure. The true flight paths are invisible, because they are hidden under the filtered tracks.

◊ At the left of the figure we see initial track formation taking place.

◊ In the centre of the figure the targets are simultaneously executing 7-g turns.

We present this as an example of what can be achieved by the Gauss filters under MCA control, and we assume that results like these can also be achieved by other filters.

Other examples of the performance of the Gauss filters under MCA control will be considered in Chapter 10.

In the down-loadable material there is a folder called ***Video_Clips***, in which you will find two sub-folders:

1. ***Doppler****:* This folder relates to a filter that we engineered, but do not discuss in this book.

2. ***TWS****:* This folder relates to the simulator mentioned in relation to Figure 1.12.

Each of these folders contains two subfolders called ***Documents*** and ***Flights.***

In ***Documents*** you will find the Readme's that explain what is going on, and should be read if you hope to understand what you are looking at.

If you run items from the ***TWS Flights*** folder you will see live versions of the kinds of flights that appear in Figure 1.12, which were for fixed-wing aircraft, as well as others for helicopters and cruise missiles performing a variety of manoeuvres involving changes in both heading and speed.

Closing comments

We have tried to keep the style of the book loose and friendly, and so hopefully it will be easy to read and enjoyable.

We would very much like to hear from you, and we welcome your comments. We would also welcome being made aware of any errors that you may find. Kindly direct your correspondence to the author at the following email address:

norman.morrison@uct.ac.za

Please include the words 'Tracking Filter' in the title of your email.

The following items are in the supplementary material

Problems for Chapter 1

Solutions to Problems for Chapter 1

Projects for Chapter 1

Appendix 1.1

Filter memory

The concept of *memory* as it relates to filters is a subject that is central to filter engineering, and in this appendix we explain what it means.

Referring to Figure A1.1.1, we assume that observations are being made on an external process.

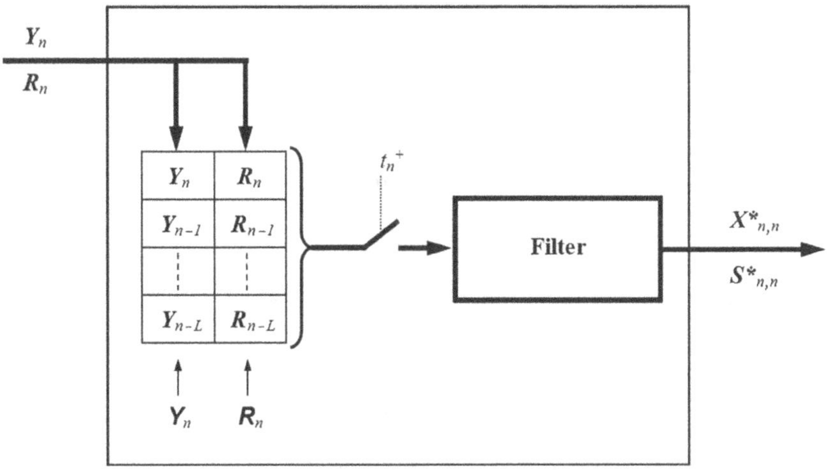

Figure A1.1.1 A filter and its input stack

The most recent observation is $\boldsymbol{Y}_n$ which could be either a scalar or a short-vector. The covariance matrix of $\boldsymbol{Y}_n$ is called $\boldsymbol{R}_n$.

The pair of quantities $(\boldsymbol{Y}_n, \boldsymbol{R}_n)$ together with their predecessors $(\boldsymbol{Y}_{n-1}, \boldsymbol{R}_{n-1})$, $(\boldsymbol{Y}_{n-2}, \boldsymbol{R}_{n-2})$, and so on, are stored in a stack. The switch is closed and the filter is cycled, and we obtain the estimate vector $\boldsymbol{X}^*_{n,n}$ and its covariance matrix $\boldsymbol{S}^*_{n,n}$.

Definition A1.1.1: The contents of the stack constitute what is in the filter's **memory**. ■■

The stack can be operated in three modes, and the filter's memory is correspondingly given the three names discussed in the following sections.

Mode 1: fixed memory

Figure A1.1.2 depicts a stack with fixed depth.

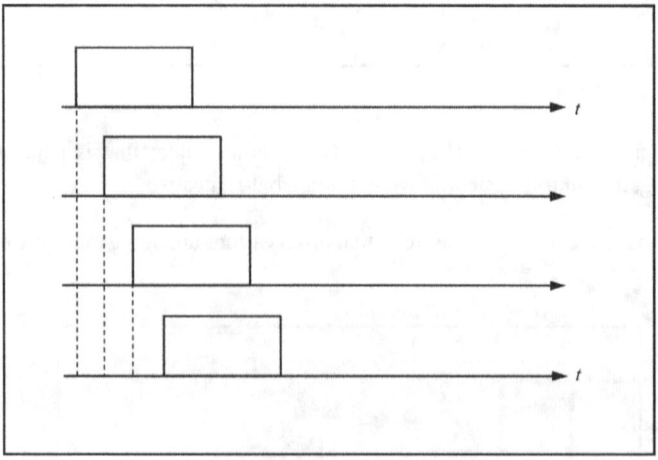

Figure A1.1.2 Fixed memory

◊ The boxes in the figure represent the stack on successive time ticks t_{n-3}, t_{n-2}, t_{n-1}, t_n, and so on.

◊ The depth of the stack is represented by the width of the box.

  ◊ The depth is fixed, and so the width never changes.

  ◊ As time progresses, the leading edge of the box moves forward and the trailing edge follows at a constant distance behind.

◊ When each new pair $(\boldsymbol{Y}, \boldsymbol{R})$ is received, the stack is pushed down, the new pair is placed at the top and the oldest pair is dropped.

◊ The contents of the stack are then processed by the filter.

When a filter is fed from a stack that operates in this way, we say that it has a *fixed memory* or a *fixed-length memory*.

The Gauss filters operate naturally with a fixed memory.

Mode 2: variable-length memory

Figure A1.1.3 depicts a stack with variable depth.

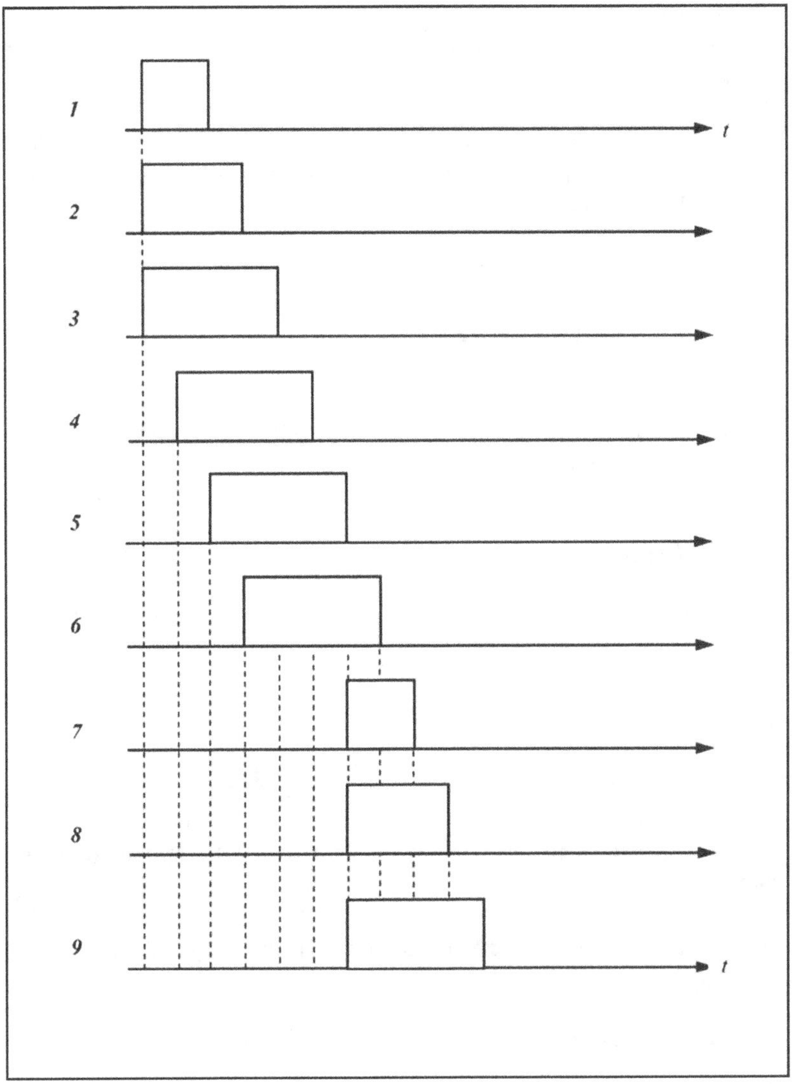

Figure A1.1.3 Variable memory

◊ The dashed lines are again the time ticks and the width of the box represents the depth of the stack.

◊ As time progresses, the leading edge of the box always moves to the right, but its trailing edge follows in an unrelated way.

  ◊ In rows 1, 2 and 3 the trailing edge remains fixed (on each time tick the newest data pair is placed at the top of the stack and nothing is dropped). The memory length is expanding.

  ◊ In rows 4, 5 and 6 the trailing edge follows at a fixed distance behind the leading edge (on each time tick the newest data pair is placed at the top of and the oldest data pair is dropped). The memory length is fixed.

  ◊ In row 7 the trailing edge has jumped forward (the newest data pair was placed at the top and the three oldest data pairs were dropped). The memory length has shortened substantially.

  ◊ In rows 7, 8 and 9 the trailing edge is again fixed (on each time tick the newest data pair is placed at the top and nothing is dropped). The memory length is expanding.

When a filter is fed from a stack that is operated in this way then we say that it has a *variable-length memory.*

The Gauss filters also operate naturally with a variable-length memory.

Mode 3: expanding memory

Figure A1.1.4 below depicts a stack with an expanding depth.

◊ The trailing edge of the box is always fixed at the time at which the filter was started.

◊ On each time tick the newest pair is placed at the top of the stack and nothing is ever dropped.

◊ As time progresses, the leading edge moves steadily forward.

When a filter is fed from a stack that is operated in this way then we say that it has an *expanding memory.*

The Gauss filters also operate naturally with an expanding memory.

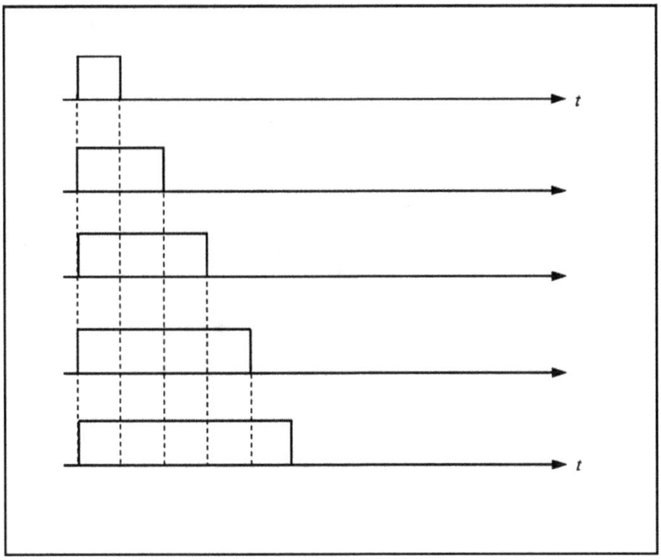

Figure A1.1.4 Expanding memory

Fading

When the observations are stage-wise uncorrelated[1] then *fading* is an attribute that can easily be superimposed onto any of the above three modes of the Gauss filters.

In Project 8.5 we consider what fading is and how it is implemented.

[1] The meaning of *stage-wise uncorrelated* is discussed in Ch. 5.

Chapter 2
Models, differential equations and transition matrices

Models in many disciplines are specified by algebraic equations. In filter engineering, they are always specified by differential equations (DEs), and in this chapter we develop the necessary background to enable us to use DEs as models.

For each DE we will see that there is a unique *transition matrix*, and it is through the transition matrix that the DE is actually implemented. Our discussion is thus about DEs and their transition matrices, and how such matrices are derived.

2.1 Linearity

Definition 2.1:

1. Let x_1 and x_2 be two scalar variables, and let a and b be constants. Then the expression

   $$ax_1 + bx_2$$

   is called a **linear combination of x_1 and x_2**. Any other combination is called a **nonlinear combination**.

   Similar definitions can be stated for m variables $x_1,\ x_2 \ldots x_m$ where m is any positive integer.

2. A **linear equation** is one in which the dependent variable is equal to a linear combination of the independent variables. ■■

Example 2.1: In the following equation the dependent variable x_1 is equal to a linear combination of the independent variables x_2 and x_3:

$$x_1 = 3x_2 + 5x_3 \tag{2.1.1}$$

It is, therefore, a linear equation. In each of the following, x_1 is equal to a nonlinear combination of x_2 and x_3, and so they are all nonlinear equations:

$$x_1 = 3x_2^2 + 5x_3 \tag{2.1.2}$$

$$x_1 = x_2 x_3 \tag{2.1.3}$$

$$x_1 = \sin(x_2) \tag{2.1.4}$$

$$x_1 = 3x_2 + 5x_3 + 1 \tag{2.1.5}$$

In the first three examples it is easy to see that the groups on the right are nonlinear combinations of x_2 and x_3. In the fourth example, the group on the right is *almost* but not quite a linear combination. In fact, it is definitely not linear because of the presence of the *1*. ■■

2.1.1 Sets of linear equations

The concept of linearity extends easily to *sets* of equations. As an example, consider

$$\begin{pmatrix} y_1 \\ y_2 \\ y_3 \end{pmatrix} = \begin{pmatrix} a_{1,1} & a_{1,2} & a_{1,3} \\ a_{2,1} & a_{2,2} & a_{2,3} \\ a_{3,1} & a_{3,2} & a_{3,3} \end{pmatrix} \begin{pmatrix} x_1 \\ x_2 \\ x_3 \end{pmatrix} \tag{2.1.6}$$

This is three equations in which the dependent variables y_1, y_2, y_3 are equal to linear combinations of the independent variables x_1, x_2, x_3, the first of them being

$$y_1 = a_{1,1} x_1 + a_{1,2} x_2 + a_{1,3} x_3 \tag{2.1.7}$$

Equation (2.1.6) is symbolized as

$$\boldsymbol{Y} = \boldsymbol{A}\boldsymbol{X} \tag{2.1.8}$$

in which $\boldsymbol{X}$ and $\boldsymbol{Y}$ are vectors and $\boldsymbol{A}$ is a matrix, and the combination is called a *linear transformation*.

Not all combinations of vectors and matrices are linear. As an example consider

$$y = \boldsymbol{X}^T \boldsymbol{A}\boldsymbol{X} \tag{2.1.9}$$

in which the dependent variable y is a scalar, $\boldsymbol{X}$ is (for example) a *2*-vector of independent variables and $\boldsymbol{A}$ is a 2×2 matrix of constants. Writing (2.1.9) in full we obtain

$$y = \begin{pmatrix} x_1 & x_2 \end{pmatrix} \begin{pmatrix} a_{1,1} & a_{1,2} \\ a_{2,1} & a_{2,2} \end{pmatrix} \begin{pmatrix} x_1 \\ x_2 \end{pmatrix} \tag{2.1.10}$$

from which

$$y = a_{1,1} x_1^2 + a_{1,2} x_1 x_2 + a_{2,1} x_2 x_1 + a_{2,2} x_2^2 \tag{2.1.11}$$

Clearly y is not equal to a linear combination of x_1 and x_2 and so (2.1.10) is not a linear equation. Expressions like (2.1.9) are called *quadratic forms* and $\boldsymbol{A}$ is called the *coefficient matrix*. Quadratic forms will occur frequently in our discussion.

2.1.2 Linear independence

A set of vectors can possess a property known as *linear independence*, and in what lies ahead an understanding of that important concept will be required. Linear independence is reviewed in Appendix 2.1 and its associated problems.

2.1.3 Linearity and differential equations

The concept of linearity extends easily to DEs.

Definition 2.2: $Dx(t)$ means $dx(t)/dt$, $D^m x(t)$ means $d^m x(t)/dt^m$, $\dot{x}(t)$ means $dx(t)/dt$, and $\ddot{x}(t)$ means $d^2x(t)/dt^2$. ■■

The following is an example of a 2nd-degree, constant-coefficient, linear, homogeneous, ordinary DE:

$$\ddot{x}(t) = ax(t) + b\dot{x}(t) \tag{2.1.12}$$

◊ The DE is 2nd-degree because the second derivative is the highest that occurs.

◊ It is constant-coefficient because the multipliers a and b are constants.

◊ It is ordinary because it contains only ordinary derivatives and no partial derivatives. Since that will always be the case in this book, we will not mention it again.

◊ It is linear because the dependent variable $\ddot{x}$ on the left is equal to a linear combination of the independent variables x and $\dot{x}$ on the right.

◊ We explain what homogeneous means by considering the following three DEs:

$$\ddot{x}(t) = ax(t) + b\dot{x}(t) \tag{2.1.13}$$

$$\ddot{x}(t) = ax(t) + b\dot{x}(t) + sin(t) \tag{2.1.14}$$

$$\ddot{x}(t) = ax(t) + b\dot{x}(t) + 2 \tag{2.1.15}$$

In (2.1.13), all functions in the DE belong to the set $S = \{x(t), \dot{x}(t), \ddot{x}(t)\}$, and membership in that set makes them all *similar*.

In (2.1.14), we see the same three terms that appeared in (2.1.13) but we also have the extra term $sin(t)$ which is not a member of S – the term $sin(t)$ is *dissimilar*.

In (2.1.15), we see again the terms that appeared in (2.1.13) but now we have the extra term *2*, which also is not a member of S – it too is dissimilar.

Definition 2.3: If we say that a linear DE is **homogeneous**, then we mean that all of its terms are similar. ■■

The dissimilar terms in (2.1.14) and (2.1.15) are called *forcing functions,* and so this means that if a linear DE is homogeneous then it contains no forcing functions.

We stated in Chapter 1 that the only types of DEs that we will consider are ones that have no forcing functions, and so all linear DEs that we consider in this book will be homogeneous.

2.1.4 Constant coefficient linear DEs

The DE in (2.1.12) can be reformulated into what we call *matrix/vector format* or *state-vector format*, as follows:

$$D\begin{pmatrix} x \\ \dot{x} \end{pmatrix} = \begin{pmatrix} 0 & 1 \\ a & b \end{pmatrix}\begin{pmatrix} x \\ \dot{x} \end{pmatrix} \qquad (2.1.16)$$

On the left we have D followed by a vector, and on the right a matrix multiplying that same vector. The first row states that $Dx = \dot{x}$ which is self-evident, and the second row that $D\dot{x} = ax + b\dot{x}$, which is what we were given in (2.1.12).

The reformulation shown in (2.1.16) is an instance of the more general form

$$DX(t) = AX(t) \qquad (2.1.17)$$

Definition 2.4: If a DE in matrix/vector format is as shown in (2.1.17), in which A is a matrix of constants, then it is said to be **constant-coefficient linear**. ■■

Definition 2.5: The vector X in (2.1.17) contains the smallest number of elements necessary to define the state of the DE in state-space at time t, and is called a **state vector**. ■■

State vectors need not contain only $x(t)$ and its derivatives. Further down we will see an example in which a state vector contains a parameter as well.

Equations (2.1.12) and (2.1.16) represent the same DE, but the format of (2.1.16) is the way in which DEs are usually stated in filter engineering. From now on, whenever discussing DEs we will try, if possible, to present them in state-vector format.

Example 2.2: The DE for 2nd-degree polynomials is

$$D^3 x(t) = 0 \qquad (2.1.18)$$

In state-vector format it becomes

$$D\begin{pmatrix} x(t) \\ \dot{x}(t) \\ \ddot{x}(t) \end{pmatrix} = \begin{pmatrix} 0 & 1 & 0 \\ 0 & 0 & 1 \\ 0 & 0 & 0 \end{pmatrix}\begin{pmatrix} x(t) \\ \dot{x}(t) \\ \ddot{x}(t) \end{pmatrix} \tag{2.1.19}$$

The first line states that $Dx(t) = \dot{x}(t)$ and the second that $D\dot{x}(t) = \ddot{x}(t)$, both of which are self-evident. The third line states that $D\ddot{x}(t) = 0$ which is what we were given in (2.1.18), and so this is the DE for second-degree polynomials in state-vector format.

We note that the matrix $\boldsymbol{A}$ in (2.1.19) has a particularly simple structure – it is all zeros, except for the super-diagonal which is all ones – and it is clear that $\boldsymbol{A}$ for polynomials of any degree will have a similar structure.

We observe that the structure of (2.1.19) is the same as (2.1.17), and so the DEs for polynomials are constant-coefficient linear. ■■

Example 2.3: The following well-known DE is used to model simple harmonic motion[1], i.e. motion made up of sines and cosines of a single frequency:

$$\ddot{x}(t) = -\omega^2 x(t) \qquad (\omega \textit{ a known nonzero constant}) \tag{2.1.20}$$

In matrix/vector format the DE becomes

$$D\begin{pmatrix} x(t) \\ \dot{x}(t) \end{pmatrix} = \begin{pmatrix} 0 & 1 \\ -\omega^2 & 0 \end{pmatrix}\begin{pmatrix} x(t) \\ \dot{x}(t) \end{pmatrix} \qquad (\omega \neq 0) \tag{2.1.21}$$

This is of the same form as (2.1.17), and so it is constant-coefficient linear. ■■

2.1.5 Time-varying linear DEs

In filter engineering we also permit the multipliers *a, b,* etc. to change with time. Consider for example the DE

$$\ddot{x} = 3(t+1)x + t^2\dot{x} \tag{2.1.22}$$

If we set *t* to a specific value in (2.1.22) then it has precisely the same form as (2.1.12), and so (2.1.22) is a linear DE for all *t*. In state-vector format it becomes

$$D\begin{pmatrix} x \\ \dot{x} \end{pmatrix} = \begin{pmatrix} 0 & 1 \\ 3(t+1) & t^2 \end{pmatrix}\begin{pmatrix} x \\ \dot{x} \end{pmatrix} \tag{2.1.23}$$

which can be written as

$$D\boldsymbol{X}(t) = \boldsymbol{A}(t)\boldsymbol{X}(t) \tag{2.1.24}$$

[1] See Reference 190.

Observe that here the matrix is shown as $A(t)$ because at least one of its elements is a function of t, whereas in (2.1.17) the matrix was shown as A because its elements were all constants.

Definition 2.6: If a DE has the structure shown in (2.1.24) in which at least one element of the matrix $A(t)$ is a function of t, then it is said to be **time-varying linear**. ■■

Example 2.4: Consider the DE

$$\ddot{x}(t) = -(t+1)^2 x(t) \tag{2.1.25}$$

In state-vector format it becomes

$$D\begin{pmatrix} x(t) \\ \dot{x}(t) \end{pmatrix} = \begin{pmatrix} 0 & 1 \\ -(t+1)^2 & 0 \end{pmatrix}\begin{pmatrix} x(t) \\ \dot{x}(t) \end{pmatrix} \tag{2.1.26}$$

This is a time-varying linear DE of the same form as (2.1.24). ■■

2.1.6 Nonlinear DEs

Consider finally the nonlinear DE

$$\ddot{x}(t) = f\left(x(t),\ \dot{x}(t)\right) \tag{2.1.27}$$

in which the argument of f is assumed to be a combination of the variables $x(t)$ and $\dot{x}(t)$ that is nonlinear. An example of such a DE would be

$$\ddot{x}(t) = \dot{x}(t) + sin\left(x(t)\right) \tag{2.1.28}$$

Writing this DE only in terms of the variables it becomes

$$\ddot{x} = \dot{x} + sin(x) \tag{2.1.29}$$

in which the variable t does not appear explicitly. By contrast, in the following two DEs the independent variable t *does* appear explicitly:

$$\ddot{x}(t) = \dot{x}(t) + sin(t) \tag{2.1.30}$$

and

$$\ddot{x}(t) = \dot{x}(t) + t\ sin\left(x(t)\right) \tag{2.1.31}$$

and this is true even when we write them in terms of only their variables $\ddot{x}$, $\dot{x}$ and x.

Thus

$$\ddot{x} = \dot{x} + sin(t) \tag{2.1.32}$$

and

$$\ddot{x} = \dot{x} + t\, sin(x) \tag{2.1.33}$$

in both of which t appears explicitly.

We spoke previously about linear DEs being either homogeneous or not. In the case of nonlinear DEs the language changes slightly, and we speak about them being either *autonomous* or not.

Definition 2.7: If a DE has a form like (2.1.27) in which the argument of f is a nonlinear combination of the state variables, and in which the independent variable t does not appear explicitly, then it is said to be an **autonomous nonlinear DE**. ■■

By this definition, (2.1.28) is an autonomous nonlinear DE, whereas (2.1.30) and (2.1.31) are non-autonomous nonlinear DEs.

Many laws in nature are the same for *now* as they were in the *past* and will be in the *future.* Therefore, t does not appear in them explicitly, and so they are expressible as autonomous DEs.

An example would be Newton's DE for simple orbital motion that appeared in Chapter 1, namely

$$D^2\boldsymbol{r} = -(\mu/r^3)\,\boldsymbol{r} \tag{2.1.34}$$

The law defined by this equation is the same for any point in time – now, past or future. The independent variable t does not appear explicitly, and so it is an autonomous nonlinear DE.

In the case of nonlinear DEs, forcing functions are always time dependent. We are restricting ourselves to DEs without forcing functions, and so in the nonlinear case they will always be autonomous.

Example 2.5: Consider the DE

$$\ddot{x}(t) = -\omega^2 x(t) \quad (\omega \textit{ is an unknown nonzero constant that is to be estimated}) \tag{2.1.35}$$

At first sight this DE appears to be the same as the one for simple harmonic motion in (2.1.20), and so we might be tempted to write it in state-vector format as we did in

(2.1.21). However, the comment in brackets after the DE requires that we introduce a further concept.

The primary objective of filter engineering is to estimate the state vector of the process being observed (see Figure 1.6). Thus, if we wish to obtain an estimate of the unknown constant ω in (2.1.35), then we will have to include it in the DEs state vector.

That in turn means that we will have to write the DE in state-vector format as follows:

$$D\begin{pmatrix} x(t) \\ \dot{x}(t) \\ \omega \end{pmatrix} = \begin{pmatrix} 0 & 1 & 0 \\ -\omega^2 & 0 & 0 \\ 0 & 0 & 0 \end{pmatrix}\begin{pmatrix} x(t) \\ \dot{x}(t) \\ \omega \end{pmatrix} \tag{2.1.36}$$

The first row states that $Dx(t) = \dot{x}(t)$, and the second row that $D\dot{x}(t) = -\omega^2 x(t)$, which is the DE that we were given. The third row states that $D\omega = 0$, making ω a constant. In this way ω has been included in the DEs state vector, and so the filter will estimate it.

However, by including ω in the state vector, we have created a problem – the term $-\omega^2 x(t)$ that appears in (2.1.36) is a nonlinear combination of two of the state variables, and so the DE is now nonlinear.

We will return to (2.1.36) on a number of occasions because it possesses some extremely useful properties. ■■

It will seldom be possible to write nonlinear DEs in the state-vector format shown in (2.1.36). However, we will always be able to write them in the general form for autonomous nonlinear DEs, namely

$$DX(t) = F(X(t)) \tag{2.1.37}$$

Definition 2.8: If a DE has the form shown in (2.1.37) in which F is a vector of combinations of the state variables, at least one of which is nonlinear, and in which t does not appear explicitly, then the DE is said to be an **autonomous nonlinear DE**. ■■

Note: The general form for non-autonomous nonlinear DEs is $DX(t) = F(X(t), t)$.

Example 2.5 (continued): For the DE in (2.1.36) the elements of the vector F would be

$$f_1\big(x(t),\ \dot{x}(t),\ \omega\big) = \dot{x}(t) \tag{2.1.38}$$

$$f_2\big(x(t),\ \dot{x}(t),\ \omega\big) = -\omega^2 x(t) \tag{2.1.39}$$

$$f_3\big(x(t),\ \dot{x}(t),\ \omega\big) = 0 \tag{2.1.40}$$

We observe that $\boldsymbol{F}$ is a 3-vector, i.e. $\boldsymbol{F} = (f_1, f_2, f_3)^T$, and that its three elements f_1, f_2, f_3 are each combinations of x, $\dot{x}$ and ω, at least one of which is nonlinear. ■■

Making use of (2.1.38), (2.1.39) and (2.1.40), we can now write equation (2.1.36) as follows:

$$D\begin{pmatrix} x(t) \\ \dot{x}(t) \\ \omega \end{pmatrix} = \begin{pmatrix} \dot{x}(t) \\ -\omega^2 x(t) \\ 0 \end{pmatrix} \tag{2.1.41}$$

◊ Observe the presence of the DEs state vector on the left of (2.1.41), even though the right-hand side is not expressible as a matrix that is multiplying the state vector. Linear DEs in state-vector format always have a matrix on the right multiplying their state vector, whereas nonlinear DEs can seldom be shown in that way. For them, we will have to content ourselves with the format shown in (2.1.37).

◊ Observe also the presence of ω in the state vector on the left of (2.1.41). This is an instance where a state vector is comprised of variables like $x(t)$ and $\dot{x}(t)$, as well as one or more parameters.

◊ Finally, observe that the DE in (2.1.41) is autonomous, because the independent variable t does not appear explicitly. This means that no forcing function is present.

2.2 The two types of models

In what lies ahead we will speak frequently about models. We raised this subject in Chapter 1, and we now elaborate further. There are two types of models:

I. The model that we believe adequately describes the physical process that we are observing. This is a construct that is external to the filter, and is called the *external* model.

II. The model on which the filter is based. This construct is internal to the filter, and is called the *filter* model.

It is essential that we think about external models and filter models as two totally separate entities.

The filter model will be expected to emulate the external model, which in turn is assumed to be an appropriate way to describe the physical process in mathematical terms.

◊ The filter model may be identical to the external model.

◊ The filter model may approximate the external model.

Either way, the filter's ability to function successfully will depend on three factors:

1. Whether or not the external model is a good representation of the physical process that is being observed.

2. Whether or not the filter model is a good emulation of the external model.

3. Whether or not we are making observations on the physical process in such a way that the filter is properly able to estimate all of the required quantities.

We defer discussion of the third item until Chapter 5, and for now consider only the first two.

2.2.1 *The external model*

Definition 2.9: An **external model** is a DE that is believed to be an appropriate way in which to describe a physical process. ■■

The external model is only implemented if we are running a simulation, and we wish to generate simulated observations, i.e. values to which appropriately selected random errors are added and are then processed by the filter.

If an analytical solution to the DE is available – as in simple satellite work – then that is the easiest way to generate the observations. However, if an analytical solution is not available – as in a more comprehensive satellite simulation – then we will have to apply numerical integration to the DE. (In the folder *Miscellaneous* in the supplementary material we have included a full discussion on numerical integration.)

In the field, where the process under observation is part of physical reality, the external model's DE is not usually implemented.

◄►

There can be more than one external model for a given physical phenomenon. These can take a number of forms, some complex and some simple, and perhaps the best way to convey what we have in mind is by two examples.

Example 2.6:* *Satellite motion (a complex external model):

◊ Implement (2.1.34), but include more than one large body, i.e. the earth, sun and moon.

◊ Include the fact that some of the large bodies may not be spherical, but have instead some arbitrary shape (such as the earth does). Specify those shapes, and include an evaluation of the gravitational forces as a function of the satellite's instantaneous position in relation to those large bodies.

◊ Include also the effects of relativity, as was done for the GPS system, or must be done for the planet Mercury if we expect the orbit in the model to precess. ■■

Example 2.7: ***Satellite motion (a simple external model):***

◊ Implement (2.1.34) for a small body moving under the effect of the gravitational force of the earth.

◊ Assume that the earth is a point mass.

◊ Ignore the effects of the sun and moon.

◊ Ignore all effects of relativity. ■■

The first external model would be much more faithful, while the second would be far easier to implement. We note that our external models are both stated as DEs.

We trust that the reader now has a better understanding of what we have in mind when we speak about an external model, and that such models can run from extremely complex to very basic.

2.2.1.1 Fundamental assumption on which this book is based

Before moving on to consider the second type of model, namely the filter model, we restate the fundamental assumption on which this book is based:

It will always be assumed that any physical phenomena that we observe are modelled by deterministic DEs that are without forcing functions – neither stochastic nor deterministic.

This is not meant to imply that stochastic DEs with or without forcing functions do not have a place in filter engineering, but simply that we do not have the space in this book to consider them.

2.2.2 The filter model

As the first step in implementing a filter, we will be required to specify the filter model.

Definition 2.10: The **filter model** is a deterministic DE on which we build a filter. ■■

The filter model will have to adequately emulate the external model of the process that we are observing. (We will have more to say about the meaning of 'adequate' in Chapter 6.) Since we will only use external models that have no forcing functions, it follows that any filter models that we use will also be without forcing functions.

We include among the possibilities for filter models the DEs for polynomials, which possess the extremely useful property of being able to approximate arbitrary external models over an appropriate time interval.

Example 2.8: We are observing a near earth satellite. We decide that Newton's equation for orbital motion in (2.1.34) will be the external model, and we make the decision to use a 2nd-degree polynomial as the filter model.

Setting aside for now the fact that the problem is three-dimensional, the filter model would then be the same as (2.1.19). ■■

In the example just considered, the external model and the filter model are not the same. However, we believe that the former can be adequately approximated by the latter over an appropriate time interval. In the next example the filter model and the external model *are* the same.

Example 2.9: As a result of carrying out an extensive investigation, two American physicists J. H. Allen and A. J. Eggers arrived at a complex autonomous nonlinear DE for modelling the re-entry of a body into the earth's atmosphere.[2] It appears in (2.2.1) in state-vector format and constitutes an external model for the re-entry phenomenon.[3]

$$D\begin{pmatrix} x_1 \\ x_2 \\ x_3 \\ x_4 \\ x_5 \\ x_6 \end{pmatrix} = \begin{pmatrix} x_4 \\ x_5 \\ x_6 \\ ce^{-kx_3}(x_4^2 + x_5^2 + x_6^2)^{1/2} x_4 \\ ce^{-kx_3}(x_4^2 + x_5^2 + x_6^2)^{1/2} x_5 \\ ce^{-kx_3}(x_4^2 + x_5^2 + x_6^2)^{1/2} x_6 - g \end{pmatrix} \tag{2.2.1}$$

Notes:

◊ $x_1, x_2, x_3, x_4, x_5, x_6$ mean $x(t), y(t), z(t), \dot{x}(t), \dot{y}(t), \dot{z}(t)$.

◊ $c = -\rho_0\alpha/2m$, where m is the body's mass and α is its drag coefficient. The quantities ρ_0 and k are constants related to atmospheric pressure.

[2] See Reference 21.

[3] There are re-entry models that are far more complex than the Allen-Eggers equation.

◊ The DE in (2.2.1) is nonlinear, as can be seen from the final three expressions in its right-hand side. It is also autonomous because the independent variable t does not appear in its statement.

◊ The gravitational quantity g appearing in the final equation is shown as a constant, but it could quite easily be written as $g(x_3)$ making it dependent on the distance from the earth's centre. However, even if left as a constant the term g still does not constitute a forcing function, as constants do for linear DEs, but is instead part of the statement of an autonomous nonlinear DE – g does not depend explicitly on the time-variable t and so it is not a forcing function.

◊ For purposes of simplification, equation (2.2.1) has been written on the assumption that the earth is flat and non-rotating. It could be reworked to include the fact that the earth is nearly elliptical and rotating.

If we decide also to use the Allen-Eggers DE as the filter model, then this would be a situation in which the filter model emulates the external model perfectly. If the external model is a good representation of what goes on in the physical world, then so will be the filter model. ■■

In the developments to come we will always be called on to specify a filter model. The filter model that we specify may or may not be the same as any of the external models that we are aware of. Our choice will be based on what we expect the filter to accomplish and what resources are available for its execution.

Unlike the external model, the filter model is always implemented – something that takes place inside the filter. As we will soon see, this is done by incorporating its associated transition matrix into the filter equations.

2.2.3 Our approach to DEs

Our approach to DEs is shown in Figure 2.1:

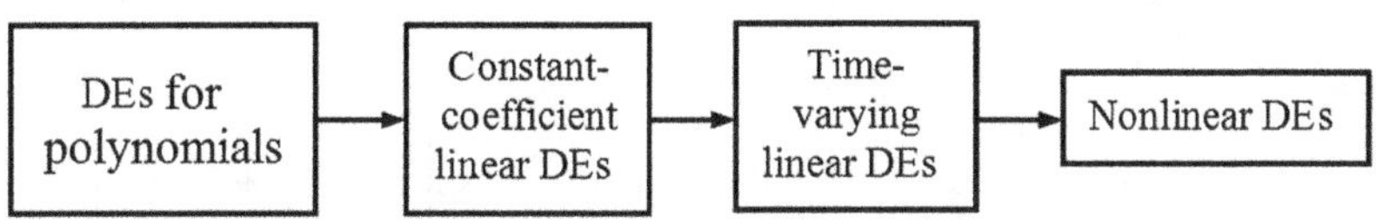

Figure 2.1 Using DEs as models

◊ First, we will consider the simplest case – the DEs for polynomials.

◊ What we learn from them will then be extended to constant-coefficient linear DEs.

◊ The next extension will be to time-varying linear DEs.

◊ Finally, we will examine the most general types – nonlinear DEs.

We will see that only the first and last – the DEs for polynomials and nonlinear DEs – are of practical interest in filter engineering, but in order to understand how to make use of nonlinear DEs we must start with the first, and then progress from there through the intermediate two.

2.3 Models based on polynomials

The simplest types of DEs for filter models are those of polynomials, which we saw earlier are a subset of constant-coefficient linear DEs.

Polynomials can be used as models over short enough time intervals with very little actual knowledge of the external model,[4] and they also give rise to the extremely compact smoothing algorithms that are discussed in Chapters 12 and 13.

Assume that our filter model is a 2nd-degree polynomial. Then from (2.1.19) we recall that its DE will be

$$D\begin{pmatrix} x(t) \\ \dot{x}(t) \\ \ddot{x}(t) \end{pmatrix} = \begin{pmatrix} 0 & 1 & 0 \\ 0 & 0 & 1 \\ 0 & 0 & 0 \end{pmatrix}\begin{pmatrix} x(t) \\ \dot{x}(t) \\ \ddot{x}(t) \end{pmatrix} \tag{2.3.1}$$

Equation (2.3.1) is the DE for 2nd-degree polynomials, and so the coefficient matrix A is 3×3. In general, for an mth-degree polynomial A would be $(m+1) \times (m+1)$, with a similar structure.

2.3.1 Notation

If we differentiate the mth-degree polynomial $x(t)$ repeatedly with respect to t and then evaluate each of the intermediate results at $t = t_n$ we obtain the following $m+1$ quantities:

$$x(t)\Big|_{t=t_n} \quad Dx(t)\Big|_{t=t_n} \quad D^2x(t)\Big|_{t=t_n} \quad \ldots\ D^jx(t)\Big|_{t=t_n} \quad \ldots\ D^mx(t)\Big|_{t=t_n} \tag{2.3.2}$$

The vertical line followed by $t = t_n$ is the usual way in which to show that $D^jx(t)$ has been evaluated at $t = t_n$. We now introduce notation by which the quantities in (2.3.2) can be represented more compactly.

[4] We have in mind here the celebrated polynomial approximation theorem by Weierstrass which states, roughly, that any reasonably smooth function can be approximated to arbitrary precision by a polynomial of sufficiently high degree. (See Reference 196.)

Definition 2.11: The abbreviation $D^j x_n$ means $d^j x(t)/dt^j$ evaluated at $t = t_n$, i.e.

$$D^j x_n \equiv d^j x(t)/dt^j \Big|_{t=t_n} \tag{2.3.3}$$

■■

As j goes from *0* to m in (2.3.3) we obtain the $m + 1$ abbreviations

$$x_n,\ \dot{x}_n,\ \ddot{x}_n,\ \ldots D^m x_n$$

and if these are now assembled into a vector we obtain

$$X(t_n) = (x,\ \dot{x},\ \ddot{x},\ \ldots D^m x)_n^T \tag{2.3.4}$$

Observe the single subscript n outside the right bracket in (2.3.4). It applies to all of the quantities inside the brackets and signifies that all of them have been evaluated at $t = t_n$.

2.3.2 The transition matrix and the transition equation

Each of the linear DEs that we will consider (polynomial, constant-coefficient linear and time-varying linear) gives rise uniquely to what are called a *transition matrix* and a *transition equation*. In filter engineering, we use them as implementations of the DE, and we now examine what they are. As a start we have the following:

Definition 2.12: The evaluation instant t_n in $\boldsymbol{X}(t_n)$ is called the **validity instant** of the state vector $\boldsymbol{X}(t_n)$. ■■

It is often the case that we are required to change the validity instant. Thus, given the model appearing in (2.3.1), let its state vector $\boldsymbol{X}(t)$ be evaluated at the two instants t_n and $t_n + \zeta$, to give the vectors $\boldsymbol{X}(t_n)$ and $\boldsymbol{X}(t_n + \zeta)$. Clearly, these must be related because they came from the same DE, and in Appendix 2.2 we show the following:

There exists a matrix $\boldsymbol{\Phi}(\zeta)$ which transforms the first vector into the second, namely:

$$\boldsymbol{X}(t_n + \zeta) = \boldsymbol{\Phi}(\zeta)\boldsymbol{X}(t_n) \tag{2.3.5}$$

We refer to this operation as *shifting the validity instant.*

Observe in (2.3.5) how multiplication of $\boldsymbol{X}(t_n)$ by $\boldsymbol{\Phi}(\zeta)$ has shifted the validity instant of $\boldsymbol{X}(t_n)$ to produce $\boldsymbol{X}(t_n + \zeta)$.

The matrix $\boldsymbol{\Phi}(\zeta)$ and its use in equation (2.3.5) occurs repeatedly in filter engineering, and we will see it often in what lies ahead.

Definition 2.13:

◊ $\boldsymbol{\Phi}(\zeta)$ is called the **transition matrix** of the selected DE.

◊ Equation (2.3.5) is called the **transition equation** of the selected DE.

◊ ζ is called the **span** of the transition. ■■

In Appendix 2.2, we show that for the DE for polynomials of degree *m,* the expression for the i, jth term of $\boldsymbol{\Phi}(\zeta)$ is given by

$$[\boldsymbol{\Phi}(\zeta)]_{i,j} = \zeta^{j-i}/(j-i)! \qquad 0 \le i, j \le m \tag{2.3.6}$$

where $1/(j-i)! = 0$ when $j - i$ is negative.

Example 2.10: Let the degree of the polynomial be $m = 3$. Then (2.3.6) gives the expressions for the elements of the transition matrix $\boldsymbol{\Phi}(\zeta)$ shown in Figure 2.2:

	$j = 0$	1	2	3
$i = 0$	1	ζ	$\zeta^2/2!$	$\zeta^3/3!$
1	0	1	ζ	$\zeta^2/2!$
2	0	0	1	ζ
3	0	0	0	1

Figure 2.2 Elements of $\boldsymbol{\Phi}(\zeta)$ for $m = 3$

■■

Example 2.11: Let the polynomial be of degree $m = 2$. Then (2.3.5) becomes

$$\begin{pmatrix} x \\ \dot{x} \\ \ddot{x} \end{pmatrix}_{t_n+\zeta} = \begin{pmatrix} 1 & \zeta & \zeta^2/2! \\ 0 & 1 & \zeta \\ 0 & 0 & 1 \end{pmatrix} \begin{pmatrix} x \\ \dot{x} \\ \ddot{x} \end{pmatrix}_{t_n} \tag{2.3.7}$$

■■

In Figure 2.3 below we show how the DE, the transition matrix and the transition equation are related.

So far we have only been considering the simplest form of models, namely those of polynomials. However, as we progress further and consider the more complex constant-coefficient and time-varying linear models, the same will always be true, namely:

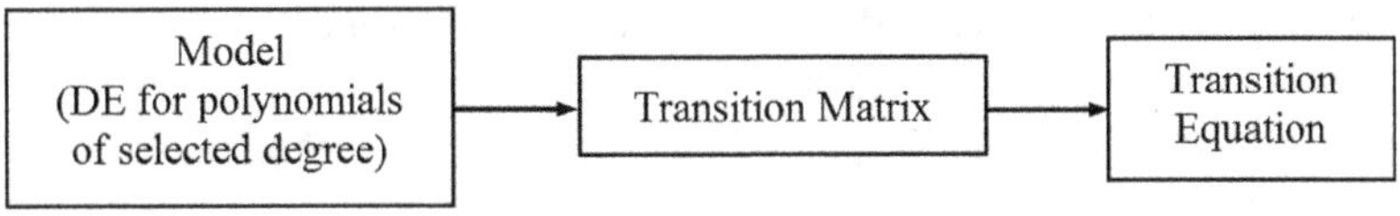

Figure 2.3 From the DE for polynomials to their transition equation

◊ The DE for the model will give rise to a unique transition matrix, which can then be used to set up a transition equation.

2.3.3 The curve implied by a transition equation

In (2.3.5), we see a linear transformation in which two vectors are related by a matrix. However, there is also a curve implied by that equation, in the following sense:

Recall that the first element of the state vector $\boldsymbol{X}(t_n + \zeta)$ is $x(t_n + \zeta)$. Then a plot of $x(t_n + \zeta)$ vs. ζ is the curve that is implied by (2.3.5).

2.3.4 The observed trajectory

Referring to Figure 2.4, every DE gives rise to infinitely many solutions depending on the choice of initial conditions.

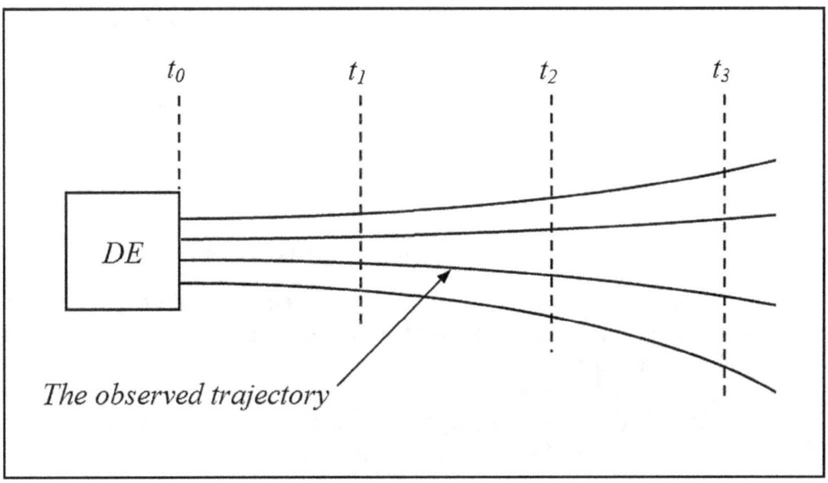

Figure 2.4 The observed trajectory

Expressed in terms of their state vectors, those solutions all have a 0th derivative, a 1st derivative and so on. In the figure we show only the 0th derivatives, each of which is known as a *trajectory*. (These are the curves that are implied by the transition equation.)

The vertical dashed lines marked t_0, t_1, t_2 and t_3 show validity instants. Starting with a state vector at any validity instant on one of the trajectories, the transition equation (2.3.5) enables us to find the state vector at any other validity instant on the *same* trajectory.

This is equivalent to numerical integration of the DE, starting from the set of initial conditions contained in the first state vector. However, the transition equation requires only a single matrix multiplication to achieve the same result, and so it is easier to use than numerical integration.

(We will see later that numerical integration cannot always be avoided.)

One of the trajectories in the figure is called the *observed trajectory,* for which we have the following definition:

Definition 2.14: The trajectory that is evolving at the time that the observations are being made, is called the **observed trajectory.** ■■

2.3.5 *Working in 3-dimensional space*

Until now we have been considering models involving only the single variable $x(t)$ and its time derivatives, whereas there are three dimensions, x, y and z, in the Cartesian coordinates in which we operate.

◊ In three dimensions, and assuming a lst-degree polynomial model, we would use the state vector and transition matrix shown in (2.3.9) below.

◊ For a 2nd-degree polynomial model, the state vector and transition matrix would be as shown in (2.3.10) below, and so on.

Notes:

1. If a submatrix is left blank as in (2.3.9) or (2.3.10) then it is to be regarded as null.
2. In (2.3.9) and (2.3.10) we are using the symbol $\boldsymbol{X}$ as the name for the overall state vector, even though it contains elements involving y and z as well. It will almost always be the case that state vectors are named $\boldsymbol{X}$.

There is a second way to represent polynomial state vectors and transition matrices in three dimensions, namely to write them as follows:

For 1st-degree:

$$\boldsymbol{X}(t_n) = \begin{pmatrix} x & y & z \\ \dot{x} & \dot{y} & \dot{z} \end{pmatrix}_n \qquad \boldsymbol{\Phi}(\zeta) = \begin{pmatrix} 1 & \zeta \\ 0 & 1 \end{pmatrix} \tag{2.3.8}$$

and for 2nd-degree:

$$\boldsymbol{X}(t_n) = \left(\begin{array}{c} x \\ \dot{x} \\ \hdashline y \\ \dot{y} \\ \hdashline z \\ \dot{z} \end{array}\right)_n \qquad \boldsymbol{\Phi}(\zeta) = \left(\begin{array}{c:c:c} \boldsymbol{G}(\zeta) & & \\ \hdashline & \boldsymbol{G}(\zeta) & \\ \hdashline & & \boldsymbol{G}(\zeta) \end{array}\right) \qquad \boldsymbol{G}(\zeta) = \begin{pmatrix} 1 & \zeta \\ 0 & 1 \end{pmatrix} \tag{2.3.9}$$

$$\boldsymbol{X}(t_n) = \begin{pmatrix} x \\ \dot{x} \\ \ddot{x} \\ \hdashline y \\ \dot{y} \\ \ddot{y} \\ \hdashline z \\ \dot{z} \\ \ddot{z} \end{pmatrix}_n \qquad \boldsymbol{\Phi}(\zeta) = \begin{pmatrix} \boldsymbol{G}(\zeta) & & \\ & \boldsymbol{G}(\zeta) & \\ & & \boldsymbol{G}(\zeta) \end{pmatrix} \qquad \boldsymbol{G}(\zeta) = \begin{pmatrix} 1 & \zeta & \zeta^2/2! \\ 0 & 1 & \zeta \\ 0 & 0 & 1 \end{pmatrix} \tag{2.3.10}$$

$$X(t_n) = \begin{pmatrix} x & y & z \\ \dot{x} & \dot{y} & \dot{z} \\ \ddot{x} & \ddot{y} & \ddot{z} \end{pmatrix}_n \qquad \Phi(\zeta) = \begin{pmatrix} 1 & \zeta & \zeta^2/2! \\ 0 & 1 & \zeta \\ 0 & 0 & 1 \end{pmatrix} \tag{2.3.11}$$

Both ways work perfectly well, but the second one has the drawback of changing the state-vectors $X(t_n)$ in (2.3.9) and (2.3.10) into the matrices appearing in (2.3.8) and (2.3.11). Throughout this book we will use (2.3.9) and (2.3.10).

2.3.6 Equally spaced observation instants

Until now we have been assuming that the observation instants are arbitrarily spaced. When they are equally spaced, then major simplifications are possible.

Definition 2.15: Let the state vector Z_n of degree m be defined as follows:

$$Z_n \equiv \left(x,\ \tau\dot{x},\ (\tau^2/2!)\ddot{x},\ \ldots(\tau^m/m!)D^m x\right)_n{}^T \tag{2.3.12}$$

in which τ is the constant interval between observation instants. Then Z_n is called a **normalized state vector**. ■■

In Appendix 2.2, we arrive at the following. Let the time instant t_n be expressed in terms of τ as follows:

$$t_n = n\tau \tag{2.3.13}$$

and let the span of the transition be $h\tau$. The dimensionless variable h – which does not have to be an integer – thus specifies the number of τ-steps, and possibly fractions thereof, that make up the span of the transition. Then:

◊ The transition matrix for the normalized state vector Z_n is

$$[\Phi(h)]_{i,j} = \binom{j}{i} h^{j-i} \quad 0 \le i, j \le m \tag{2.3.14}$$

where

$$\binom{j}{i} = 0 \text{ when } j < i \text{ and } h^0 = 1 \text{ for } h \text{ positive, negative or zero.} \tag{2.3.15}$$

◊ The transition equation for Z is

$$Z_{n+h} = \Phi(h) Z_n \tag{2.3.16}$$

We note from (2.3.14) that $\boldsymbol{\Phi}$ is independent of τ and depends only on h. ■■

Example 2.12: Let the polynomial degree be $m = 2$ and let the span of the transition be $h\tau$ in which $h = 1$. Then (2.3.16) becomes

$$\begin{pmatrix} z \\ \dot{z} \\ \ddot{z} \end{pmatrix}_{n+1} = \begin{pmatrix} 1 & 1 & 1 \\ 0 & 1 & 2 \\ 0 & 0 & 1 \end{pmatrix} \begin{pmatrix} z \\ \dot{z} \\ \ddot{z} \end{pmatrix}_{n} \tag{2.3.17}$$

■■

We note that when $h = 1$ the transition matrix for $\boldsymbol{Z}$ is simply Pascal's triangle, appropriately positioned as an upper-triangular matrix.

Example 2.13: Let the degree be $m = 2$ and assume that we are making a *1*-step retrodiction, and so $h = -1$. Then (2.3.16) becomes

$$\begin{pmatrix} z \\ \dot{z} \\ \ddot{z} \end{pmatrix}_{n-1} = \begin{pmatrix} 1 & -1 & 1 \\ 0 & 1 & -2 \\ 0 & 0 & 1 \end{pmatrix} \begin{pmatrix} z \\ \dot{z} \\ \ddot{z} \end{pmatrix}_{n} \tag{2.3.18}$$

■■

The transition matrices in (2.3.17) and (2.3.18) should be inverses (because the one undoes what the other accomplishes), and it can easily be verified that in fact they are.

In Chapters 12 and 13, we discuss the polynomial filters, and we make extensive use of the vector $\boldsymbol{Z}_n$ shown in (2.3.12) and its transition matrix $\boldsymbol{\Phi}(h)$ in (2.3.14). In those chapters, we refer to $\boldsymbol{Z}_n$ as the *normalized* state vector, and to $\boldsymbol{X}_n$ in (2.3.4) as the *denormalized* state vector.

Given the normalized state vector $\boldsymbol{Z}_n$, we can *denormalize* it as follows. First, we rewrite (2.3.12) as follows:

$$\boldsymbol{Z}_n = \boldsymbol{B}(\tau)\boldsymbol{X}_n \tag{2.3.19}$$

in which $\boldsymbol{B}(\tau)$ is the diagonal matrix whose ith diagonal element is

$$[\boldsymbol{B}(\tau)]_{i,i} = \tau^i/i! \qquad 0 \le i \le m \tag{2.3.20}$$

Thus

$$\boldsymbol{B}(\tau) = diag(\tau^0/0!,\ \tau^1/1!,\ \tau^2/2! \ldots \tau^m/m!) \tag{2.3.21}$$

It then follows from (2.3.19) that

$$\boldsymbol{X}_n = \boldsymbol{B}(\tau)^{-1}\boldsymbol{Z}_n \tag{2.3.22}$$

in which

$$B(\tau)^{-1} = diag(0!/\tau^0,\ 1!/\tau^1,\ 2!/\tau^2 \dots m!/\tau^m) \tag{2.3.23}$$

giving us

$$X_n = (z,\ \dot{z}/\tau,\ 2!\ddot{z}/\tau^2,\ \dots m!D^m z/\tau^m)_n{}^T \tag{2.3.24}$$

■■

2.4 Models based on constant-coefficient linear DEs

The general form of the constant-coefficient linear DE was discussed in Section 2.1 where it appeared as

$$DX(t) = AX(t) \tag{2.4.1}$$

Other than when A has the structure for polynomials similar to (2.3.1), and perhaps also for sinusoids of a single frequency, models based on (2.4.1) are seldom used in filter engineering. However, some extremely important results can be obtained from (2.4.1) which then also apply to the polynomial models of the preceding section, and so we examine it further.

The general solution to (2.4.1) is the following:

$$X(t) = exp(tA)X(0) \tag{2.4.2}$$

in which $exp(tA)$ is the matrix exponential of tA defined as follows:

$$exp(tA) \equiv I + tA + t^2A^2/2! + \cdots + t^mA^m/m! + \cdots \tag{2.4.3}$$

Note: Many of the properties of $exp(tA)$ are the same as those of the scalar exponential $exp(ta)$. In particular:

$$\begin{aligned} &exp(t_1A)exp(t_2A) = exp\big((t_1 + t_2)A\big) \\ &\big(exp(tA)\big)^{-1} = exp(-tA) \\ &D\,exp(tA) = A\,exp(tA) \end{aligned} \tag{2.4.4}$$

However,

$$exp(tA)exp(tB) = exp\big(t(A + B)\big) \tag{2.4.5}$$

if and only if A and B commute. ■■

To see that (2.4.2) is the solution to (2.4.1), we differentiate the former and end up with the latter. Thus

$$DX(t) = D\big(exp\,(tA)X(0)\big) = A\,exp\,(tA)X(0) = AX(t) \tag{2.4.6}$$

proving that (2.4.2) is in fact the solution to (2.4.1). From (2.4.2) it now follows that

$$\begin{aligned} X(t+\zeta) = exp\,\big((t+\zeta)A\big)X(0) &= exp(\zeta A)exp(tA)X(0) \\ &= exp(\zeta A)X(t) \end{aligned} \tag{2.4.7}$$

The first and last terms in this equation show that

$$X(t+\zeta) = exp(\zeta A)X(t) \tag{2.4.8}$$

This is a transition equation of the same form as (2.3.5) which means that the transition matrix for every constant-coefficient linear DE is related to its A-matrix by

$$\Phi(\zeta) = exp(\zeta A) \tag{2.4.9}$$

This result then also applies to polynomials because their DEs are constant coefficient linear.

2.4.1 One way to derive transition matrices

Equation (2.4.9) gives us one way to derive the transition matrices for polynomials.

Example 2.14: From (2.3.1), the matrix A in the DE for second degree polynomials is

$$A = \begin{pmatrix} 0 & 1 & 0 \\ 0 & 0 & 1 \\ 0 & 0 & 0 \end{pmatrix} \tag{2.4.10}$$

Forming the matrix exponential of ζA we obtain

$$\begin{aligned} &exp(\zeta A) = I + \zeta A + \zeta^2 A^2/2! + \cdots \\ &= \begin{pmatrix} 1 & 0 & 0 \\ 0 & 1 & 0 \\ 0 & 0 & 1 \end{pmatrix} + \zeta \begin{pmatrix} 0 & 1 & 0 \\ 0 & 0 & 1 \\ 0 & 0 & 0 \end{pmatrix} + \zeta^2/2! \begin{pmatrix} 0 & 0 & 1 \\ 0 & 0 & 0 \\ 0 & 0 & 0 \end{pmatrix} \end{aligned} \tag{2.4.11}$$

at which point the power series terminates because all higher powers of A are null, and so

$$exp(\zeta A) = \begin{pmatrix} 1 & \zeta & \zeta^2/2! \\ 0 & 1 & \zeta \\ 0 & 0 & 1 \end{pmatrix} \tag{2.4.12}$$

This is identical to the expression for $\boldsymbol{\Phi}(\zeta)$ in (2.3.7). ■■

With the benefit of the knowledge gained from (2.4.9), we can now examine the following properties of the polynomial transition matrices. We see intuitively that the ones on the left must be true, and we can now also see the exponential structure behind them on the right.

$$\boldsymbol{\Phi}(0) = \boldsymbol{I} \qquad exp(0\boldsymbol{A}) = \boldsymbol{I} \tag{2.4.13}$$

$$\boldsymbol{\Phi}(\zeta)^{-1} = \boldsymbol{\Phi}(-\zeta) \qquad exp(\zeta\boldsymbol{A})^{-1} = exp(-\zeta\boldsymbol{A}) \tag{2.4.14}$$

$$\boldsymbol{\Phi}(\alpha)\boldsymbol{\Phi}(\beta) = \boldsymbol{\Phi}(\alpha+\beta) \qquad exp(\alpha\boldsymbol{A})exp(\beta\boldsymbol{A}) = exp\big((\alpha+\beta)\boldsymbol{A}\big) \tag{2.4.15}$$

2.4.2 Every transition matrix is nonsingular

From (2.4.14) it appears that $\boldsymbol{\Phi}(\zeta)$ has an inverse for all ζ. This is a property that applies to all transition matrices, which we state as follows:

Theorem 2.1: Every transition matrix is nonsingular. ■■

A proof of this theorem is provided in Appendix 2.5 for the transition matrices for constant-coefficient linear DEs. In the next section, we will provide the proof for the transition matrices for time-varying linear DEs.

2.4.3 A general way to find transition matrices for constant-coefficient linear DEs

We start with the transition equation

$$\boldsymbol{X}(t_n+\zeta) = \boldsymbol{\Phi}(\zeta)\boldsymbol{X}(t_n) \tag{2.4.16}$$

Then the following must be true:

$$\partial\big[\boldsymbol{X}(t_n+\zeta)\big]_i \big/ \partial\big[\boldsymbol{X}(t_n)\big]_j = \big[\boldsymbol{\Phi}(\zeta)\big]_{i,j} \tag{2.4.17}$$

in which $\left[X(t_n+\zeta)\right]_i$ is the ith element of $X(t_n+\zeta)$, and $\left[X(t_n)\right]_j$ is the jth element of $X(t_n)$.

Example 2.15: Applying (2.4.17) to the transition equation in (2.3.7) we obtain the following:

$$\partial\left[X(t_n+\zeta)\right]_1/\partial\left[X(t_n)\right]_1 = 1 = \left[\Phi(\zeta)\right]_{1,1} \tag{2.4.18}$$

$$\partial\left[X(t_n+\zeta)\right]_1/\partial\left[X(t_n)\right]_2 = \zeta = \left[\Phi(\zeta)\right]_{1,2} \tag{2.4.19}$$

$$\partial\left[X(t_n+\zeta)\right]_1/\partial\left[X(t_n)\right]_3 = \zeta^2/2 = \left[\Phi(\zeta)\right]_{1,3} \tag{2.4.20}$$

and so on. ■■

We write (2.4.17) symbolically as follows:

$$\Phi(\zeta) = \partial X(t+\zeta)/\partial X(t) \tag{2.4.21}$$

Equation (2.4.21) provides us with a general way to find the transition matrix for any constant-coefficient linear DE. The complete discussion is given in Appendix 2.6, where we use it to re-derive the result shown earlier in (2.3.7) for polynomials, as well as the following result:

Example 2.16: The DE for simple harmonic motion that we considered in Example 2.3 is the following:

$$D\begin{pmatrix} x(t) \\ \dot{x}(t) \end{pmatrix} = \begin{pmatrix} 0 & 1 \\ -\omega^2 & 0 \end{pmatrix}\begin{pmatrix} x(t) \\ \dot{x}(t) \end{pmatrix} \qquad (\omega \neq 0) \tag{2.4.22}$$

for which the transition matrix is (see Problem 2.22)

$$\Phi(\zeta) = \begin{pmatrix} \cos(\omega\zeta) & (1/\omega)\sin(\omega\zeta) \\ -\omega\sin(\omega\zeta) & \cos(\omega\zeta) \end{pmatrix} \tag{2.4.23}$$

This matrix will be used in a number of places in later chapters. ■■

2.4.4 The DE governing a transition matrix

The transition matrix $\boldsymbol{\Phi}(\zeta)$ is a function of the span-variable ζ which is related to the time-variable t, and so there should be a DE for $\boldsymbol{\Phi}(\zeta)$. If we differentiate (2.4.9) with respect to ζ we obtain the following:

$$d/d\zeta\, \boldsymbol{\Phi}(\zeta) = d/d\zeta\, exp(\zeta \boldsymbol{A}) = \boldsymbol{A}\, exp(\zeta \boldsymbol{A}) = \boldsymbol{A}\boldsymbol{\Phi}(\zeta) \tag{2.4.24}$$

This shows that the DE for $\boldsymbol{\Phi}(\zeta)$ is

$$d/d\zeta\, \boldsymbol{\Phi}(\zeta) = \boldsymbol{A}\boldsymbol{\Phi}(\zeta) \tag{2.4.25}$$

We now write (2.4.1) and (2.4.25) side by side as follows,

$$D\boldsymbol{X}(t) = \boldsymbol{A}\boldsymbol{X}(t) \qquad d/d\zeta\, \boldsymbol{\Phi}(\zeta) = \boldsymbol{A}\boldsymbol{\Phi}(\zeta) \tag{2.4.26}$$

Comparison of these equations shows that the two DEs are the same. This gives us

Theorem 2.2: The DE for the transition matrix $\boldsymbol{\Phi}(\zeta)$ is the same as the DE for the state vector $\boldsymbol{X}(t)$ on which it operates. ■■

We will see in the next section that this theorem also applies to time-varying linear DEs. It will play a critical role when we come to models based on nonlinear DEs.

2.5 Models based on time-varying linear DEs

Like the constant-coefficient linear DEs in the preceding section, time-varying linear DEs are seldom used as models in filter engineering. However, in the next section we will see that they provide us with the key to the use of nonlinear DEs as models, and so it is necessary that we examine them.

The general form of the time-varying linear DE was given in Section 2.1 as

$$D\boldsymbol{X}(t) = \boldsymbol{A}(t)\boldsymbol{X}(t) \tag{2.5.1}$$

Even though this has a superficial resemblance to the constant-coefficient linear DE that appeared in (2.4.1), the matrix $\boldsymbol{A}(t)$ contains one or more time-varying elements whereas the matrix $\boldsymbol{A}$ in (2.4.1) contained only constants. As a result, the solution here no longer has the exponential form that appeared in (2.4.2).

To see that this is so, we start from the assumption that the solution to (2.5.1) is of the same exponential form, namely:

$$\boldsymbol{X}(t) = exp(t\boldsymbol{A}(t))\boldsymbol{X}(0) \tag{2.5.2}$$

which we differentiate, obtaining

$$DX(t) = D\big(exp(tA(t))X(0)\big) = exp\big(tA(t)\big)D\big(tA(t)\big)X(0)$$

$$= exp\big(tA(t)\big)\big(A(t) + tDA(t)\big)X(0) \tag{2.5.3}$$

but the presence of the term $tDA(t)$ now forces us to conclude that (2.5.2) is, in fact, *not* the solution to (2.5.1).

However, by virtue of the linearity of (2.5.1), and under fairly relaxed conditions on the matrix $A(t)$, it can be shown[5] that there exists a matrix $P(t)$ for which both of the following are true:

1. The solution to (2.5.1) is

$$X(t) = P(t)X(0) \tag{2.5.4}$$

2. $P(t)$ is nonsingular for all t. ■■

Starting from (2.5.4) we then have

$$X(t_n) = P(t_n)X(0) \tag{2.5.5}$$

and also

$$X(t_n + \zeta) = P(t_n + \zeta)X(0) \tag{2.5.6}$$

Using the fact that $P(t)$ is non-singular we now combine these two equations, obtaining

$$X(t_n + \zeta) = P(t_n + \zeta)P(t_n)^{-1}X(t_n) \tag{2.5.7}$$

Define

$$\Phi(t_n + \zeta, t_n) \equiv P(t_n + \zeta)P(t_n)^{-1} \tag{2.5.8}$$

in which the double argument of $\Phi(t_n + \zeta, t_n)$ signifies that its elements are functions of the two variables $t_n + \zeta$ and t_n. Then (2.5.7) becomes

$$X(t_n + \zeta) = \Phi(t_n + \zeta, t_n)X(t_n) \tag{2.5.9}$$

This is now seen to be a transition equation similar to (2.3.5). From (2.5.8) we also see that

[5] See Reference 15.

$$\left(\boldsymbol{\Phi}(t_n+\zeta\,,t_n)\right)^{-1}=\boldsymbol{P}(t_n)\boldsymbol{P}(t_n+\zeta)^{-1} \tag{2.5.10}$$

and because $\boldsymbol{P}(t)$ is nonsingular, it follows from this equation that $\left(\boldsymbol{\Phi}(t_n+\zeta\,,\,t_n)\right)^{-1}$ exists, i.e. that $\boldsymbol{\Phi}(t_n+\zeta\,,t_n)$ is nonsingular.

This completes the proof of Theorem 2.1 which we stated in the preceding section, namely that all transition matrices are nonsingular.

2.5.1 Comparison with constant-coefficient linear DEs

There are crucial differences between (2.5.9) and the transition equations that appeared earlier for polynomials and other constant-coefficient linear DEs:

◊ In the earlier case the transition matrix $\boldsymbol{\Phi}(\zeta)$ had only the single argument ζ which was the span of the transition, and $\boldsymbol{\Phi}$ did not depend on the instant at which the transition was initiated. Moreover, $\boldsymbol{\Phi}(\zeta)$ had the exponential structure that appeared in (2.4.9).

◊ Here the transition matrix $\boldsymbol{\Phi}(t_n+\zeta\,,\,t_n)$ has two arguments, showing that it is dependent both on the span of the transition ζ as well as on the instant t_n at which the transition is initiated. (This is a consequence of the fact that the model is time-varying.) Furthermore, the exponential structure of (2.4.9) does not apply.

2.5.2 Obtaining the transition matrix $\boldsymbol{\Phi}(t_n+\zeta\,,t_n)$

We now address the question of how exactly one obtains the matrix $\boldsymbol{\Phi}(t_n+\zeta\,,t_n)$. First, we observe that (2.5.9) can also be written as follows:

$$\boldsymbol{X}(t_m)=\boldsymbol{\Phi}(t_m,t_n)\boldsymbol{X}(t_n) \tag{2.5.11}$$

in which we have used

$$t_m=t_n+\zeta \tag{2.5.12}$$

Setting $m=n$ in (2.5.11), it follows immediately that $\boldsymbol{\Phi}$ must satisfy

$$\boldsymbol{\Phi}(t_n\,,\,t_n)=\boldsymbol{I} \tag{2.5.13}$$

This not-surprising result states that

$$\boldsymbol{\Phi}(t_n+0\,,\,t_n)\,\boldsymbol{X}(t_n)=\boldsymbol{X}(t_n) \tag{2.5.14}$$

Returning now to (2.5.8), if we try to obtain $\Phi(t_n + \zeta, t_n)$ from that equation, then we will have to find $P(t)$ and its inverse for any time-varying linear DE that we might encounter – something that would constitute an insurmountable problem.

Instead, the following is a much more practical approach. In Appendix 2.3, we show that the matrix Φ for time-varying linear DEs satisfies the same DE as the vector X in (2.5.1) on which it operates, i.e. that the following two equations hold true:

$$DX(t) = A(t)X(t) \tag{2.5.15}$$

and

$$d/d\zeta\, \Phi(t_n + \zeta, t_n) = A(t_n + \zeta)\Phi(t_n + \zeta, t_n) \tag{2.5.16}$$

Theorem 2.2, namely that the DE for a transition matrix is the same as the DE for the state vector $X(t)$ on which it operates, thus applies to time-varying linear DEs as well.

Given the DE in (2.5.16), it then follows that we can obtain $\Phi(t_n + \zeta, t_n)$ by numerical integration. For initial conditions we must use (2.5.13).

We have now mentioned numerical integration on a number of occasions. This is a topic that is central to tracking filter engineering, and in the folder ***Miscellaneous*** in the supplementary material there is a file called ***Numerical_Integration*** in which the topic is fully explored.

In the next section we show how the results derived here for time-varying linear DEs can be extended to nonlinear DEs, making it possible to use them also as models.

2.6 Models based on nonlinear DEs

Nonlinear DEs are perhaps the most extensive class of external models, and so it is essential that we be able to use them in tracking filter engineering. Accordingly, assume that the model is the autonomous nonlinear DE

$$DX(t) = F\big(X(t)\big) \tag{2.6.1}$$

in which F is a vector of combinations of the elements of the state vector X, one or more of which is nonlinear, and in which the independent variable t does not appear explicitly.

To make the discussion in this section easier to follow, we will work an example in stages, accompanying each of the theoretical developments by a further development of the example.

Example 2.17: A body is moving in one dimension through a viscous medium, in which the drag force is proportional to the square of the body's speed. This is modelled by the algebraic equation

$$f_d = {}^1\!/\!_2\, c_d \rho v^2 \tag{2.6.2}$$

in which f_d is the drag force, c_d is the body's drag coefficient, ρ is the medium's density and v is the body's speed. Using Newton's second law then gives us the DE

$$mD\dot{x}(t) = -{}^1\!/\!_2\, \rho c_d \left(\dot{x}(t)\right)^2 \tag{2.6.3}$$

in which the minus sign on the right shows that the term is a drag force rather than a propulsive one. The DE in (2.6.3) now constitutes the external model for the process that we are considering. We note that it is autonomous.

Define the constant k by

$$k \equiv -{}^1\!/\!_2\, \rho c_d / m \tag{2.6.4}$$

Then (2.6.3) becomes

$$D\dot{x}(t) = k\left(\dot{x}(t)\right)^2 \tag{2.6.5}$$

in which k is always negative. Reorganizing (2.6.5) into state-vector format, we arrive at the following:

$$D\begin{pmatrix} x(t) \\ \dot{x}(t) \end{pmatrix} = \begin{pmatrix} \dot{x}(t) \\ k\left(\dot{x}(t)\right)^2 \end{pmatrix} \tag{2.6.6}$$

This is of the same form as (2.6.1), in which the vector $\boldsymbol{F}$ is as follows:

$$\boldsymbol{F} = \begin{pmatrix} f_1\left(x(t), \dot{x}(t)\right) \\ f_2\left(x(t), \dot{x}(t)\right) \end{pmatrix} = \begin{pmatrix} \dot{x}(t) \\ k\left(\dot{x}(t)\right)^2 \end{pmatrix} \tag{2.6.7}$$

■■

In the preceding sections in this chapter, the DEs that we considered were linear. That gave us the ability to derive a transition matrix which then enabled us to replace the DE by a transition equation that had the following form:

◊ For polynomials and constant-coefficient linear DEs:

$$\boldsymbol{X}(t_n + \zeta) = \boldsymbol{\Phi}(\zeta)\boldsymbol{X}(t_n) \tag{2.6.8}$$

◊ For time-varying-linear DEs:

$$X(t_n + \zeta) = \Phi(t_n + \zeta, t_n)X(t_n) \tag{2.6.9}$$

We are now considering nonlinear DEs for our models, but we are immediately confronted with a problem: Nonlinear DEs do not have transition matrices, and if we do not have a transition matrix then we cannot set up a transition equation.

At first sight we might then conclude that we cannot use nonlinear DEs as models, but there is a procedure that enables us to get around this obstacle.

That procedure is called the *method of local linearization*, and here's how it works.

2.6.1 The method of local linearization

Depending on the initial conditions, the DE (2.6.1) gives rise to infinitely many trajectories and their state vectors. However, there will be one particular trajectory that the filter will attempt to estimate from the observations.

Let the true state vector of that trajectory be $X(t) = (x_1(t),\ x_2(t),\ \ldots x_m(t))^T$ in which the first element, namely $x_1(t)$, was earlier called the *observed trajectory*.

In Figure 2.5, we show $x_1(t)$ as the dashed line with the label $X(t)$, and the observations that are made on it as the series of small black dots. Observe how the observations follow the observed trajectory with only minor deviations due to the observation errors.

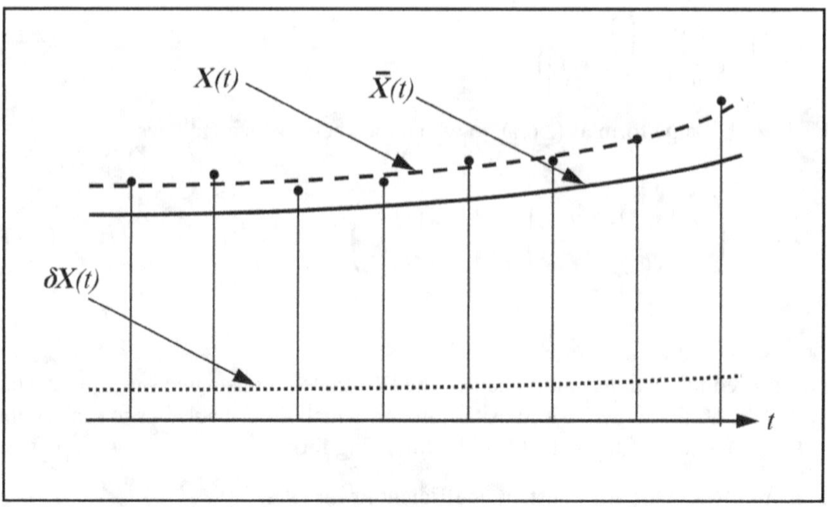

Figure 2.5 The observed trajectory and the nominal trajectory

We now assume that there is a *known* trajectory with state vector $\bar{X}(t)$, that has the following two properties:

1. $\bar{X}(t)$ satisfies the same DE as $X(t)$, namely (2.6.1).
2. $\bar{X}(t)$ is close to $X(t)$.

In Figure 2.5, we show $\bar{X}(t)$ lying close to $X(t)$, and in order to reinforce the fact that it is *known*, we show it as a solid black line.

We call the first element of $\bar{X}(t)$ a *nominal trajectory,* and in Chapter 9 we discuss how $\bar{X}(t)$ is actually obtained.

By the assumption that $X(t)$ and $\bar{X}(t)$ are close, it must be true that

$$X(t) = \bar{X}(t) + \delta X(t) \tag{2.6.10}$$

in which $\delta X(t)$ is a vector whose elements are small in relation to the corresponding elements of either $X(t)$ or $\bar{X}(t)$.

The vector $\delta X(t)$ is called a *perturbation vector*, and in the figure we show its first element as the (small) dotted function just above the time line.

In what now follows we focus our attention on $\delta X(t)$, and it will appear as though we have lost interest in $X(t)$. However, that is not the case. Because $\bar{X}(t)$ in (2.6.10) is fully known, whatever we learn about $\delta X(t)$ will then also give us knowledge about $X(t)$.

From (2.6.10) we see that $\delta X(t)$ is given by

$$\delta X(t) = X(t) - \bar{X}(t) \tag{2.6.11}$$

We now set out to find the DE that governs $\delta X(t)$ and we find that it is *time-varying linear*. That being the case, we are able to derive a transition matrix that can be used to create a transition equation for $\delta X(t)$, and the obstacle referred to earlier will have been bypassed. The entire procedure is depicted in Figure 2.6 below.

In order to avoid breaking the flow of the discussion we derive the DE that governs $\delta X(t)$ in Appendix 2.4, where the following is obtained:

$$D\delta X(t) = A\big(\bar{X}(t)\big)\,\delta X(t) \tag{2.6.12}$$

in which the matrix $A\big(\bar{X}(t)\big)$ is defined as follows:

Definition 2.16: Let X and F in (2.6.1) be m-vectors. Then $A\big(\bar{X}(t)\big)$ is the $m \times m$ matrix called the **DEs sensitivity matrix** whose i, jth element is

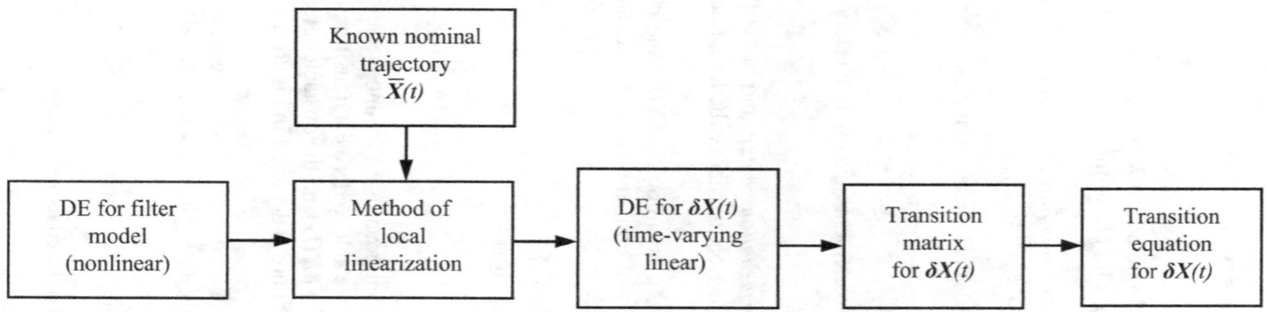

Figure 2.6 From nonlinear model to transition equation

$$[A(\bar{X}(t))]_{i,j} = \partial f_i(x_1, \ \ldots x_m)/\partial x_j|_{\bar{X}(t)} \qquad 1 \le i, j \le m \tag{2.6.13}$$

in which $f_i(x_1, \ldots x_m)$ is the ith function in the vector $\boldsymbol{F}$, and x_j is the jth element of the vector $\boldsymbol{X}$.

As shown in (2.6.13), final evaluation of the partial derivatives in the matrix $A(\bar{X}(t))$ must be carried out using the numerical values contained in $\bar{X}(t)$.

Stated in words: The i, jth element of the DEs sensitivity matrix $A(\bar{X}(t))$ is the partial derivative of the ith element of $\boldsymbol{F}$ with respect to the jth element of $\boldsymbol{X}$, evaluated using the elements of $\bar{X}(t)$.

Because the vectors $\boldsymbol{X}$ and $\boldsymbol{F}$ have the same dimension, it follows that the matrix $A(\bar{X}(t))$ will always be square. ■■

Example 2.17 (continued): The vector of functions $\boldsymbol{F}$ for the DE for *1*-dimensional motion through a viscous medium in (2.6.6) was as follows:

$$\boldsymbol{F} \equiv \begin{pmatrix} f_1(x(t), \ \dot{x}(t)) \\ f_2(x(t), \ \dot{x}(t)) \end{pmatrix} = \begin{pmatrix} \dot{x}(t) \\ k(\dot{x}(t))^2 \end{pmatrix} \tag{2.6.14}$$

Forming the matrix A by the rule stated in (2.6.13) we obtain:

$$A(\bar{X}(t)) = \begin{pmatrix} \partial f_1/\partial x & \partial f_1/\partial \dot{x} \\ \partial f_2/\partial x & \partial f_2/\partial \dot{x} \end{pmatrix}_{\bar{X}(t)} = \begin{pmatrix} 0 & 1 \\ 0 & 2k\dot{x}(t) \end{pmatrix}_{\bar{X}(t)} \tag{2.6.15}$$

The DE for $\boldsymbol{\delta X}(t)$ in state-vector format then becomes

$$D\begin{pmatrix} \delta x(t) \\ \delta \dot{x}(t) \end{pmatrix} = \begin{pmatrix} 0 & 1 \\ 0 & 2k\dot{x}(t) \end{pmatrix}_{\bar{X}(t)} \begin{pmatrix} \delta x(t) \\ \delta \dot{x}(t) \end{pmatrix} \tag{2.6.16}$$

which is seen to be time-varying linear. The subscripted $\bar{X}(t)$ after the matrix on the right signifies that for each value of t, the value of $\dot{x}(t)$ is that of $\bar{\dot{x}}(t)$ in $\bar{X}(t)$. ■■

2.6.2 *Using the results from time-varying linear DEs*

We have shown that the DE for $\boldsymbol{\delta X}(t)$ that the method of local linearization gives us is time-varying linear, and so all of the discussion in Section 2.5 regarding those types of DEs now applies here, with only the following minor differences:

◊ There, the coefficient matrix $A(t)$ for time-varying linear DEs depended directly on t.

Here, the matrix $A\left(\bar{X}(t)\right)$ appearing in (2.6.12), while time-dependent, is not dependent directly on t. Instead it depends on the nominal trajectory $\bar{X}(t)$ which is dependent on t, and for this reason we write it as $A\left(\bar{X}(t)\right)$ rather than $A(t)$.

◊ There, we showed the transition matrix as $\boldsymbol{\Phi}(t_n+\zeta\,, t_n)$.

Here, the transition matrix is derived using the nominal trajectory $\bar{X}(t)$ and so we will show it as $\boldsymbol{\Phi}(t_n+\zeta\,, t_n; \bar{X})$.

◊ There, we obtained a transition equation for $X(t)$ of the form

$$X(t_n+\zeta) = \boldsymbol{\Phi}(t_n+\zeta\,, t_n)\,X(t_n) \tag{2.6.17}$$

in which the transition matrix $\boldsymbol{\Phi}(t_n+\zeta\,, t_n)$ was obtained by numerical integration of the DE

$$d/d\zeta\;\boldsymbol{\Phi}(t_n+\zeta\,, t_n) = A(t_n+\zeta)\;\boldsymbol{\Phi}(t_n+\zeta\,, t_n) \tag{2.6.18}$$

starting from the initial condition $\boldsymbol{\Phi}(t_n\,, t_n) = \boldsymbol{I}$.

Here, in exactly the same way, we have a transition equation for $\delta X(t)$ of the form

$$\delta X(t_n+\zeta) = \boldsymbol{\Phi}(t_n+\zeta\,, t_n; \bar{X})\,\delta X(t_n) \tag{2.6.19}$$

in which the transition matrix $\boldsymbol{\Phi}(t_n+\zeta\,, t_n; \bar{X})$ must be obtained by numerical integration of the DE

$$d/d\zeta\;\boldsymbol{\Phi}(t_n+\zeta\,, t_n; \bar{X}) = A\left(\bar{X}(t_n+\zeta)\right)\boldsymbol{\Phi}(t_n+\zeta\,, t_n; \bar{X}) \tag{2.6.20}$$

starting from initial conditions

$$\boldsymbol{\Phi}(t_n\,, t_n; \bar{X}) = \boldsymbol{I} \tag{2.6.21}$$

In order to be able to fill in the values in $A\left(\bar{X}(t_n+\zeta)\right)$ during the numerical integration of (2.6.20), we will also have to integrate the DE for $\bar{X}(t)$.

This means that we will be integrating two DEs, the DE for $\boldsymbol{\Phi}$ in (2.6.20) using as initial conditions (2.6.21), as well as the nonlinear DE

$$DX(t) = F\left(X(t)\right) \tag{2.6.22}$$

that we started with in (2.6.1), using as initial conditions the known state vector $\bar{X}(t_n)$ of the nominal trajectory.

All of the equations that have brought us to this stage of our development are listed below. In the files contained in ***Miscellaneous\Numerical_Integration*** we discuss in fine detail how the numerical integration of the two DEs (2.6.23) and (2.6.27) is actually carried out.

$$DX(t) = F(X(t)) \qquad (X \text{ and } F \text{ are } m\text{-vectors}) \tag{2.6.23}$$

$$X(t) = \bar{X}(t) + \delta X(t) \tag{2.6.24}$$

$$D\delta X(t) = A(\bar{X}(t))\delta X(t) \qquad (A \text{ is } m \times m) \tag{2.6.25}$$

$$[A(\bar{X}(t))]_{i,j} = \partial f_i(x_1, \dots x_m)/\partial x_j \Big|_{\bar{X}(t)} \qquad 1 \le i,j \le m \tag{2.6.26}$$

$$d/d\zeta\, \Phi(t_n+\zeta, t_n; \bar{X}) = A(\bar{X}(t_n+\zeta))\Phi(t_n+\zeta, t_n; \bar{X}) \tag{2.6.27}$$

$$\Phi(t_n, t_n; \bar{X}) = I \qquad (\Phi \text{ is } m \times m) \tag{2.6.28}$$

$$\delta X(t_n+\zeta) = \Phi(t_n+\zeta, t_n; \bar{X})\, \delta X(t_n) \tag{2.6.29}$$

Example 2.17 (continued): Returning to the DE for *1*-dimensional motion through a viscous medium, we have seen that

$$A(\bar{X}(t)) = \begin{pmatrix} 0 & 1 \\ 0 & 2k\dot{x}(t) \end{pmatrix}_{\bar{X}(t)} \tag{2.6.30}$$

Then from (2.6.20) the DE for the transition matrix $\Phi(t_n+\zeta, t_n; \bar{X})$ will be

$$d/d\zeta \begin{pmatrix} \varphi_{1,1} & \varphi_{1,2} \\ \varphi_{2,1} & \varphi_{2,2} \end{pmatrix} = \begin{pmatrix} 0 & 1 \\ 0 & 2k\dot{x}(t_n+\zeta) \end{pmatrix}_{\bar{X}(t)} \begin{pmatrix} \varphi_{1,1} & \varphi_{1,2} \\ \varphi_{2,1} & \varphi_{2,2} \end{pmatrix} \tag{2.6.31}$$

with initial conditions

$$\varphi_{1,1} = 1, \ \varphi_{1,2} = 0, \ \varphi_{2,1} = 0, \ \varphi_{2,2} = 1 \tag{2.6.32}$$

The DE for the nominal trajectory $\bar{X}(t)$ that is required to evaluate the elements of the first matrix on the right of (2.6.31) is

$$D\begin{pmatrix} x(t) \\ \dot{x}(t) \end{pmatrix} = \begin{pmatrix} \dot{x}(t) \\ k\left(\dot{x}(t)\right)^2 \end{pmatrix} \tag{2.6.33}$$

with initial conditions given by the known state vector of the nominal trajectory $\bar{X}(t)$ at time t_n, namely

$$\bar{X}(t_n) = \begin{pmatrix} \bar{x}(t) \\ \dot{\bar{x}}(t) \end{pmatrix}_{t_n} \tag{2.6.34}$$

■■

2.6.3 Summary

◊ For models based on polynomials and constant-coefficient linear DEs, the transition matrix $\mathbf{\Phi}(\zeta)$ has only one argument, namely ζ. Thus $\mathbf{\Phi}$ depends only on the span of the transition. It does not depend on the starting time of the transition, and it also does not depend on the trajectory along which the transition is taking place.

For such models we can always find the analytical form of the $\mathbf{\Phi}$ matrix, and so numerical integration is not required.

◊ For models based on time-varying linear DEs the transition matrix $\mathbf{\Phi}(t_n + \zeta, t_n)$ has two arguments, namely $t_n + \zeta$ and t_n. Thus $\mathbf{\Phi}$ depends both on the span of the transition as well as on the starting time of the transition. It does not depend on the trajectory along which the transition is taking place.

For time-varying linear models we do not know the analytical form of $\mathbf{\Phi}$, and so numerical integration of the DE (2.5.16) will be required.

◊ When using nonlinear DEs as models, the transition matrix $\mathbf{\Phi}(t_n + \zeta, t_n; \bar{X})$ for the perturbation vector $\delta X(t)$ has three arguments, namely $t_n + \zeta$, t_n and $\bar{X}$. Thus $\mathbf{\Phi}$ depends on the span of the transition, on the starting time of the transition and on the trajectory along which the transition is taking place.

For nonlinear models we do not know the analytical form of $\mathbf{\Phi}$, and so numerical integration of (2.6.20) will always be required. This will also require numerical integration of the DE (2.6.1) for $\bar{X}(t)$, and so when the model is nonlinear, we will almost always have to use numerical integration on two DEs.

2.6.4 Examples of analytical solutions

We have just stated that when the model is nonlinear, then the two DEs (2.6.20) and (2.6.1) will almost always have to be integrated numerically to obtain the transition matrix.

When the analytical solution to the nonlinear DE can be found – a situation which is admittedly rare – then we can sometimes also find the analytical expression for the transition matrix.

Examples are as follows:

1. In Example 2.5, we saw that the DE for simple harmonic motion in which ω is to be estimated, is as follows:

$$D\begin{pmatrix} x(t) \\ \dot{x}(t) \\ \omega \end{pmatrix} = \begin{pmatrix} \dot{x}(t) \\ -\omega^2 x(t) \\ 0 \end{pmatrix} \tag{2.6.35}$$

This DE is nonlinear. Deriving its sensitivity matrix $A\big(\bar{X}(t)\big)$ using (2.6.13) we obtain

$$A\big(\bar{X}(t)\big) = \begin{pmatrix} 0 & 1 & 0 \\ -\omega^2 & 0 & -2\omega x \\ 0 & 0 & 0 \end{pmatrix}_{\bar{X}(t)} \tag{2.6.36}$$

Then the DE for the perturbation vector for (2.6.35) is as follows:

$$D\begin{pmatrix} \delta x(t) \\ \delta \dot{x}(t) \\ \delta\omega \end{pmatrix} = A\big(\bar{X}(t)\big)\begin{pmatrix} \delta x(t) \\ \delta \dot{x}(t) \\ \delta\omega \end{pmatrix} \tag{2.6.37}$$

in which $A(\bar{X}(t))$ is given by (2.6.36). Equation (2.6.37) has the associated transition matrix $\Phi(t_n+\zeta\,,\,t_n;\bar{X})$ that satisfies the following DE:

$$d/d\zeta\,\Phi(t_n+\zeta\,,\,t_n;\bar{X}) = A(\bar{X}(t_n+\zeta))\,\Phi(t_n+\zeta\,,\,t_n;\bar{X}) \tag{2.6.38}$$

with initial conditions

$$\Phi(t_n\,,\,t_n;\bar{X}) = I \tag{2.6.39}$$

In Problem 3.10 in the next chapter we ask you to show that the analytical solution to (2.6.35) is as follows:

$$\begin{aligned} x(t) &= \bar{x}_n cos(\bar{\omega}t) + (\bar{\dot{x}}_n/\bar{\omega})sin(\bar{\omega}t) \\ \dot{x}(t) &= -\bar{x}_n\bar{\omega}sin(\bar{\omega}t) + \bar{\dot{x}}_n cos(\bar{\omega}t) \\ \omega(t) &= \bar{\omega} \end{aligned} \tag{2.6.40}$$

in which $\bar{x}_n$, $\bar{\dot{x}}_n$ and $\bar{\omega}$ are the values contained in the nominal trajectory

$$\bar{X}_n = (\bar{x},\ \bar{\dot{x}},\ \bar{\omega})_n^T \tag{2.6.41}$$

Based on the trajectory given by (2.6.40), the analytical solution to (2.6.38) subject to (2.6.39) is also derived in Problem 3.10. The result is shown in equation (2.6.42) lower down.

The parameters $\bar{x}_n$, $\bar{\dot{x}}_n$ and $\bar{\omega}$ in (2.6.41) are the same three *fixed* values that specify the nominal trajectory in (2.6.40).

We observe that t_n does not appear in the right of (2.6.42), even though it appears in two places on its left. Instead t_n comes into the right of (2.6.42) *implicitly*, by whatever is the validity instant of the nominal trajectory $\bar{X}_n$ in (2.6.41). ■■

It has been our experience that some people have difficulty comprehending what the matrix $\Phi(t_n+\zeta\,,\,t_n;\bar{X})$ really means, and fully understanding how the DE in (2.6.38) generates that matrix.

$$\boldsymbol{\Phi}(t_n+\zeta\,,\,t_n;\bar{\boldsymbol{X}}) = \begin{pmatrix} \cos(\bar{\omega}\zeta) & (1/\bar{\omega})\sin(\bar{\omega}\zeta) & \begin{array}{l} -\bar{x}_n\zeta\sin(\bar{\omega}\zeta) \\ -(\dot{\bar{x}}_n/\bar{\omega}^2)\left(\sin(\bar{\omega}\zeta)-\bar{\omega}\zeta\cos(\bar{\omega}\zeta)\right) \end{array} \\ -\bar{\omega}\sin(\bar{\omega}\zeta) & \cos(\bar{\omega}\zeta) & \begin{array}{l} -\bar{x}_n\left(\sin(\bar{\omega}\zeta)+\bar{\omega}\zeta\cos(\bar{\omega}\zeta)\right) \\ -\dot{\bar{x}}_n\zeta\sin(\bar{\omega}\zeta) \end{array} \\ 0 & 0 & 1 \end{pmatrix} \tag{2.6.42}$$

If you are having similar problems, then please take the necessary time to work through Problem 3.10. Any mystery regarding the meaning of $\boldsymbol{\Phi}(t_n+\zeta\,,\,t_n;\,\bar{\boldsymbol{X}})$ and of its DE in (2.6.38) should then be dispelled, and after that you will have no difficulty understanding how numerical integration is used in the nonlinear case to derive the transition matrix $\boldsymbol{\Phi}$ that goes with any other $\boldsymbol{A}\big(\bar{\boldsymbol{X}}(t)\big)$.

2. The DE for simple satellite motion appearing in (1.5.1), namely

$$D^2\boldsymbol{r} = -(\mu/r^3)\boldsymbol{r} \tag{2.6.43}$$

has an analytical solution that is well known, which is expressed in terms of the six orbital parameters called the Keplerians. We discuss the DE and its solution in the documentation for the program ***14_Orbit_Track***.

The derivation of the matrix $\boldsymbol{A}\big(\bar{\boldsymbol{X}}(t)\big)$ associated with (2.6.43) is carried out in Problem 2.20. The associated transition matrix $\boldsymbol{\Phi}(t_n+\zeta\,,\,t_n;\,\bar{\boldsymbol{X}})$ for ***14_Orbit_Track*** is then obtained by numerical integration of (2.6.20).

Example 2.18: We return to the Allen-Eggers re-entry equation considered in Example 2.9. The *6*-vector of functions $\boldsymbol{F}(\boldsymbol{X})$ is shown on the right of (2.2.1) in which the DEs state vector is

$$\boldsymbol{X}_n = (x_1\,,\,x_2\,,\,x_3\,,\,x_4\,,\,x_5\,,\,x_6)_n{}^T \equiv (x\,,\,y\,,\,z\,,\,\dot{x}\,,\,\dot{y}\,,\,\dot{z})_n{}^T \tag{2.6.44}$$

Define the perturbation vector

$$\boldsymbol{\delta X}(t) \equiv \big(\delta x_1(t)\,,\,\delta x_2(t)\,,\,\delta x_3(t)\,,\,\delta x_4(t)\,,\,\delta x_5(t)\,,\,\delta x_6(t)\big)^T \tag{2.6.45}$$

Then the DE for $\boldsymbol{\delta X}(t)$ is

$$D\boldsymbol{\delta X}(t) = \boldsymbol{A}\big(\bar{\boldsymbol{X}}(t)\big)\,\boldsymbol{\delta X}(t) \tag{2.6.46}$$

in which the 6×6 sensitivity matrix $\boldsymbol{A}\big(\bar{\boldsymbol{X}}(t)\big)$ is obtained by partial differentiation of the expressions in (2.2.1), according to the rule stated in (2.6.13). The result is shown in (2.6.47) below.

$$A(\bar{X}(t)) = \left(\begin{array}{c:c:c:c:c:c} 0 & 0 & 0 & 1 & 0 & 0 \\ \hdashline 0 & 0 & 0 & 0 & 1 & 0 \\ \hdashline 0 & 0 & 0 & 0 & 0 & 1 \\ \hdashline 0 & 0 & kce^{-kx_3} vx_4 & -ce^{-kx_3} \frac{v^2+x_4{}^2}{v} & -ce^{-kx_3} \frac{x_4x_5}{v} & -ce^{-kx_3} \frac{x_4x_6}{v} \\ \hdashline 0 & 0 & kce^{-kx_3} vx_5 & -ce^{-kx_3} \frac{x_4x_5}{v} & -ce^{-kx_3} \frac{v^2+x_5{}^2}{v} & -ce^{-kx_3} \frac{x_5x_6}{v} \\ \hdashline 0 & 0 & kce^{-kx_3} vx_6 & -ce^{-kx_3} \frac{x_4x_6}{v} & -ce^{-kx_3} \frac{x_5x_6}{v} & -ce^{-kx_3} \frac{v^2+x_6}{v} \end{array} \right)_{X = \bar{X}(t)}$$

Notes: $c = \rho_0 \alpha / 2m$ in which m is the body's mass and α is the body's drag coefficient.

The constants ρ_0 and k are related to atmospheric pressure. (2.6.47)

$v = (x_4{}^2 + x_5{}^2 + x_6{}^2)^{1/2}$

Notes:

1. The transition equation for $\delta X(t)$ in (2.6.46) will be

$$\delta X(t_n + \zeta) = \Phi(t_n + \zeta, t_n; \bar{X}) \delta X(t_n) \tag{2.6.48}$$

The matrix $\Phi(t_n + \zeta, t_n; \bar{X})$ in this equation is obtained by numerical integration of the set of *36* coupled DEs contained in

$$d/d\zeta\, \Phi(t_n + \zeta, t_n; \bar{X}) = A\big(\bar{X}(t_n + \zeta)\big)\Phi(t_n + \zeta, t_n; \bar{X}) \tag{2.6.49}$$

The matrix $A\big(\bar{X}(t)\big)$ appears in (2.6.47).

The initial conditions for (2.6.49) are

$$\Phi(t_n, t_n; \bar{X}) = I \tag{2.6.50}$$

2. In order to integrate (2.6.49) we require values for $\bar{X}(t)$, and so we will also have to integrate the Allen-Eggers equation (2.2.1).

The initial conditions for this second integration are the numbers contained in the known nominal-trajectory state vector, namely

$$\bar{X}_n = (\bar{x}_1, \bar{x}_2, \bar{x}_3, \bar{x}_4, \bar{x}_5, \bar{x}_6)_n{}^T \tag{2.6.51}$$

3. In Chapter 9, we consider how the nominal trajectory vector $\bar{X}_n$ is obtained.

4. In the files contained in ***Miscellaneous\Numerical_Integration*** we examine in fine detail how numerical integration is carried out. ■■

◄►

The methodology discussed in this section was used in the following four programs, all of which involve nonlinear DEs:

11_Gauss_Newton_A, ***12_Gauss_Newton_B***, ***13_Gauss_Newton_C***, ***14_Orbit_Track***

◊ In the first three of these programs, we made use of the analytical solutions appearing in (2.6.40) and (2.6.42).

◊ In ***14_Orbit_Track*** we used the analytical solution to the DE in (2.6.43) and numerical integration for the transition matrix.

Further discussion is given in the documentation that accompanies the programs.

◆

2.7 Numerical partial differentiation

Partial differentiation is required in at least three places in filter engineering:

◊ Forming the $A(\bar{X})$ matrix defined in (2.6.13) in this chapter.

◊ Forming the $M(\bar{X})$ matrix defined in (3.3.15) in Chapter 3.

◊ Forming the Jacobi matrix of a nonlinear transformation in order to transform covariance matrices, as discussed in Section 5.1.

When partial derivatives are required, then the best approach is to obtain their analytical expressions using pencil and paper. However, from time to time we may find ourselves in a situation where that is so tedious that we wish to perform partial differentiation numerically with the assistance of a computer.

Alternatively, even if we have obtained the analytical expressions using pencil and paper, it could be reassuring to use a computer to validate those expressions.

See the file in the supplementary material entitled

Miscellaneous\Numerical_Partial_Differentiation

for a complete discussion on numerical partial differentiation.

The following items are in the supplementary material

Problems for Chapter 2

Solutions to Problems for Chapter 2

Projects for Chapter 2

Appendix 2.1

Linear independence of a set of vectors

Definition A2.1.1: Let $S = \{\boldsymbol{X}_1, \boldsymbol{X}_2, \ldots \boldsymbol{X}_m\}$ be a set of two or more vectors all with the same number of elements, and let $\alpha_1 \boldsymbol{X}_1 + \alpha_2 \boldsymbol{X}_2 + \ldots + \alpha_m \boldsymbol{X}_m$ be a linear combination of those vectors, where $\alpha_1, \alpha_2, \ldots \alpha_m$ are constants. Let $\boldsymbol{\emptyset}$ be the zero vector.

Then S is a **linearly independent set** if the following is true only when all α s are zero.

$$\alpha_1 \boldsymbol{X}_1 + \alpha_2 \boldsymbol{X}_2 + \ldots + \alpha_m \boldsymbol{X}_m = \boldsymbol{\emptyset} \qquad \text{(A2.1.1)}$$

On the other hand, if (A2.1.1) is true when one or more of the α s are nonzero then S is a **linearly dependent set.** ■■

Equation (A2.1.1) is often called the *testing function* for linear independence.

In Problems 2.21a to 2.21e, we ask you to apply Definition A2.1.1 to various sets of vectors in order to determine whether or not they are linearly independent.

Appendix 2.2

The polynomial transition matrices

We derive the general expressions for the polynomial transition matrices. Assume first that the sampling instants t_n are arbitrarily spaced and let $x(t)$ be a polynomial of degree m. Then by Taylor's theorem:

$$x(t_n + \zeta) = x(t_n) + \zeta\, \dot{x}(t_n) + \zeta^2/2!\, \ddot{x}(t_n) + \ldots + \zeta^m/m!\, D^m x(t_n)$$
$$= \sum_{j=0}^{m} (\zeta^j/j!)\, D^j x(t_n) \qquad \text{(A2.2.1)}$$

in which the series terminates because all higher derivatives are zero. By repeatedly differentiating (A2.2.1) with respect to ζ we then also have

$$d^i x(t_n + \zeta)/d\zeta^i = \sum_{j=i}^{m} (\zeta^{j-i}/(j-i)!) D^j x(t_n) \qquad 0 \le i \le m \qquad \text{(A2.2.2)}$$

Define the transition matrix $\boldsymbol{\Phi}(\zeta)$ whose i, jth term is

$$[\boldsymbol{\Phi}(\zeta)]_{i,j} \equiv \zeta^{j-i}/(j-i)! \qquad 0 \le i, j \le m \qquad \text{(A2.2.3)}$$

where by definition $1/(j-i)! = 0$ when $j < i$.

Define also the vector $\boldsymbol{X}$ whose jth element is

$$[\boldsymbol{X}(t)]_j \equiv D^j x(t) \qquad 0 \le i \le m \qquad \text{(A2.2.4)}$$

Then (A2.2.2) is equivalent to the transition equation

$$\boldsymbol{X}(t_n + \zeta) = \boldsymbol{\Phi}(\zeta)\, \boldsymbol{X}(t_n) \qquad \text{(A2.2.5)}$$

This shows that (A2.2.3) is the definition for the transition matrix $\boldsymbol{\Phi}(\zeta)$ for polynomials when the sample spacing is arbitrary.

Suppose next that the sampling instants are equally spaced, and let the separation between those instants be τ seconds. Then

$$x(t_n) = x(n\tau) \tag{A2.2.6}$$

and if we let $\zeta = h\tau$ then

$$x(t_n + \zeta) = x(n\tau + h\tau) = x\big((n+h)\tau\big) \tag{A2.2.7}$$

Returning to (A2.2.2) we replace $x(t_n+\zeta)$ by $x\big((n+h)\tau\big)$ and $x(t_n)$ by $x(n\tau)$, obtaining

$$\begin{aligned} D^i x\big((n+h)\tau\big) &= \sum_{j=i}^{m} \big((h\tau)^{j-i}/(j-i)!\big)\, D^j x(n\tau) \\ &= i!/\tau^i \sum_{j=i} j!/\big(i!(j-i)!\big) h^{j-i} \tau^j / j!\, D^j x(n\tau) \\ &= i!/\tau^i \sum_{j=i}^{m} \binom{j}{i} h^{j-i} \big(\tau^j / j!\big)\, D^j x(n\tau) \end{aligned} \tag{A2.2.8}$$

Then

$$\tau^i / i!\, D^i x\big((n+h)\tau\big) = \sum_{j-i}^{m} \binom{j}{i} h^{j-i} \big(\tau^j / j!\big)\, D^j x(n\tau) \qquad 0 \le i \le m \tag{A2.2.9}$$

Define the vector $\boldsymbol{Z}_n$ by

$$\boldsymbol{Z}_n \equiv \big(x,\ \ \tau\dot{x},\ \ (\tau^2/2!)\ddot{x}, \ldots (\tau^m/m!) D^m x\big)_n^{\boldsymbol{T}} \tag{A2.2.10}$$

and the transition matrix $\boldsymbol{\Phi}(h)$ by

$$[\boldsymbol{\Phi}(h)]_{i,j} \equiv \binom{j}{i} h^{j-i} \qquad 0 \le i, j \le m \tag{A2.2.11}$$

where

$$\binom{j}{i} = 0 \text{ when } j < i. \tag{A2.2.12}$$

Then (A2.2.9) is equivalent to the transition equation

$$\boldsymbol{Z}_{n+h} = \boldsymbol{\Phi}(h)\, \boldsymbol{Z}_n \tag{A2.2.13}$$

From (A2.2.11) we see that $\boldsymbol{\Phi}$ is now independent of τ and depends only on the dimensionless variable h that specifies the number of τ-steps (and fractions thereof) that make up the span of the transition.

We have thus shown that (A2.2.11) is the definition for the transition matrix $\boldsymbol{\Phi}(\zeta)$ for polynomials when the samples are equally spaced and the state vector has the normalized form appearing in (A2.2.10).

In Chapters 12 and 13 where we discuss the polynomial filters, we will make extensive use of the normalized vector $\boldsymbol{Z}_n$ in (A2.2.10) and its transition matrix $\boldsymbol{\Phi}$ in (A2.2.11).

Appendix 2.3

Derivation of the DE for the transition matrix $\mathbf{\Phi}(t_n+\zeta\,,\,t_n)$

The general form for time-varying linear DEs is

$$DX(t) = A(t)X(t) \tag{A2.3.1}$$

We showed in Section 2.5 that there exists a transition matrix $\mathbf{\Phi}(t_n+\zeta\,,\,t_n)$ for (A2.3.1) such that

$$X(t_n+\zeta\,) = \mathbf{\Phi}(t_n+\zeta\,,\,t_n)\,X(t_n) \tag{A2.3.2}$$

We now derive the DE for $\mathbf{\Phi}(t_n+\zeta\,,\,t_n)$ in two ways.

Method 1:

We rewrite (A2.3.2) as follows:

$$\mathbf{\Phi}(t_n+\zeta\,,\,t_n)\,X(t_n) = X(t_n+\zeta) \tag{A2.3.3}$$

Differentiation with respect to ζ then gives

$$d/d\zeta\,\mathbf{\Phi}(t_n+\zeta\,,\,t_n)\,X(t_n) = d/d\zeta X(t_n+\zeta\,) \tag{A2.3.4}$$

Using (A2.3.1) this continues as

$$= A(t_n+\zeta\,)\,X(t_n+\zeta\,) \tag{A2.3.5}$$

Finally, using (A2.3.2) on the right of (A2.3.5), we continue as

$$= A(t_n+\zeta\,)\mathbf{\Phi}(t_n+\zeta\,,\,t_n)\,X(t_n) \tag{A2.3.6}$$

We have thus shown that

$$d/d\zeta\,\mathbf{\Phi}(t_n+\zeta\,,\,t_n)\,X(t_n) = A(t_n+\zeta\,)\mathbf{\Phi}(t_n+\zeta\,,\,t_n)\,X(t_n) \tag{A2.3.7}$$

However, $X(t_n)$ is an arbitrary vector and so it must be true that

$$d/d\zeta\,\boldsymbol{\Phi}(t_n+\zeta\,,\,t_n) = \boldsymbol{A}(t_n+\zeta\,)\boldsymbol{\Phi}(t_n+\zeta\,,\,t_n) \tag{A2.3.8}$$

which is the required DE for $\boldsymbol{\Phi}$. ■■

Method 2:

Let $[\boldsymbol{X}(t_n+\,\zeta\,)]_i$ be the ith element of $\boldsymbol{X}(t_n+\zeta\,)$, and $[\boldsymbol{X}(t_n)]_j$ be the jth element of $\boldsymbol{X}(t_n)$.

Starting from (A2.3.2) we then have

$$\partial[\boldsymbol{X}(t_n+\zeta\,)]_i/\partial[\boldsymbol{X}(t_n)]_j = [\boldsymbol{\Phi}(t_n+\zeta\,,\,t_n)]_{i\,,\,j} \tag{A2.3.9}$$

We symbolize this as

$$\boldsymbol{\Phi}(t_n+\zeta\,,\,t_n) = \partial\boldsymbol{X}(t_n+\zeta\,)/\partial\boldsymbol{X}(t_n) \tag{A2.3.10}$$

Differentiation with respect to ζ gives

$$\begin{aligned} d/d\zeta\,\boldsymbol{\Phi}(t_n+\zeta\,,\,t_n) &= d/d\zeta\left(\partial\boldsymbol{X}(t_n+\zeta\,)/\partial\boldsymbol{X}(t_n)\right) \\ &= \partial(d/d\zeta\boldsymbol{X}(t_n+\zeta\,))/\partial\boldsymbol{X}(t_n) \end{aligned}$$

in which we have assumed that it is permissible to interchange the order of differentiation. Using (A2.3.1) this continues as

$$\begin{aligned} &= \partial\left(\boldsymbol{A}(t_n+\zeta)\boldsymbol{X}(t_n+\zeta\,)\right)/\partial\boldsymbol{X}(t_n) \\ &= \boldsymbol{A}(t_n+\zeta\,)\partial\boldsymbol{X}(t_n+\zeta\,)/\partial\boldsymbol{X}(t_n) \\ &= \boldsymbol{A}(t_n+\zeta\,)\boldsymbol{\Phi}(t_n+\zeta\,,\,t_n) \end{aligned} \tag{A2.3.11}$$

We have thus shown again that $\boldsymbol{\Phi}$ satisfies

$$d/d\zeta\,\boldsymbol{\Phi}(t_n+\zeta\,,\,t_n) = \boldsymbol{A}(t_n+\zeta\,)\,\boldsymbol{\Phi}(t_n+\zeta\,,\,t_n) \tag{A2.3.12}$$

■■

Appendix 2.4

The method of local linearization

We derive the DE that governs the perturbation vector $\boldsymbol{\delta X}(t)$ defined in (2.6.11) and we show that it is time-varying linear. Starting from (2.6.11) we have

$$\boldsymbol{\delta X}(t) = \boldsymbol{X}(t) - \bar{\boldsymbol{X}}(t) \tag{A2.4.1}$$

Then

$$\begin{aligned} D\boldsymbol{\delta X}(t) &= D\boldsymbol{X}(t) - D\bar{\boldsymbol{X}}(t) \\ &= \boldsymbol{F}(\boldsymbol{X}(t)) - \boldsymbol{F}(\bar{\boldsymbol{X}}(t)) \\ &= \boldsymbol{F}(\bar{\boldsymbol{X}}(t) + \boldsymbol{\delta X}(t)) - \boldsymbol{F}(\bar{\boldsymbol{X}}(t)) \end{aligned} \tag{A2.4.2}$$

This gives us the following expression for the derivative of the perturbation vector:

$$D\boldsymbol{\delta X}(t) = \boldsymbol{F}(\bar{\boldsymbol{X}}(t) + \boldsymbol{\delta X}(t)) - \boldsymbol{F}(\bar{\boldsymbol{X}}(t)) \tag{A2.4.3}$$

We now reduce the expression on the right of (A2.4.3), and that will lead us to the DE that governs $\boldsymbol{\delta X}(t)$.

For simplicity in what follows we will assume that both $\boldsymbol{\delta X}(t)$ and $\boldsymbol{F}(\bar{\boldsymbol{X}}(t))$ are 2-vectors. We will also temporarily drop the (t) after each of the variables and so equation (A2.4.3) becomes

$$D\begin{pmatrix} \delta x_1 \\ \delta x_2 \end{pmatrix} = \begin{pmatrix} f_1(\bar{x}_1 + \delta x_1, \bar{x}_2 + \delta x_2) \\ f_2(\bar{x}_1 + \delta x_1, \bar{x}_2 + \delta x_2) \end{pmatrix} - \begin{pmatrix} f_1(\bar{x}_1, \bar{x}_2) \\ f_2(\bar{x}_1, \bar{x}_2) \end{pmatrix} \tag{A2.4.4}$$

Expanding the first matrix on the right of (A2.4.4) using Taylor's theorem and simplifying (see Problem 2.23) results in the following:

$$D\begin{pmatrix} \delta x_1(t) \\ \delta x_2(t) \end{pmatrix} = \begin{pmatrix} D_{x_1} f_1(x_1(t), x_2(t)) & D_{x_2} f_1(x_1(t), x_2(t)) \\ D_{x_1} f_2(x_1(t), x_2(t)) & D_{x_2} f_2(x_1(t), x_2(t)) \end{pmatrix}_{\bar{\boldsymbol{X}}(t)} \begin{pmatrix} \delta x_1(t) \\ \delta x_2(t) \end{pmatrix} \tag{A2.4.5}$$

in which we have reintroduced the (t), and so the dependence on time is clearly evident.

Note 1: D_{x_1} means the partial derivative with respect to x_1, etc.

Note 2: The subscript $\bar{X}(t)$ appearing in (A2.4.5) means that the functions $x_1(t)$ and $x_2(t)$ in the partial derivatives are evaluated using the elements of the vector $\bar{X}(t)$.

We started from the nonlinear DE (2.6.1). We assumed that we knew of a nominal trajectory that also satisfied (2.6.1) and differed from it by the perturbation vector $\boldsymbol{\delta X}(t)$. Then the DE for $\boldsymbol{\delta X}(t)$ appears in (A2.4.5), which we note is linear.

In general, the form of that DE is

$$D\boldsymbol{\delta X}(t) = \boldsymbol{A}(\bar{\boldsymbol{X}}(t))\ \boldsymbol{\delta X}(t) \tag{A2.4.6}$$

in which the matrix $\boldsymbol{A}(\bar{\boldsymbol{X}}(t))$ is defined as follows:

Definition A2.4.1: Let $\boldsymbol{X}$ and $\mathbf{F}$ in (2.6.1) be m-vectors. Then $\boldsymbol{A}(\bar{\boldsymbol{X}}(t))$ in (A2.4.6) is the $m \times m$ matrix called the **DE's sensitivity matrix** whose i, jth element is

$$[\boldsymbol{A}(\bar{\boldsymbol{X}}(t))]_{i,j} = \partial f_i(x_1, \dots x_m)/\partial x_j \Big|_{\bar{\boldsymbol{X}}(t)} \qquad 1 \le i, j \le m \tag{A2.4.7}$$

in which $f_i(x_1, \dots x_m)$ is the ith function in the vector $\boldsymbol{F}$ and x_j is the jth element of the vector $\boldsymbol{X}$, evaluated using the elements of $\bar{\boldsymbol{X}}(t)$. ■■

Stated in words: The i, jth element of the DE sensitivity matrix $\boldsymbol{A}(\bar{\boldsymbol{X}}(t))$ is the partial derivative of the ith element of $\boldsymbol{F}$ with respect to the jth element of $\boldsymbol{X}$, evaluated using the elements of $\bar{\boldsymbol{X}}(t)$.

We refer you to Problems 2.24 and 2.25 where we ask you to apply the above.

Appendix 2.5

Proof of Theorem 2.1: Every transition matrix $\mathbf{\Phi}(\zeta)$ is nonsingular

For constant-coefficient linear DEs (which includes polynomials), we know from (2.4.9) that

$$\mathbf{\Phi}(\zeta) = exp(\zeta A) \tag{A2.5.1}$$

Consider now the following property for matrix exponentials:

$$exp\Big((\zeta A) + (-\zeta A)\Big) = exp(\zeta A)\, exp(-\zeta A) \tag{A2.5.2}$$

(This is true because the matrices ζA and $-\zeta A$ commute.)

The left side of (A2.5.2) is $exp(0) = \boldsymbol{I}$, and the right side is $\mathbf{\Phi}(\zeta)\,\mathbf{\Phi}(-\zeta)$, showing that

$$\mathbf{\Phi}(\zeta)\,\mathbf{\Phi}(-\zeta) = \boldsymbol{I} \tag{A2.5.3}$$

This means that the matrices $\mathbf{\Phi}(\zeta)$ and $\mathbf{\Phi}(-\zeta)$ are each other's inverses, proving that $\mathbf{\Phi}(\zeta)$ has an inverse and so it is non-singular.

Appendix 2.6

A general way to find transition matrices

Equation (2.4.9) for the transition matrix for constant coefficient linear DEs was as follows:

$$\boldsymbol{\Phi}(\zeta) = exp(\zeta \boldsymbol{A}) \tag{A2.6.1}$$

This told us that $\boldsymbol{\Phi}(\zeta)$ is the exponential of ζ times the $\boldsymbol{A}$ matrix of the DE when written in state-vector format. However, if we wish to use (A2.6.1) to find $\boldsymbol{\Phi}(\zeta)$ then we must find the precise form of $exp(\zeta \boldsymbol{A})$.

Although that was not much of a problem in the simple polynomial example that we considered in Example 2.15, it could be a bit more time consuming in other situations as the following shows.

Example A2.6.1: In Example 2.3, we considered the DE for simple harmonic motion when the frequency variable ω is known (and so it is not included in the state vector), namely

$$D\begin{pmatrix} x(t) \\ \dot{x}(t) \end{pmatrix} = \begin{pmatrix} 0 & 1 \\ -\omega^2 & 0 \end{pmatrix}\begin{pmatrix} x(t) \\ \dot{x}(t) \end{pmatrix} \qquad (\omega \neq 0) \tag{A2.6.2}$$

From this we see that

$$\boldsymbol{A} = \begin{pmatrix} 0 & 1 \\ -\omega^2 & 0 \end{pmatrix} \tag{A2.6.3}$$

If we now try to obtain the transition matrix for this DE by forming the exponential power series for this $\boldsymbol{A}$, we find that it no longer truncates as it did in (2.4.11). Instead, here's what happens when we raise $\boldsymbol{A}$ to successive powers:

$$A^2 = \begin{pmatrix} -\omega^2 & 0 \\ 0 & -\omega^2 \end{pmatrix} \quad A^3 = \begin{pmatrix} 0 & -\omega^2 \\ \omega^4 & 0 \end{pmatrix} \quad A^4 = \begin{pmatrix} \omega^4 & 0 \\ 0 & \omega^4 \end{pmatrix} \tag{A2.6.4}$$

and this never truncates. In Problem 2.12e, we ask you to sum the exponential power series which this leads to.

There is a more direct way to find the expression for the transition matrix of a constant-coefficient linear DE *when the analytical solution to the DE is known.*

Starting with the transition equation, we have:

$$\boldsymbol{X}(t+\zeta) = \boldsymbol{\Phi}(\zeta)\boldsymbol{X}(t) \tag{A2.6.5}$$

Let $[\boldsymbol{X}(t+\zeta)]_i$ be the ith element of $\boldsymbol{X}(t+\zeta)$ and $[\boldsymbol{X}(t)]_j$ be the jth element of $\boldsymbol{X}(t)$. Then from (A2.6.5) it follows immediately that

$$\partial[\boldsymbol{X}(t+\zeta)]_i / \partial[\boldsymbol{X}(t)]_j = [\boldsymbol{\Phi}(\zeta)]_{i,j} \tag{A2.6.6}$$

Equation (A2.6.6) is written symbolically as follows:

$$\boldsymbol{\Phi}(\zeta) = \partial\boldsymbol{X}(t+\zeta)/\partial\boldsymbol{X}(t) \tag{A2.6.7}$$

Using equation (A2.6.7) to find transition matrices

Solving constant-coefficient linear DEs is a straight-forward process. Textbooks on differential equations show how that is done and we will assume, given such a DE, that its general solution has been obtained.

We now show how we can start from that general solution and arrive at the associated transition matrix by the use of (A2.6.7).

Example A2.6.2: Consider the DE for second degree motion along the x-axis, namely

$$D^3x(t) = 0 \tag{A2.6.8}$$

Deriving the general solution we arrive at

$$x(t) = a_0 + a_1 t + a_2 t^2 \tag{A2.6.9}$$

We could now proceed to derive the transition matrix for (A2.6.8) starting from the general solution in (A2.6.9) but that would involve us in unnecessary algebraic complexity. To simplify matters we instead make use of the following fact.

> For a constant-coefficient linear DE there is only a single form of the transition matrix which can then operate on *any* trajectory generated by that DE.

Accordingly, we will select a convenient trajectory generated by the DE and use it as the starting point for the derivation of the transition matrix.

However, care must be taken to ensure that, when making that selection, we make one that does indeed come from the DE under consideration and not from a simpler one. By this we mean the following:

The DE $\ddot{x}(t) = -2x(t) - 3\dot{x}(t)$ has the general solution $x(t) = a_1 e^{-t} + a_2 e^{-2t}$.

If we now select the particular solution $x(t) = e^{-t}$ as our starting point for finding the transition matrix then we will not succeed, because this particular solution arises from a different DE, namely $\dot{x}(t) = x$, and not from the one that we are considering. A correct particular solution to select would be $x(t) = e^{-t} + e^{-2t}$. This is the subject of Problem 2.10.

Returning to the problem of finding the transition matrix for the DE in (A2.6.8), for convenience, we now select a simple particular solution by letting $a_0 = a_1 = 0$, $a_2 = 1$ in (A2.6.9), obtaining

$$\begin{aligned} x(t) &= t^2 \\ \dot{x}(t) &= 2t \\ \ddot{x}(t) &= 2 \end{aligned} \tag{A2.6.10}$$

This then gives the following three results:

$$\begin{aligned} x(t+\zeta) &= (t+\zeta)^2 = t^2 + 2t\,\zeta + \zeta^2 = x(t) + \zeta\dot{x}(t) + (\zeta^2/2)\,\ddot{x}(t) \\ \dot{x}(t+\zeta) &= 2(t+\zeta) = 2t + 2\zeta = \dot{x}(t) + \ddot{x}(t)\zeta \\ \ddot{x}(t+\zeta) &= \ddot{x}(t) \end{aligned} \tag{A2.6.11}$$

Observe how we have eliminated the variable t and expressed $x(t+\zeta)$, $\dot{x}(t+\zeta)$ and $\ddot{x}(t+\zeta)$ in terms of $x(t)$, $\dot{x}(t)$ and $\ddot{x}(t)$, which is the essential step in finding $\boldsymbol{\Phi}(\zeta)$.

We now apply (A2.6.6) to the terms in (A2.6.11), and obtain the following nine expressions:

$$\left.\begin{aligned} \varphi_{1,1}(\zeta) &= \partial x(t+\zeta)/\partial x(t) = 1 \\ \varphi_{1,2}(\zeta) &= \partial x(t+\zeta)/\partial \dot{x}(t) = \zeta \\ \varphi_{1,3}(\zeta) &= \partial x(t+\zeta)/\partial \ddot{x}(t) = \zeta^2/2! \end{aligned}\right\} \tag{A2.6.12}$$

$$\left.\begin{aligned} \varphi_{2,1}(\zeta) &= \partial \dot{x}(t+\zeta)/\partial x(t) = 0 \\ \varphi_{2,2}(\zeta) &= \partial \dot{x}(t+\zeta)/\partial \dot{x}(t) = 1 \\ \varphi_{2,3}(\zeta) &= \partial \dot{x}(t+\zeta)/\partial \ddot{x}(t) = \zeta \end{aligned}\right\} \tag{A2.6.13}$$

$$\left.\begin{aligned} \varphi_{3,1}(\zeta) &= \partial \ddot{x}(t+\zeta)/\partial x(t) = 0 \\ \varphi_{3,2}(\zeta) &= \partial \ddot{x}(t+\zeta)/\partial \dot{x}(t) = 0 \\ \varphi_{3,3}(\zeta) &= \partial \ddot{x}(t+\zeta)/\partial \ddot{x}(t) = 1 \end{aligned}\right\} \tag{A2.6.14}$$

Not surprisingly, after assembling the transition matrix according to (A2.6.7) the result is

$$\mathbf{\Phi}(\zeta) = \begin{pmatrix} 1 & \zeta & \zeta^2/2! \\ 0 & 1 & \zeta \\ 0 & 0 & 1 \end{pmatrix} \tag{A2.6.15}$$

■■

In Problems 2.12, 2.13 and 2.14, we ask you to apply the above method to a number of constant-coefficient linear DEs in order to obtain their transition matrices.

In Problem 2.22, we ask you to show that the DE

$$\ddot{x}(t) = -\omega^2 x(t) \qquad (\omega \text{ a known nonzero constant}) \tag{A2.6.16}$$

has the following transition matrix:

$$\mathbf{\Phi}(\zeta) = \begin{pmatrix} \cos(\omega\zeta) & 1/\omega \sin(\omega\zeta) \\ -\omega \sin(\omega\zeta) & \cos(\omega\zeta) \end{pmatrix} \tag{A2.6.17}$$

Chapter 3

Observation schemes

In Chapter 2 we discussed *models* and how they are specified by DEs and their transition matrices.

In this chapter we examine the way in which *observations* are made.

We will find that four cases can arise, and that for each case there is a matrix called $\boldsymbol{T}$.

Then in Chapter 8 we will see that the Gauss filters are based on $\boldsymbol{T}$ in the follo wing way:

1. First, a matrix $\boldsymbol{W}$ is created from $\boldsymbol{T}$ as follows:

   $$\boldsymbol{W} = (\boldsymbol{T}^T \boldsymbol{R}_Y^{-1} \boldsymbol{T})^{-1} \boldsymbol{T}^T \boldsymbol{R}_Y^{-1}$$

2. An estimate vector $\boldsymbol{X}^*$ is then obtained from $\boldsymbol{W}$ and the observation vector $\boldsymbol{Y}$ according to the equation

   $$\boldsymbol{X}^* = \boldsymbol{W}\boldsymbol{Y}$$

3. Finally, the covariance matrix of $\boldsymbol{X}^*$ is computed according to

   $$\boldsymbol{S}^* = \boldsymbol{W}\boldsymbol{R}_Y\boldsymbol{W}^T$$

Thus in order to build a Gauss filter, we are first required to build the matrix $\boldsymbol{T}$.

This chapter is devoted to that matrix, to how it comes about and to the four versions that it can assume.

3.1 How filtering works

In Figure 3.1 we see the following:

1. The observed trajectory has true state vector $\boldsymbol{X}$.
2. Observations are made by an observation instrument.

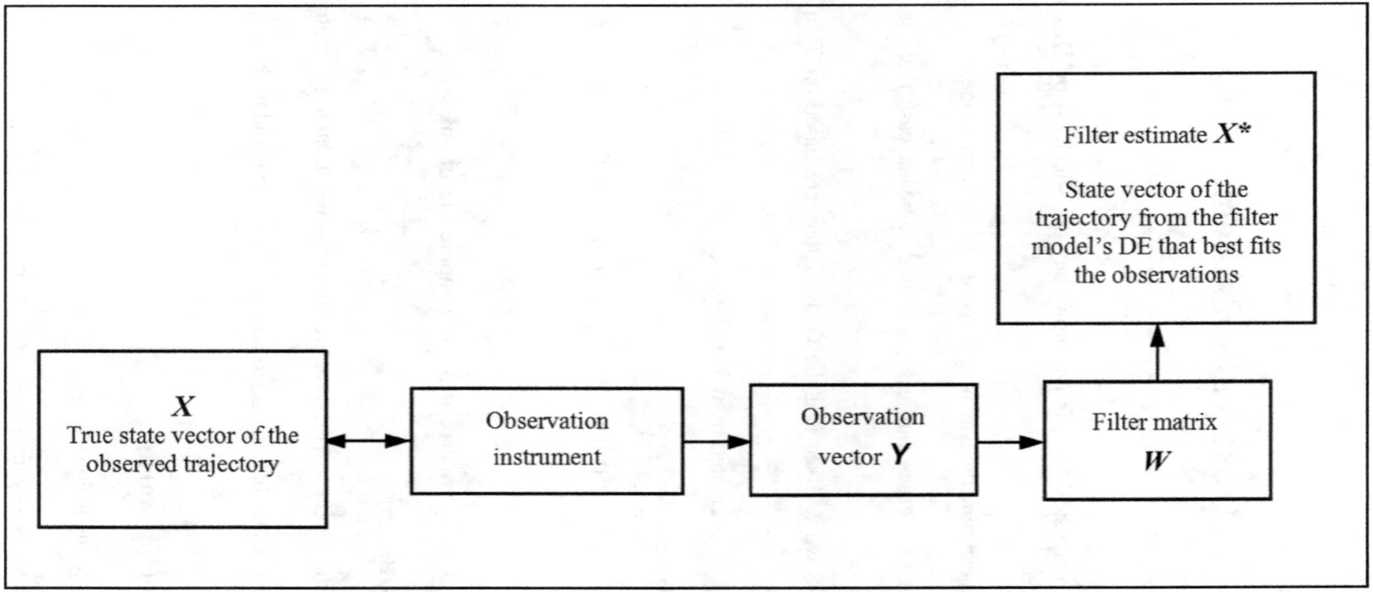

Figure 3.1 From the state vector of the observed trajectory to the filter estimate

3. The observations become the inputs to the filter.
4. The filter processes the observations and derives an estimate of $\boldsymbol{X}$ that is called $\boldsymbol{X}^*$.

   It does that by finding the trajectory from the filter model's DE that best fits the observations.

   In Chapter 8 we will specify what *best* means.

3.1.1 The true state vector $\boldsymbol{X}$

The true state vector $\boldsymbol{X}$ was mentioned on a number of occasions in the first two chapters and we have now mentioned it again, yet up to now we haven't specified the DE for which $\boldsymbol{X}$ is the state vector.

One's first guess might be that $\boldsymbol{X}$ is the state vector of the external model's DE, but that is incorrect. $\boldsymbol{X}$ is always the state vector of the filter model's DE (unless we specifically state otherwise).

To see why this is so, consider the following:

The external model's DE is not implemented anywhere in the filter. That DE is for the most part only conceptual, and serves as a reference against which we compare the filter model's DE. The only time that the external model's DE is actually implemented is in a simulation in order to generate values to which errors are added, the results then serving as simulated observations.

Regarding for which DE $\boldsymbol{X}$ is the state vector, if the filter model is the same as the external model then $\boldsymbol{X}$ is the state vector of either.

However, if the filter model differs from the external model, then each will have its own state vector. (Their dimensions do not even have to be the same.)

If we now make $\boldsymbol{X}$ the state vector of the external model, then we will have to build the filter using the external model, and that is not what we want to do. What we want, and as shown in Figure 3.1, is for the filter to find the trajectory from its own DE that best fits the observations. To accomplish that, we must build the filter using the *filter* model's DE.

Accordingly, $\boldsymbol{X}$ must always be thought of as the error-free state vector of the filter model's DE.

3.1.2 The observation equations

When observations are made on a physical process, rules are established by which those observations are made. For example, in the case of radar, some of the rules cover how the target's slant-range, azimuth and elevation angles will be measured.

Such rules form part of what we call the *observation scheme*, and they are *implicitly* included in the filter.

In addition to those rules, there are two items that must be *explicitly* included in the filter.

◊ ***The filter model's DE:*** The filter model's DE is included by including the transition matrix $\boldsymbol{\Phi}$ that is derived from that DE.

◊ ***The observation equations:*** These equations specify the relationships between the observations and $\boldsymbol{X}$, the true state vector of the observed trajectory. The observation equations are included by including an observation matrix called $\boldsymbol{M}$.

The way in which these two matrices are actually included in the filter is by using them to build the matrix $\boldsymbol{T}$ which is then included in the filter equations.

◊ This chapter is devoted to the way in which $\boldsymbol{\Phi}$ and $\boldsymbol{M}$ are used to create $\boldsymbol{T}$.

◊ In Chapters 8 and 9, we will see how $\boldsymbol{T}$ is included in the filter.

3.1.3 The four cases

In Chapter 2 we spoke about *models* that are either linear or nonlinear. In this chapter, we will see that there are *observation equations* that can be either linear or nonlinear.

Considering filter models and observation equations as pairs, there are then the four cases shown in Figure 3.2.

		Filter model	
		Linear	Nonlinear
Observation equations	Linear	***Case 1***	***Case 3***
	Nonlinear	***Case 2***	***Case 4***

Figure 3.2 The four cases

At the top of the figure we see the filter model and its two possibilities, linear and nonlinear, and at the left we see the observation equations with their two possibilities, also linear and nonlinear. The centre portion of the figure shows each of the four possible cases, namely ***Case 1***, ***Case 2***, ***Case 3*** and ***Case 4***.

In filter engineering all four cases can and do occur.

3.2 *Case 1:* Linear filter model, linear observation equation

We are considering ***Case 1*** of the four possibilities in Figure 3.2, for which both the filter model and observation equation are linear.

Let $\boldsymbol{Y}_n$ be a k-vector of observations that are made at time t_n, i.e.

$$\boldsymbol{Y}_n = (y_1, y_2, \ldots y_k)_n^T \tag{3.2.1}$$

in which k could be *1*, *2* or more.

Definition 3.1: $\boldsymbol{Y}_n$ is called the **observation vector**. The instant t_n at which the observations are made – as specified by a real-time clock – is called the **time stamp** of $\boldsymbol{Y}_n$. Together with the observations themselves, the time stamp forms an intrinsic part of $\boldsymbol{Y}_n$. ■■

Let the vector of errors in $\boldsymbol{Y}_n$ be

$$\boldsymbol{N}_n = (\nu_1, \nu_2, \ldots \nu_k)_n^T \tag{3.2.2}$$

Reminder: The letter $\boldsymbol{N}$ is the upper-case Greek letter *Nu*, and so its elements are the lower-case Greek letter *nu*, namely ν.

Let $\boldsymbol{X}_n$ be the true state vector of the observed trajectory at time t_n and let it be an m-vector. Then for ***Case 1*** the observation vector $\boldsymbol{Y}_n$ is related to $\boldsymbol{X}_n$ as follows:

$$\boldsymbol{Y}_n = \boldsymbol{M}\boldsymbol{X}_n + \boldsymbol{N}_n \tag{3.2.3}$$

in which $\boldsymbol{M}$ is a $k \times m$ matrix of constants (usually only *1*'s and *0*'s).

Example 3.1: Let $\boldsymbol{X}_n = (x, \dot{x}, \ddot{x})_n^T$ and assume that we are observing only x. Then $k = 1$ and $m = 3$, and (3.2.3) would be

$$y_n = (1 \quad 0 \quad 0)\begin{pmatrix} x \\ \dot{x} \\ \ddot{x} \end{pmatrix}_n + \nu_n \tag{3.2.4}$$

Equation (3.2.4) constitutes what we call the *observation equation(s)* in which $\boldsymbol{M}$ is the 1×3 matrix of *1's* and *0's* that is multiplying $\boldsymbol{X}_n$. ■■

Example 3.2: Let $\boldsymbol{X}_n = (x, \dot{x}, y, \dot{y}, z, \dot{z})_n^T$ and assume that we are observing x, y and z. In this case $k = 3$ and $m = 6$, and (3.2.3) would be

$$\begin{pmatrix} y_1 \\ y_2 \\ y_3 \end{pmatrix}_n = \begin{pmatrix} 1 & 0 & 0 & 0 & 0 & 0 \\ 0 & 0 & 1 & 0 & 0 & 0 \\ 0 & 0 & 0 & 0 & 1 & 0 \end{pmatrix} \begin{pmatrix} x \\ \dot{x} \\ y \\ \dot{y} \\ z \\ \dot{z} \end{pmatrix}_n + \begin{pmatrix} v_1 \\ v_2 \\ v_3 \end{pmatrix}_n \tag{3.2.5}$$

Then the observation equations are contained in (3.2.5) and $\boldsymbol{M}$ is the 3×6 matrix of *1's* and *0's* appearing in that equation. ■■

Definition 3.2: Equation (3.2.3) that links $\boldsymbol{Y}_n$ and $\boldsymbol{X}_n$ constitutes the **observation equation(s)** and $\boldsymbol{M}$ is called the **observation matrix.** ■■

If $\boldsymbol{M}$ changes with n then we symbolize it as $\boldsymbol{M}_n$ and (3.2.3) becomes

$$\boldsymbol{Y}_n = \boldsymbol{M}_n \boldsymbol{X}_n + \boldsymbol{N}_n \tag{3.2.6}$$

We will return to (3.2.6) shortly, and we will see that it is not very different from (3.2.3).

3.2.1 Sequences of observation vectors

In the filtering algorithms that we use, an estimate of the vector $\boldsymbol{X}_n$ will always be derived from a *sequence* of $L+1$ observation vectors that have the time stamps $t_n, t_{n-1}, \ldots t_{n-L}$.

Let those $L+1$ observation vectors be called $\boldsymbol{Y}_n, \boldsymbol{Y}_{n-1}, \ldots \boldsymbol{Y}_{n-L}$ and let the associated $L+1$ observation equations be as follows:

$$\boldsymbol{Y}_{n-k} = \boldsymbol{M}\boldsymbol{X}_{n-k} + \boldsymbol{N}_{n-k} \qquad 0 \le k \le L \tag{3.2.7}$$

Concatenation then gives

$$\begin{pmatrix} Y_n \\ \hdashline Y_{n-1} \\ \hdashline \vdots \\ \hdashline Y_{n-L} \end{pmatrix} = \begin{pmatrix} MX_n \\ \hdashline MX_{n-1} \\ \hdashline \vdots \\ \hdashline MX_{n-L} \end{pmatrix} + \begin{pmatrix} N_n \\ \hdashline N_{n-1} \\ \hdashline \vdots \\ \hdashline N_{n-L} \end{pmatrix} \tag{3.2.8}$$

The $\boldsymbol{M}$ matrices appearing in (3.2.8) are assumed to be all the same. However, if that were not the case and they changed with *n*, then nothing would be lost, and we would simply use the symbol $\boldsymbol{M}_n$ in place of $\boldsymbol{M}$. (Of course if $\boldsymbol{M}$ changes, then the corresponding subvectors $\boldsymbol{Y}$ and $\boldsymbol{N}$ might also change.)

We are in ***Case 1***, and so we are assuming that the filter model is linear.

In filter engineering we seldom encounter time-varying linear models, and so we will assume further that the DE of the filter model is constant-coefficient linear and that its transition matrix is $\boldsymbol{\Phi}(\zeta)$.

Note: In the unlikely situation in which the filter model's DE is time-varying linear, the reader should have no difficulty in making the necessary extensions. Keep in mind that numerical integration will then be required in order to derive the transition matrix $\boldsymbol{\Phi}(t_n + \zeta, t_n)$ for various values of ζ.

Let ζ be the (negative) quantity $t_{n-k} - t_n$. Then the vectors $\boldsymbol{X}_{n-k}$ and $\boldsymbol{X}_n$ will be related as follows:

$$\boldsymbol{X}_{n-k} = \boldsymbol{\Phi}(t_{n-k} - t_n)\boldsymbol{X}_n \tag{3.2.9}$$

Using (3.2.9) in (3.2.8) then gives

$$\begin{pmatrix} Y_n \\ \hdashline Y_{n-1} \\ \hdashline \vdots \\ \hdashline Y_{n-L} \end{pmatrix} = \begin{pmatrix} MX_n \\ \hdashline M\Phi(t_{n-1} - t_n)X_n \\ \hdashline \vdots \\ \hdashline M\Phi(t_{n-L} - t_n)X_n \end{pmatrix} + \begin{pmatrix} N_n \\ \hdashline N_{n-1} \\ \hdashline \vdots \\ \hdashline N_{n-L} \end{pmatrix} \tag{3.2.10}$$

We now factor out X_n from the matrix on the right, obtaining

$$\begin{pmatrix} Y_n \\ \hdashline Y_{n-1} \\ \hdashline \vdots \\ \hdashline Y_{n-L} \end{pmatrix} = \begin{pmatrix} M \\ \hdashline M\Phi(t_{n-1} - t_n) \\ \hdashline \vdots \\ \hdashline M\Phi(t_{n-L} - t_n) \end{pmatrix} X_n + \begin{pmatrix} N_n \\ \hdashline N_{n-1} \\ \hdashline \vdots \\ \hdashline N_{n-L} \end{pmatrix} \tag{3.2.11}$$

In (3.2.11), the first and last vectors are given the names Y_n and N_n and the matrix in the centre is called T_n. Thus:

$$\mathsf{Y}_n \equiv \begin{pmatrix} Y_n \\ \hdashline Y_{n-1} \\ \hdashline \vdots \\ \hdashline Y_{n-L} \end{pmatrix} \qquad \mathsf{N}_n \equiv \begin{pmatrix} N_n \\ \hdashline N_{n-1} \\ \hdashline \vdots \\ \hdashline N_{n-L} \end{pmatrix} \qquad \mathsf{T}_n \equiv \begin{pmatrix} M \\ \hdashline M\Phi(t_{n-1} - t_n) \\ \hdashline \vdots \\ \hdashline M\Phi(t_{n-L} - t_n) \end{pmatrix} \tag{3.2.12}$$

◊ Observe that we are using *sans-serif* letters for Y, N and T. This signifies that multiple time stamps are involved.

◊ Observe also that the subscripts of the vectors Y_n and N_n are the same as those of their most recent subvectors, namely Y_n and N_n.

We will refer to vectors like Y_n and N_n which contain subvectors with multiple time stamps as *long-vectors*, and to vectors such as Y_n, N_n and X_n that have only a single time stamp as *short-vectors*. A rectangular matrix like T_n that has more rows than columns is called a *high-matrix*, and one that has more columns than rows is called a *wide-matrix*.

Definition 3.3:

1. The long-vectors Y_n and N_n in (3.2.12) are called the **total observation vector** and the **total error vector**.

2. The high-matrix T_n is called the **total observation matrix**, and for ***Case 1*** it is shown in (3.2.12).

■■

Equation (3.2.11) can now be written as

$$\boldsymbol{Y}_n = \boldsymbol{T}_n \boldsymbol{X}_n + \boldsymbol{N}_n \tag{3.2.13}$$

We call (3.2.13) the **total observation equation**. Observe that if $\boldsymbol{N}_n$ is removed, then $\boldsymbol{Y}_n$ is related to $\boldsymbol{X}_n$ by a linear transformation, which is why we refer to (3.2.13) as a *linear* observation equation.

Equation (3.2.13) and the definition of $\boldsymbol{T}_n$ in (3.2.12) are part of the rules by which the observations were made. If those observations are to be submitted to a filter for the extraction of an estimate, then the rules will have to be conveyed to the filter so that it can make sense of them. The way in which that is done is by incorporating the matrix $\boldsymbol{T}$ into the filtering algorithm. We will return to that later.

3.2.2 The two matrices that are used to build $\boldsymbol{T}$

Taking another look at $\boldsymbol{T}_n$ in (3.2.12) we see that it is constructed from two matrices.

◊ The transition matrix $\boldsymbol{\Phi}$ that uniquely specifies the filter model.

◊ The matrix $\boldsymbol{M}$ that specifies how the observations $\boldsymbol{Y}_{n-k}$ are related to the state vector $\boldsymbol{X}$.

We note that numerical integration will not be required in order to derive the matrix $\boldsymbol{\Phi}$ appearing in (3.2.12), because the filter model here has been assumed to be constant-coefficient linear and so $\boldsymbol{\Phi}$ can always be written in explicit form (see Chapter 2). However, in two of the cases further down the models will be nonlinear, for which numerical integration *will* be required to obtain $\boldsymbol{\Phi}$.

3.2.3 The need for linear independence

It is clear from the above that $\boldsymbol{T}_n$ will always be high, and so it has long column vectors.

We must ensure that those column vectors are linearly independent because if not, our progress at a later stage will be blocked.

If we are observing the process properly, then the columns of $\boldsymbol{T}_n$ *will* be linearly independent, but if not, then they may not be. We defer further discussion of this important topic to Section 5.6 where we examine it in detail.

3.3 *Case 4:* Nonlinear filter model, nonlinear observation equation

We are considering ***Case 4*** of the four possibilities in Figure 3.2, for which both the filter model and observation equation are nonlinear. ***Case 4*** is the most challenging, and once it is understood, ***Cases 2*** and ***3*** will be easy.

Example 3.3: An example of nonlinear observation equations are the ones used in radar systems. Referring to Figure 3.3, the point ***P*** has radar coordinates ρ, θ, ψ, where ρ is called the slant-range, θ is the elevation angle and ψ the azimuth angle.[1]

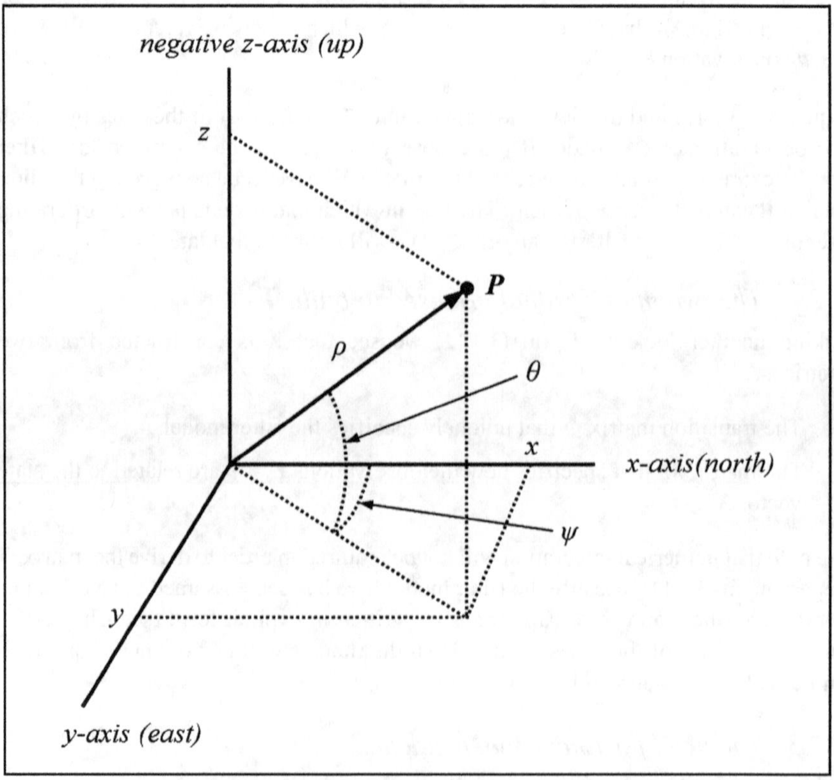

Figure 3.3 NED coordinates

◊ The Cartesian coordinates of ***P*** are shown as x, y, z.

◊ The azimuth angle ψ is measured in the x, y plane starting from the x-axis, which is typically north, and increases positively as we move towards the y-axis, which is east.

◊ In order for the Cartesian axes to form a right-handed coordinate system, the positive z-axis is down. Such a system is referred to as *north-east-down* or *NED*.

(See Appendix 3.1 where we discuss the east-north-up (ENU) system.)

[1] Mathematicians usually refer to these as polar coordinates.

◊ The elevation angle θ is measured vertically from the x, y plane to the $\boldsymbol{P}$ vector. The positive z-axis is down, and has an elevation value of $\theta = \pi/2$ *rads*. As it appears in the figure, the value of θ is negative because $\boldsymbol{P}$ lies in the region where z is negative.

We accordingly have the following relationships:

$$\begin{aligned} x &= \rho\, cos(\theta)\, cos(\psi) \\ y &= \rho\, cos(\theta)\, sin(\psi) \\ z &= \rho\, sin(\theta) \end{aligned} \tag{3.3.1}$$

The inverse set of equations to (3.3.1) is as follows:

$$\begin{aligned} \rho &= (x^2 + y^2 + z^2)^{1/2} \\ \theta &= tan^{-1}\left(z/(x^2 + y^2)^{1/2}\right) \quad (\textit{2-quadrant arctan}) \\ \psi &= tan^{-1}(y/x) \qquad\qquad (\textit{4-quadrant arctan}) \end{aligned} \tag{3.3.2}$$

As an example, let the state vector of the filter model be

$$\boldsymbol{X}_n = (x, \dot{x}, y, \dot{y}, z, \dot{z})_n^T \tag{3.3.3}$$

and assume that the radar is observing ρ, θ and ψ. Then, in place of (3.2.5) the observation equations now become

$$\boldsymbol{Y}_n = \begin{pmatrix} \rho \\ \theta \\ \psi \end{pmatrix}_n = \begin{pmatrix} (x^2 + y^2 + z^2)^{1/2} \\ tan^{-1}\left(z/(x^2 + y^2)^{1/2}\right) \\ tan^{-1}(y/x) \end{pmatrix}_n + \begin{pmatrix} v_1 \\ v_2 \\ v_3 \end{pmatrix}_n \tag{3.3.4}$$

Equation (3.3.4) is seen to be of the form

$$\boldsymbol{Y}_n = \boldsymbol{G}(\boldsymbol{X}_n) + \boldsymbol{N}_n \tag{3.3.5}$$

in which $\boldsymbol{G}$ is a vector of combinations of the elements of $\boldsymbol{X}_n$, one or more of which is nonlinear, and so the observation equations in (3.3.5) are nonlinear. ■■

Example 3.4: A satellite is being observed by radar. For present purposes we will use (3.3.4) as the observation equations. (In actuality, the observation equations, while nonlinear, would be considerably more complex than (3.3.4.) See the documentation for ***14_ Orbit_ Track***.)

A simplified filter model would be Newton's DE in (1.5.1) that governs the motion of a small object under the influence of a large central body, namely

$$D^2\boldsymbol{r} = -(\mu/r^3)\boldsymbol{r} \tag{3.3.6}$$

In Problem 2.20 we ask you to resolve (3.3.6) into its Cartesian components and obtain the following:

$$D\begin{pmatrix} x_1 \\ x_2 \\ x_3 \\ x_4 \\ x_5 \\ x_6 \end{pmatrix} = \begin{pmatrix} x_2 \\ -\mu x_1/r^3 \\ x_4 \\ -\mu x_3/r^3 \\ x_6 \\ -\mu x_5/r^3 \end{pmatrix} \tag{3.3.7}$$

The quantities in (3.3.7) have the following definitions:

$$(x_1, x_2, x_3, x_4, x_5, x_6) \equiv (x, \dot{x}, y, \dot{y}, z, \dot{z}) \quad r \equiv (x^2 + y^2 + z^2)^{1/2} \tag{3.3.8}$$

The DE in (3.3.7) is clearly nonlinear because of the presence of r^3 in the three denominators on the right, and so this is a nonlinear filter model of the form

$$D\boldsymbol{X}(t) = \boldsymbol{F}(\boldsymbol{X}(t)) \tag{3.3.9}$$

Taken together with the nonlinear observation equations in (3.3.5), this is an example of ***Case 4***. ■■

3.3.1 *Local linearization applied to observation equations*

Recall that in Section 2.6 we applied the method of local linearization to an equation of the form shown in (3.3.9), and arrived at a linear DE governing the perturbation vector $\delta\boldsymbol{X}(t)$. We now apply the same method to (3.3.5) and arrive at a linear observation equation that relates the perturbation vectors $\delta\boldsymbol{Y}_n$ and $\delta\boldsymbol{X}_n$.

As we did in Section 2.6, we again assume the existence of a known nominal trajectory $\bar{\boldsymbol{X}}(t)$ that is close to the true trajectory $\boldsymbol{X}(t)$ and which is governed by the same DE, namely (3.3.9).

Then $\boldsymbol{X}(t)$ and $\bar{\boldsymbol{X}}(t)$ are related by

$$\boldsymbol{X}(t) = \bar{\boldsymbol{X}}(t) + \delta\boldsymbol{X}(t) \tag{3.3.10}$$

The perturbation vector $\boldsymbol{\delta X}(t)$ in this equation is comprised of time-dependent functions which are small in relation to the corresponding elements of $\bar{\boldsymbol{X}}(t)$.

Using the nominal trajectory $\bar{\boldsymbol{X}}(t)$ we showed in Appendix 2.4 how the DE for the perturbation vector $\boldsymbol{\delta X}(t)$ can be derived. Because that DE was linear, it led us to (2.6.19), namely the transition equation

$$\boldsymbol{\delta X}(t_n + \zeta) = \boldsymbol{\Phi}(t_n + \zeta, t_n; \bar{\boldsymbol{X}})\boldsymbol{\delta X}(t_n) \tag{3.3.11}$$

In Sections 2.5 and 2.6 we provided the method for deriving the transition matrix $\boldsymbol{\Phi}(t_n + \zeta, t_n; \bar{\boldsymbol{X}})$ that appears in (3.3.11). This involved numerical integration of its DE.

Note: The details for obtaining $\boldsymbol{\Phi}$ by using numerical integration are discussed fully in the files contained in ***Miscellaneous\Numerical_Integration***.

Starting from the nonlinear observation equation (3.3.5), we now consider what must be done to arrive at an observation equation that is similarly linear. We state a few preliminary items here, but in order not to disrupt the developments we have moved the main discussion to Problem 3.6.

In that problem we start by forming what is called the *simulated observation vector* $\bar{\boldsymbol{Y}}_n$ as follows:

$$\bar{\boldsymbol{Y}}_n = \boldsymbol{G}(\bar{\boldsymbol{X}}_n) \tag{3.3.12}$$

in which $\boldsymbol{G}$ is the vector of functions that appears in (3.3.5). Thus $\bar{\boldsymbol{Y}}_n$ is what we would observe in the absence of errors if $\bar{\boldsymbol{X}}_n$ was the state vector of the observed trajectory.

The simulated observation vector $\bar{\boldsymbol{Y}}_n$ is then subtracted from the actual observation vector $\boldsymbol{Y}_n$ to give the *observation perturbation vector*

$$\boldsymbol{\delta Y}_n = \boldsymbol{Y}_n - \bar{\boldsymbol{Y}}_n \tag{3.3.13}$$

The end result of the derivation in Problem 3.6 is that $\boldsymbol{\delta Y}_n$ is related to $\boldsymbol{\delta X}_n$ as follows:

$$\boldsymbol{\delta Y}_n = \boldsymbol{M}(\bar{\boldsymbol{X}}_n)\,\boldsymbol{\delta X}_n + \boldsymbol{N}_n \tag{3.3.14}$$

Equation (3.3.14) then constitutes the required linearized observation equation that connects the perturbation vectors $\boldsymbol{\delta Y}_n$ and $\boldsymbol{\delta X}_n$.

Observe that (3.3.14) has a structure that is similar to (3.2.3), except that here we are talking about perturbation vectors, whereas there we were talking about whole vectors, and here we have the matrix $\boldsymbol{M}(\bar{\boldsymbol{X}}_n)$, whereas there the matrix was $\boldsymbol{M}$.

In Problem 3.6 we show that the matrix $\boldsymbol{M}(\bar{\boldsymbol{X}}_n)$ in (3.3.14) has a definition that is analogous to that of $\boldsymbol{A}\left(\bar{\boldsymbol{X}}(t)\right)$ in (2.6.13), and is as follows:

Definition 3.4: Let $\boldsymbol{G}(\boldsymbol{X}_n)$ in (3.3.5) be a k-vector of functions of the elements of the m-vector $\boldsymbol{X}_n$.

Then $\boldsymbol{M}(\bar{\boldsymbol{X}}_n)$ in (3.3.14) is the $k \times m$ matrix called the **observation sensitivity matrix** whose i, jth element is

$$\left[\boldsymbol{M}(\bar{\boldsymbol{X}}_n)\right]_{i,j} = \partial g_i(x_1, \ldots x_m)/\partial x_j \Big|_{\bar{\boldsymbol{X}}_n} \qquad 1 \le i \le k, \quad 1 \le j \le m \tag{3.3.15}$$

in which $g_i(x_1, \ldots x_m)$ is the ith function in the vector $\boldsymbol{G}(\boldsymbol{X}_n)$, and x_j is the jth element of the state vector $\boldsymbol{X}_n$.

Stated in words: The i, jth element of the matrix $\boldsymbol{M}(\bar{\boldsymbol{X}}_n)$ is the partial derivative of the ith element of $\boldsymbol{G}$ with respect to the jth element of $\boldsymbol{X}$, evaluated using the elements of $\bar{\boldsymbol{X}}_n$. ■■

Example 3.5: Let the state vector of the observed trajectory be

$$\boldsymbol{X}_n = (x_1, x_2, x_3, x_4, x_5, x_6)_n^T \equiv (x, y, z, \dot{x}, \dot{y}, \dot{z})_n^T \tag{3.3.16}$$

and let the observation vector in radar-coordinates be $\boldsymbol{Y}_n = (\rho, \theta, \psi)^T$ shown in (3.3.4). Then $k = 3$ and $m = 6$.

In order to form the 3×6 matrix $\boldsymbol{M}(\bar{\boldsymbol{X}}_n)$ by the rule stated in (3.3.15), we must apply partial differentiation to each of the three expressions in (3.3.2). From the first of these we obtain:

$$\begin{aligned}
\partial g_1/\partial x_1 &= \partial\rho/\partial x = \partial\,(x^2 + y^2 + z^2)^{1/2}/\partial x = x/\rho = x_1/\rho \\
\partial g_1/\partial x_2 &= \partial\rho/\partial y = \partial\,(x^2 + y^2 + z^2)^{1/2}/\partial y = y/\rho = x_2/\rho \\
\partial g_1/\partial x_3 &= \partial\rho/\partial z = \partial\,(x^2 + y^2 + z^2)^{1/2}/\partial z = z/\rho = x_3/\rho \\
\partial g_1/\partial x_4 &= \partial\rho/\partial \dot{x} = \partial\,(x^2 + y^2 + z^2)^{1/2}/\partial \dot{x} = 0 \\
\partial g_1/\partial x_5 &= \partial\rho/\partial \dot{y} = \partial\,(x^2 + y^2 + z^2)^{1/2}/\partial \dot{y} = 0 \\
\partial g_1/\partial x_6 &= \partial\rho/\partial \dot{z} = \partial\,(x^2 + y^2 + z^2)^{1/2}/\partial \dot{z} = 0
\end{aligned} \tag{3.3.17}$$

Then these six results form the first row of $\boldsymbol{M}(\bar{\boldsymbol{X}}_n)$. In the same way we differentiate each of θ and ψ partially with respect to each of $x, y, z, \dot{x}, \dot{y}, \dot{z}$, obtaining finally the result shown in (3.3.18).

$$\boldsymbol{M}(\bar{\boldsymbol{X}}_n) \left(\begin{array}{cccccc} x_1/\rho & x_2/\rho & x_3/\rho & 0 & 0 & 0 \\ -x_1 x_3/\rho^2 s & -x_2 x_3/\rho^2 s & s/\rho^2 & 0 & 0 & 0 \\ -x_2/s^2 & x_1/s^2 & 0 & 0 & 0 & 0 \end{array} \right)_{\bar{\boldsymbol{X}}_n} \tag{3.3.18}$$

in which $\rho = (x^2 + y^2 + z^2)^{1/2}$ and $s = (x^2 + y^2)^{1/2}$

All the equation that have brought us to this stage of our development are listed below.

$$D\boldsymbol{X}(t) = \boldsymbol{F}(\boldsymbol{X}(t)) \qquad (\boldsymbol{X} \text{ and } \boldsymbol{F} \text{ are } m\text{-vectors}) \tag{3.3.19}$$

$$\boldsymbol{X}(t) = \overline{\boldsymbol{X}}(t) + \boldsymbol{\delta X}(t) \tag{3.3.20}$$

$$D\boldsymbol{\delta X}(t) = \boldsymbol{A}(\overline{\boldsymbol{X}}(t))\boldsymbol{\delta X}(t) \qquad (\boldsymbol{A} \text{ is } m \times m) \tag{3.3.21}$$

$$[A(X(t))]_{i,j} = \partial f_i(x_1, \ldots\ x_m)/\partial x_j \Big|_{\overline{X}(t)} \qquad 1 \leq i, j \leq m \tag{3.3.22}$$

$$d/d\zeta\, \boldsymbol{\Phi}(t_n+\zeta, t_n\,; \overline{\boldsymbol{X}}) = \boldsymbol{A}(\overline{\boldsymbol{X}}(t_n+\zeta))\, \boldsymbol{\Phi}(t_n+\zeta, t_n\,; \overline{\boldsymbol{X}}) \tag{3.3.23}$$

$$\boldsymbol{\Phi}(t_n, t_n\,; \overline{\boldsymbol{X}}) = \boldsymbol{I} \tag{3.3.24}$$

$$\boldsymbol{\delta}\mathbf{X}(t_m) = \boldsymbol{\Phi}(t_m, t_n\,; \overline{\boldsymbol{X}})\boldsymbol{\delta X}(t_n) \qquad (\boldsymbol{\Phi} \text{ is } m \times m) \tag{3.3.25}$$

$$\boldsymbol{Y}_n = \boldsymbol{G}(\boldsymbol{X}_n) + \boldsymbol{N}_n \qquad (\boldsymbol{Y} \text{ and } \boldsymbol{G} \text{ are } k\text{-vectors}) \tag{3.3.26}$$

$$\overline{\boldsymbol{Y}}_n = \boldsymbol{G}(\overline{\boldsymbol{X}}_n) \tag{3.3.27}$$

$$\boldsymbol{\delta Y}_n = \boldsymbol{Y}_n - \overline{\boldsymbol{Y}}_n \tag{3.3.28}$$

$$\boldsymbol{\delta Y}_n = \boldsymbol{M}(\overline{\boldsymbol{X}}_n)\,\boldsymbol{\delta X}_n + \boldsymbol{N}_n \qquad (\boldsymbol{M} \text{ is } k \times m) \tag{3.3.29}$$

$$[\boldsymbol{M}(\overline{\boldsymbol{X}}_n)]_{i,j} = \partial g_i(x_1, \ldots\ x_m)/\partial x_j \Big|_{\overline{\boldsymbol{X}}_n} \qquad 1 \leq i \leq k, \quad 1 \leq j \leq m \tag{3.3.30}$$

3.3.2 Sequences of observations

We now assume that we have obtained a sequence of $L+1$ observation vectors $\boldsymbol{Y}_{n-k}(0 \leq k \leq L)$ with time stamps $t_n, t_{n-1}, \ldots t_{n-L}$. Then we have $L+1$ equations similar to (3.3.14), and by concatenation we obtain

$$\begin{pmatrix} \delta\boldsymbol{Y}_n \\ \hline \delta\boldsymbol{Y}_{n-1} \\ \hline \vdots \\ \hline \delta\boldsymbol{Y}_{n-L} \end{pmatrix} = \begin{pmatrix} \boldsymbol{M}(\bar{\boldsymbol{X}}_n)\,\delta\boldsymbol{X}_n \\ \hline \boldsymbol{M}(\bar{\boldsymbol{X}}_{n-1})\,\delta\boldsymbol{X}_{n-1} \\ \hline \vdots \\ \hline \boldsymbol{M}(\bar{\boldsymbol{X}}_{n-L})\,\delta\boldsymbol{X}_{n-L} \end{pmatrix} + \begin{pmatrix} \boldsymbol{N}_n \\ \hline \boldsymbol{N}_{n-1} \\ \hline \vdots \\ \hline \boldsymbol{N}_{n-L} \end{pmatrix} \tag{3.3.31}$$

From (2.6.19) the transition equation for $\delta\boldsymbol{X}_n$ is

$$\delta\boldsymbol{X}_m = \boldsymbol{\Phi}\,(t_m, t_n; \bar{\boldsymbol{X}}\,)\,\delta\boldsymbol{X}_n \tag{3.3.32}$$

in which $\boldsymbol{\Phi}(t_m, t_n; \bar{\boldsymbol{X}}\,)$ is derived by integrating the DE in (2.6.20). Using (3.3.32) in (3.3.31) then gives

$$\begin{pmatrix} \delta\boldsymbol{Y}_n \\ \hline \delta\boldsymbol{Y}_{n-1} \\ \hline \vdots \\ \hline \delta\boldsymbol{Y}_{n-L} \end{pmatrix} = \begin{pmatrix} \boldsymbol{M}(\bar{\boldsymbol{X}}_n)\delta\boldsymbol{X}_n \\ \hline \boldsymbol{M}(\bar{\boldsymbol{X}}_{n-1})\,\boldsymbol{\Phi}\,(t_{n-1}, t_n; \bar{\boldsymbol{X}}\,)\,\delta\bar{\boldsymbol{X}}_n \\ \hline \vdots \\ \hline \boldsymbol{M}(\bar{\boldsymbol{X}}_{n-L})\,\boldsymbol{\Phi}\,(t_{n-L}, t_n; \bar{\boldsymbol{X}}\,)\,\delta\bar{\boldsymbol{X}}_n \end{pmatrix} + \begin{pmatrix} \boldsymbol{N}_n \\ \hline \boldsymbol{N}_{n-1} \\ \hline \vdots \\ \hline \boldsymbol{N}_{n-L} \end{pmatrix} \tag{3.3.33}$$

We now factor out $\delta\boldsymbol{X}_n$ from the matrix on the right, obtaining

$$\begin{pmatrix} \delta\boldsymbol{Y}_n \\ \hline \delta\boldsymbol{Y}_{n-1} \\ \hline \vdots \\ \hline \delta\boldsymbol{Y}_{n-L} \end{pmatrix} = \begin{pmatrix} \boldsymbol{M}(\bar{\boldsymbol{X}}_n) \\ \hline \boldsymbol{M}(\bar{\boldsymbol{X}}_{n-1})\,\boldsymbol{\Phi}\,(t_{n-1}, t_n; \bar{\boldsymbol{X}}\,) \\ \hline \vdots \\ \hline \boldsymbol{M}(\bar{\boldsymbol{X}}_{n-L})\,\boldsymbol{\Phi}\,(t_{n-L}, t_n; \bar{\boldsymbol{X}}\,) \end{pmatrix} \delta\boldsymbol{X}_n + \begin{pmatrix} \boldsymbol{N}_n \\ \hline \boldsymbol{N}_{n-1} \\ \hline \vdots \\ \hline \boldsymbol{N}_{n-L} \end{pmatrix} \tag{3.3.34}$$

Definition 3.5: The **total observation matrix** for ***Case 4*** is as follows:

$$\boldsymbol{T}(\bar{X}_n) \equiv \begin{pmatrix} \boldsymbol{M}(\bar{X}_n) \\ \hdashline \boldsymbol{M}(\bar{X}_{n-1})\,\boldsymbol{\Phi}(t_{n-1}, t_n; \bar{X}) \\ \hdashline \vdots \\ \hdashline \boldsymbol{M}(\bar{X}_{n-L})\,\boldsymbol{\Phi}(t_{n-L}, t_n; \bar{X}) \end{pmatrix} \tag{3.3.35}$$

■■

This is the form that the $\boldsymbol{T}$ matrix assumes for ***Case 4*** in which both the filter model and the observation equation are nonlinear. We note that $\boldsymbol{T}$ depends on the nominal trajectory $\bar{X}_n$, and to make that clear we write it as $\boldsymbol{T}(\bar{X}_n)$.

Using the definition appearing in (3.3.35) we now write (3.3.34) as

$$\delta \boldsymbol{Y}_n = \boldsymbol{T}(\bar{X}_n)\,\delta X_n + N_n \tag{3.3.36}$$

in which the vectors $\delta \boldsymbol{Y}_n$ and $\boldsymbol{N}_n$ are long, and the matrix $\boldsymbol{T}(\bar{X}_n)$ is high. Observe that if $\boldsymbol{N}_n$ is removed, then $\delta \boldsymbol{Y}_n$ is related to δX_n by a linear transformation.

As we did with (3.2.13), we refer to (3.3.36) as the total observation equation. In (3.2.13), the whole vectors $\boldsymbol{Y}_n$ and X_n were linearly related, whereas here it is the perturbation vectors $\delta \boldsymbol{Y}_n$ and $\delta \boldsymbol{X}_n$ that are linearly related.

Details of the numerical integration for obtaining the transition matrices $\boldsymbol{\Phi}(t_{n-k}, t_n; \bar{X})$ in (3.3.35) are discussed in the files contained in the folder

Miscellaneous\Numerical_Integration.

Once again we point out that the long column vectors of $\boldsymbol{T}(\bar{X}_n)$ in (3.3.35) must be linearly independent or else our progress will be blocked later on. We defer discussion of this important topic to Section 5.6.

3.4 *Case 3:* Nonlinear filter model, linear observation equation

We are now considering ***Case 3*** of the four possibilities in Figure 3.2 for which the filter model is nonlinear and the observation equation is linear.

Example 3.6: In the previous example we considered a small body in orbit around a spherical earth. The observation vector was of the form

$$\boldsymbol{Y}_{r,n} = (\rho, \theta, \psi)_n^T \tag{3.4.1}$$

and the filter model was the nonlinear DE shown in (3.3.7). Note that we have now added a subscripted r to $\boldsymbol{Y}_{r,n}$ in (3.4.1) to show that it is a vector of observations in radar coordinates.

It is often possible to transform the raw observations $\boldsymbol{Y}_{r,n}$ from radar to Cartesian coordinates, before they are processed by the filter. (This transformation is done externally to the filter and precedes it.)

The equations that perform the transformation are those stated earlier in (3.3.1), namely:

$$\begin{aligned} x_n &= \rho_n \cos(\theta_n) \cos(\psi_n) \\ y_n &= \rho_n \cos(\theta_n) \sin(\psi_n) \\ z_n &= \rho_n \sin(\theta_n) \end{aligned} \tag{3.4.2}$$

Assembling x_n, y_n and z_n into a *3*-vector then gives us the vector of observations

$$\boldsymbol{Y}_{c,n} = (x, y, z)_n^T \tag{3.4.3}$$

where the subscript c designates Cartesian coordinates.

If the true state vector is $\boldsymbol{X}_n = (x, y, z, \dot{x}, \dot{y}, \dot{z})_n^T$ then the vector of observations $\boldsymbol{Y}_{c,n}$ in (3.4.3) would be related to $\boldsymbol{X}_n$ as follows:

$$\begin{pmatrix} y_1 \\ y_2 \\ y_3 \end{pmatrix}_{c,n} = \begin{pmatrix} 1 & 0 & 0 & 0 & 0 & 0 \\ 0 & 1 & 0 & 0 & 0 & 0 \\ 0 & 0 & 1 & 0 & 0 & 0 \end{pmatrix} \begin{pmatrix} x \\ y \\ z \\ \dot{x} \\ \dot{y} \\ \dot{z} \end{pmatrix}_n + \begin{pmatrix} v_x \\ v_y \\ v_z \end{pmatrix}_{c,n} \tag{3.4.4}$$

Equation (3.4.4) has the general form

$$\boldsymbol{Y}_n = \boldsymbol{M}\boldsymbol{X}_n + \boldsymbol{N}_n \tag{3.4.5}$$

and under these circumstances we would have the nonlinear filter model in (3.3.7) and the linear observation equation in (3.4.5), which is the situation for ***Case 3***. ■■

As we did in ***Case 4*** above, we again assume the existence of a known nominal trajectory $\bar{\boldsymbol{X}}(t)$ that is close to the observed trajectory $\boldsymbol{X}(t)$, and which is governed by the same DE as $\boldsymbol{X}(t)$, namely (3.3.9).

We now form the simulated observation vector $\bar{\boldsymbol{Y}}_n$ defined by

$$\bar{\boldsymbol{Y}}_n = \boldsymbol{M}\bar{\boldsymbol{X}}_n \tag{3.4.6}$$

Subtracting (3.4.6) from (3.4.5) we obtain

$$\boldsymbol{Y}_n - \bar{\boldsymbol{Y}}_n = \boldsymbol{M}\boldsymbol{X}_n + \boldsymbol{N}_n - \boldsymbol{M}\bar{\boldsymbol{X}}_n = \boldsymbol{M}(\boldsymbol{X}_n - \bar{\boldsymbol{X}}_n) + \boldsymbol{N}_n = \boldsymbol{M}\,\delta\boldsymbol{X}_n + \boldsymbol{N}_n \tag{3.4.7}$$

Now define

$$\delta\boldsymbol{Y}_n \equiv \boldsymbol{Y}_n - \bar{\boldsymbol{Y}}_n \tag{3.4.8}$$

Then (3.4.7) becomes

$$\delta\boldsymbol{Y}_n = \boldsymbol{M}\,\delta\boldsymbol{X}_n + \boldsymbol{N}_n \tag{3.4.9}$$

Repeating this for each of the observation instants $t_n, t_{n-1}, \ldots t_{n-L}$ gives us the sequence of $L+1$ perturbation observation equations

$$\delta\boldsymbol{Y}_{n-k} = \boldsymbol{M}\,\delta\boldsymbol{X}_{n-k} + \boldsymbol{N}_{n-k} \quad 0 \le k \le L \tag{3.4.10}$$

from which, by concatenation,

$$\begin{pmatrix} \delta\boldsymbol{Y}_n \\ \hdashline \delta\boldsymbol{Y}_{n-1} \\ \hdashline \vdots \\ \hdashline \delta\boldsymbol{Y}_{n-L} \end{pmatrix} = \begin{pmatrix} \boldsymbol{M}\,\delta\boldsymbol{X}_n \\ \hdashline \boldsymbol{M}\,\delta\boldsymbol{X}_{n-1} \\ \hdashline \vdots \\ \hdashline \boldsymbol{M}\,\delta\boldsymbol{X}_{n-L} \end{pmatrix} + \begin{pmatrix} \boldsymbol{N}_n \\ \hdashline \boldsymbol{N}_{n-1} \\ \hdashline \vdots \\ \hdashline \boldsymbol{N}_{n-L} \end{pmatrix} \tag{3.4.11}$$

We are in ***Case 3*** for which the filter model is nonlinear, and so the transition equation for the perturbation vector $\delta\boldsymbol{X}_n$ is as shown in (3.3.11). Using that transition equation in (3.4.11) we obtain

$$\begin{pmatrix} \delta Y_n \\ \text{------} \\ \delta Y_{n-1} \\ \text{------} \\ \vdots \\ \text{------} \\ \delta Y_{n-L} \end{pmatrix} = \begin{pmatrix} M\,\delta X_n \\ \text{------------------} \\ M\,\Phi(t_{n-1}, t_n; \bar{X})\delta X_n \\ \text{------------------} \\ \vdots \\ \text{------------------} \\ M\,\Phi(t_{n-L}, t_n; \bar{X})\delta X_n \end{pmatrix} + \begin{pmatrix} N_n \\ \text{------} \\ N_{n-1} \\ \text{------} \\ \vdots \\ \text{------} \\ N_{n-L} \end{pmatrix} \tag{3.4.12}$$

and finally, factoring out δX_n gives the result

$$\begin{pmatrix} \delta Y_n \\ \text{------} \\ \delta Y_{n-1} \\ \text{------} \\ \vdots \\ \text{------} \\ \delta Y_{n-L} \end{pmatrix} = \begin{pmatrix} M \\ \text{------------------} \\ M\,\Phi(t_{n-1}, t_n; \bar{X}) \\ \text{------------------} \\ \vdots \\ \text{------------------} \\ M\,\Phi(t_{n-L}, t_n; \bar{X}) \end{pmatrix} \delta X_n + \begin{pmatrix} N_n \\ \text{------} \\ N_{n-1} \\ \text{------} \\ \vdots \\ \text{------} \\ N_{n-L} \end{pmatrix} \tag{3.4.13}$$

Definition 3.6: The **total observation matrix** for ***Case 3*** is as follows:

$$T(\bar{X}_n) \equiv \begin{pmatrix} M \\ \text{------------------} \\ M\,\Phi(t_{n-1}, t_n; \bar{X}) \\ \text{------------------} \\ \vdots \\ \text{------------------} \\ M\,\Phi(t_{n-L}, t_n; \bar{X}) \end{pmatrix} \tag{3.4.14}$$

■■

This is the form that $\boldsymbol{T}$ assumes for ***Case 3***, in which the filter model is nonlinear and the observation equation is linear. We note that $\boldsymbol{T}(\bar{X}_n)$ depends on the nominal trajectory $\bar{X}$.

Using the definition appearing in (3.4.14) we are now able to write (3.4.13) as

$$\boldsymbol{\delta Y}_n = \boldsymbol{T}(\bar{\boldsymbol{X}}_n)\,\boldsymbol{\delta X}_n + \boldsymbol{N}_n \tag{3.4.15}$$

in which $\boldsymbol{\delta Y}_n$ is the total observation perturbation vector, and $\boldsymbol{N}_n$ is the total error vector. We refer to (3.4.15) as the total observation equation, in which the perturbation vectors $\boldsymbol{\delta Y}_n$ and $\boldsymbol{\delta X}_n$ are seen to be linearly related.

As with ***Case 4*** above, we will require numerical integration in order to derive values for the transition matrices $\boldsymbol{\Phi}(t_{n-L}, t_n; \bar{\boldsymbol{X}})$ appearing in (3.4.14). This is discussed in detail in the files contained in the folder

Miscellaneous\Numerical_Integration.

The long column vectors of $\boldsymbol{T}(\bar{\boldsymbol{X}}_n)$ in (3.4.14) must be linearly independent, or else our progress will be blocked later on. Discussion of this topic is deferred to Section 5.6.

3.5 *Case 2:* Linear filter model, nonlinear observation equation

We are now on the last of the four possibilities in Figure 3.2, namely ***Case 2*** for which the filter model is linear and the observation equation is nonlinear.

Example 3.7*:* Consider an aircraft that is flying in the neighbourhood of a CW (continuous wave) transmitter.

◊ The Doppler shift in the backscatter from the aircraft is observed by a number of receivers – say *24* – that have been distributed more or less uniformly over a circular region whose radius is about *40 km*.

◊ Each of the receivers also measures the bearing of the aircraft relative to its own location.

The equation for the Doppler shift f_d as observed at the kth receiver is as follows:

$$f_d(x, y, z)_k = K_d\left(\frac{(x - x_0)\dot{x} + (y - y_0)\dot{y} + (z - z_0)\dot{z}}{\left((x - x_0)^2 + (y - y_0)^2 + (z - z_0)^2\right)^{1/2}} + \frac{(x - x_k)\dot{x} + (y - y_k)\dot{y} + (z - z_k)\dot{z}}{\left((x - x_k)^2 + (y - y_k)^2 + (z - z_k)^2\right)^{1/2}}\right) \tag{3.5.1}$$

in which x, y, z are the coordinates of the aircraft, x_0, y_0, z_0 are the coordinates of the transmitter, x_k, y_k, z_k are the coordinates of the kth receiver and K_d is a constant that is related to the wavelength of the energy being transmitted.

The equation for the bearing of the target as observed by the kth receiver is

$$\psi_k(x, y, z) = tan^{-1}((y - y_k)/(x - x_k)) \quad (-\pi < \psi_k \leq \pi) \tag{3.5.2}$$

Assume that for each Doppler observation from each receiver there is also a bearing observation, and at time t_n let all of the Doppler and bearing observations from all of the receivers be concatenated into an observation vector that we call $\boldsymbol{Y}_n$.

◊ Both (3.5.1) and (3.5.2) contain nonlinear combinations of the variables in the state vector

$$\boldsymbol{X}_n = (x, \dot{x}, \ddot{x}, \ldots y, \dot{y}, \ddot{y}, \ldots z, \dot{z}, \ddot{z}, \ldots)$$

and so the vector of observations $\boldsymbol{Y}_n$ will be related to the state vector $\boldsymbol{X}_n$ by an equation that is nonlinear.

◊ The flight path of the target is assumed to be a *3*-dimensional polynomial, and so the filter model is linear.

Then this is an example of a linear filter model and a set of nonlinear observation equations, which is the situation for ***Case 2***. ■■

Even though the filter model is linear, we must still work with perturbation vectors because the observation equations are nonlinear. Accordingly, we start from

$$\boldsymbol{X}_n = \bar{\boldsymbol{X}}_n + \delta\boldsymbol{X}_n \tag{3.5.3}$$

in which $\bar{\boldsymbol{X}}_n$ is a known nominal trajectory that is assumed to be close to the observed trajectory $\boldsymbol{X}_n$ and $\delta\boldsymbol{X}_n$ is a vector of perturbations. Then

$$\delta\boldsymbol{X}_n = \boldsymbol{X}_n - \bar{\boldsymbol{X}}_n \tag{3.5.4}$$

The nominal trajectory $\bar{\boldsymbol{X}}$ is governed by the same linear DE as the vector $\boldsymbol{X}$, and so they both have the same transition matrix. Let it be called $\boldsymbol{\Phi}$. We now multiply both sides of (3.5.4) by $\boldsymbol{\Phi}(t_m - t_n)$, obtaining

$$\boldsymbol{\Phi}(t_m - t_n)\delta\boldsymbol{X}_n = \boldsymbol{\Phi}(t_m - t_n)(\boldsymbol{X}_n - \bar{\boldsymbol{X}}_n) = \boldsymbol{X}_m - \bar{\boldsymbol{X}}_m = \delta\boldsymbol{X}_m \tag{3.5.5}$$

Comparison of the first and last terms in this equation shows that the perturbation vector $\delta\boldsymbol{X}_n$ also has $\boldsymbol{\Phi}(t_m - t_n)$ as its transition matrix, and so we have the transition equation

$$\delta\boldsymbol{X}_m = \boldsymbol{\Phi}(t_m - t_n)\delta\boldsymbol{X}_n \tag{3.5.6}$$

We are in *Case 2* in which the observation equation is nonlinear, and so we must use the equation for the observation perturbation vector that appeared in (3.3.14), namely

$$\delta \boldsymbol{Y}_n = \boldsymbol{M}(\bar{\boldsymbol{X}}_n)\,\delta \boldsymbol{X}_n + \boldsymbol{N}_n \tag{3.5.7}$$

Then concatenation of $L+1$ such equations gives

$$\begin{pmatrix} \delta \boldsymbol{Y}_n \\ \hline \delta \boldsymbol{Y}_{n-1} \\ \hline \vdots \\ \hline \delta \boldsymbol{Y}_{n-L} \end{pmatrix} = \begin{pmatrix} \boldsymbol{M}(\bar{\boldsymbol{X}}_n)\,\delta \boldsymbol{X}_n \\ \hline \boldsymbol{M}(\bar{\boldsymbol{X}}_{n-1})\,\delta \boldsymbol{X}_{n-1} \\ \hline \vdots \\ \hline \boldsymbol{M}(\bar{\boldsymbol{X}}_{n-L})\,\delta \boldsymbol{X}_{n-L} \end{pmatrix} + \begin{pmatrix} \boldsymbol{N}_n \\ \hline \boldsymbol{N}_{n-1} \\ \hline \vdots \\ \hline \boldsymbol{N}_{n-L} \end{pmatrix} \tag{3.5.8}$$

We now apply (3.5.6) and factor out $\delta \boldsymbol{X}_n$, obtaining

$$\begin{pmatrix} \delta \boldsymbol{Y}_n \\ \hline \delta \boldsymbol{Y}_{n-1} \\ \hline \vdots \\ \hline \delta \boldsymbol{Y}_{n-L} \end{pmatrix} = \begin{pmatrix} \boldsymbol{M}(\bar{\boldsymbol{X}}_n) \\ \hline \boldsymbol{M}(\bar{\boldsymbol{X}}_{n-1})\boldsymbol{\Phi}(t_{n-1} - t_n) \\ \hline \vdots \\ \hline \boldsymbol{M}(\bar{\boldsymbol{X}}_{n-L})\boldsymbol{\Phi}(t_{n-L} - t_n) \end{pmatrix} \delta \boldsymbol{X}_n + \begin{pmatrix} \boldsymbol{N}_n \\ \hline \boldsymbol{N}_{n-1} \\ \hline \vdots \\ \hline \boldsymbol{N}_{n-L} \end{pmatrix} \tag{3.5.9}$$

Definition 3.7: The **total observation matrix** for ***Case 2*** is as follows:

$$\boldsymbol{T}(\bar{\boldsymbol{X}}_n) \equiv \begin{pmatrix} \boldsymbol{M}(\bar{\boldsymbol{X}}_n) \\ \hline \boldsymbol{M}(\bar{\boldsymbol{X}}_{n-1})\boldsymbol{\Phi}(t_{n-1} - t_n) \\ \hline \vdots \\ \hline \boldsymbol{M}(\bar{\boldsymbol{X}}_{n-L})\boldsymbol{\Phi}(t_{n-L} - t_n) \end{pmatrix} \tag{3.5.10}$$

■■

This is the form that $\boldsymbol{T}$ assumes for ***Case 2*** in which the filter model is linear and the observation equation is nonlinear. We note that $\boldsymbol{T}(\bar{\boldsymbol{X}}_n)$ depends on the nominal trajectory $\bar{\boldsymbol{X}}_n$.

Using the definition in (3.5.10) enables us to write (3.5.9) as

$$\delta \boldsymbol{Y}_n = \boldsymbol{T}(\bar{X}_n)\delta X_n + \boldsymbol{N}_n \tag{3.5.11}$$

in which $\delta \boldsymbol{Y}_n$ is the total observation perturbation vector and $\boldsymbol{N}_n$ is the total error vector.

In (3.5.11) we have arrived at a total observation equation in which the perturbation vectors $\delta \boldsymbol{Y}_n$ and $\delta \boldsymbol{X}_n$ are linearly related.

As with ***Case 1***, we will not require numerical integration in order to derive the transition matrices $\boldsymbol{\Phi}(t_{n-k} - t_n)$ appearing in (3.5.10), because they can be written in explicit form (see Chapter 2).

It is essential that the column vectors of the high-matrix $\boldsymbol{T}(\bar{X}_n)$ in (3.5.10) be linearly independent. As with ***Cases 1, 2*** and ***4***, discussion of this topic is deferred to Section 5.6.

3.6 Summary

We have considered the four possible cases, and have shown that we can always arrive at a linear total observation equation. For ***Case 1***, it was

$$\boldsymbol{Y}_n = \boldsymbol{T}_n X_n + \boldsymbol{N}_n \tag{3.6.1}$$

and for ***Cases 2, 3*** and ***4*** it was

$$\delta \boldsymbol{Y}_n = \boldsymbol{T}(\bar{X}_n)\,\delta X_n + \boldsymbol{N}_n \tag{3.6.2}$$

Each of the four cases has its own definition for $\boldsymbol{T}$ which is shown in Figure 3.4 below.

- ◊ The total observation equation for ***Case 1***, when the observation equation(s) and the filter model are linear, is shown above in (3.6.1). We note that it involves whole vectors rather than perturbation vectors. The $\boldsymbol{T}$ matrix is shown in Figure 3.4 below, where we see that it does not depend on a nominal trajectory.

- ◊ The total observation equation for ***Cases 2, 3*** and ***4***, when nonlinearities are present either in the filter model or the observation equation(s) or in both, is as shown in (3.6.2). We note that it involves perturbation vectors. The $\boldsymbol{T}(\bar{X}_n)$ matrices for those cases are also shown in Figure 3.4, where we see that they depend on a nominal trajectory $\bar{X}$.

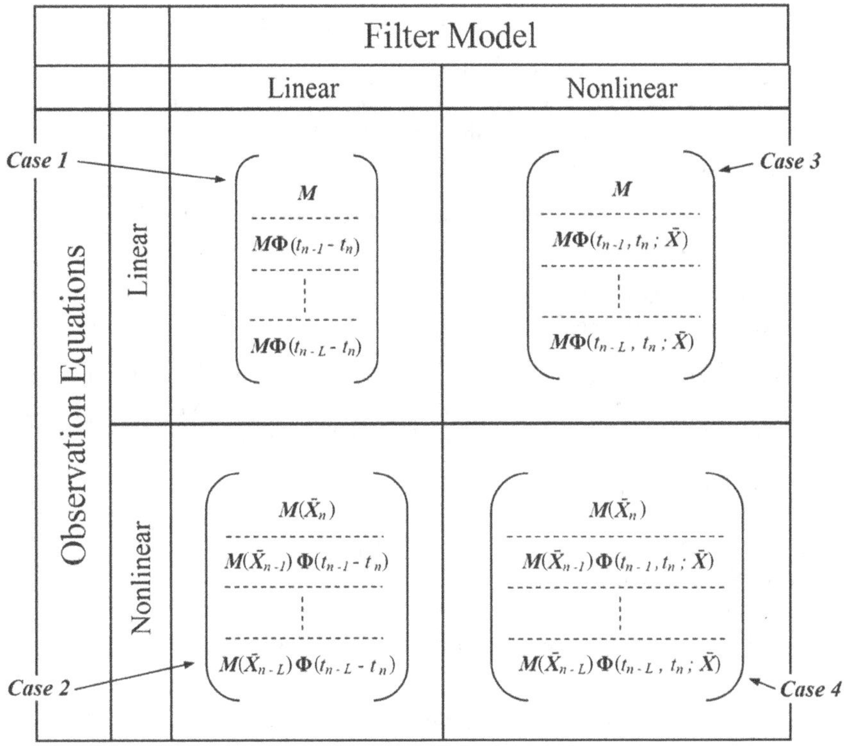

*Figure 3.4 The four **T** matrices*

◊ In ***Cases 3*** and ***4*** we will require numerical integration in order to derive the transition matrices $\Phi(t_{n-1}, t_n; \bar{X})$. This is discussed in detail in the files contained in the folder

Miscellaneous\Numerical_Integration.

◊ In Chapter 9, we discuss how a nominal trajectory is obtained for the three cases involving nonlinearities.

3.7 Incorporating a *T* matrix into the filter

Each of the ***T*** matrices that we derived above was based on the filter model and the observation equations.

Once the observations have been made they become the inputs to the filter, and in order for the latter to be able to operate on them correctly, it is essential that the details of the filter model and the observation equations be built into the filter.

The way in which that is done is by including a $\boldsymbol{T}$ matrix in the filtering algorithm.

Referring back to Figure 1.7 in which we showed Version 1 of the Gauss–Aitken filter, we see that the filter matrix $\boldsymbol{W}$ is created out of two components, a $\boldsymbol{T}$ matrix and the covariance matrix $\boldsymbol{R_Y}$ of the total observation vector $\boldsymbol{Y}$, and we now examine what that accomplishes.

We have spoken in the preceding two chapters about the external model and the filter model, and about the fact that they are two separate constructs.

The external model is selected by the filter engineer out of a possible set of alternatives, and it could be complex or simple (see discussion in Chapter 2).

The filter model is likewise selected by the filter engineer, and it could be the same as, or different from, the external model, but if bias errors are to be avoided then it must successfully emulate the external model (see Chapter 6).

Either way (same or different):

◊ The filter model is specified through whatever transition matrix $\boldsymbol{\Phi}$ appears in the $\boldsymbol{T}$ matrix that is incorporated into the filter, and if the filter model is identical to the external model, then the transition matrix $\boldsymbol{\Phi}$ appearing in $\boldsymbol{T}$ will be the same as the transition matrix associated with the DE of the external model.

In the event that the transition matrix $\boldsymbol{\Phi}$ appearing in $\boldsymbol{T}$ is different from the transition matrix of the external model, then this means that we have elected to use a filter model that differs from the external mode.

◊ The details of how the true state vector $\boldsymbol{X}$ was observed are conveyed to the filter by the observation matrix $\boldsymbol{M}$ that appears in $\boldsymbol{T}$.

All of this is made clear by the following two examples. In the first of them the filter model and the external model are the same, and in the second they are different.

Example 3.8: In Example. 3.4, we considered a near-earth satellite being observed by radar, and we arrived at the following $\boldsymbol{T}$ matrix appearing in (3.7.1) below.

$$\boldsymbol{T}(\bar{\boldsymbol{X}}_n) \equiv \begin{pmatrix} \boldsymbol{M}(\bar{\boldsymbol{X}}_n) \\ \hdashline \boldsymbol{M}(\bar{\boldsymbol{X}}_{n-1})\boldsymbol{\Phi}(t_{n-1}, t_n; \bar{\boldsymbol{X}}) \\ \hdashline \vdots \\ \hdashline \boldsymbol{M}(\bar{\boldsymbol{X}}_{n-L})\boldsymbol{\Phi}(t_{n-L}, t_n; \bar{\boldsymbol{X}}) \end{pmatrix} \tag{3.7.1}$$

Recall that both the filter model and the observation equation were nonlinear, and so we were in ***Case 4***.

The filter model was the nonlinear DE for orbital motion appearing in (3.3.7) in which the true state vector was

$$\boldsymbol{X} = (x, y, z, \dot{x}, \dot{y}, \dot{z})^T \tag{3.7.2}$$

This is a *6*-vector, and when the matrix $\boldsymbol{M}(\bar{\boldsymbol{X}})$ is derived using the rule given in (3.3.15), it is done using partial differentiation with respect to each of the elements of this vector. Because there are *3* equations in (3.3.4), it follows that $\boldsymbol{M}(\bar{\boldsymbol{X}})$ will be 3×6. The resulting matrix $\boldsymbol{M}(\bar{\boldsymbol{X}})$ appeared in (3.3.18).

The transition matrix $\boldsymbol{\Phi}(t_{n-1}, t_n; \bar{\boldsymbol{X}})$ for the *6*-vector in (3.7.2) is 6×6, and so the matrices $\boldsymbol{M}(\bar{\boldsymbol{X}}_m)$ and $\boldsymbol{\Phi}(t_{n-1}, t_n; \bar{\boldsymbol{X}})$ are conformable, meaning that they can be multiplied together. Accordingly, if this $\boldsymbol{T}$ matrix is included in the filter, then it will be possible to evaluate those products and the filter can be cycled.

Processing of the observations by the filter is then being done on the assumption that the filter model is identical to the external model because the transition matrix $\boldsymbol{\Phi}(t_{n-1}, t_n; \bar{\boldsymbol{X}})$ that appears in $\boldsymbol{T}(\bar{\boldsymbol{X}}_n)$ is the same as the transition matrix for the external model. ■■

Example 3.9: In this example we make the decision that the filter model will not be the same as the orbital external model, which is nonlinear. Instead of using (3.3.7) as the filter model we will use one based on 2nd-degree polynomials whose state vector and transition matrix appear in (2.3.10). The filter model will then be linear.

There are two reasons why we might wish to do this:

◊ Polynomials have the property of being able to fit themselves to trajectories without us knowing the equations for the latter.

◊ Polynomials free us from the need for numerical integration to derive the transition matrix.

Note: When we use a filter model that differs from the external model, the possibility arises that we might introduce bias errors. Bias errors are considered in Chapter 6.

The filter model will now be the linear DE for 2nd-degree polynomials, rather than the nonlinear one appearing in (3.3.7), and from (2.3.10) its state vector will be the *9*-vector

$$\boldsymbol{X} = (x, \dot{x}, \ddot{x}, y, \dot{y}, \ddot{y}, z, \dot{z}, \ddot{z})^T \tag{3.7.3}$$

The transition matrix will be the 9×9 matrix $\boldsymbol{\Phi}$ that also appears in (2.3.10).

The matrix $\boldsymbol{M}(\bar{\boldsymbol{X}}_n)$ must be derived using the rule given in (3.3.15) applied to (3.3.4).

Partial differentiation must now be performed with respect to the elements of the state vector appearing in (3.7.3), rather than the one in (3.7.2), and so $\boldsymbol{M}(\bar{\boldsymbol{X}}_n)$ will be 3×9.

The filter model is now linear but the observation equation remains nonlinear. We have moved from ***Case 4*** to ***Case 2***, and so the $\boldsymbol{T}$ matrix will have the structure

$$\boldsymbol{T}(\bar{\boldsymbol{X}}_n) \equiv \begin{pmatrix} \boldsymbol{M}(\bar{\boldsymbol{X}}_n) \\ \hdashline \boldsymbol{M}(\bar{\boldsymbol{X}}_{n-1})\boldsymbol{\Phi}(t_{n-1} - t_n) \\ \hdashline \vdots \\ \hdashline \boldsymbol{M}(\bar{\boldsymbol{X}}_{n-L})\boldsymbol{\Phi}(t_{n-L} - t_n) \end{pmatrix} \tag{3.7.4}$$

Because $\boldsymbol{M}(\bar{\boldsymbol{X}}_n)$ is now 3×9 and $\boldsymbol{\Phi}$ is 9×9, the matrices $\boldsymbol{M}(\bar{\boldsymbol{X}}_m)$ and $\boldsymbol{\Phi}(t_m - t_n)$ appearing in (3.7.4) are again conformable and can be multiplied. The matrix $\boldsymbol{T}(\bar{\boldsymbol{X}}_n)$ can be evaluated and again the matrix can be cycled. ■■

3.7.1 Conclusion

The filter model will be implemented when the $\boldsymbol{T}$ matrix is included in the equations of the filter. It could be the same as, or different from, the external model.

In the same way, the observation equation(s) will become part of the filtering algorithm when $\boldsymbol{T}$ is included in the filter algorithm.

All of the results derived in this chapter will be needed when we discuss the Gauss–Aitken and Gauss–Newton filters in Chapters 8 and 9.

The following items are in the supplementary material

Problems for Chapter 3

Solutions to Problems for Chapter 3

Projects for Chapter 3

Appendix 3.1
ENU (east–north–up) coordinates

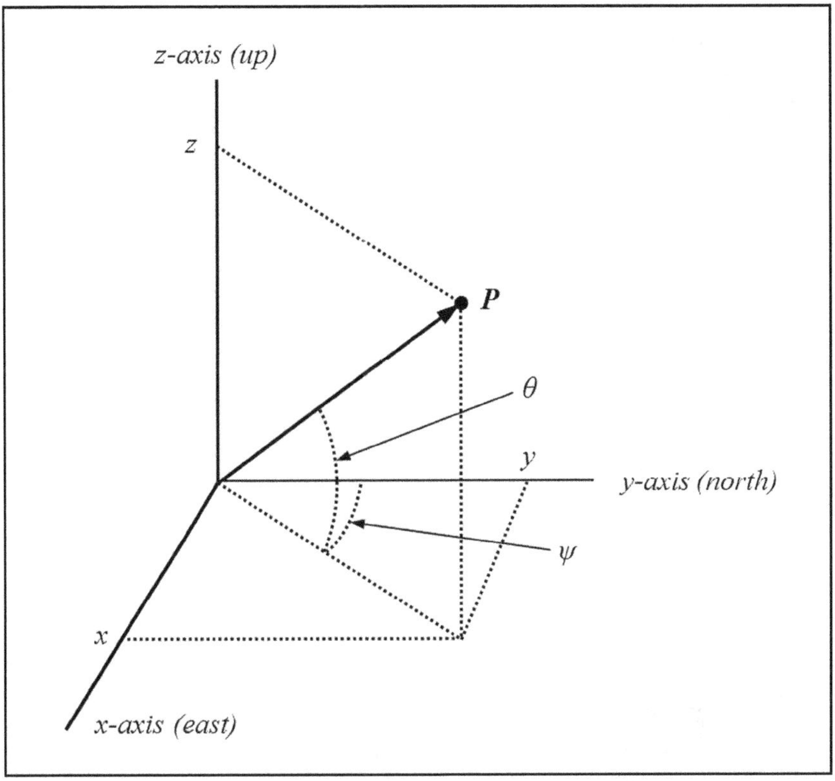

Figure A3.1.1 ENU coordinates

In *ENU* or *east-north-up* coordinates, the azimuth angle ψ is measured counter-clockwise in the x,y-plane starting from the y-axis, which is typically *north*, and increases positively as we move towards the x-axis which is *east*. In order that the Cartesian axes form a right-handed coordinate system, the positive z-axis is *up*.

The elevation angle θ is measured vertically from the x, y plane to the $\boldsymbol{P}$-vector. The positive z-axis has an elevation angle $\theta = \pi/2$ rads. As shown in Figure A3.1.1 the value of θ is positive because $\boldsymbol{P}$ lies in the region where z is positive.

We then have the following relationships:

$$\begin{aligned} x &= \rho \, cos(\theta) \, sin(\psi) \\ y &= \rho \, cos(\theta) \, cos(\psi) \\ z &= \rho \, sin(\theta) \end{aligned} \tag{A3.1.1}$$

The inverse set of equations is as follows:

$$\begin{aligned} \rho &= \left(x^2 + y^2 + z^2\right)^{1/2} \\ \theta &= \tan^{-1}\left(\frac{z}{(x^2 + y^2)^{1/2}}\right) \quad (2\text{-}quadrant\ arctan) \\ \psi &= \tan^{-1}\left(\frac{x}{y}\right) \quad (4\text{-}quadrant\ arctan) \end{aligned} \tag{A3.1.2}$$

Chapter 4

Random vectors and covariance matrices – theory

Random vectors and covariance matrices are the basic building blocks of filter engineering. In this chapter we review some of their properties, and in the next we consider how they are used.

In the first and second sections of this chapter we discuss random vectors and their covariance matrices, and at first sight the reader may feel that the material has been well covered in first courses in probability and statistics. However, that is not the case. It is here that we lay the foundation for the split between the two types of covariance matrices – *supposed* and *actual* – and between what is theoretical and covered in most such courses, and what is empirical and more often encountered in Monte-Carlo simulations.

In the third and fourth sections of the chapter we discuss the positive-definite property of certain matrices, a concept that is often not covered in introductory linear algebra and which plays a key role in filter engineering.

4.1 Random variables

Random variables (RVs) are of two types – continuous and discrete. In this book we are concerned only with the former and so when we use the abbreviation *RV* we are always referring to *continuous* random variables.

RVs are characterized by what are known as their *probability density functions* (PDFs). We use the following standard notation:

◊ All RVs will be represented initially by upper case italic letters, e.g. X, Y, Z etc.

◊ All PDFs will be represented by the symbol f, e.g. $f(x)$ or $f(x, y)$.

◊ To distinguish between the PDFs of X and Y we attach subscripts to f. Thus $f_X(x)$ is the PDF of X and $f_Y(y)$ is the PDF of Y. In the first of these we are using an upper-case X for the name of the RV and a lower-case x for the independent variable of f, but that is only for convenience. We could just as well have written the PDF of X as $f_X(u)$, and the PDF of Y as $f_Y(v)$.

◊ The PDF of the *2*-dimensional RV (X, Y) will be shown as $f_{X,Y}(x, y)$.

Definition 4.1: The **1-dimensional random variable** X is an infinite population of real numbers lying on the real line $-\infty < x < \infty$ and distributed in such a way that its PDF is $f_X(x)$. Such a random variable is said to be *univariate* and its PDF is a *univariate* PDF.

The **2-dimensional random variable** (X, Y) is an infinite population of real number pairs lying on the real plane $-\infty < x, y < \infty$ and distributed in such a way that its PDF is $f_{X,Y}(x, y)$. Such a random variable is said to be *bivariate* and its PDF is a *bivariate* PDF. ■■

Definition 4.2: Let a sequence of values $x_1, x_2, \ldots x_k$ be selected randomly from the RV X whose PDF is $f_X(x)$. Then each of those values is called a **draw**. ■■

A similar definition can be given for draws from the bivariate RV (X, Y) whose PDF is $f_{X,Y}(x, y)$, in which case the draws would be real-number pairs of the form $(x, y)_1, (x, y)_2, \ldots (x, y)_k$.

If an increasingly large number of draws from X is assembled into a histogram and the width of the histogram blocks is steadily reduced, then the histogram will tend towards a plot of the PDF $f_X(x)$. The illustration in Figure 4.1 comes from the program ***04_Monte_Carlo*** and shows this convergence taking place.

The solid black curve in the figure is a plot of $f_X(x)$ and the narrow vertical rectangles make up the histogram. In this case the number of draws was ten million.

In the same way, a 2-dimensional histogram consisting of an increasingly large number of draws from the RV (X, Y) will tend towards the PDF surface $z = f_{X,Y}(x, y)$.

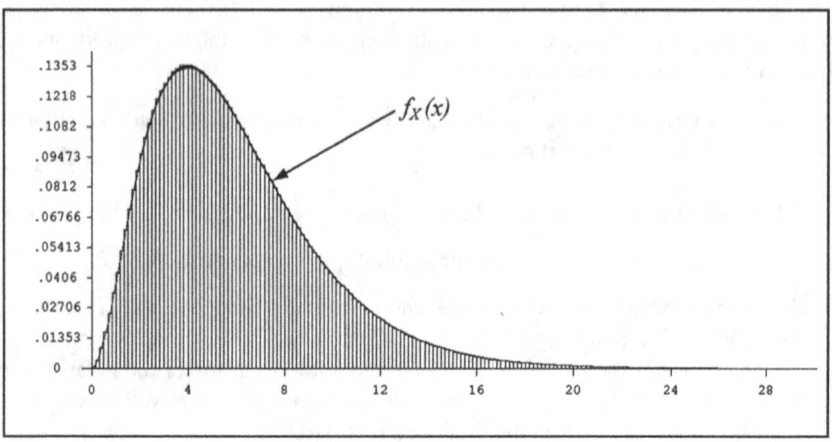

Figure 4.1 A PDF and a histogram

4.1.1 Averages and expectations

The average value of X is a scalar that is defined as follows:

$$Ave\{X\} \equiv \lim_{M\to\infty} 1/M \sum_{r=1}^{M} x_r \tag{4.1.1}$$

in which x_r is the rth draw from X.

Equation (4.1.1) is an empirical definition[1] in the sense that it leads to an actionable procedure. Thus, suppose that a Monte-Carlo generates M successive draws from X. Then we can estimate $Ave\{X\}$ as follows:

$$Ave\{X\} \approx 1/M \sum_{r=1}^{M} x_r \tag{4.1.2}$$

Equation (4.1.2) is an algorithm that can be easily implemented. The value that it produces is approximate in relation to (4.1.1), but its accuracy increases with M. We note that (4.1.2) involves only the most basic of arithmetic operations, namely addition and division.

Starting from (4.1.1), by using the much more sophisticated operation of integration and the concept of PDFs, mathematicians have devised a *formula* for the average value of X called the *expectation* of X, which is as follows:

$$\mathcal{E}\{X\} \equiv \int x f_X(x)dx \tag{4.1.3}$$

Note: The limits of integration in (4.1.3) and all integrals below are from $-\infty$ to $+\infty$.

We referred to (4.1.1) as an *empirical* statement, and we call (4.1.3) a *theoretical* statement.

If the integral appearing in (4.1.3) is evaluated, then the resulting value will be the same as the one obtained from (4.1.1). However, (4.1.1) is the *actual* average whereas (4.1.3) is a formula that is *supposed* to yield that average, and they should not be regarded as one and the same.

◀▶

[1] *Empirical:* Capable of being confirmed, verified or disproved by observation or experiment. (Merriam-Webster).

Let the univariate random variable X have the PDF $f_X(x)$ and the bivariate random variable (X, Y) have the PDF $f_{X,Y}(x, y)$. Then we also have the following empirical/theoretical pairs:

$$Ave\{X^2\} \equiv \lim_{M\to\infty} 1/M \sum_{r=1}^{M} x_r^2 \qquad \mathcal{E}\{X^2\} \equiv \int x^2 f_X(x)dx \tag{4.1.4}$$

$$Ave\{XY\} \equiv \lim_{M\to\infty} 1/M \sum_{r=1}^{M} x_r y_r \quad \mathcal{E}\{XY\} \equiv \iint x\,y\,f_{X,Y}(x,y)dx\,dy \tag{4.1.5}$$

Corresponding to $Ave\,\{X^2\}$ and $Ave\,\{XY\}$ in the above equations are the following two estimation algorithms:

$$Ave\{X^2\} \approx 1/M \sum_{r=1}^{M} x_r{}^2 \tag{4.1.6}$$

and

$$Ave\{XY\} \approx 1/M \sum_{r=1}^{M} x_r y_r \tag{4.1.7}$$

In (4.1.5) and (4.1.7) the expressions XY, xy and $x_r y_r$ are in each case the *products* of the two quantities involved.

Let a and b be constants. Then it is clear from the above definitions that the following pairs of empirical/theoretical statements must also be true:

$$Ave\{aX\} = a\,Ave\,\{X\} \qquad \mathcal{E}\{aX\} = a\mathcal{E}\{X\} \tag{4.1.8}$$

$$Ave\{X+Y\} = Ave\,\{X\} + Ave\{Y\} \quad \mathcal{E}\{X+Y\} = \mathcal{E}\{X\} + \mathcal{E}\{Y\} \tag{4.1.9}$$

$$Ave\{(aX)(bY)\} = ab\,Ave\{XY\} \quad \mathcal{E}\{(aX)(bY)\} = ab\mathcal{E}\{XY\} \tag{4.1.10}$$

4.1.2 Variance and standard deviation

We now temporarily discontinue consideration of the empirical *Ave* expressions and focus our attention on the theoretical formulas that use $\mathcal{E}$. However, the reader should bear in mind that for each formula involving $\mathcal{E}$ there is always a corresponding expression involving *Ave.*

If $\mathcal{E}\{X\}$ is subtracted from each member of the population X, then we obtain the RV X - $\mathcal{E}\{X\}$.

Definition 4.3: The **variance** of X is defined as

$$\sigma_X{}^2 \equiv \mathcal{E}\{(X - \mathcal{E}\{X\})^2\} \tag{4.1.11}$$

■■

Clearly the variance of any random variable will be a nonnegative number and the only way in which it could be zero is if all members of the population are equal to $\mathcal{E}\{X\}$, something that we can exclude as being of no interest. Thus the variance of any random variable is always positive.

It will often be the case that X is zero-mean, i.e. that

$$\mathcal{E}\{X\} = 0 \tag{4.1.12}$$

in which case its variance becomes simply

$$\sigma_X{}^2 = \mathcal{E}\{X^2\} \tag{4.1.13}$$

The positive square root of the variance is called the *standard deviation,* and is written as σ_X. Thus:

$$\sigma_X = (\sigma_X{}^2)^{1/2} \tag{4.1.14}$$

4.1.3 Covariance matrices – actual and supposed

Covariance matrices will be considered further down in full detail, but for now we make the following comments.

In Chapter 1 we spoke about the filter covariance matrix $\boldsymbol{S}^*$ and the actual estimation-error covariance matrix $\boldsymbol{S}_{actual}$ and we defined ECM consistency as the situation for which $\boldsymbol{S}^*$ is a good representation of $\boldsymbol{S}_{actual}$. We also stated that in Chapter 7 we will specify precisely what we mean by 'a good representation'.

The two matrices $\boldsymbol{S}^*$ and $\boldsymbol{S}_{actual}$ appeared in Figure 1.8 that is reproduced here for convenience as Figure 4.2:

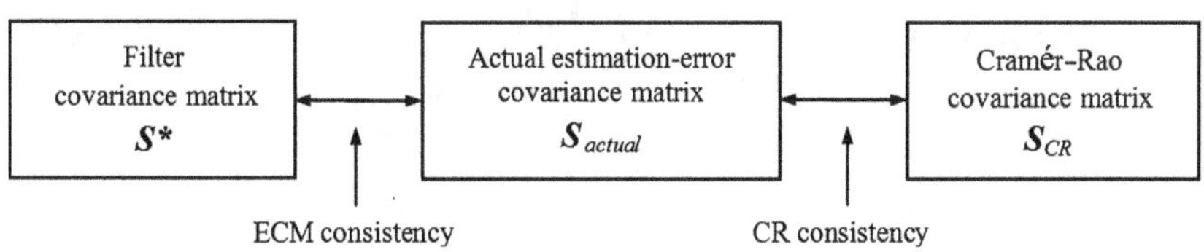

Figure 4.2 ECM and CR consistency

We will see further down that the theoretical concept $\mathcal{E}$ leads us to the filter covariance matrix $\boldsymbol{S}^*$ on the left, and the empirical concept *Ave* leads us to the covariance matrix $\boldsymbol{S}_{actual}$ in the centre.

4.2 Random vectors and covariance matrices

4.2.1 Random vectors

Up to now the symbols X and Y have been used to represent scalars RVs. We now use the bold symbols $\boldsymbol{X}$ and $\boldsymbol{Y}$ to represent vectors whose elements are scalar RVs. Such vectors are called *random vectors*. As examples:

$$\boldsymbol{X} = (X_1, X_2, X_3)^T \quad \text{and} \quad \boldsymbol{Y} = (Y_1, Y_2, Y_3)^T \tag{4.2.1}$$

in which the vector elements X and Y are scalar random variables.

We will simplify the notation by using the more usual *lower-case* letters for the vector elements as follows:

$$\boldsymbol{X} = (x_1, x_2, x_3)^T \quad \text{and} \quad \boldsymbol{Y} = (y_1, y_2, y_3)^T \tag{4.2.2}$$

in which the symbols x and y no longer represent draws, but are now scalar random variables.

Definition 4.4: The **expectations of the random vectors $\boldsymbol{X}$** and $\boldsymbol{Y}$ in (4.2.2) are as follows:

$$\mathcal{E}\{\boldsymbol{X}\} \equiv \left(\mathcal{E}\{x_1\},\ \mathcal{E}\{x_2\},\ \mathcal{E}\{x_3\}\right)^T \quad \text{and} \quad \mathcal{E}\{\boldsymbol{Y}\} \equiv \left(\mathcal{E}\{y_1\},\ \mathcal{E}\{y_2\},\ \mathcal{E}\{y_3\}\right)^T \tag{4.2.3}$$

4.2.2 Random matrices

Consider now the product $\boldsymbol{XY}^T$. By the usual rules of matrix multiplication:

$$\boldsymbol{XY}^T = \begin{pmatrix} x_1 \\ x_2 \\ x_3 \end{pmatrix} (y_1 \quad y_2 \quad y_3) = \begin{pmatrix} x_1y_1 & x_1y_2 & x_1y_3 \\ x_2y_1 & x_2y_2 & x_2y_3 \\ x_3y_1 & x_3y_2 & x_3y_3 \end{pmatrix} \tag{4.2.4}$$

The elements of the matrix on the right of (4.2.4) are products of the form $x_i y_j$, each of which is a univariate RV. A matrix whose elements are univariate RVs is called a *random matrix*.

Definition 4.5: The **expectation of the random matrix XY^T** is as follows:

$$\mathcal{E}\{XY^T\} \equiv \begin{pmatrix} \mathcal{E}\{x_1 y_1\} & \mathcal{E}\{x_1 y_2\} & \mathcal{E}\{x_1 y_3\} \\ \mathcal{E}\{x_2 y_1\} & \mathcal{E}\{x_2 y_2\} & \mathcal{E}\{x_2 y_3\} \\ \mathcal{E}\{x_3 y_1\} & \mathcal{E}\{x_3 y_2\} & \mathcal{E}\{x_3 y_3\} \end{pmatrix} \tag{4.2.5}$$

■■

From the definition of $\mathcal{E}$ in (4.1.5) we see that each of the elements in $\mathcal{E}\{XY^T\}$ is a double integral. Thus

$$\mathcal{E}\{x_i y_j\} = \iint xy f_{x_i, y_j}(x, y) dx\, dy \tag{4.2.6}$$

Definition 4.6: Let x_i and y_j be zero-mean scalar random variables. Then

(a) $\mathcal{E}\{x_i y_j\}$ is called the **covariance** of x_i and y_j.

(b) If

$$\mathcal{E}\{x_i y_j\} = 0 \tag{4.2.7}$$

then we say that the RVs x_i and y_j are **uncorrelated**, and if

$$\mathcal{E}\{x_i y_j\} \neq 0 \tag{4.2.8}$$

then we say that the two RVs are **correlated.** ■■

We note that a more general definition of x_i and y_j being uncorrelated is the following:

Definition 4.7: The scalar random variables x_i and y_j are said to be **uncorrelated** if

$$\mathcal{E}\{x_i y_j\} = \mathcal{E}\{x_i\}\mathcal{E}\{y_j\} \tag{4.2.9}$$

■■

It then follows that if either x_i or y_j is zero-mean, that (4.2.9) becomes

$$\mathcal{E}\{x_i y_j\} = 0 \tag{4.2.10}$$

which is the same as (4.2.7).

Corresponding to (4.2.6) is $Ave\{x_i\, y_j\}$ that can be estimated by

$$Ave\{x_i y_j\} \approx 1/M \sum_{r=1}^{M} x_{i,\,r}\, y_{j,\,r} \tag{4.2.11}$$

This statement tells us that the average of the random matrix $\boldsymbol{X}\boldsymbol{Y}^T$ can be estimated by

$$Ave\{\boldsymbol{X}\boldsymbol{Y}^T\} \approx 1/M \sum_{r=1}^{M} \boldsymbol{X}_r \boldsymbol{Y}_r^{\,T} \tag{4.2.12}$$

in which $\boldsymbol{X}_r$ and $\boldsymbol{Y}_r$ are the rth draws from $\boldsymbol{X}$ and $\boldsymbol{Y}$.

4.2.3 Covariance matrices

Let $\boldsymbol{N}$ be a random vector. As an example, for the 3-dimensional case,

$$\boldsymbol{N} = (\nu_1, \nu_2, \nu_3)^T \tag{4.2.13}$$

in which each of the elements ν_1, ν_2, ν_3 is a scalar random variable. Forming $\boldsymbol{N} - \mathcal{E}\{\boldsymbol{N}\}$ we obtain the random vector $\boldsymbol{N} - \mathcal{E}\{\boldsymbol{N}\}$.

Definition 4.8: The **covariance matrix** of the vector $\boldsymbol{N}$ is defined as follows:

$$\boldsymbol{R}_N \equiv \mathcal{E}\{(\boldsymbol{N} - \mathcal{E}\{\boldsymbol{N}\})(\boldsymbol{N} - \mathcal{E}\{\boldsymbol{N}\})^T\} \tag{4.2.14}$$

■■

It will usually be the case that $\boldsymbol{N}$ is zero-mean, i.e. that

$$\mathcal{E}\{\boldsymbol{N}\} = \boldsymbol{\emptyset} \tag{4.2.15}$$

In that case we have

Definition 4.9: The **covariance matrix** of a zero-mean random vector $\boldsymbol{N}$ is defined as follows:

$$\boldsymbol{R}_N \equiv \mathcal{E}\{\boldsymbol{N}\boldsymbol{N}^T\} \tag{4.2.16}$$

■■

For example, if $\boldsymbol{N}$ is a zero-mean *3*-vector, then

$$\boldsymbol{R}_N = \begin{pmatrix} \mathcal{E}\{\nu_1{}^2\} & \mathcal{E}\{\nu_1\nu_2\} & \mathcal{E}\{\nu_1\nu_3\} \\ \mathcal{E}\{\nu_2\nu_1\} & \mathcal{E}\{\nu_2{}^2\} & \mathcal{E}\{\nu_2\nu_3\} \\ \mathcal{E}\{\nu_3\nu_1\} & \mathcal{E}\{\nu_3\nu_2\} & \mathcal{E}\{\nu_3{}^2\} \end{pmatrix} \tag{4.2.17}$$

from which we see also that

$$\boldsymbol{R}_N = \begin{pmatrix} \sigma_{\nu_1}{}^2 & \mathcal{E}\{\nu_1\nu_2\} & \mathcal{E}\{\nu_1\nu_3\} \\ \mathcal{E}\{\nu_2\nu_1\} & \sigma_{\nu_2}{}^2 & \mathcal{E}\{\nu_2\nu_3\} \\ \mathcal{E}\{\nu_3\nu_1\} & \mathcal{E}\{\nu_3\nu_2\} & \sigma_{\nu_3}{}^2 \end{pmatrix} \tag{4.2.18}$$

The diagonal terms of $\boldsymbol{R}_N$ are then the variances of the elements of $\boldsymbol{N}$, and the off-diagonal terms are the covariances.

Observe that a covariance matrix is always real and symmetric – real because its elements are real, and symmetric because its covariance terms satisfy

$$\mathcal{E}\{\nu_i\nu_j\} = \mathcal{E}\{\nu_j\nu_i\} \tag{4.2.19}$$

From Definition 4.6(b) it also follows that the covariance matrix of a vector of uncorrelated random variables will be diagonal.

4.2.4 Covariance matrices – theoretical and actual

Each of the above statements regarding covariance matrices was a theoretical statement since they all used the expectation operator $\mathcal{E}$. Accordingly, we will refer to the covariance matrices in Definitions 4.8 and 4.9 as *theoretical* covariance matrices, for which we use the symbol $\boldsymbol{R}_{N,\ theoretical}$.

Using the empirical *Ave* concept we also have the following:

Definition 4.10: The **actual covariance matrix** $\boldsymbol{R}_{N,\ actual}$ of a zero-mean random vector $\boldsymbol{N}$ is defined as follows:

$$\boldsymbol{R}_{N,\ actual} \equiv Ave\{\boldsymbol{N}\boldsymbol{N}^T\} = \lim_{M\to\infty} 1/M \sum_{r=1}^{M} \boldsymbol{N}_r\boldsymbol{N}_r{}^T \tag{4.2.20}$$

■■

In theory $\boldsymbol{R}_{N,\ theoretical}$ and $\boldsymbol{R}_{N,\ actual}$ are numerically equal. However, $\boldsymbol{R}_{N,\ theoretical}$ is a theoretical concept whereas $\boldsymbol{R}_{N,\ actual}$ is empirical, and leads to the following estimation algorithm:

$$\boldsymbol{R}_{N,\ actual} \approx 1/M \sum_{r=1}^{M} \boldsymbol{N}_r \boldsymbol{N}_r^{\,T} \tag{4.2.21}$$

Equation (4.2.21) tells us that if draws of the zero-mean random vector $\boldsymbol{N}$ are generated by a Monte-Carlo then we can use them to form the products $\boldsymbol{N}_r \boldsymbol{N}_r^{\,T}$, and by adding M of these together and finally dividing the sum by M we obtain an estimate of the *actual* covariance matrix of $\boldsymbol{N}$.

The algorithm in (4.2.21) for estimating actual covariance matrices is used in a number of the programs, and in Figure 4.3 we show a matrix that was generated by the program ***09_Random_Vectors***:

```
The theoretical covariance matrix is

  3                 -1
 -1                  3

The actual covariance matrix of the 1000000 random vectors
that were generated is

  3.0003196        -1.0005974
 -1.0005974         3.0055888
```

Figure 4.3 Theoretical and actual covariance matrices from ***09_Random_Vectors***

Observe the presence of the two constructs – the *theoretical covariance* matrix that came from a formula that could be evaluated almost instantaneously by the computer, and the *actual covariance matrix* that was estimated by making one million draws from a random vector.

4.2.5 The correlation matrix

If we pre- and post-multiply the covariance matrix $\boldsymbol{R}_N$ in (4.2.18) by the following diagonal matrix

$$\boldsymbol{D} \equiv \begin{pmatrix} 1/\sigma_{v_1} & 0 & 0 \\ 0 & 1/\sigma_{v_2} & 0 \\ 0 & 0 & 1/\sigma_{v_3} \end{pmatrix} \tag{4.2.22}$$

we obtain what is called the *correlation matrix* of $\boldsymbol{N}$, namely

$$\boldsymbol{R}'_N = \begin{pmatrix} 1 & \rho(\nu_1, \nu_2) & \rho(\nu_1, \nu_3) \\ \rho(\nu_2, \nu_1) & 1 & \rho(\nu_2, \nu_3) \\ \rho(\nu_3, \nu_1) & \rho(\nu_3, \nu_2) & 1 \end{pmatrix} \tag{4.2.23}$$

The off-diagonal quantities appearing in $\boldsymbol{R}'_N$ have the following meanings:

$$\rho(\nu_i, \nu_j) \equiv \frac{\mathcal{E}\{\nu_i, \nu_j\}}{\sigma_{\nu_i}\sigma_{\nu_j}} \tag{4.2.24}$$

These quantities are called the *correlation coefficients,* and it can be shown that correlation coefficients always lie in the range ± 1. (See problem 4.7.)

When $\rho(\nu_i, \nu_j) = 0$ then the random variables ν_i and ν_j are uncorrelated, and when $\rho(\nu_i, \nu_j) = \pm 1$ then they are linearly dependent in the mathematical sense (see Appendix 2.1). If two random variables are linearly dependent in this way then we say that they are *perfectly correlated.*

Linear dependence of random variables will be considered again further down.

It can be shown (see Problem 4.9) that if two or more elements of $\boldsymbol{N}$ are perfectly correlated, then its correlation matrix will be singular, and hence also that its covariance matrix $\boldsymbol{R}_N$ will be singular. In that case the determinant of $\boldsymbol{R}_N$ will be zero.

Starting from being uncorrelated, the elements of $\boldsymbol{N}$ can approach this singular condition, and as that convergence takes place the determinant will tend to zero, making it increasingly difficult to invert $\boldsymbol{R}_N$ with any accuracy.

4.2.6 Uncorrelated random vectors

Let $\boldsymbol{X}$ and $\boldsymbol{Y}$ be two zero-mean random vectors from which we form the matrix $\mathcal{E}\{\boldsymbol{X}\boldsymbol{Y}^T\}$. As an example, let $\boldsymbol{X}$ be a 2-vector and $\boldsymbol{Y}$ a 3-vector. Then

$$\mathcal{E}\{\boldsymbol{X}\boldsymbol{Y}^T\} \equiv \begin{pmatrix} \mathcal{E}\{x_1 y_1\} & \mathcal{E}\{x_1 y_2\} & \mathcal{E}\{x_1 y_3\} \\ \mathcal{E}\{x_2 y_1\} & \mathcal{E}\{x_2 y_2\} & \mathcal{E}\{x_2 y_3\} \end{pmatrix} \tag{4.2.25}$$

***Definition 4.11**:* If every element of the matrix $\mathcal{E}\{\boldsymbol{X}\boldsymbol{Y}^T\}$ is zero, we say that $\boldsymbol{X}$ and $\boldsymbol{Y}$ are **uncorrelated random vectors**. ■■

4.2.7 Linear transformation of a random vector

Let $\boldsymbol{N}$ be a zero-mean random k-vector with theoretical covariance matrix $\boldsymbol{R}_{N,\ theoretical}$ and let $\boldsymbol{A}$ be an $m \times k$ matrix of constants. Let $\boldsymbol{N}'$ be the random m-vector formed from $\boldsymbol{N}$ as follows:

$$\boldsymbol{N}' = \boldsymbol{A}\boldsymbol{N} \tag{4.2.26}$$

Equation (4.2.26) is depicted in Figure 4.4 in which $m < k$, and so $\boldsymbol{A}$ is wide.

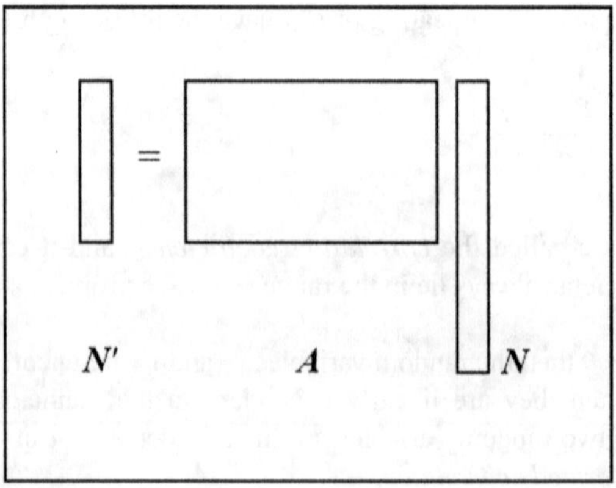

Figure 4.4 $\boldsymbol{N}' = \boldsymbol{A}\boldsymbol{N}$

Similar figures could be created in which $m = k$ ($\boldsymbol{A}$ is square) and $m > k$ ($\boldsymbol{A}$ is high).

Because $\boldsymbol{N}$ is zero-mean it follows that

$$\mathcal{E}\{\boldsymbol{N}'\} = \mathcal{E}\{\boldsymbol{A}\boldsymbol{N}\} = \boldsymbol{A}\mathcal{E}\{\boldsymbol{N}\} = \boldsymbol{\emptyset} \tag{4.2.27}$$

and so $\boldsymbol{N}'$ is also zero-mean. Then the theoretical covariance matrix of $\boldsymbol{N}'$ will be

$$\begin{aligned}\boldsymbol{R}_{N',\ theoretical} \equiv \mathcal{E}\{\boldsymbol{N}'\boldsymbol{N}'^T\} &= \mathcal{E}\left\{(\boldsymbol{A}\boldsymbol{N})(\boldsymbol{A}\boldsymbol{N})^T\right\} = \mathcal{E}\{\boldsymbol{A}\boldsymbol{N}\boldsymbol{N}^T\boldsymbol{A}^T\} \\ &= \boldsymbol{A}\mathcal{E}\{\boldsymbol{N}\boldsymbol{N}^T\}\boldsymbol{A}^T = \boldsymbol{A}\boldsymbol{R}_{N,\ theoretical}\boldsymbol{A}^T\end{aligned} \tag{4.2.28}$$

This expression gives us the following theorem that we will use frequently in what lies ahead.

Theorem 4.1*: The linear transformation of a random vector and its theoretical covariance matrix.*

Let $\boldsymbol{N}$ be a zero-mean random k-vector with theoretical covariance matrix $\boldsymbol{R}_{N,\ theoretical}$ and let $\boldsymbol{A}$ be an $m \times k$ matrix of constants. Let $\boldsymbol{N}'$ be the random m-vector formed by

$$N' = AN \tag{4.2.29}$$

Then N' is also zero-mean and the theoretical covariance matrix of N' is

$$R_{N',\ theoretical} = AR_{N,\ theoretical}A^T \tag{4.2.30}$$

■■

We draw the reader's attention to (4.2.29) and (4.2.30). The first equation, in which there is a single matrix A on the left, shows how a random vector N is transformed, and the second equation, in which A appears on the left and A^T appears on the right, shows how its covariance matrix is transformed.

Theorem 4.1 was derived using the $\mathcal{E}$ operator in (4.2.28), and so it gave us a relationship between the two theoretical matrices $R_{N',\ theoretical}$ and $R_{N,\ theoretical}$. As one would expect, the theorem has the following counterpart using the *Ave* operation, which gives us a relationship between the two actual matrices $R_{N',\ actual}$ and $R_{N,\ actual}$.

Theorem 4.1b: ***The linear transformation of a random vector and its actual covariance matrix.***

Let N_r be the rth draw of a zero-mean random k-vector N being generated by a Monte-Carlo with actual covariance matrix $R_{N,\ actual}$ and let A be an $m \times k$ matrix of constants. Let N'_r be the random m-vector formed by

$$N'_r = AN_r \tag{4.2.31}$$

Then N'_r is a draw from a random vector N' that is zero-mean, and whose actual covariance matrix is $AR_{N,\ actual}A^T$. ■■

Proof that N' is zero-mean is similar to (4.2.27). To prove that the actual covariance matrix of N' is $AR_{N,\ actual}A^T$ we proceed as follows:

$$\begin{aligned}
R_{N',\ actual} &= Ave\{N'N'^T\} = \underset{M\to\infty}{Lim}\ 1/M\sum_{r=1}^{M} N'_r N'^T_r \\
&= \underset{M\to\infty}{Lim}\ 1/M\sum_{r=1}^{M}(AN_r)(AN_r)^T \\
&= \underset{M\to\infty}{Lim}\ 1/M\sum_{r=1}^{M} AN_r N_r^T A^T \\
&= \underset{M\to\infty}{Lim}\ A\left(1/M\sum_{r=1}^{M} N_r N_r^T\right)A^T \\
&= A\left(\underset{M\to\infty}{Lim}\ 1/M\sum_{r=1}^{M} N_r N_r^T\right)A^T = AR_{N,\ actual}A^T
\end{aligned} \tag{4.2.32}$$

Comparison of the first and last items in (4.2.32) shows that

$$\boldsymbol{R}_{N',\, actual} = \boldsymbol{A}\boldsymbol{R}_{N,\, actual}\boldsymbol{A}^T \tag{4.2.33}$$

■■

The matrices $\boldsymbol{R}_{N',\, theoretical}$ obtained in Theorem 4.1 and $\boldsymbol{R}_{N',\, actual}$ obtained in Theorem 4.1b should be equal. However, they are not one and the same.

We will see in the next chapter that Theorem 4.1 and $\boldsymbol{R}_{N',\, theoretical}$ lead us to the filter covariance matrix $\boldsymbol{S}^*$, whereas Theorem 4.1b and $\boldsymbol{R}_{N',\, actual}$ lead us to the actual estimation-error covariance matrix $\boldsymbol{S}_{actual}$ (See Figure 4.2.).

4.2.8 The covariance matrices of $\boldsymbol{X}$ and $\boldsymbol{N_X}$

Let the random vector $\boldsymbol{X}$ be partitioned as follows:

$$\boldsymbol{X} = \mathcal{E}\{\boldsymbol{X}\} + \boldsymbol{N_X} \tag{4.2.34}$$

in which $\boldsymbol{N_X}$ is zero-mean, i.e.

$$\mathcal{E}\{\boldsymbol{N_X}\} = \boldsymbol{\emptyset} \tag{4.2.35}$$

We have therefore split $\boldsymbol{X}$ into the sum of its mean component $\mathcal{E}\{\boldsymbol{X}\}$ and its zero-mean component $\boldsymbol{N_X}$.

Let the covariance matrix of $\boldsymbol{N_X}$ be $\boldsymbol{R_{N_X}}$. Then by Definition 4.8, the covariance matrix of $\boldsymbol{X}$ will be

$$\boldsymbol{R_X} = \mathcal{E}\{(\boldsymbol{X} - \mathcal{E}\{\boldsymbol{X}\})(\boldsymbol{X} - \mathcal{E}\{\boldsymbol{X}\})^T\} = \mathcal{E}\{\boldsymbol{N_X}\boldsymbol{N_X}^T\} = \boldsymbol{R_{N_X}} \tag{4.2.36}$$

giving us the result

$$\boldsymbol{R_X} = \boldsymbol{R_{N_X}} \tag{4.2.37}$$

This shows that the covariance matrix of a nonzero-mean random vector $\boldsymbol{X}$ is the same as the covariance matrix of its zero-mean component $\boldsymbol{N_X}$.

Thus, if we have an estimate vector $\boldsymbol{X}^*$ whose error vector is $\boldsymbol{N}^*$ and if the covariance matrix of $\boldsymbol{N}^*$ is $\boldsymbol{S}^*$, then the covariance matrix of $\boldsymbol{X}^*$ is also $\boldsymbol{S}^*$.

4.2.9 The covariance matrix of the sum or difference of uncorrelated vectors

Let $\boldsymbol{X}$ and $\boldsymbol{Y}$ be two random vectors that have been partitioned as follows:

$$\boldsymbol{X} = \mathcal{E}\{\boldsymbol{X}\} + \boldsymbol{N_X} \quad \text{and} \quad \boldsymbol{Y} = \mathcal{E}\{\boldsymbol{Y}\} + \boldsymbol{N_Y} \tag{4.2.38}$$

in which the random vectors $\boldsymbol{N}_X$ and $\boldsymbol{N}_Y$ are zero-mean. Let $\boldsymbol{N}_X$ and $\boldsymbol{N}_Y$ be uncorrelated, i.e.

$$\mathcal{E}\{\boldsymbol{N}_X \boldsymbol{N}_Y^T\} = \boldsymbol{\emptyset} \tag{4.2.39}$$

and let the random vector $\boldsymbol{Z}$ be the sum or difference of $\boldsymbol{X}$ and $\boldsymbol{Y}$. Thus

$$\boldsymbol{Z} = \boldsymbol{X} \pm \boldsymbol{Y} = \big(\mathcal{E}\{\boldsymbol{X}\} \pm \mathcal{E}\{\boldsymbol{Y}\}\big) + (\boldsymbol{N}_X \pm \boldsymbol{N}_Y) \tag{4.2.40}$$

Then it follows that

$$\boldsymbol{Z} = \mathcal{E}\{\boldsymbol{Z}\} + \boldsymbol{N}_Z \tag{4.2.41}$$

in which

$$\mathcal{E}\{\boldsymbol{Z}\} = \mathcal{E}\{\boldsymbol{X}\} \pm \mathcal{E}\{\boldsymbol{Y}\} \quad \text{and} \quad \boldsymbol{N}_Z = \boldsymbol{N}_X \pm \boldsymbol{N}_Y \tag{4.2.42}$$

By Definition 4.8 the covariance matrix of $\boldsymbol{Z}$ will be

$$\begin{aligned} \boldsymbol{R}_Z &= \mathcal{E}\{\big(\boldsymbol{Z} - \mathcal{E}\{\boldsymbol{Z}\}\big)\big(\boldsymbol{Z} - \mathcal{E}\{\boldsymbol{Z}\}\big)^T\} \\ &= \mathcal{E}\{\boldsymbol{N}_Z \boldsymbol{N}_Z^T\} = \mathcal{E}\{(\boldsymbol{N}_X \pm \boldsymbol{N}_Y)(\boldsymbol{N}_X \pm \boldsymbol{N}_Y)^T\} \\ &= \mathcal{E}\{\boldsymbol{N}_X \boldsymbol{N}_X^T\} + \mathcal{E}\{\boldsymbol{N}_Y \boldsymbol{N}_Y^T\} \pm \mathcal{E}\{\boldsymbol{N}_X \boldsymbol{N}_Y^T\} \pm \mathcal{E}\{\boldsymbol{N}_Y \boldsymbol{N}_X^T\} \end{aligned} \tag{4.2.43}$$

The last two terms are null because $\boldsymbol{N}_X$ and $\boldsymbol{N}_Y$ were assumed to be uncorrelated, and so

$$\boldsymbol{R}_Z = \boldsymbol{R}_{N_X} + \boldsymbol{R}_{N_Y} \tag{4.2.44}$$

This gives us the following:

Theorem 4.2: Let

$$\boldsymbol{Z} = \boldsymbol{X} \pm \boldsymbol{Y} \tag{4.2.45}$$

where

$$\boldsymbol{X} = \mathcal{E}\{\boldsymbol{X}\} + \boldsymbol{N}_X \quad \text{and} \quad \boldsymbol{Y} = \mathcal{E}\{\boldsymbol{Y}\} + \boldsymbol{N}_Y \tag{4.2.46}$$

and where $\boldsymbol{N}_X$ and $\boldsymbol{N}_Y$ are uncorrelated and zero-mean, with covariance matrices $\boldsymbol{R}_{N_X}$ and $\boldsymbol{R}_{N_Y}$. Then the covariance matrix of $\boldsymbol{Z}$ is

$$\boldsymbol{R_Z} = \boldsymbol{R_{N_X}} + \boldsymbol{R_{N_Y}} = \boldsymbol{R_X} + \boldsymbol{R_Y} \tag{4.2.47}$$

■■

This result tells us that if we form the sum or difference of two random vectors $\boldsymbol{X}$ and $\boldsymbol{Y}$ whose zero-mean components are uncorrelated, then the covariance matrix of that sum or difference is the sum of the covariance matrices of $\boldsymbol{X}$ and $\boldsymbol{Y}$.

Finally, we note that if $\boldsymbol{X}$ and $\boldsymbol{Y}$ in (4.2.34) and (4.2.35) are zero-mean, i.e. if

$$\boldsymbol{X} = \boldsymbol{N_X} \quad \text{and} \quad \boldsymbol{Y} = \boldsymbol{N_Y} \tag{4.2.48}$$

in which $\boldsymbol{N_X}$ and $\boldsymbol{N_Y}$ are uncorrelated, and if we form

$$\boldsymbol{Z} = \boldsymbol{X} \pm \boldsymbol{Y} \tag{4.2.49}$$

then it must also be true that

$$\boldsymbol{R_Z} = \boldsymbol{R_X} + \boldsymbol{R_Y} \tag{4.2.50}$$

Thus regardless of whether or not $\boldsymbol{X}$ and $\boldsymbol{Y}$ are zero mean, if $\boldsymbol{N_X}$ and $\boldsymbol{N_Y}$ are uncorrelated then

$$\boldsymbol{R_Z} = \boldsymbol{R_X} + \boldsymbol{R_Y} \tag{4.2.51}$$

■■

4.2.10 Correlation and independence

Definition 4.12: Let A and B be two random events and let $A \cap B$ be the random event *both A and B*. Let their probabilities be $p(A)$, $p(B)$ and $p(A \cap B)$. Then if

$$p(A \cap B) = p(A)p(B) \tag{4.2.52}$$

we say that the events A and B are **statistically independent**, or more simply **independent**.[2] ■■

Statistical independence can also be defined for random variables. Thus:

Definition 4.13: Let the random variables X and Y have univariate PDFs $f_X(x)$ and $f_Y(y)$, and let their bivariate PDF be $f_{X,Y}(x, y)$. Then if

$$f_{X,Y}(x, y) = f_X(x) f_Y(y) \tag{4.2.53}$$

we say that X and Y are **statistically independent**, or more simply **independent.** ■■

[2] Statistical independence should not be confused with linear independence. Nor should independent variables in the statistical sense be confused with x being the independent variable in $f(x)$.

When two random variables are independent, then they are also uncorrelated. Thus for random variables from any PDF $f_{X,Y}(x,y)$, if (4.2.53) is true then the following is also always true:

$$\mathcal{E}\{X,Y\} = \mathcal{E}\{X\}\mathcal{E}\{Y\} \tag{4.2.54}$$

However, the reverse may not be true, i.e. (4.2.54) may or may not imply (4.2.53).

In the Gaussian case, however, (4.2.54) does imply (4.2.53). By this we mean that if the random variables X and Y are jointly Gaussian and they are uncorrelated, then they are also independent.

4.2.11 Linear transformation of Gaussian random vectors

We close this section by stating a fact that is well known, and for which we will not provide a proof.

Let the random vectors $\boldsymbol{X}$ and $\boldsymbol{Y}$ be related by the linear transformation

$$\boldsymbol{X} = \boldsymbol{A}\boldsymbol{Y} \tag{4.2.55}$$

in which $\boldsymbol{Y}$ is multivariate Gaussian with covariance matrix $\boldsymbol{R_Y}$. Then $\boldsymbol{X}$ is also multivariate Gaussian, and by Theorem 4.1 its covariance matrix will be

$$\boldsymbol{R_X} = \boldsymbol{A}\boldsymbol{R_Y}\boldsymbol{A}^T \tag{4.2.56}$$

■■

4.3 The positive-definite property of covariance matrices

We now show that all covariance matrices considered in this book possess what is called the *positive-definite property*. Further down we will discuss the related *positive-semidefinite property* that applies to certain other matrices that are also considered in the book.

4.3.1 The positive-definite property

Let $\boldsymbol{N}$ be an error vector whose covariance matrix is $\boldsymbol{R}$ (where we have temporarily dropped the subscript $\boldsymbol{N}$), and let the scalar quantity a be formed as follows:

$$a = \boldsymbol{c}^T\boldsymbol{R}\boldsymbol{c} \tag{4.3.1}$$

The vector $\boldsymbol{c}$ in this equation is any nonzero vector of constants of the same dimension as $\boldsymbol{N}$. We stress the words *any* and *nonzero* in relation to $\boldsymbol{c}$, and point out that infinitely many scalars a can be generated from $\boldsymbol{R}$ in this way.

Keeping in mind that $\boldsymbol{N}^T\boldsymbol{c} = \boldsymbol{c}^T\boldsymbol{N}$ (because these are both scalars), we then have

$$\begin{aligned} a &= \boldsymbol{c}^T\boldsymbol{R}\boldsymbol{c} = \boldsymbol{c}^T\mathcal{E}\{\boldsymbol{N}\boldsymbol{N}^T\}\boldsymbol{c} = \mathcal{E}\{\boldsymbol{c}^T\boldsymbol{N}\boldsymbol{N}^T\boldsymbol{c}\} \\ &= \mathcal{E}\{(\boldsymbol{c}^T\boldsymbol{N})(\boldsymbol{c}^T\boldsymbol{N})\} = \mathcal{E}\{(\boldsymbol{c}^T\boldsymbol{N})^2\} \end{aligned} \tag{4.3.2}$$

Thus (4.3.1) is the same as

$$a = \mathcal{E}\{(\boldsymbol{c}^T\boldsymbol{N})^2\} \tag{4.3.3}$$

Now $(\boldsymbol{c}^T\boldsymbol{N})^2$ is a scalar random variable that can be either positive or zero, but it cannot be negative. Thus all of the infinitely many a's must satisfy

$$a = \mathcal{E}\{(\boldsymbol{c}^T\boldsymbol{N})^2\} \geq 0 \tag{4.3.4}$$

and the only way in which the equality can be achieved for any of the a's is if, for some nonzero vector $\boldsymbol{c}$,

$$\boldsymbol{c}^T\boldsymbol{N} = 0 \quad \text{with probability } 1 \tag{4.3.5}$$

(See footnote.[3])

The vectors $\boldsymbol{c}$ and $\boldsymbol{N}$ can in general be of any dimension, but for simplicity we will assume that they are *3*-vectors. Thus

$$\boldsymbol{c} = (c_1, c_2, c_3)^T \quad \text{and} \quad \boldsymbol{N} = (\nu_1, \nu_2, \nu_3)^T \tag{4.3.6}$$

Then (4.3.5) is the same as

$$c_1\nu_1 + c_2\nu_2 + c_3\nu_3 = 0 \tag{4.3.7}$$

and so the only way in which the equality in (4.3.4) can be achieved for any of the a's is if (4.3.7) is true for some nonzero vector $\boldsymbol{c}$ and every draw from $\boldsymbol{N}$.

We have stated that $\boldsymbol{N}$ is a vector whose elements are errors, and this fact has implications for (4.3.5). To better understand what they are, we consider an example.

Example 4.1: $\boldsymbol{c}$ is assumed to be a nonzero vector and so at least one of c_1, c_2, c_3 in (4.3.7) must be nonzero. Let c_1 be nonzero. Then (4.3.7) can be written as

[3] When every member of a population possesses a particular attribute, then we say that the population possesses that attribute with probability *1*.

$$\nu_1 = -(c_2/c_1)\nu_2 - (c_3/c_1)\nu_3 \tag{4.3.8}$$

This equation means that one of the following must be true:

◊ The first of the errors in $\boldsymbol{N}$ – namely ν_1 – is always zero (if $c_2 = c_3 = 0$).

◊ The first of the errors in $\boldsymbol{N}$ is always a linear combination of the other two (if either or both c_2 and c_3 are nonzero).

It is certainly possible that such error vectors might exist, but we will consider both of these situations to be highly contrived, and will exclude them. That being the case, (4.3.7) will never be true and so none of the a's will be zero. ■■

Starting once more from (4.3.7) we now observe the following:

If we think of the elements of $\boldsymbol{N}$ – namely ν_1, ν_2, ν_3 – as constructs that can be either linearly independent or dependent, then (4.3.7) can be regarded as the testing function for their linear independence (see Appendix 2.1).

Thus, if the elements of $\boldsymbol{N}$ are linearly independent, then the only way in which (4.3.7) can be true for all $\boldsymbol{N}$ is if all elements of $\boldsymbol{c}$ are zero.

However, by assumption, $\boldsymbol{c}$ is a nonzero vector, and so we have arrived at the following:

◊ If the elements of the random vector $\boldsymbol{N}$ form a linearly independent set and $\boldsymbol{c}$ is a nonzero vector of constants, then $\boldsymbol{c}^T\boldsymbol{N}$ will be nonzero with probability 1.

But if $\boldsymbol{c}^T\boldsymbol{N}$ is always nonzero then $(\boldsymbol{c}^T\boldsymbol{N})^2$ is always positive, and so from (4.3.3) all scalars a in (4.3.1) will be positive.

We have thus proved the following:

***Theorem 4.3**:* Let $\boldsymbol{R}$ be the covariance matrix of the error vector $\boldsymbol{N}$ whose elements form a linearly independent set with probability 1, and let $\boldsymbol{c}$ be any nonzero vector of constants. Then

$$\boldsymbol{c}^T\boldsymbol{R}\boldsymbol{c} > 0 \tag{4.3.9}$$

■■

This deceptively simple property of $\boldsymbol{R}$ is given a special name, and an astonishing number of consequences flow from it. (See for example, References 123, 125, 129, 130, 132.)

***Definition 4.14**:* If a real symmetric matrix $\boldsymbol{R}$ always satisfies

$$\boldsymbol{c}^T\boldsymbol{R}\boldsymbol{c} > 0 \tag{4.3.10}$$

in which $\boldsymbol{c}$ is any nonzero vector of constants, then we say that $\boldsymbol{R}$ is **positive-definite**. ■■

Note: The positive-definite property is the extension to higher dimensions of the positive property for scalars. Thus for r a positive scalar and c any nonzero scalar constant, equation (4.3.10) becomes $crc = rc^2 > 0$.

◄►

The positive-definite property applies only to real symmetric matrices, and many textbooks call it the *symmetric* positive-definite property. That name seemed a bit too cumbersome for us, and so we elected to drop the word 'symmetric'. However, it is always to be understood that

◊ Only real symmetric matrices can be positive-definite.

◊ If a matrix is deemed to be positive-definite, then it must be real and symmetric.

Throughout the rest of this book we will always assume that the elements of each and every error vector that we encounter form a linearly independent set with probability *1*, and so all of their covariance matrices will be positive-definite.

4.3.2 Geometric interpretation

The following is a very useful geometric interpretation of the positive-definite property:

Let x, y, z be the usual Cartesian variables in *3* dimensions, let $\boldsymbol{c}$ be a *2*-vector whose elements are any values of x and y, i.e. $\boldsymbol{c} = (x, y)^T$, and let $\boldsymbol{R}$ be a 2×2 positive-definite matrix.

Then in Appendix 4.1 we show that the algebraic equation

$$z = \boldsymbol{c}^T \boldsymbol{R} \boldsymbol{c} = \begin{pmatrix} x & y \end{pmatrix} \begin{pmatrix} r_{1,1} & r_{1,2} \\ r_{1,2} & r_{2,2} \end{pmatrix} \begin{pmatrix} x \\ y \end{pmatrix}$$

$$= r_{1,1}x^2 + 2r_{1,2}xy + r_{2,2}y^2 \tag{4.3.11}$$

represents a parabolic surface like the one shown in Figure 4.5 below.

The horizontal cross-sections of the surface are ellipses whose axes of symmetry are possibly rotated with respect to the x and y axes, and the vertical cross-sections of the surface are parabolas.

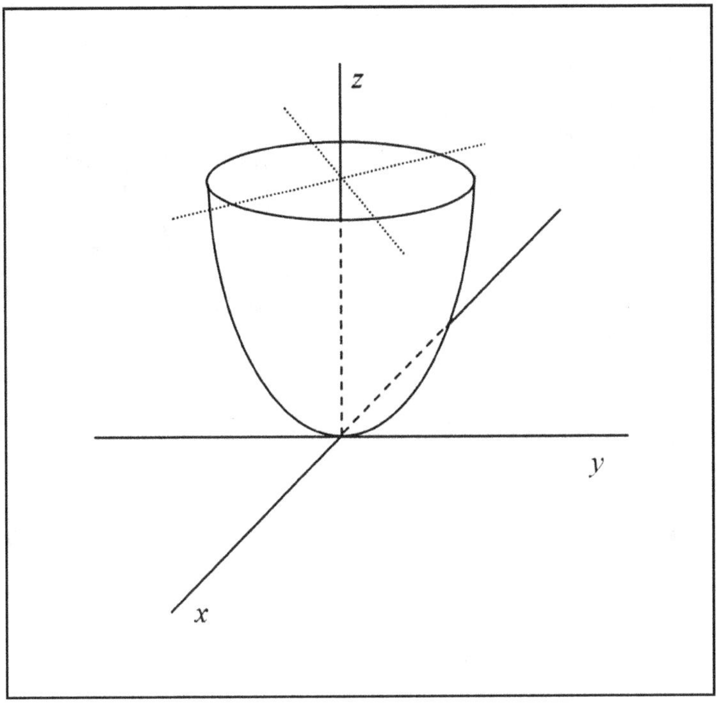

Figure 4.5 A surface corresponding to a positive-definite matrix

The surface touches the x, y plane at the origin when $\boldsymbol{c} = (0, 0)$, and lies above the plane if $\boldsymbol{c}$ is any nonzero vector. In higher dimensions (4.3.11) represents sets of ellipsoids and hyper-ellipsoids nested about the origin, whose sizes grow at a quadratic rate with z. ■■

4.3.3 Three worked examples

Example 4.2: Diagonal covariance matrices.

A radar system has error variances that are assembled into a matrix as follows:

$$\boldsymbol{R} = \begin{pmatrix} \sigma_\rho^2 & 0 & 0 \\ 0 & \sigma_\theta^2 & 0 \\ 0 & 0 & \sigma_\psi^2 \end{pmatrix} \tag{4.3.12}$$

Then $\boldsymbol{R}$ is positive-definite, and hence so is every diagonal matrix of positive numbers.

Proof: Let $\boldsymbol{c} = (c_1, c_2, c_3)^T$ be any nonzero 3-vector. Then

$$\boldsymbol{c}^T\boldsymbol{Rc} = \begin{pmatrix} c_1 & c_2 & c_3 \end{pmatrix} \begin{pmatrix} \sigma_\rho^2 & 0 & 0 \\ 0 & \sigma_\theta^2 & 0 \\ 0 & 0 & \sigma_\psi^2 \end{pmatrix} \begin{pmatrix} c_1 \\ c_2 \\ c_3 \end{pmatrix}$$

$$= c_1^2\sigma_\rho^2 + c_2^2\sigma_\theta^2 + c_3^2\sigma_\psi^2 > 0 \tag{4.3.13}$$

This proves that $\boldsymbol{R}$ is positive-definite. ■■

Example 4.3: The following matrix is positive-definite:

$$\boldsymbol{R} = \begin{pmatrix} 1 & -2 \\ -2 & 5 \end{pmatrix} \tag{4.3.14}$$

Proof: Let $\boldsymbol{c} = (c_1, c_2)^T$ be any nonzero 2-vector. Then

$$\boldsymbol{c}^T\boldsymbol{Rc} = c_1^2 - 4c_1c_2 + 5c_2^2 = (c_1 - 2c_2)^2 + c_2^2 \tag{4.3.15}$$

Thus $\boldsymbol{c}^T\boldsymbol{Rc}$ is the sum of two squares which means that it cannot be negative, and the only way in which it can be zero is if both c_1 and c_2 are zero. But that would make $\boldsymbol{c}$ the zero 2-vector which has been excluded, and so for any nonzero $\boldsymbol{c}$ we must have

$$\boldsymbol{c}^T\boldsymbol{Rc} > 0 \tag{4.3.16}$$

This proves that $\boldsymbol{R}$ is positive-definite. ■■

Example 4.4: The following matrix is *not* positive-definite:

$$\boldsymbol{B} = \begin{pmatrix} 1 & -2 \\ -2 & 4 \end{pmatrix} \tag{4.3.17}$$

■■

Proof: Let $\boldsymbol{c} = (c_1, c_2)^T$ be any nonzero 2-vector. Then

$$\boldsymbol{c}^T\boldsymbol{Bc} = c_1^2 - 4c_1c_2 + 4c_2^2 = (c_1 - 2c_2)^2 \tag{4.3.18}$$

Thus $\boldsymbol{c}^T\boldsymbol{B}\boldsymbol{c}$ is equal to a square and so it cannot be negative. However, if we let $c_1 = 2c_2$ where both c_1 and c_2 are nonzero, then $\boldsymbol{c}$ is a nonzero vector but

$$\boldsymbol{c}^T\boldsymbol{B}\boldsymbol{c} = 0 \tag{4.3.19}$$

This proves that $\boldsymbol{B}$ is not positive-definite. ■■

4.4 Properties of positive-definite matrices

The following properties will be invoked throughout the rest of the book. For proofs see Problems 4.18 to 4.25.

1. The sum of two positive-definite matrices is positive-definite.

2. The eigenvalues of a positive-definite matrix are all positive, and its determinant is positive.

3. If the eigenvalues of a real symmetric matrix are all positive, then it is positive-definite.

4. Every positive-definite matrix is nonsingular, i.e. it has an inverse.

5. The inverse of a positive-definite matrix is positive-definite.

6. The diagonal elements of a positive-definite matrix are all positive, and all of its principal submatrices[4] are positive-definite.

7. Every positive-definite matrix $\boldsymbol{A}$ has a positive-definite square-root matrix $\boldsymbol{B}$ such that $\boldsymbol{B}^2 = \boldsymbol{A}$. We write this as $\boldsymbol{B} = \boldsymbol{A}^{1/2}$.

Definition 4.15: Let $\boldsymbol{A}$, $\boldsymbol{R}$ and $\boldsymbol{S}$ be square matrices of the same size where $\boldsymbol{A}$ is nonsingular, and let them be related as follows:

$$\boldsymbol{R} = \boldsymbol{A}^T\boldsymbol{S}\boldsymbol{A} \tag{4.4.1}$$

Then $\boldsymbol{R}$ and $\boldsymbol{S}$ are said to be related by a **congruence transformation**. ■■

8. The positive-definite property is preserved under congruence, by which we mean that if $\boldsymbol{A}$ is non-singular then $\boldsymbol{R}$ in (4.4.1) is positive-definite if and only if $\boldsymbol{S}$ is positive-definite.

[4] A principal submatrix is a square submatrix whose diagonal elements lie on the diagonal of the matrix.

Property 8 provides us with an easy way to generate positive-definite matrices. Thus:

◊ Fill an $m \times m$ matrix $\boldsymbol{A}$ with random numbers. (It will almost certainly not be symmetric.)

◊ Verify that $\boldsymbol{A}$ is nonsingular by evaluating its determinant. In the unlikely event that $det(\boldsymbol{A}) = 0$ then reject it and start again.

◊ Multiply $\boldsymbol{A}$ by $\boldsymbol{A}^T$ to form $\boldsymbol{R} = \boldsymbol{A}^T\boldsymbol{A}$.

Then $\boldsymbol{R}$ is positive-definite.

***Proof*:** $\boldsymbol{R}^T = \boldsymbol{A}^T\boldsymbol{A} = \boldsymbol{R}$ and so $\boldsymbol{R}$ is symmetric. Moreover, $\boldsymbol{R} = \boldsymbol{A}^T\boldsymbol{A} = \boldsymbol{A}^T\boldsymbol{I}\boldsymbol{A}$ which is a congruence transformation on the positive-definite matrix $\boldsymbol{I}$, and so $\boldsymbol{R}$ is positive-definite. ■■

Property 8 has an important implication, namely the fact that if a matrix is positive-definite and we interchange a given pair of rows and then interchange the same pair of columns, the resulting matrix is also positive definite. Proof is given in Problem 4.25.

***Example 4.5*:** The following matrix is positive-definite:

$$\boldsymbol{R}_1 = \begin{pmatrix} 1 & 2 & 1 & 3 \\ 2 & 5 & -3 & 7 \\ 1 & -3 & 28 & -4 \\ 3 & 7 & -4 & 15 \end{pmatrix} \tag{4.4.2}$$

Interchanging rows *1* and *2* gives the intermediate result

$$\boldsymbol{R}_2 = \begin{pmatrix} 2 & 5 & -3 & 7 \\ 1 & 2 & 1 & 3 \\ 1 & -3 & 28 & -4 \\ 3 & 7 & -4 & 15 \end{pmatrix} \tag{4.4.3}$$

which is no longer symmetric. Interchanging columns *1* and *2* then gives

$$\boldsymbol{R}_3 = \begin{pmatrix} 5 & 2 & -3 & 7 \\ 2 & 1 & 1 & 3 \\ -3 & 1 & 28 & -4 \\ 7 & 3 & -4 & 15 \end{pmatrix} \tag{4.4.4}$$

which is once more symmetric, and $\boldsymbol{R}_3$ is also positive-definite. ■■

4.4.1 The rank of a matrix

To proceed further we require an understanding of the concept of *rank.*[5]

Definition 4.16: Let $\boldsymbol{A}$ be a matrix of constants with m rows, all of which are linearly independent. Then we say that $\boldsymbol{A}$ has **row-rank** m or $\boldsymbol{A}$ has **full row-rank**.

If one of the rows of $\boldsymbol{A}$ is a linear combination of the other $m-1$ rows then we say that $\boldsymbol{A}$ has **row-rank** $m-1$ or $\boldsymbol{A}$ has a **row-rank defect of** *1,* and if two of its rows are linear combinations of the other $m-2$, we say that $\boldsymbol{A}$ has **row-rank** $m-2$ or a **row-rank defect of** *2,* and so on.

Similar definitions can be stated for the columns of $\boldsymbol{A}$, in which case we speak about the **column-rank** of $\boldsymbol{A}$ and its **column-rank defect.** ■■

The expressions 'the rows of $\boldsymbol{A}$ are linearly independent' and '$\boldsymbol{A}$ has full row-rank' mean the same thing, but *rank* is a more compact concept and also enables us to speak about *rank defect,* i.e. when the rows are not linearly independent.

It is shown in elementary courses on linear algebra that the row-rank of any matrix is equal to its column rank.

Example 4.6: Let $\boldsymbol{A}$ be a nonsingular $m \times m$ matrix. Then the row-rank and column-rank of $\boldsymbol{A}$ are both full and are equal to m, and if $\boldsymbol{A}$ is singular then the row-rank and column-rank of $\boldsymbol{A}$ are both less than m. ■■

[5] While *rank* is usually covered in first courses on linear algebra, we feel that a brief review of the concept would be helpful. See also References 16, 27, 28, 49.

Example 4.7: Let the matrices $\boldsymbol{A}_1, \boldsymbol{A}_2, \boldsymbol{A}_3$ and $\boldsymbol{A}_4$ be as follows:

$$A_1 = \begin{pmatrix} 1 & 2 & 5 \\ 1 & 3 & 4 \end{pmatrix} \quad A_2 = \begin{pmatrix} 1 & 2 & 3 \\ 2 & 4 & 6 \end{pmatrix}$$
$$A_3 = \begin{pmatrix} 1 & 2 \\ 3 & 2 \\ 1 & 3 \end{pmatrix} \qquad A_4 = \begin{pmatrix} 1 & 2 \\ 2 & 4 \\ 4 & 8 \end{pmatrix} \tag{4.4.5}$$

◊ The rows of $\boldsymbol{A}_1$ are linearly independent and so it has row-rank *2*, i.e. full row-rank. Its column-rank is also *2* and it has a column-rank defect of *1*. (Any of its columns is a linear combination of the other two.) Its row-rank is equal to its column rank.

◊ The rows of $\boldsymbol{A}_2$ are not linearly independent (its second row is a multiple of its first) and so it has less than full row-rank. It has row-rank *1* and a row-rank defect of *1*. It has column-rank *1* and a column-rank defect of *2* (any two of its columns are multiples of the remaining one). Its row-rank is equal to its column-rank.

◊ The columns of $\boldsymbol{A}_3$ are linearly independent and so it has column-rank *2*, i.e. full column-rank. It has row-rank *2* and a row-rank defect of *1* (any of its rows is a linear combination of the other two). Its row-rank is equal to its column-rank.

◊ The columns of $\boldsymbol{A}_4$ are not linearly independent (either of its columns is a multiple of the other) and so it has less than full column-rank. It has column-rank *1* and a column-rank defect of *1*. It has row-rank *1* and a row-rank defect of *2* (any two of its rows are multiples of the remaining one). Its row-rank is equal to its column-rank.

■■

4.4.2 Three important theorems regarding positive-definite matrices

In Chapter 8 when we derive the equations for the Gauss filters we will require the following:

Theorem 4.4: Let $\boldsymbol{T}$ be a $k \times m$ matrix where $k > m$ (i.e. $\boldsymbol{T}$ is high) and let $\boldsymbol{R}$ be a positive-definite $k \times k$ matrix. Then $\boldsymbol{T}^T\boldsymbol{R}^{-1}\boldsymbol{T}$ is nonsingular if and only if $\boldsymbol{T}$ has full column rank.

■■

Proof is given in Problem 4.26.

Theorem 4.5: Let $\boldsymbol{W}$ be an $m \times k$ matrix where $k > m$ (i.e. $\boldsymbol{W}$ is wide) and let $\boldsymbol{R_Y}$ be a positive-definite $k \times k$ matrix. Then $\boldsymbol{W R_Y W}^T$ is positive-definite if and only if $\boldsymbol{W}$ has full row rank. ■■

Proof is given in Problem 4.27.

Note that Theorem 4.4 speaks about the triple product $\boldsymbol{T}^T\boldsymbol{R}^{-1}\boldsymbol{T}$ being *nonsingular*, whereas Theorem 4.5 speaks about the triple product $\boldsymbol{W R_Y W}^T$ being *positive-definite,* which is a far more demanding condition.

Note also that Theorem 4.5 is a generalization of Property 8 in which the matrix $\boldsymbol{W}$ in the triple product $\boldsymbol{W R_Y W}^T$ was square and nonsingular and so the triple product was a congruence transformation. Here the matrix $\boldsymbol{W}$ is wide and of full row-rank, and so the triple product is seen to be a *generalized congruence transformation*.

Theorem 4.1 stated that if $\boldsymbol{N}' = \boldsymbol{A}\boldsymbol{N}$ then $\boldsymbol{R}_{N'} = \boldsymbol{A}\boldsymbol{R}_N\boldsymbol{A}^T$. The theorem is true regardless of the dimensions of the matrix $\boldsymbol{A}$. However, by Theorem 4.5, if $\boldsymbol{R}_N$ is positive-definite then $\boldsymbol{R}_{N'}$ will be positive-definite if and only if $\boldsymbol{A}$ has full row rank.

Throughout the rest of the book, whenever Theorem 4.1 is applied, it will always be the case that the matrix $\boldsymbol{A}$ (or whatever it is called at that time) will have full row rank, and that the matrix $\boldsymbol{R}_N$ (or whatever it is called at that time) will be positive-definite, and so the matrix $\boldsymbol{R}_{N'}$ emerging from the transformation will be positive-definite.

Definition 4.17: Let $\boldsymbol{x}$ be an m-vector. Then $\| \boldsymbol{x} \|$ means the Euclidean length of x, i.e $\| \boldsymbol{x} \| \equiv (x_1^2 + x_2^2 + \cdots + x_m^2)^{1/2}$. ■■

The following is an important statement that can be made about a quadratic form on a positive-definite matrix:

Theorem 4.6: Let $\boldsymbol{x}^T\boldsymbol{B}\boldsymbol{x}$ be a quadratic form on a positive-definite matrix $\boldsymbol{B}$. Then $\boldsymbol{x}^T\boldsymbol{B}\boldsymbol{x}$ increases like $k\| \boldsymbol{x} \|^2$ in which k is a constant.

Proof is given in Problem 4.28. ■■

From Theorem 4.6 we see that if the length of $\boldsymbol{x}$ grows, then $\boldsymbol{x}^T\boldsymbol{B}\boldsymbol{x}$ grows at a rate that is proportional to the length of $\boldsymbol{x}$ squared. We note that the length of $\boldsymbol{x}$ can grow even if only one of its elements grows.

4.5 Final items

4.5.1 Positive-semidefinite matrices

The defining condition for the positive-definite property was as follows:

$$\boldsymbol{c}^T\boldsymbol{R}\boldsymbol{c} > 0 \tag{4.5.1}$$

in which $\boldsymbol{c}$ is any nonzero vector, and the inequality is 'strictly-greater-than'.

However, it is also possible for a real symmetric matrix $\boldsymbol{B}$ to satisfy a similar relationship but with the inequality being 'greater-than-or-equal-to', namely

$$\boldsymbol{c}^T\boldsymbol{B}\boldsymbol{c} \geq 0 \tag{4.5.2}$$

in which $\boldsymbol{c}$ is again any nonzero vector. What this means is that the quadratic form in (4.5.2) is in general positive, but that there exist specific sets of (nonzero) vectors $\boldsymbol{c}$ for which it is zero.

Such a matrix is called *positive-semidefinite* for which we have the following definition.

Definition 4.18: A real symmetric matrix $\boldsymbol{B}$ is **positive-semidefinite** if for all vectors $\boldsymbol{a}$

$$\boldsymbol{a}^T\boldsymbol{B}\boldsymbol{a} \geq 0 \tag{4.5.3}$$

and there exist nonzero vectors $\boldsymbol{c}$ such that

$$\boldsymbol{c}^T\boldsymbol{B}\boldsymbol{c} = 0 \tag{4.5.4}$$

■■

From this definition it follows that if a matrix is positive-semidefinite, then it cannot also be positive-definite, and *vice-versa*. Thus the sets of all positive-definite and all positive-semidefinite matrices are disjoint.

Example 4.8: Show that the following matrix is positive-semidefinite:

$$\boldsymbol{B} = \begin{pmatrix} 1 & -2 \\ -2 & 4 \end{pmatrix} \tag{4.5.5}$$

Proof: We examined $\boldsymbol{B}$ earlier in Example 4.4 where we saw that

$$\boldsymbol{c}^T\boldsymbol{B}\boldsymbol{c} = (c_1 - 2c_2)^2 \tag{4.5.6}$$

By assumption, the vector $\boldsymbol{c}$ is nonzero, and so c_1 and c_2 cannot both be zero. In general therefore, $(c_1 - 2c_2)^2$ is positive.

However, if $c_1 = 2c_2$ then $\boldsymbol{c}^T\boldsymbol{B}\boldsymbol{c}$ is zero, even though c_1 and c_2 are not both zero. Moreover, that will also be true for the entire set of vectors $\boldsymbol{u} = k\boldsymbol{c}$ where k is any nonzero number.

This means that, for $\boldsymbol{c}$ any nonzero vector and for the matrix $\boldsymbol{B}$ in (4.5.5),

$$\boldsymbol{c}^T\boldsymbol{B}\boldsymbol{c} \geq 0 \tag{4.5.7}$$

showing that $\boldsymbol{B}$ is positive-semidefinite. ■■

Some of the properties of positive-semidefinite matrices are similar to those of positive-definite matrices, but here the symbol $\geq$ appears, instead of the symbol $>$.

4.5.2 Properties of positive-semidefinite matrices

As with the properties of positive-definite matrices discussed earlier, the following properties of positive-semidefinite matrices will also be invoked. Proofs are provided in Problems 4.29 to 4.32.

9. The eigenvalues of a positive-semidefinite matrix are either positive or zero, with at least one being zero, and its determinant is equal to zero.

10. If one or more of the eigenvalues of a real symmetric matrix are zero and the others are positive, then it is positive-semidefinite.

11. The diagonal elements of a positive-semidefinite matrix are all greater than or equal to zero, and its principal sub-matrices are either positive-semidefinite or positive-definite.

12. If a diagonal element of a positive-semidefinite matrix is zero, then all elements in the same row and column are zero.

The next item follows immediately from Property 12.

13. If the diagonal elements of a positive-semidefinite matrix are all zero, then all elements of the matrix are zero, and so it is the null matrix.

14. The positive-semidefinite property is preserved under congruence, by which we mean that if $\boldsymbol{A}$ is non-singular then $\boldsymbol{R}$ in (4.4.1) is positive-semidefinite if and only if $\boldsymbol{S}$ is positive-semidefinite.

4.5.3 Important facts from linear algebra

As a reminder to the reader we state a few important facts (without proof) from elementary linear algebra. (See e.g. Refs. 27 or 28.)

◊ All eigenvalues of a real symmetric matrix are real.

◊ The eigenvectors of distinct eigenvalues of a real symmetric matrix are mutually orthogonal.

◊ The eigenvectors of a real symmetric $n \times n$ matrix are linearly independent and can form a basis for n-space.

◊ An orthogonal matrix $\boldsymbol{Q}$ is one for which

$$\boldsymbol{Q}^T = \boldsymbol{Q}^{-1} \tag{4.5.8}$$

◊ The determinant of a matrix is equal to the product of its eigenvalues.

◊ The trace of a matrix is equal to the sum of its eigenvalues.

The following theorem will be used frequently:

Theorem 4.7: Let $\boldsymbol{B}$ be any real symmetric matrix. Then there exists an orthogonal matrix $\boldsymbol{Q}$ composed of the orthonormalized eigenvectors of $\boldsymbol{B}$ such that

$$\boldsymbol{Q}^T\boldsymbol{B}\boldsymbol{Q} = \boldsymbol{\Lambda} \tag{4.5.9}$$

where $\boldsymbol{\Lambda}$ is the diagonal matrix comprised of the eigenvalues of $\boldsymbol{B}$ arrayed in the same order as the corresponding eigenvectors in $\boldsymbol{Q}$. ■■

The following items are in the supplementary material

Problems for Chapter 4

Solutions to Problems for Chapter 4

Projects for Chapter 4

Appendix 4.1

Geometric interpretation of the positive-definite property

For convenience we repeat Figure 4.5:

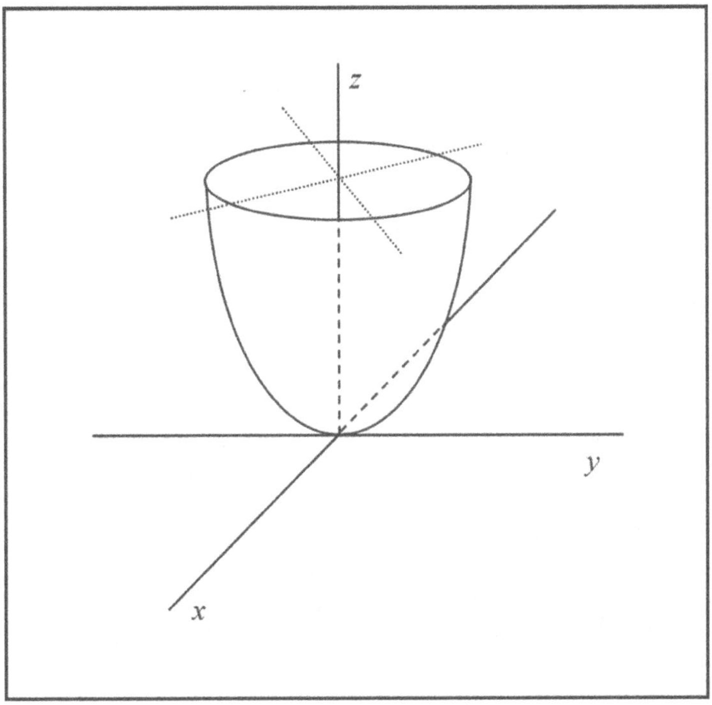

Figure A4.1.1 A surface corresponding to a positive definite matrix

Consider the positive-definite 2×2 matrix:

$$\boldsymbol{R} = \begin{pmatrix} r_{1,1} & r_{1,2} \\ r_{1,2} & r_{2,2} \end{pmatrix} \tag{A4.1.1}$$

and the vector $c = (x, y)^T$. We form $\boldsymbol{c}^T\boldsymbol{R}\boldsymbol{c}$ and set it equal to the variable z. Then

$$z = r_{1,1}x^2 + 2\,r_{1,2}\,x\,y + r_{2,2}\,y^2 \tag{A4.1.2}$$

Equation (A4.1.2) represents a surface in 3-space, with the following properties:

◊ If $\boldsymbol{c}$ is the zero vector then $z = 0$, and so the surface passes through the origin.

◊ If $\boldsymbol{c}$ is any nonzero vector then $z > 0$ (because $\boldsymbol{R}$ is positive-definite) and so, other than the point at the origin, the surface lies above the x, y plane.

◊ Setting $z = k$, where k is a positive constant, gives the equation for horizontal cross sections of the surface as

$$r_{1,1}x^2 + 2\,r_{1,2}\,x\,y + r_{2,2}\,y^{\,2} = k \tag{A4.1.3}$$

It is shown in first courses on linear algebra that if the constants $r_{1,1}$, $r_{1,2}$ and $r_{2,2}$ in (A4.1.3) come from a real symmetric matrix $\boldsymbol{R}$ whose eigenvalues are all positive, then (A4.1.3) represents an ellipse with its principal axes possibly rotated away from the x- and y-axes, but with their intersection at the origin.

The eigenvalues of $\boldsymbol{R}$ *are* all positive (because $\boldsymbol{R}$ is positive-definite) and so all horizontal cross sections of the surface in the figure are ellipses.

◊ We now let

$$y = mx \tag{A4.1.4}$$

Then this represents a plane that is perpendicular to the x, y plane which contains the z axis, and the parameter m determines the azimuthal orientation of that plane.

Substitution of (A4.1.4) into (A4.1.2) gives the equation for vertical cross sections of the surface as

$$\begin{aligned} z &= r_{1,1}x^2 + 2r_{1,2}\,x(m\,x) + r_{2,2}\,(mx)^2 \\ &= (r_{1,1} + 2mr_{1,2} + m^2 r_{2,2})\,x^2 = kx^2 \end{aligned} \tag{A4.1.5}$$

in which k is positive (because $z > 0$). Consider now the following:

As we move away from the origin in the x, y plane along the line $y = mx$, the distance from the origin is given by

$$\|\,\boldsymbol{c}\,\| \equiv (x^2 + y^2)^{1/2} = x(1 + m^2)^{1/2} \tag{A4.1.6}$$

from which

$$x = \| \boldsymbol{c} \| / (1 + m^2)^{1/2} \tag{A4.1.7}$$

Using this in (A4.1.5) gives

$$z = Ax^2 = \left(k/(1 + m^2)\right) \| \boldsymbol{c} \|^2 = k(m) \| \boldsymbol{c} \|^2 \tag{A4.1.8}$$

This means that there exists a positive constant $k(m) \equiv A/(1 + m^2)$ that is independent of $\| \boldsymbol{c} \|$ such that

$$z = \boldsymbol{c}^T \boldsymbol{R} \boldsymbol{c} = k(m) \| \boldsymbol{c} \|^2 \tag{A4.1.9}$$

This is the equation to a parabola that passes through the origin, and so all vertical cross sections of the surface produced by cutting it with a vertical plane are parabolas.

We have thus shown that, as we depart radially in any direction from the origin, the quadratic form $z = \boldsymbol{c}^T \boldsymbol{R} \boldsymbol{c}$ grows at a rate that is proportional the square of the distance from the origin, where the growth constant $k(m)$ depends only on the azimuthal direction in which we are departing, and not on $\| \boldsymbol{c} \|$.

Thus any 2×2 positive-definite matrix $\boldsymbol{R}$ can be associated with a parabolic surface of the form shown in Figure A4.1.1.

Similar demonstrations are possible for positive-definite matrices of any size.

Chapter 5

Random vectors and covariance matrices in filter engineering

In this chapter we explore the role that random vectors and covariance matrices play in filter engineering. For now we are making the assumption that all errors are zero-mean. In Chapter 6 we will investigate the situation where biases are present and the errors are not zero-mean.

5.1 Transformation of random vectors and covariance matrices

In Theorem 4.1 we saw how a random vector and its covariance matrix pass through a linear transformation. However, in many practical cases the transformations involved are nonlinear, and in what now follows we examine such situations.

As the vehicle for our analysis we will use the NED coordinate system (Figure 3.3), and although we are considering a specific example, the techniques that we develop are perfectly general.

5.1.1 The transformation equations

We are given the radar vector $\boldsymbol{P}_R$ and its diagonal covariance matrix $\boldsymbol{R}_{\boldsymbol{P}_R}$ namely

$$\boldsymbol{P}_R \equiv (\rho, \theta, \psi)^T \quad \text{and} \quad \boldsymbol{R}_{\boldsymbol{P}_R} \equiv diag(\sigma_\rho^2, \sigma_\theta^2, \sigma_\psi^2) \tag{5.1.1}$$

and we wish to find the Cartesian image vector $\boldsymbol{P}_C \equiv (x, y, z)^T$ and its associated covariance matrix $\boldsymbol{R}_{\boldsymbol{P}_C}$ assuming NED coordinates.

We recall from (3.3.1) that $\boldsymbol{P}_C$ and $\boldsymbol{P}_R$ in NED coordinates are related as follows:

$$\begin{aligned} x &= \rho\, cos(\theta)\, cos(\psi) \\ y &= \rho\, cos(\theta)\, sin(\psi) \\ z &= \rho\, sin(\theta) \end{aligned} \tag{5.1.2}$$

Then (5.1.2) are the required equations that transform $\boldsymbol{P}_R$ to $\boldsymbol{P}_C$ and we observe that they are nonlinear.

We can also transform in the reverse direction. Thus, solving for ρ, θ and ψ from (5.1.2) we obtain

$$\begin{aligned} \rho &= (x^2 + y^2 + z^2)^{1/2} \\ \theta &= \tan^{-1}\left(z/(x^2 + y^2)^{1/2}\right) \quad \text{(2-quadrant arctan)} \\ \psi &= \tan^{-1}(y/x) \quad \text{(4-quadrant arctan)} \end{aligned} \tag{5.1.3}$$

Definition 5.1: A nonlinear transformation, in which either set of coordinates can be uniquely transformed into the other set, is called a **nonsingular** nonlinear transformation. ■■

By this definition, the nonlinear transformation associated with NED coordinates is nonsingular.

Example 5.1: A second example of a nonsingular nonlinear transformations occurs in the ENU coordinates discussed in Appendix 3.1, namely:

$$\begin{aligned} x &= \rho \cos(\theta) \sin(\psi) \\ y &= \rho \cos(\theta) \cos(\psi) \\ z &= \rho \sin(\theta) \end{aligned} \tag{5.1.4}$$

for which the inverse set of equations is as follows:

$$\begin{aligned} \rho &= (x^2 + y^2 + z^2)^{1/2} \\ \theta &= \tan^{-1}\left(z/(x^2 + y^2)^{1/2}\right) \quad \text{(2-quadrant arctan)} \\ \psi &= \tan^{-1}(x/y) \quad \text{(4-quadrant arctan)} \end{aligned} \tag{5.1.5}$$

■■

Example 5.2: An example of a *singular* nonlinear transformation is the following: Starting from the transformation in (5.1.2) we obtain slant-range and azimuth as follows:

$$\begin{aligned} \rho &= (x^2 + y^2 + z^2)^{1/2} \\ \psi &= \tan^{-1}(y/x) \quad \text{(4-quadrant arctan)} \end{aligned} \tag{5.1.6}$$

However, given the values for slant-range and azimuth (ρ and ψ) appearing in (5.1.6) it is not possible to derive the Cartesian coordinates (x, y, z) of the object. The transformation from Cartesian coordinates to slant-range and azimuth is therefore singular. ■■

5.1.2 *Transforming the covariance matrix in a nonlinear transformation*

Returning now to (5.1.2), suppose that the quantities ρ, θ, ψ change by small amounts $\delta\rho, \delta\theta, \delta\psi$ (called differentials). Then the quantities x, y, z will change by differentials $\delta x, \delta y, \delta z$ as follows:

$$\begin{aligned} \delta x &= (\partial x/\partial\rho)\,\delta\rho + (\partial x/\partial\theta)\,\delta\theta + (\partial x/\partial\psi)\,\delta\psi \\ \delta y &= (\partial y/\partial\rho)\,\delta\rho + (\partial y/\partial\theta)\,\delta\theta + (\partial y/\partial\psi)\,\delta\psi \\ \delta z &= (\partial z/\partial\rho)\,\delta\rho + (\partial z/\partial\theta)\,\delta\theta + (\partial z/\partial\psi)\,\delta\psi \end{aligned} \tag{5.1.7}$$

These equations are only approximate as long as the differentials are of finite size, but they become increasingly more accurate as the differentials tend to zero. Restating (5.1.7) in matrix/vector format we obtain

$$\begin{pmatrix} \delta x \\ \delta y \\ \delta z \end{pmatrix} = \begin{pmatrix} \partial x/\partial\rho & \partial x/\partial\theta & \partial x/\partial\psi \\ \partial y/\partial\rho & \partial y/\partial\theta & \partial y/\partial\psi \\ \partial z/\partial\rho & \partial z/\partial\theta & \partial z/\partial\psi \end{pmatrix} \begin{pmatrix} \delta\rho \\ \delta\theta \\ \delta\psi \end{pmatrix} \tag{5.1.8}$$

Definition 5.2: The matrix of partial derivatives in (5.1.8) is called the **Jacobi matrix** of the transformation.[1] ■■

We now assume that the differentials $\delta\rho, \delta\theta, \delta\psi$ are, in fact, the *errors* in ρ, θ, ψ, and rename them $\nu_\rho, \nu_\theta, \nu_\psi$. Then the differentials $\delta x, \delta y, \delta z$ will be the corresponding errors in x, y, z, which we call ν_x, ν_y, ν_z.

Using these symbols, equation (5.1.8) becomes

$$\begin{pmatrix} \nu_x \\ \nu_y \\ \nu_z \end{pmatrix} = \begin{pmatrix} \partial x/\partial\rho & \partial x/\partial\theta & \partial x/\partial\psi \\ \partial y/\partial\rho & \partial y/\partial\theta & \partial y/\partial\psi \\ \partial z/\partial\rho & \partial z/\partial\theta & \partial z/\partial\psi \end{pmatrix} \begin{pmatrix} \nu_\rho \\ \nu_\theta \\ \nu_\psi \end{pmatrix} \tag{5.1.9}$$

This equation shows (to first order) that the errors in Cartesian coordinates are related to the errors in radar coordinates by a linear transformation based on the Jacobi matrix. That in turn means that if the errors in radar coordinates are Gaussian, then the errors in Cartesian coordinates can be regarded as Gaussian.

Before proceeding further we pause to form the partial derivatives in the Jacobi matrix for NED coordinates in (5.1.9). Let it be called $\boldsymbol{G}$, i.e.

[1] See Reference 191.

$$\boldsymbol{G} \equiv \begin{pmatrix} \partial x/\partial\rho & \partial x/\partial\theta & \partial x/\partial\psi \\ \partial y/\partial\rho & \partial y/\partial\theta & \partial y/\partial\psi \\ \partial z/\partial\rho & \partial z/\partial\theta & \partial z/\partial\psi \end{pmatrix} \tag{5.1.10}$$

Carrying out the indicated partial differentiation on the three functions in (5.1.2) we obtain

$$\boldsymbol{G} = \begin{pmatrix} cos(\theta)\,cos(\psi) & -\rho\,sin(\theta)\,cos(\psi) & -\rho\,cos(\theta)\,sin(\psi) \\ cos(\theta)\,sin(\psi) & -\rho\,sin(\theta)\,sin(\psi) & \rho\,cos(\theta)\,cos(\psi) \\ sin(\theta) & \rho\,cos(\theta) & 0 \end{pmatrix} \tag{5.1.11}$$

It is easily verified that

$$det(\boldsymbol{G}) = -\rho^2 cos\,(\theta) \tag{5.1.12}$$

from which we see that the determinant of $\boldsymbol{G}$ will be zero if $\rho = 0$ or $\theta = \pm\pi/2$.

The first of these conditions can be excluded for physical reasons (the object being observed will never have slant range zero), and the second, for which the object being observed is directly overhead (or underhead), can be regarded as a singularity and also excluded.

Other than that, the determinant of (5.1.11) will be nonzero, and so the Jacobi matrix $\boldsymbol{G}$ can be considered to be nonsingular.

Equation (5.1.9) can now be written as

$$\boldsymbol{N}_C = \boldsymbol{G}\boldsymbol{N}_R \tag{5.1.13}$$

in which

$$\boldsymbol{N}_C \equiv (\nu_x, \nu_y, \nu_z)^T \text{ and } \boldsymbol{N}_R \equiv (\nu_\rho, \nu_\theta, \nu_\psi)^T \tag{5.1.14}$$

These are the error vectors in Cartesian and radar coordinates.

Suppose we had instead started out from (5.1.3). Then in the same way we would have obtained the equation

$$\boldsymbol{N}_R = \boldsymbol{H}\boldsymbol{N}_C \tag{5.1.15}$$

in which the Jacobi matrix was

$$H = \begin{pmatrix} \partial\rho/\partial x & \partial\rho/\partial y & \partial\rho/\partial z \\ \partial\theta/\partial x & \partial\theta/\partial y & \partial\theta/\partial z \\ \partial\psi/\partial x & \partial\psi/\partial y & \partial\psi/\partial z \end{pmatrix} \tag{5.1.16}$$

If we now compare (5.1.13) and (5.1.15) then we see that $\boldsymbol{G}$ and $\boldsymbol{H}$ must be each other's inverses, namely

$$G^{-1} = H \tag{5.1.17}$$

or, in terms of their Jacobi matrices,

$$\begin{pmatrix} \partial x/\partial\rho & \partial x/\partial\theta & \partial x/\partial\psi \\ \partial y/\partial\rho & \partial y/\partial\theta & \partial y/\partial\psi \\ \partial z/\partial\rho & \partial z/\partial\theta & \partial z/\partial\psi \end{pmatrix}^{-1} = \begin{pmatrix} \partial\rho/\partial x & \partial\rho/\partial y & \partial\rho/\partial z \\ \partial\theta/\partial x & \partial\theta/\partial y & \partial\theta/\partial z \\ \partial\psi/\partial x & \partial\psi/\partial y & \partial\psi/\partial z \end{pmatrix} \tag{5.1.18}$$

This equation shows that if a transformation is nonsingular, then the associated Jacobi matrix has an inverse. It also shows that if we require the inverse of $\boldsymbol{G}$ in functional form, then we can obtain it more easily by forming the matrix $\boldsymbol{H}$ starting from (5.1.3).

Consider now (5.1.13). We are assuming that the radar errors are zero mean. It then follows that the Cartesian errors will also be zero mean.

In equation (5.1.13) we see the random vector $\boldsymbol{N}_R$ passing through a linear transformation to produce the vector $\boldsymbol{N}_C$. Let the covariance matrix of $\boldsymbol{N}_R$ be as follows:

$$R_{N_R} = \begin{pmatrix} \sigma_\rho^2 & 0 & 0 \\ 0 & \sigma_\theta^2 & 0 \\ 0 & 0 & \sigma_\psi^2 \end{pmatrix} \tag{5.1.19}$$

and let the covariance matrix of $\boldsymbol{N}_C$ be called $\boldsymbol{R}_{N_C}$. Then it follows from Theorem 4.1 that

$$R_{N_C} = G R_{N_R} G^T \tag{5.1.20}$$

We have thus shown that under the nonsingular nonlinear NED transformation from radar to Cartesian coordinates, the error vectors and their covariance matrices transform as follows:

$$N_C = G N_R \tag{5.1.21}$$

$$R_{N_C} = G R_{N_R} G^T \tag{5.1.22}$$

in which $\boldsymbol{G}$ is the Jacobi matrix appearing in (5.1.11). Writing (5.1.22) in full we obtain the result shown below in equation (5.1.25).

In Example 4.2 we showed that $\boldsymbol{R}_{\boldsymbol{N}_R}$ was positive-definite, and in (5.1.22) we see that $\boldsymbol{R}_{\boldsymbol{N}_C}$ and $\boldsymbol{R}_{\boldsymbol{N}_R}$ are related by a congruence transformation (because $\boldsymbol{G}$ is nonsingular). Since the positive-definite property is preserved under congruence (Property 8), it follows that $\boldsymbol{R}_{\boldsymbol{N}_C}$ in (5.1.22) will also be positive-definite.

We can generalize the above result as follows: Let two estimates $\boldsymbol{P}^*$ and $\boldsymbol{Q}^*$ that are related by a nonsingular nonlinear transformation have error vectors $\boldsymbol{N}_{\boldsymbol{P}^*}$ and $\boldsymbol{N}_{\boldsymbol{Q}^*}$. Then those error vectors satisfy

$$\boldsymbol{N}_{\boldsymbol{P}^*} = \boldsymbol{G}\boldsymbol{N}_{\boldsymbol{Q}^*} \tag{5.1.23}$$

in which $\boldsymbol{G}$ is the Jacobi matrix of the transformation, and to first order their covariance matrices satisfy

$$\boldsymbol{R}_{\boldsymbol{N}_{\boldsymbol{P}^*}} = \boldsymbol{G}\boldsymbol{R}_{\boldsymbol{N}_{\boldsymbol{Q}^*}}\boldsymbol{G}^T \tag{5.1.24}$$

5.1.3 *Three final items*

1. The matrix $\boldsymbol{G}$ that appears in (5.1.21) and (5.1.22) contains the variables ρ, θ and ψ, which, from the context of the above analysis, are the exact values.

   However, in the field we do not know those exact values. All that we have there are what the observation instrument (e.g. the radar) has observed, and while those observations are hopefully close to the exact values, they are not precisely equal to those values.

   Accordingly, in the field we will have to use the observed values when evaluating $\boldsymbol{G}$, and acknowledge that, in addition to the above analysis being true only to first order, a further source of error is present because of the differences between the exact and observed values of ρ, θ and ψ.

   By contrast, when running a simulation we have a choice between the exact values and the observed values (as seen by the observation instrument). The details of the situation would dictate which to use.

2. Referring to equation (5.1.17), if we require $\boldsymbol{G}^{-1}$ in functional form and we obtain it by deriving $\boldsymbol{H}$, then we could easily find that we have obtained $\boldsymbol{G}^{-1}$ expressed in the wrong coordinate system. In Problem 5.2 we consider how to prevent this from happening.

$$\boldsymbol{R}_{N_C} = \begin{pmatrix} \mathcal{E}\{x^2\} & \mathcal{E}\{xy\} & \mathcal{E}\{xz\} \\ \mathcal{E}\{xy\} & \mathcal{E}\{y^2\} & \mathcal{E}\{yz\} \\ \mathcal{E}\{xz\} & \mathcal{E}\{yz\} & \mathcal{E}\{z^2\} \end{pmatrix} = \boldsymbol{G}\boldsymbol{R}_{N_R}\boldsymbol{G}^T$$

$$= \begin{pmatrix} cos(\theta)\,cos(\psi) & -\rho\,sin(\theta)\,cos(\psi) & -\rho\,cos(\theta)\,sin(\psi) \\ cos(\theta)\,sin(\psi) & -\rho\,sin(\theta)\,sin(\psi) & \rho\,cos(\theta)\,cos(\psi) \\ sin(\theta) & \rho\,cos(\theta) & 0 \end{pmatrix} \begin{pmatrix} \sigma_\rho^2 & 0 & 0 \\ 0 & \sigma_\theta^2 & 0 \\ 0 & 0 & \sigma_\psi^2 \end{pmatrix} \begin{pmatrix} cos(\theta)\,cos(\psi) & -\rho\,sin(\theta)\,cos(\psi) & -\rho\,cos(\theta)\,sin(\psi) \\ cos(\theta)\,sin(\psi) & -\rho\,sin(\theta)\,sin(\psi) & \rho\,cos(\theta)\,cos(\psi) \\ sin(\theta) & \rho\,cos(\theta) & 0 \end{pmatrix}^T \tag{5.1.25}$$

Transforming $\boldsymbol{R}_{N_R}$ to $\boldsymbol{R}_{N_C}$ for NED radar coordinates

3. In the above discussion it was assumed that the nonlinear transformations are in the form of a set of explicit algebraic equations (see (5.1.4) and (5.1.5)). However, that may not always be the case and one or more of the equations may be implicit.

   In satellite work, for example, the transformation from Keplerians to Cartesians (see the documentation for ***14_Orbit_Track***) makes use of Kepler's equation

$$M = E - e\,sin(E) \tag{5.1.26}$$

   This is an implicit expression for E in which M and e are known, and solving for E is then done by Newton's method. Obtaining partial derivatives by hand under such circumstances would be either difficult or impossible.

   Alternately, even if all of the equations in the nonlinear transformation are explicit, they may be extremely complex, as for example in the transformation from Cartesians to Keplerians (see the documentation for ***14_Orbit_Track***), and we may not want to make the extensive effort to find the partial derivatives by hand that make up the Jacobi matrix.

   If either of these situations is present, we refer you to Project 5.2.

5.1.4 Summary

◊ The NED and ENU transformations from radar to Cartesian coordinates are nonsingular nonlinear transformations.

◊ The NED transformation is shown in (5.1.2), and its Jacobi matrix, $\boldsymbol{G}$, is shown in (5.1.11).

◊ The NED inverse transformation is shown in (5.1.3), and its Jacobi matrix, $\boldsymbol{H}$, is shown in (5.1.16).

◊ Both $\boldsymbol{G}$ and $\boldsymbol{H}$ can be regarded as nonsingular and they satisfy $\boldsymbol{GH} = \boldsymbol{I}$.

◊ To first order accuracy, the error vectors $\boldsymbol{N}_C$ and $\boldsymbol{N}_R$ and their covariance matrices $\boldsymbol{R}_{\boldsymbol{N}_C}$ and $\boldsymbol{R}_{\boldsymbol{N}_R}$ are related as shown in (5.1.21) and (5.1.22).

◊ The complete NED covariance matrix transformation is shown in (5.1.25).

◊ If the covariance matrix $\boldsymbol{R}_{\boldsymbol{N}_R}$ is positive-definite then $\boldsymbol{R}_{\boldsymbol{N}_C}$ will also be positive-definite, and *vice versa*.

◊ If the radar errors are zero-mean, then to first order the Cartesian errors will also be zero-mean.

◊ If the radar errors are Gaussian, then to first order the errors in Cartesian coordinates are also Gaussian.

◊ In practice we do not know the exact values of (ρ, θ, ψ) in the transformation equations (5.1.2) and (5.1.25). In the field we must therefore use the values obtained from the observation instrument, which include the observation errors. When running a simulation we have a choice between the exact values and the observed values. ■■

The techniques that we developed above for transforming a random vector and its covariance matrix from one coordinate system to another, in which the transformation is nonlinear, has many applications in filter engineering, and in Appendix 5.1 we explore them further.

5.2 The covariance matrix of the observation errors

In this section we identify a number of properties of the observation-error covariance matrix $\boldsymbol{R}_{\mathsf{Y}_n}$.

Suppose we make a sequence of observation vectors, the most recent of which is the short-vector called $\boldsymbol{Y}_n$. Let the vector of random errors in $\boldsymbol{Y}_n$ be called $\boldsymbol{N}_n$ and assume for simplicity that they are both *2*-vectors, namely:

$$\boldsymbol{Y}_n = \begin{pmatrix} y_1 \\ y_2 \end{pmatrix}_n \qquad \boldsymbol{N}_n = \begin{pmatrix} \nu_1 \\ \nu_2 \end{pmatrix}_n \tag{5.2.1}$$

Assume also that similar vectors of observations were made at the L previous instants $t_{n-1}, \ldots t_{n-L}$. We now concatenate the observation and error vectors, obtaining two long-vectors as follows:

$$\mathsf{Y}_n = \begin{pmatrix} \boldsymbol{Y}_n \\ \text{------} \\ \boldsymbol{Y}_{n-1} \\ \text{------} \\ \vdots \\ \text{------} \\ \boldsymbol{Y}_{n-L} \end{pmatrix} \qquad \mathsf{N}_n = \begin{pmatrix} \boldsymbol{N}_n \\ \text{-------} \\ \boldsymbol{N}_{n-1} \\ \text{-------} \\ \vdots \\ \text{-------} \\ \boldsymbol{N}_{n-L} \end{pmatrix} \tag{5.2.2}$$

As we did in Chapter 3, we are using the *sans-serif* symbols Y_n and N_n as the names for long-vectors. In that chapter we referred to Y_n and N_n as the total observation vector and the total error vector, and we now add the following:

Definition 5.3: Let $\boldsymbol{R}_{\mathsf{Y}_n}$ be the covariance matrix of Y_n. Then $\boldsymbol{R}_{\mathsf{Y}_n}$ is called the **total observation covariance matrix.** ■■

In general L will be a fairly large integer, but in order to save trees we will assume that there have been only two observation instants t_n and t_{n-1}, and so $L = 1$. We will also assume that the error vectors $\boldsymbol{N}_n$ and $\boldsymbol{N}_{n-1}$ are both zero mean.

Forming $\boldsymbol{R}_{\boldsymbol{Y}_n}$ we then obtain

$$\boldsymbol{R}_{\boldsymbol{Y}_n} = \mathcal{E}\{\boldsymbol{N}_n \boldsymbol{N}_n{}^T\} = \mathcal{E}\begin{pmatrix} \boldsymbol{N}_n \\ \hdashline \boldsymbol{N}_{n-1} \end{pmatrix}\begin{pmatrix} \boldsymbol{N}_n{}^T & \vdots & \boldsymbol{N}_{n-1}{}^T \end{pmatrix}$$

$$= \mathcal{E}\begin{pmatrix} \nu_{1,n} \\ \nu_{2,n} \\ \hdashline \nu_{1,n-1} \\ \nu_{2,n-1} \end{pmatrix}\begin{pmatrix} \nu_{1,n} & \nu_{2,n} & \vdots & \nu_{1,n-1} & \nu_{2,n-1} \end{pmatrix} \tag{5.2.3}$$

The final expression in (5.2.3) multiplies out as shown in (5.2.4) on next page.

We now make a careful examination of the matrix $\boldsymbol{R}_{\boldsymbol{Y}_n}$ in (5.2.4):

◊ Overall, it contains $L + 1 = 2$ time stamps, namely t_n and t_{n-1}.

◊ The upper left principal submatrix is 2×2 because $\boldsymbol{N}_n$ is a 2-vector, and it contains only the single time stamp t_n.

Its on-diagonal terms are the variances of $\nu_{1,n}$ and $\nu_{2,n}$ and its two off-diagonal terms are their covariances, which are equal.

This submatrix is symmetric.

◊ The lower right principal submatrix is also 2×2 because $\boldsymbol{N}_{n-1}$ is a 2-vector. (If it had been a 3-vector, then the matrix would have been 3×3.) Again we see that it contains only a single time stamp which is now t_{n-1}.

Its on-diagonal terms are the variances of $\nu_{1,n-1}$ and $\nu_{2,n-1}$ and its off-diagonal terms are their covariances, which are equal.

This submatrix is also symmetric.

◊ The off-diagonal submatrices are 2×2. (However, if $\boldsymbol{N}_n$ had been a 2-vector and $\boldsymbol{N}_{n-1}$ a 3-vector, then one of the off-diagonal submatrices would have been 2×3 and the other 3×2.)

$$\boldsymbol{R}_{\mathbf{Y}_n} = \left(\begin{array}{cc:cc} \mathcal{E}\{\nu_{1,n}{}^2\} & \mathcal{E}\{\nu_{1,n}\nu_{2,n}\} & \mathcal{E}\{\nu_{1,n}\nu_{1,n-1}\} & \mathcal{E}\{\nu_{1,n}\nu_{2,n-1}\} \\ \mathcal{E}\{\nu_{2,n}\nu_{1,n}\} & \mathcal{E}\{\nu_{2,n}{}^2\} & \mathcal{E}\{\nu_{2,n}\nu_{1,n-1}\} & \mathcal{E}\{\nu_{2,n}\nu_{2,n-1}\} \\ \hdashline \mathcal{E}\{\nu_{1,n-1}\nu_{1,n}\} & \mathcal{E}\{\nu_{1,n-1}\nu_{2,n}\} & \mathcal{E}\{\nu_{1,n-1}{}^2\} & \mathcal{E}\{\nu_{1,n-1}\nu_{2,n-1}\} \\ \mathcal{E}\{\nu_{2,n-1}\nu_{1,n}\} & \mathcal{E}\{\nu_{2,n-1}\nu_{2,n}\} & \mathcal{E}\{\nu_{2,n-1}\nu_{1,n-1}\} & \mathcal{E}\{\nu_{2,n-1}{}^2\} \end{array}\right) \tag{5.2.4}$$

Each term in the off-diagonal submatrices contains both time stamps t_n and t_{n-1}.

The upper right submatrix is equal to the transpose of the lower left submatrix. They contain only covariances, all four of which are different.

These submatrices are asymmetric.

There are two possibilities that occur frequently.

1. The observations made at time t_n are uncorrelated with those made at time t_{n-1}. Then $\boldsymbol{R}_{\boldsymbol{Y}_n}$ will have the form

$$\boldsymbol{R}_{\boldsymbol{Y}_n} =$$

$$\left(\begin{array}{cc|cc} \mathcal{E}\{\nu_{1,n}{}^2\} & \mathcal{E}\{\nu_{1,n}\nu_{2,n}\} & & \\ \mathcal{E}\{\nu_{2,n}\nu_{1,n}\} & \mathcal{E}\{\nu_{2,n}{}^2\} & & \\ \hline & & \mathcal{E}\{\nu_{1,n-1}{}^2\} & \mathcal{E}\{\nu_{1,n-1}\nu_{2,n-1}\} \\ & & \mathcal{E}\{\nu_{2,n-1}\nu_{1,n-1}\} & \mathcal{E}\{\nu_{2,n-1}{}^2\} \end{array}\right) \tag{5.2.5}$$

Each on-diagonal block now has only a single time subscript, and all off-diagonal blocks are null. The on-diagonal blocks are symmetric.

When this condition is present we say that the observations are *stage-wise uncorrelated* for which we have the following definition:

Definition 5.4: Let $\boldsymbol{Y}_n$ and $\boldsymbol{Y}_m$ be two vectors of observations made at any different time instants t_n and t_m and let their error vectors $\boldsymbol{N}_n$ and $\boldsymbol{N}_m$ satisfy

$$\mathcal{E}\{\boldsymbol{N}_n\boldsymbol{N}_m{}^T\} = \boldsymbol{\emptyset} \tag{5.2.6}$$

Then we say that $\boldsymbol{Y}_n$ and $\boldsymbol{Y}_m$ are **stage-wise uncorrelated.** ■■

This situation occurs very frequently and we will see it often in the developments that lie ahead.

Note: A stage is a time instant, and so *stage-wise* means from time-instant to time-instant (referring to the times at which the observations are made). Thus the term stage-wise uncorrelated means that errors from different time instants are uncorrelated.

2. In addition to the above, if the errors in the observations made at time t_n are also uncorrelated with each other, and those made at time t_{n-1} are likewise uncorrelated with each other, then $\boldsymbol{R}_{\boldsymbol{Y}_n}$ will have the form

$$\boldsymbol{R}_{\boldsymbol{Y}_n} = \left(\begin{array}{cc|cc} \mathcal{E}\{\nu_{1,n}{}^2\} & & & \\ & \mathcal{E}\{\nu_{2,n}{}^2\} & & \\ \hline & & \mathcal{E}\{\nu_{1,n-1}{}^2\} & \\ & & & \mathcal{E}\{\nu_{2,n-1}{}^2\} \end{array}\right) \tag{5.2.7}$$

Definition 5.5: When the total covariance matrix $\boldsymbol{R}_{\boldsymbol{Y}_n}$ has the diagonal structure shown in (5.2.7) we say that the observation errors are stage-wise uncorrelated as well as **locally uncorrelated.** ■■

5.3 The covariance matrix of the estimation errors

We continue to assume that all errors are zero-mean.

We are observing a physical process and we are estimating the state vector of the observed trajectory $\boldsymbol{X}_n$. The estimate vector is called $\boldsymbol{X}^*_{n,n}$.

*5.3.1 The actual estimation-error vector N**

Definition 5.6: The **actual estimation-error vector** $N^*_{n,n}$ is defined as follows:

$$\boldsymbol{N}^*_{n,n} = \boldsymbol{X}^*_{n,n} - \boldsymbol{X}_n \tag{5.3.1}$$

■■

In the field we cannot evaluate (5.3.1) because we do not know $\boldsymbol{X}_n$. However, when running a Monte-Carlo simulation both items on the right of (5.3.1) are known and so multiple draws of the random vector $\boldsymbol{N}^*_{n,n}$ can be generated by using this equation.

◊ $\boldsymbol{N}^*_{n,n}$ has a covariance matrix which in Section 1.8 was called the actual estimation-error covariance matrix. This is now symbolized as $\boldsymbol{S}_{actual,n,n}$ and further down we will give a precise definition of $\boldsymbol{S}_{actual,n,n}$ and discuss how it can be estimated.[2]

[2] Note the absence of a star in $\boldsymbol{S}_{actual,n,n}$. That is because this quantity is not normally produced by the filter.

◊ There is also a formula for the covariance matrix of $\boldsymbol{N}^*_{n,n}$ which will be used to create what we have called the filter covariance matrix $\boldsymbol{S}^*_{n,n}$.

Always keep in mind that $\boldsymbol{S}^*_{n,n}$ is generated by a *formula* whereas $\boldsymbol{S}_{actual,n,n}$ is *actual* and they are not the same. While it is to be hoped that $\boldsymbol{S}^*_{n,n}$ is a good representation of $\boldsymbol{S}_{actual,n,n}$ that might not always be the case.[3]

5.3.2 The filter matrix W

The total observation equation is (see (3.2.13))

$$\boldsymbol{Y}_n = \boldsymbol{T}_n \boldsymbol{X}_n + \boldsymbol{N}_n \tag{5.3.2}$$

in which $\boldsymbol{N}_n$ is the vector of random errors in the total observation vector $\boldsymbol{Y}_n$. The covariance matrix of $\boldsymbol{Y}_n$ was discussed in the previous section and is called $\boldsymbol{R}_{\boldsymbol{Y}_n}$.

It will *always* be the case that we obtain the estimate vector $\boldsymbol{X}^*_{n,n}$ by means of a linear transformation on $\boldsymbol{Y}_n$. Thus our filters will all have the basic structure

$$\boldsymbol{X}^*_{n,n} = \boldsymbol{W}_n \boldsymbol{Y}_n \tag{5.3.3}$$

This applies to every filter that we develop in this book, and although (5.3.3) may not be the way in which a filter is actually implemented, it implicitly always lies behind every implementation.

Definition 5.7: The matrix $\boldsymbol{W}_n$ in (5.3.3) is called the **filter matrix.** ▪▪

The dimension of $\boldsymbol{Y}_n$ will always be greater than the dimension of $\boldsymbol{X}^*_{n,n}$ and so $\boldsymbol{W}_n$ will always be wide.

Clearly, filter matrices will have to possess some very specific properties. Later on we will identify others, but one of them is as follows:

◊ Every filter matrix $\boldsymbol{W}_n$ used in (5.3.3) must produce an estimate vector $\boldsymbol{X}^*_{n,n}$ that is an *unbiased estimate* of $\boldsymbol{X}_n$, i.e. an estimate $\boldsymbol{X}^*_{n,n}$ which satisfies

$$\mathcal{E}\{\boldsymbol{X}^*_{n,n}\} = \boldsymbol{X}_n \tag{5.3.4}$$

or equivalently, the errors in $\boldsymbol{X}^*_{n,n}$ must be zero-mean, namely

$$\mathcal{E}\{\boldsymbol{N}^*_{n,n}\} = \boldsymbol{\emptyset} \tag{5.3.5}$$

[3] Think of $\boldsymbol{S}_{actual}$ as a bottle of wine that is *there,* standing on the table, and of $\boldsymbol{S}^*$ as a set of instructions for how to make such a bottle of wine. The former is *actual* whereas the latter is *supposed,* and if you follow the instructions precisely, the outcome will supposedly be such a bottle of wine. Clearly the two are not the same.

The constraint on $\boldsymbol{W}_n$ which ensures that (5.3.4) and (5.3.5) are both true is as follows:

$$\boldsymbol{W}_n \boldsymbol{T}_n = \boldsymbol{I} \tag{5.3.6}$$

In this equation, $\boldsymbol{T}_n$ is the total observation matrix discussed in Chapter 3 and $\boldsymbol{I}$ is the identity matrix of the same order as $\boldsymbol{X}^*_{n,n}$. Equation (5.3.6) is called the *exactness constraint* and will be derived in Chapter 6.

There is an immediate consequence to (5.3.6). In Chapter 6 we also derive the following:

Theorem 6.2: Let the matrices $\boldsymbol{W}_n$ and $\boldsymbol{T}_n$ satisfy

$$\boldsymbol{W}_n \boldsymbol{T}_n = \boldsymbol{I} \tag{5.3.7}$$

Then $\boldsymbol{W}_n$ has full row-rank and $\boldsymbol{T}_n$ has full column-rank. ■■

Since all filter matrices $\boldsymbol{W}_n$ that we use in (5.3.3) will satisfy (5.3.6), it follows that they will all have full row-rank.

In Appendix 5.2 we consider the row-rank of $\boldsymbol{W}_n$ from another perspective.

5.3.3 The filter covariance matrix $\boldsymbol{S}^*$

Starting from (5.3.1) we now have:

$$\begin{aligned} \boldsymbol{N}^*_{n,n} &= \boldsymbol{X}^*_{n,n} - \boldsymbol{X}_n = \boldsymbol{W}_n \mathbf{Y}_n - \boldsymbol{X}_n \\ &= \boldsymbol{W}_n(\boldsymbol{T}_n \boldsymbol{X}_n + \mathbf{N}_n) - \boldsymbol{X}_n = \boldsymbol{W}_n \boldsymbol{T}_n \boldsymbol{X}_n + \boldsymbol{W}_n \mathbf{N}_n - \boldsymbol{X}_n \end{aligned} \tag{5.3.8}$$

and by (5.3.6) this reduces to

$$\boldsymbol{N}^*_{n,n} = \boldsymbol{W}_n \mathbf{N}_n \tag{5.3.9}$$

This equation shows how the actual estimation-error vector $\boldsymbol{N}^*_{n,n}$ is related to the total observation-error vector $\mathbf{N}_n$ under the assumption that the filter is producing an unbiased estimate vector $\boldsymbol{X}^*_{n,n}$.

From (5.3.9) and Theorem 4.1 the formula for the filter covariance matrix of $\boldsymbol{N}^*_{n,n}$ will be as follows:

$$\boldsymbol{S}^*_{n,n} = \boldsymbol{W}_n \mathbf{R}_{\mathbf{Y}_n} \boldsymbol{W}_n{}^T \tag{5.3.10}$$

It is to be hoped that $\boldsymbol{S}^*_{n,n}$ is a good representation of $\boldsymbol{S}_{actual,n,n}$. However, regardless of whether or not that is true, the filter equations will be as follows:

$$X^*_{n,n} = W_n Y_n \tag{5.3.11}$$

$$S^*_{n,n} = W_n R_{Y_n} W_n{}^T \tag{5.3.12}$$

These two equations show how the estimate vector $X^*_{n,n}$ and the filter covariance matrix $S^*_{n,n}$ are related, through the filter matrix W_n, to the total observation vector Y_n and its covariance matrix R_{Y_n}.

We note that R_{Y_n} will always be positive-definite and that the wide-matrix W_n will always have full row-rank. The conditions for Theorem 4.5 are thus satisfied, and so $S^*{}_{n,n}$ will always be positive-definite.

5.3.4 Error/covariance-matrix consistency

In Chapter 1 we introduced the concept of ECM consistency, meaning that the filter covariance matrix $S^*_{n,n}$ is a good representation of the actual estimation-error covariance matrix $S_{actual,\,n,\,n}$ (see Figure 1.8).

In Chapter 7 we discuss three tests to determine if ECM consistency is present or absent, and we also state what we mean by 'a good representation'.

◊ If $S^*_{n,n}$ is a good representation of $S_{actual,\,n,\,n}$ then the filter will be ECM consistent and those tests will pass.

◊ If $S^*_{n,n}$ is not a good representation of $S_{actual,\,n,\,n}$ then the filter will be ECM inconsistent – a condition which is a fatal – and they will fail.

How could ECM inconsistency arise? There are a number of ways:

1. The most likely causes are programming errors, theoretical mistakes and unacceptable approximations. The three ECM consistency tests in Chapter 7 will make us aware that such problems exist, after which they must be appropriately addressed. (In Chapter 6 we discuss approximations and their possible effects on ECM consistency.)
2. A second way is for the filtering algorithm – even though correctly engineered and correctly programmed – to malfunction spontaneously. As is well known, the extended Kalman filters are vulnerable to this possibility, and it is for that reason that an appropriately-selected Q matrix is always included when running them. (The same is also true for the Swerling filter.) In Chapter 11 we discuss this matter further.
3. A third way in which ECM inconsistency can arise is as follows:

   Take a look at the definition of $S^*_{n,n}$ in (5.3.10). There are two items on the right of that equation, namely W_n and R_{Y_n}.

We create $\boldsymbol{W}_n$ and *we* are using it in the algorithm. Thus the filter matrix $\boldsymbol{W}_n$ is completely known to us under all conditions – field and simulation.

However, the matrix $\boldsymbol{R}_{\boldsymbol{Y}_n}$ is supposedly the covariance matrix of the observation vector $\boldsymbol{Y}_n$.

- ◊ In a simulation $\boldsymbol{R}_{\boldsymbol{Y}_n}$ is fully known, and then, if we exclude the spontaneous instability of the Kalman and Swerling filters, by everything that we know from mathematics and statistics, $\boldsymbol{S}^*_{n,n}$ should be a good formula for $\boldsymbol{S}_{actual,n,n}$.

- ◊ In the field, however, we are basing our knowledge of $\boldsymbol{R}_{\boldsymbol{Y}_n}$ on what the engineers who built the observation instrument have told us (see Figure 1.6) and if they are wrong then what we are using for $\boldsymbol{R}_{\boldsymbol{Y}_n}$ could be in error. In that case $\boldsymbol{S}^*_{n,n}$ might not be a good representation of $\boldsymbol{S}_{actual,n,n}$ and the filter could be ECM inconsistent (even though in the field we won't know it).

The three ECM-consistency tests that we discuss in Chapter 7 can only be used in simulations. ECM inconsistency in the field is by definition impossible to detect because we cannot know $\boldsymbol{X}_n$, but hopefully what we learn from the simulations will continue to apply in the field.

5.3.5 The complete filter

Definition 5.8: Equations (5.3.11) and (5.3.12) constitute a **complete filter.** ■■

Any filters that we construct should always be complete, by providing both an estimate vector and a filter covariance matrix.

5.4 Ways to generate a prediction

This is a frequently-occurring requirement in filter engineering which poses a number of problems that must be carefully addressed. We have placed the full discussion in Project 5.4.

5.5 The actual estimation-error covariance matrix

In this section we restate what we mean by the *actual estimation-error covariance matrix* $\boldsymbol{S}_{actual,n,n}$ and we also comment on how we can estimate it.

Starting from (4.2.20) we have the following:

Definition 5.9: The **actual estimation-error covariance matrix** is defined as

$$\boldsymbol{S}_{actual,n,n} \equiv \lim_{M\to\infty} 1/M \sum_{r=1}^{M} \boldsymbol{N}^*_{n,n,r} \boldsymbol{N}^{*T}_{n,n,r} \tag{5.5.1}$$

in which $N^*_{n,n,r}$ is the rth draw of the actual estimation-error vector $N^*_{n,n}$ defined in (5.3.1). ■■

Based on this definition we can estimate $S_{actual,n,n}$ by use of the algorithm

$$S_{actual,n,n} \approx 1/M \sum_{r=1}^{M} N^*_{n,n,r} \, N^{*\,T}_{n,n,r} \tag{5.5.2}$$

Equation (5.5.2) assumes that repeated draws of $N^*_{n,n}$ are being generated by a Monte-Carlo, and for each such draw the matrix product $N^*_{n,n} \, N^{*T}_{n,n}$ is formed. These are added together until M of them have been accumulated, giving us the quantity

$$\sum_{r=1}^{M} N^*_{n,n,r} \, N^{*\,T}_{n,n,r} \tag{5.5.3}$$

This sum is then divided by M which gives the required estimate of $S_{actual,n,n}$.

Using a Monte-Carlo in which there are M repeated runs – where $M = 10{,}000$ or $100{,}000$ or even more – we can, in this way, obtain estimates of $S_{actual,n,n}$ to a high degree of accuracy.

In order to evaluate $S_{actual,n,n}$ in (5.5.2) at a selected point B on a selected flight path, for example, it is necessary that the Monte-Carlo be repeatedly restarted at point A and terminated at the point B where the draws of $N^*_{n,n}$ are calculated.

Thus (5.5.2) should not be thought of as a way in which to obtain ongoing and continuous readouts of $S_{actual,n,n,}$ but is instead a way in which to estimate its value at a specific pre-selected point on a pre-selected flight during a simulation.

To obtain ongoing and continuous readouts of estimates of $S_{actual,n,n}$ in the field, we must instead use the covariance matrix $S^*_{n,n}$ which is evaluated on every cycle of the filter, and then depend on the ECM consistency of the filter to ensure that these matrices are good representations of the corresponding matrices $S_{actual,n,n}$, which for practical reasons cannot be evaluated continuously.

We refer you to the programs ***04_Monte_Carlo*** and ***12_Gauss_Newton_B***, both of which involve the estimation of actual estimation-error covariance matrices using the algorithm appearing in (5.5.2).

*5.5.1 Comment regarding N**

The vector $N^*_{n,n}$ in (5.3.1) is the *actual* vector of errors in $X^*_{n,n}$ and can always be derived in a simulation because there we know X_n. From repeated draws of $N^*_{n,n}$ we can then estimate $S_{actual,n,n}$ using (5.5.2).

This is the key fact behind the matrix-to-matrix ECM-consistency test discussed in Chapter 7 and the matrix-to-matrix CR-consistency tests discussed in Chapter 9.

5.6 The column-rank of the $\boldsymbol{T}$ matrices

In Chapter 3 we stated repeatedly that the columns of the high-matrix $\boldsymbol{T}$ must be linearly independent, i.e. that $\boldsymbol{T}$ must have full column-rank, or else further progress will be blocked.

The column-rank of the $\boldsymbol{T}$ matrices is a topic that goes to the very heart of filter engineering, and the way in which we construct our observation schemes.

The complete discussion of this topic is straightforward but lengthy, and rather than including it in the printed part of the book, we have placed it in Project 5.3.

In Chapter 8 it will be assumed that the material in that project has been read and fully absorbed.

The following items are in the supplementary material

Problems for Chapter 5

Solutions to Problems for Chapter 5

Projects for Chapter 5

Appendix 5.1

Nonlinear transformation of random vectors and covariance matrices

Given the following *linear* relationship between two zero-mean random vectors $\boldsymbol{Z}$ and $\boldsymbol{X}$

$$\boldsymbol{Z} = \boldsymbol{G}\boldsymbol{X} \tag{A5.1.1}$$

by Theorem 4.1 we can then say that their covariance matrices $\boldsymbol{R_Z}$ and $\boldsymbol{R_X}$ are related as follows:

$$\boldsymbol{R_Z} = \boldsymbol{G}\boldsymbol{R_X}\boldsymbol{G}^T \tag{A5.1.2}$$

By Property 8, if $\boldsymbol{G}$ is non-singular and $\boldsymbol{R_X}$ is positive-definite, then $\boldsymbol{R_Z}$ will also be positive-definite.

Suppose instead that we have the following nonlinear relationship between $\boldsymbol{Z}$ and $\boldsymbol{X}$:

$$\boldsymbol{Z} = \boldsymbol{F}(\boldsymbol{X}) \tag{A5.1.3}$$

in which $\boldsymbol{F}$ is a vector of functions of the variables $\boldsymbol{X} = (x_1, x_2, \ldots x_m)^T$, at least one of which is non-linear, and we wish to find the relationship between their covariance matrices. Following the developments in Section 5.1, here's how we would proceed.

Assume that the elements of $\boldsymbol{X}$ in (A5.1.3) change by small amounts $\boldsymbol{\delta X}$. Then the elements of $\boldsymbol{Z}$ will change correspondingly by amounts

$$\boldsymbol{\delta Z} = \boldsymbol{G}(\boldsymbol{X})\,\boldsymbol{\delta X} \tag{A5.1.4}$$

in which the Jacobi matrix $\boldsymbol{G}$ has the following definition:

Definition A5.1.1: Let $\boldsymbol{F}(\boldsymbol{X})$ be the vector of non-linear functions in (A5.1.3) and let $f_i(x_1, \ldots x_m)$ be the ith element of the vector $\boldsymbol{F}(\boldsymbol{X})$. Then the i, jth element of the matrix $\boldsymbol{G}(\boldsymbol{X})$ in (A5.1.4) is given by

$$[\boldsymbol{G}(\boldsymbol{X}(t))]_{i,j} = \partial f_i(x_1, \ldots x_m) \big/ \partial x_j \quad 1 \le i, j \le m \tag{A5.1.5}$$

■■

If $\boldsymbol{\delta X}$ is the vector of errors in $\boldsymbol{X}$, then $\boldsymbol{\delta Z}$ will be the corresponding vector of errors in $\boldsymbol{Z}$, from which it follows by Theorem 4.1 that their covariance matrices will be related as follows:

$$\boldsymbol{R_Z} = \boldsymbol{G}(\boldsymbol{X})\boldsymbol{R_X}\boldsymbol{G}(\boldsymbol{X})^T \tag{A5.1.6}$$

If $\boldsymbol{G}(\boldsymbol{X})$ is nonsingular and $\boldsymbol{R_X}$ is positive-definite, then $\boldsymbol{R_Z}$ will also be positive-definite.

By the use of either (A5.1.2) or (A5.1.6) we are thus able to transform covariance matrices from one coordinate system to another, regardless of whether the transformation is linear or non-linear.

In Problem 5.11 we apply the results that were developed above.

Appendix 5.2

Linear independence of the rows of $\boldsymbol{W}$

There are at least two reasons why an acceptable filter matrix $\boldsymbol{W}$ must have full row-rank.

Reason 1

This is a plausibility argument.

For simplicity let $\boldsymbol{W}$ be $3\times k$, i.e $\boldsymbol{W}$ consists of three row-vectors of dimension k, and since $\boldsymbol{W}$ must always be wide it follows that $k > 3$. Let those row-vectors be called $\boldsymbol{w}_1$, $\boldsymbol{w}_2$ and $\boldsymbol{w}_3$. Then $\boldsymbol{X}^* = \boldsymbol{W}\,\mathbf{Y}$ in (5.3.3) becomes

$$\boldsymbol{X}^* = \begin{pmatrix} x_1^* \\ x_2^* \\ x_3^* \end{pmatrix} = \begin{pmatrix} \boldsymbol{w}_1 \\ \boldsymbol{w}_2 \\ \boldsymbol{w}_3 \end{pmatrix} \mathbf{Y} \tag{A5.2.1}$$

in which each of $\boldsymbol{w}_1$, $\boldsymbol{w}_2$ and $\boldsymbol{w}_3$ is a row-vector of dimension k.

Suppose now that $\boldsymbol{W}$ has less than full row-rank, and so $\boldsymbol{w}_1$, $\boldsymbol{w}_2$ and $\boldsymbol{w}_3$ are linearly dependent. As an example, let

$$\boldsymbol{w}_3 = \alpha \boldsymbol{w}_1 + \beta \boldsymbol{w}_2 \tag{A5.2.2}$$

Then this would mean that x_3^* in (A5.2.1) is being derived as follows:

$$x_3^* = \boldsymbol{w}_3 \mathbf{Y} = (\alpha \boldsymbol{w}_1 + \beta \boldsymbol{w}_2)\mathbf{Y} = \alpha \boldsymbol{w}_1 \mathbf{Y} + \beta \boldsymbol{w}_2 \mathbf{Y} = \alpha x_1^* + \beta x_2^* \tag{A5.2.3}$$

and so the third element of $\boldsymbol{X}^*$ is now a linear combination of the first two.

Even though mathematically possible, it makes no sense to create a filter that produces one estimate which is a linear combination of the other two – it would be far simpler for the filter to produce the two linearly independent estimates and thereafter to create the third from them.

That being the case, the rows of an acceptable filter matrix $\boldsymbol{W}$ will be linearly independent and so it will have full row-rank. ∎∎

Reason 2

This is a mathematical proof that all acceptable filter matrices $\boldsymbol{W}$ will have full row-rank.

In Chapter 6 we derive the following relationship between $\boldsymbol{W}$ and $\boldsymbol{T}$:

$$\boldsymbol{W}\boldsymbol{T} = \boldsymbol{I} \tag{A5.2.4}$$

This is called the *exactness constraint* and it ensures that if the observation vector $\boldsymbol{Y}$ is unbiased then the estimate vector $\boldsymbol{X}^*$ will be unbiased, and so we will only use $\boldsymbol{W}$ matrices that satisfy it.

In Appendix 6.2 we then prove the following:

Theorem 6.2: Let the matrices $\boldsymbol{W}$ and $\boldsymbol{T}$ satisfy

$$\boldsymbol{W}\boldsymbol{T} = \boldsymbol{I} \tag{A5.2.5}$$

Then $\boldsymbol{W}$ has full row-rank and $\boldsymbol{T}$ has full column-rank. ∎∎

Since we will only use $\boldsymbol{W}$ matrices that satisfy the exactness constraint, it follows that they will always have full row-rank.

Chapter 6

Bias errors

6.1 Introduction

In Chapter 5 we examined the errors in the observations and the resulting errors in the estimates, and everything that we developed there was based on errors which were assumed to be zero-mean. Such errors were referred to as *random* errors.

In this chapter we examine the second type of errors that might arise in filtering systems. These are known as *bias errors*, sometimes also called *systematic errors*. These are *non*zero-mean.

In Chapter 5 we made use of the following proposition without deriving it:

If the observation errors are zero-mean, then the filter will produce unbiased estimates if and only if the filter matrix $\boldsymbol{W}$ and the $\boldsymbol{T}$ matrix satisfy

$$\boldsymbol{W}\boldsymbol{T} = \boldsymbol{I} \tag{6.1.1}$$

This equation was called the *exactness constraint* and it plays a key role in filter engineering.

We also showed that when the exactness constraint is satisfied and the observation errors are zero-mean, then the errors in the estimate will also be zero-mean, and will be related to the observation errors by

$$N^*_{random} = \boldsymbol{W}\boldsymbol{N}_{random} \tag{6.1.2}$$

In this chapter we will derive the exactness constraint, and we will also derive the companion equation to (6.1.2) for the situation where the observation errors are nonzero-mean, i.e. when bias errors are present in $\boldsymbol{Y}$. We will see that those bias errors give rise to bias errors in the estimate according to the equation

$$N^*_{bias} = \boldsymbol{W}\boldsymbol{N}_{bias} \tag{6.1.3}$$

6.2 The ensemble

Imagine the following: We assemble a collection of infinitely-many filtering systems – all of them identical save for one feature that we will specify shortly.

Such a collection is called an *ensemble,* and the imagined undertaking by which we construct it is called a *thought experiment.*[1]

As an example, suppose that we are observing a satellite in orbit about the earth. Our imagined ensemble would then consist of infinitely many identical radars all positioned at precisely the same location, each of them making observations on the satellite which are then processed by its own filter. The latter is identical to all other filters in the ensemble, and each filter estimates the satellite's orbital parameters in the same way.

In Figure 6.1 we show the ensemble that corresponds to the system in Figure 1.6.

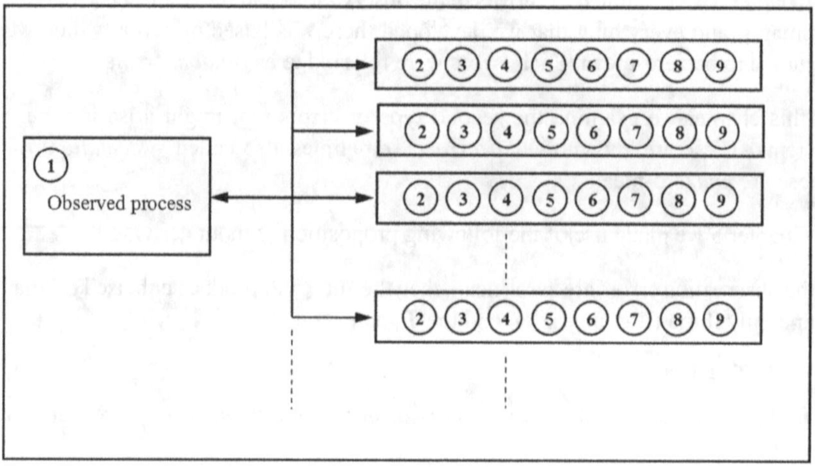

Figure 6.1 Ensemble of filtering systems

The numbers in the circles in Figure 6.1 have the same meanings that they did in Figure 1.6, and so each of the long rectangular blocks represents a complete filtering system – from observation instrument in (2) to filter covariance matrix $\boldsymbol{S}^*$ in (9).

The errors in such an ensemble will be of two types – *random* errors and *bias* errors. We consider them one at a time.

1. The random errors:

    ◊ The observation vectors $\boldsymbol{Y}$ will contain error vectors that differ randomly from one system in the ensemble to another at each time instant, but such error vectors are similar in the sense that they are all drawn from the same multivariate PDF.

[1] In German it is called a *gedankenexperiment.*

◊ These vectors of random errors in $\boldsymbol{Y}$ give rise to the error vectors $\boldsymbol{N}^*_{random}$ (the random errors in $\boldsymbol{X}^*$) which also differ randomly from one system in the ensemble to another at each time instant, but they are also all similar in that they are drawn from the same multivariate PDF.

2. The bias errors:

◊ Each vector of observations $\boldsymbol{Y}$ will potentially contain a vector of what are called *bias errors* that are *literally* the same for all systems in the ensemble at each time instant.

◊ These bias errors in the observations then give rise to bias errors in the estimates which are also all the same across the ensemble.

◊ It is also possible for bias errors to be present in the estimates, even when there are no bias errors in the observations.

In Chapter 5 we examined the random errors in both the observations and the estimates, and from time to time we will add to the comments made there. However, in this chapter our primary focus will be on bias errors – what they are and how they come about.

In Chapters 7 and 10 we will consider how bias errors can often be detected.

6.3 Bias errors in the observations

We consider first the bias errors in the observations.

In Chapters 3 and 5 we made use of the total observation equation

$$\boldsymbol{Y} = \boldsymbol{T}\boldsymbol{X} + \boldsymbol{N} \tag{6.3.1}$$

In this equation $\boldsymbol{Y}$ is the total observation vector, $\boldsymbol{T}$ is the total observation matrix, $\boldsymbol{X}$ is the true state vector of the observed process and $\boldsymbol{N}$ is the vector of errors in $\boldsymbol{Y}$.

In Chapter 5 we assumed that the elements of $\boldsymbol{N}$ are random variables that are zero-mean, and so

$$\mathcal{E}\{\boldsymbol{N}\} = \emptyset \tag{6.3.2}$$

However, it is possible under certain circumstances that the observations will be offset by constant amounts, and it is these constant amounts that are called the *bias errors*.

As an example, suppose that the device that measures the azimuth angle of a radar antenna has been mounted incorrectly by *0.1 degrees*. Then, in addition to the zero-mean errors in the azimuth angles, there will also be a constant offset of this amount. This offset would constitute a bias error.

Under such circumstances the total error vector $\boldsymbol{N}$ would consist of *two* parts, namely

◊ A random error component that we call $\boldsymbol{N}_{random}$. When considered across the ensemble at a single time instant, this vector is zero-mean.

◊ A bias error component that we call $\boldsymbol{N}_{bias}$. When considered across the ensemble at a single time instant, this vector is the same for all members of the ensemble.

We will therefore write the error vector $\boldsymbol{N}$ as

$$\boldsymbol{N} = \boldsymbol{N}_{random} + \boldsymbol{N}_{bias} \tag{6.3.3}$$

and the total observation equation in (6.3.1) as

$$\boldsymbol{Y} = \boldsymbol{TX} + \boldsymbol{N}_{random} + \boldsymbol{N}_{bias} \tag{6.3.4}$$

Taking the expectation of (6.3.3) we obtain

$$\mathcal{E}\{\boldsymbol{N}\} = \boldsymbol{N}_{bias} \tag{6.3.5}$$

and so when bias errors are present, the total error vector $\boldsymbol{N}$ has the bias vector $\boldsymbol{N}_{bias}$ as its mean.

6.3.1 Calibration and bore-sighting

When observation instruments are put into service, every effort is made to reduce the bias vector N_{bias} to zero, or at least to acceptably small levels.

In radar engineering this procedure is referred to as *bore-sighting*, and in Project 6.4, where we discuss bore-sighting in slightly greater detail, we point out that it involves a number of activities:

◊ The radar's physical location must be carefully established.

◊ The antenna must be carefully aligned in relation to north, east and down.

◊ The values that the radar puts out must be carefully calibrated:

  ◊ The radar's slant-range outputs must be compared to the physical range between its own location and one or more reflecting objects that have been placed at distant points.

  ◊ The same must be done for the radar's azimuth and elevation outputs.

However, even with perfect bore-sighting, there is another potential source of bias errors in the observations.

The observation vectors $Y_n, Y_{n-1}, \ldots Y_{n-L}$ that make up the total observation vector $\boldsymbol{Y}$ are made directly on the observed trajectory. It is we who then introduce the concept of a filter model in the form of a DE, as our way of modelling what we are observing.

If the DE of the filter model that we select differs from the DE of the external process, then by *our* standards (i.e. by our filter model) the observations could be biased. Thus in addition to the assumption that bore-sighting has been properly carried out, we must also be sure to select a filter model which is a good representation of the external process.

Example 6.1: The external model is Newton's DE for orbital motion

$$D^2 \boldsymbol{r} = -(\mu/r^3)\boldsymbol{r} \tag{6.3.6}$$

and the filter model is the same DE.

Then bias errors could be present because of poor bore-sighting, but they cannot be present because of a mismatch between external and filter models since the two are the same. ■■

Example 6.2: The external model is Newton's DE for orbital motion

$$D^2 \boldsymbol{r} = -(\mu/r^3)\boldsymbol{r} \tag{6.3.7}$$

and the filter model is the *3*-dimensional DE for 2nd-degree polynomials.

This situation could arise if we made the decision to approximate the orbit by 2nd-degree polynomials over a limited time interval.

Bias errors could now be present, *both* because of poor bore-sighting *and* because there is a mismatch between the external and filter models. ■■

Example 6.3: The external model is the set of equations for a point object moving in *3*-space with constant radial acceleration (i.e. on a circular flight path), and the filter model is the *3*-dimensional DE for 2nd-degree polynomials.

This situation arises in Chapter 10 where we discuss the Master Control Algorithm (MCA). We will see that the MCA uses either the *3*-dimensional DE for 1st-degree polynomials or the *3*-dimensional DE for 2nd-degree polynomials for tracking manoeuvring targets.

Bias errors could now be present *both* because of poor bore sighting *and* because there is a mismatch between the external and filter models. ■■

◀▶

In Appendix 6.1 we cite a few paragraphs from Reference 4 in which Gauss, in his own words, speaks about calibration, and about the random and bias errors in the observations. His remarkable insights make for fascinating reading.

6.4 Errors in the estimate

In the preceding section we focused our attention on errors in the observations. Consider now the errors in the estimate.

When the total observation vector $\boldsymbol{Y}$ is submitted to a filter, the estimate vector $\boldsymbol{X}^*$ that emerges will contain an error vector that we call $\boldsymbol{N}^*$. As with the observations considered earlier, the errors that make up $\boldsymbol{N}^*$ will be of two kinds:

◊ ***Random errors in the estimate:*** There will be a random component that we call $\boldsymbol{N}^*_{random}$. These vary for each member of the ensemble, but taken as a set across the ensemble at each time instant they all share the same PDF which is zero-mean.

  In Chapter 5 we derived the equation by which $\boldsymbol{N}^*_{random}$ is produced from $\boldsymbol{N}_{random}$, namely

$$\boldsymbol{N}^*_{random} = \boldsymbol{W}\boldsymbol{N}_{random} \tag{6.4.1}$$

◊ ***Bias errors in the estimate:*** There could also be a bias component that we call $\boldsymbol{N}^*_{bias}$ which is the same for all members of the ensemble. Lower down we will derive the equation by which $\boldsymbol{N}^*_{bias}$ is produced from $\boldsymbol{N}_{bias}$.

6.4.1 The random errors in the estimate

We discussed the random errors in the estimate in Chapter 5, and we examined how they are caused by the random errors in the observations. Everything that we did there was based on the assumption that the observation errors were zero-mean.

We can summarize what we derived there, as follows:

In Figure 6.2 we show a simplified block diagram of a filtering system.

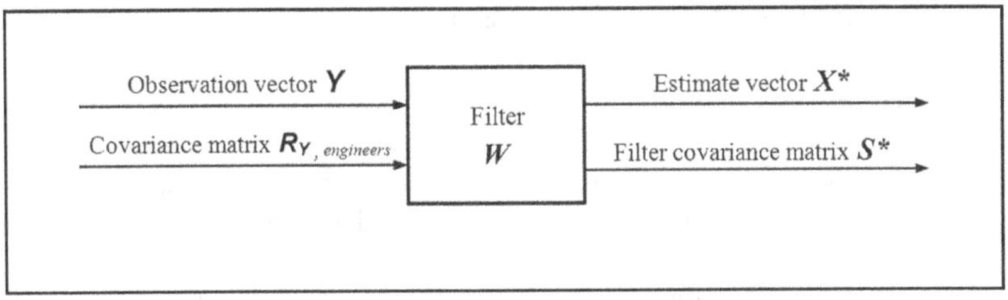

Figure 6.2 Filter inputs and outputs

The observation vector $\boldsymbol{Y}$ produces the estimate vector $\boldsymbol{X}^*$, and the random errors in $\boldsymbol{Y}$ produce the random error vector N^*_{random} that is contained in $\boldsymbol{X}^*$.

The random errors in $\boldsymbol{Y}$ have an actual covariance matrix $\boldsymbol{R}_{\boldsymbol{Y},\ actual}$ that is unknown to us except in a simulation. In its place we use $\boldsymbol{R}_{\boldsymbol{Y},\ engineers}$ that is provided by the engineers who built the observation instrument – a matrix that must be accepted by the filter engineers as unquestionably correct.

In Chapter 5 we saw that the formula for how $\boldsymbol{R}_{\boldsymbol{Y},\ engineers}$ is transformed into the filter covariance matrix $\boldsymbol{S}^*$ is as follows:

$$\boldsymbol{S}^* = \boldsymbol{W}\boldsymbol{R}_{\boldsymbol{Y},\ engineers}\,\boldsymbol{W}^T \tag{6.4.2}$$

This equation, which was based on Theorem 4.1, will be discussed further in Chapters 8 and 9.

What we would *really* like is for the filter to provide us with successive values of the actual estimation-error covariance matrix $\boldsymbol{S}_{actual}$. However, it takes a Monte-Carlo to obtain a single estimate of that matrix at a specific point in time, and so it is impossible to obtain such successive values.

In the field we are therefore forced to use the stream of matrices $\boldsymbol{S}^*$ emanating from the filter according to (6.4.2) as the representations of successive values of $\boldsymbol{S}_{actual}$. We must therefore hope that the filter is always ECM consistent, thereby ensuring that those $\boldsymbol{S}^*$ matrices are good representations of $\boldsymbol{S}_{actual}$.

6.4.2 The bias errors in the estimate

In equation (6.3.3), we saw that the observation error vector $\boldsymbol{N}$ is potentially comprised of two components, the random errors $\boldsymbol{N}_{random}$ and the bias error $\boldsymbol{N}_{bias}$.

Further down we will derive the equation by which $\boldsymbol{N}_{bias}$ gives rise to bias errors in the estimate. However, bias errors in the estimate can arise even when there are none in the observations, from two other possible sources.

1. The filter matrix $\boldsymbol{W}$ is not properly constructed.

2. The filter model that we are using differs from the external model.

The first item will be examined in the next section, where we show that the matrix $\boldsymbol{W}$ must satisfy an important constraint if it is not to introduce bias errors in the estimate.

The second item was mentioned earlier and will be examined again in Section 6.7, where we derive the expression for the bias errors that can arise when the filter model differs from the external model.

6.5 The exactness constraint

The filters that we consider in this book all have the general form

$$\boldsymbol{X}^*_{n,n} = \boldsymbol{W}_n \boldsymbol{Y}_n \tag{6.5.1}$$

As stated in Chapter 5, equation (6.5.1) may not always be the way in which a filter is actually implemented, but it implicitly always lies behind every implementation.

In (6.5.1) the dimension of $\boldsymbol{Y}_n$ is always greater than the dimension of $\boldsymbol{X}^*_{n,n}$ and so the filter matrix $\boldsymbol{W}_n$ is always wide.

Any such matrices $\boldsymbol{W}_n$ that we use will have to possess some very specific properties. We will identify others in later chapters, but one of them is as follows:

Assuming that the errors in $\boldsymbol{Y}_n$ are zero-mean, we expect that the estimate vector $\boldsymbol{X}^*_{n,n}$ in (6.5.1) shall be an unbiased estimate of $\boldsymbol{X}_n$, i.e. that

$$\mathcal{E}\{\boldsymbol{X}^*_{n,n}\} = \boldsymbol{X}_n \tag{6.5.2}$$

There is a constraint on $\boldsymbol{W}_n$ which ensures that (6.5.2) will be true, which we state as follows:

Theorem 6.1: Let $\boldsymbol{Y}_n$ be related to $\boldsymbol{X}_n$ by the total observation equation

$$\boldsymbol{Y}_n = \boldsymbol{T}_n \boldsymbol{X}_n + \boldsymbol{N}_n \tag{6.5.3}$$

in which the errors in $\boldsymbol{Y}_n$ are zero-mean, i.e.

$$\mathcal{E}\{\boldsymbol{N}_n\} = \emptyset \tag{6.5.4}$$

Let the estimate vector $\boldsymbol{X}^*_{n,n}$ be derived from $\boldsymbol{Y}_n$ by the filter equation

$$\boldsymbol{X}^*_{n,n} = \boldsymbol{W}_n \boldsymbol{Y}_n \tag{6.5.5}$$

Then $X^*_{n,n}$ will be an unbiased estimate of X_n if and only if

$$W_n T_n = I \tag{6.5.6}$$

Proof: Combining (6.5.3) and (6.5.5) gives

$$X^*_{n,n} = W_n (T_n X_n + N_n) = W_n T_n X_n + W_n N_n \tag{6.5.7}$$

Taking the expectation then gives

$$\mathcal{E}\{X^*_{n,n}\} = W_n T_n X_n + W_n \mathcal{E}\{N_n\} \tag{6.5.8}$$

and by (6.5.4) this reduces to

$$\mathcal{E}\{X^*_{n,n}\} = W_n T_n X_n \tag{6.5.9}$$

Then $\mathcal{E}\{X^*_{n,n}\}$ will be equal to X_n if and only if

$$W_n T_n = I \tag{6.5.10}$$

■■

Equation (6.5.10) applies to every filter that we may create, and it is one of the pillars on which we build the Gauss filters in later chapters.

We note that both W_n and T_n in (6.5.10) are subscripted with an n. This means that as T_n changes when the filter is cycled then W_n must change accordingly so that (6.5.10) is always satisfied. We return to this in Chapter 8.

Definition 6.2: Equation (6.5.10) is called the **exactness constraint**. ■■

The following is an important consequence to the exactness constraint:

Theorem 6.2: Let the matrices W and T satisfy

$$WT = I \tag{6.5.11}$$

Then W must have full row-rank and T full column-rank. ■■

Proof of Theorem 6.2 is given in Appendix 6.2.

Since we will always use W matrices that satisfy the exactness constraint, it follows that they will always have full row-rank.

6.6 The infinitely many W matrices that satisfy the exactness constraint

There are infinitely many matrices $\boldsymbol{W}$ that satisfy the exactness constraint relative to a given $\boldsymbol{T}$, and we now consider how they arise. First, we restate the exactness constraint in its transposed form as follows:

$$\boldsymbol{T}^T \boldsymbol{W}^T = \boldsymbol{I} \tag{6.6.1}$$

In (6.6.1) the matrix $\boldsymbol{T}^T$ can be regarded as known and the matrix $\boldsymbol{W}^T$ as unknown.

The matrices $\boldsymbol{W}$ are $m \times k$ and $\boldsymbol{T}$ is $k \times m$ with $k > m$ and as a simple example let $k = 3$ and $m = 2$. Then (6.6.1) becomes

$$\begin{pmatrix} t_{1,1} & t_{2,1} & t_{3,1} \\ t_{1,2} & t_{2,2} & t_{3,2} \end{pmatrix} \begin{pmatrix} w_{1,1} & w_{2,1} \\ w_{1,2} & w_{2,2} \\ w_{1,3} & w_{2,3} \end{pmatrix} = \begin{pmatrix} 1 & 0 \\ 0 & 1 \end{pmatrix} \tag{6.6.2}$$

This can now be rearranged as follows:

$$\begin{pmatrix} t_{1,1} & t_{2,1} & t_{3,1} & 0 & 0 & 0 \\ 0 & 0 & 0 & t_{1,1} & t_{2,1} & t_{3,1} \\ t_{1,2} & t_{2,2} & t_{3,2} & 0 & 0 & 0 \\ 0 & 0 & 0 & t_{1,2} & t_{2,2} & t_{3,2} \end{pmatrix} \begin{pmatrix} w_{1,1} \\ w_{1,2} \\ w_{1,3} \\ w_{2,1} \\ w_{2,2} \\ w_{2,3} \end{pmatrix} = \begin{pmatrix} 1 \\ 0 \\ 0 \\ 1 \end{pmatrix} \tag{6.6.3}$$

Equation (6.6.3) comprises four equations for the six unknowns in $\boldsymbol{W}$, and so it is under-constrained. There will therefore be infinitely many solutions for $\boldsymbol{W}$ with the following structure[2]:

$$\boldsymbol{W} = \boldsymbol{G}_p + a\boldsymbol{G}_{h_1} + b\boldsymbol{G}_{h_2} \tag{6.6.4}$$

◊ The matrices $\boldsymbol{G}_p$, $\boldsymbol{G}_{h_1}$ and $\boldsymbol{G}_{h_2}$ in (6.6.4) are made up of fixed constants that depend on the elements of $\boldsymbol{T}$, and in this case they are 2×3.

[2] See e.g. Ref. 133.

◊ $\boldsymbol{G}_p$ is called the *particular solution*.

◊ $\boldsymbol{G}_{h_1}$ and $\boldsymbol{G}_{h_2}$ make up what is called the *homogeneous solution*.

◊ There are $mk - m^2$ matrices in the homogeneous solution. In this case $m = 2$ and $k = 3$ and so $mk - m^2 = 2$. Thus there are two matrices in the homogeneous solution.

◊ The constants a and b can assume any values on the real line.

In this way we see that there are infinitely many matrices $\boldsymbol{W}$, all of them satisfying the exactness constraint relative to the given $\boldsymbol{T}$, and by Theorem 6.2, all of them have full row-rank which in this example is $m = 2$.

In Chapters 8 and 9, we will return to the infinitely-many matrices $\boldsymbol{W}$ that satisfy the exactness constraint relative to a given $\boldsymbol{T}$, and we will see there that there exists *one* particular $\boldsymbol{W}$ among them called $\boldsymbol{W}_{minvar}$ which lies at the very heart of tracking filters – be they Gauss–Newton, Swerling or Kalman.

6.7 When the observation vector is biased

In going from (6.5.8) to (6.5.9) we assumed that the errors in the observation vector $\boldsymbol{Y}_n$ were zero-mean, and so we used $\mathcal{E}\{\boldsymbol{N}_n\} = \boldsymbol{\emptyset}$. However, if a bias vector is present in $\boldsymbol{Y}_n$ then it will give rise to a bias vector in the estimate as follows:

Combining (6.5.5) with (6.3.4) gives

$$\begin{aligned} \boldsymbol{X}^*_{n,n} &= \boldsymbol{W}_n \boldsymbol{Y}_n = \boldsymbol{W}_n (\boldsymbol{T}_n \boldsymbol{X}_n + \boldsymbol{N}_{random,n} + \boldsymbol{N}_{bias,n}) \\ &= \boldsymbol{W}_n \boldsymbol{T}_n \boldsymbol{X}_n + \boldsymbol{W}_n \boldsymbol{N}_{random,n} + \boldsymbol{W}_n \boldsymbol{N}_{bias,n} \end{aligned} \tag{6.7.1}$$

Taking the expectation, and recalling that $\mathcal{E}\{\boldsymbol{N}_{random,n}\} = \boldsymbol{\emptyset}$ gives

$$\mathcal{E}\{\boldsymbol{X}^*_{n,n}\} = \boldsymbol{W}_n \boldsymbol{T}_n \boldsymbol{X}_n + \boldsymbol{W}_n \boldsymbol{N}_{bias,n} \tag{6.7.2}$$

and when the exactness constraint is satisfied this reduces further to

$$\mathcal{E}\{\boldsymbol{X}^*_{n,n}\} = \boldsymbol{X}_n + \boldsymbol{W}_n \boldsymbol{N}_{bias,n} \tag{6.7.3}$$

Thus if the observations contain a bias, then even with the exactness constraint on $\boldsymbol{W}$ being satisfied, the estimate $\boldsymbol{X}^*$ will contain the following bias vector

$$\boldsymbol{N}^*_b = \boldsymbol{W} \boldsymbol{N}_{bias,n} \tag{6.7.4}$$

Accordingly, we should not think of the exactness constraint as the only requirement for an unbiased estimate – we also require that the observations be unbiased.

6.7.1 Biased observations and the test for goodness-of-fit

In Chapter 1 we mentioned eight performance monitors, one of which is the goodness-of-fit (GOF) test. The full discussion of the GOF test takes place in Chapter 10 and its appendices, but for now we make a few preliminary comments.

The GOF test evaluates the fit between the observation vector $\mathbf{Y}$ and the curve implied by the estimate vector X^*, and if the fit is good then the test will pass.

However, passing only the GOF test is not sufficient to ensure that the estimate vector X^* is unbiased. The vector $\mathbf{Y}$ must also be unbiased.

To understand this better, consider the following (admittedly extreme) example in which there is a positive bias of $90°$ in every azimuth measurement obtained from a radar.

Then a target which is actually due north will be reported as being due east, and even though the GOF test passes we will have no way of knowing that this bias is present unless we attempt to engage the target.

Clearly then, for us to be able to say that X^* is an unbiased estimate, *two* conditions must be present:

1. The GOF test must pass.
2. The total observation vector must be unbiased.

We will discuss the GOF test fully in Chapter 10 and the ways in which it can be used. We will also examine how bias errors can affect the GOF test as well as how they can affect the *3*-sigma and Chi-squared tests that are discussed in Chapter 7.

6.8 The filter model differs from the external model

In Chapter 3 we considered the matrix $\mathbf{T}$ which relates the total observation vector $\mathbf{Y}$ to the true state vector X as follows:

$$\mathbf{Y} = \mathbf{T}X + \mathbf{N} \tag{6.8.1}$$

In Chapter 8 we will see that the matrix $\mathbf{T}$ is used to build the filter matrix W according to

$$W = (\mathbf{T}^T\mathbf{R}_\mathbf{Y}^{-1}\mathbf{T})^{-1}\mathbf{T}^T\mathbf{R}_\mathbf{Y}^{-1} \tag{6.8.2}$$

We stated above that if the filter model differs from the external model, then bias errors in the estimate can arise, even when there are none in the observations.

There are situations in which, by deliberate choice, the filter model *will* differ from the external model, and in that case the matrix $\boldsymbol{T}$ that is based on the transition matrix for the filter model (as in Chapter 3) will differ from the matrix $\boldsymbol{T}$ that could be based on the transition matrix for the external model.

Consider the following two examples.

Example 6.4: This situation was considered in Example 6.2. The external model is the DE for orbital motion, and the filter model is the (*3*-dimensional) DE for 2nd-degree polynomials. The $\boldsymbol{T}$ matrix would then be based on the transition matrix of the latter.

This situation could arise if we made the decision to approximate the orbit by 2nd-degree polynomials over an acceptable time interval.

In the program ***14_Orbit_Track*** this is exactly what is done, so that high-data-rate observations can be prefiltered by the polynomial filters discussed in Chapters 12 and 13.

This conserves machine resources, and we then need only cycle the main filters – Gauss–Newton, Swerling or Kalman – from time and to time, rather than at the high data-rate of the radar's prf.

Prefiltering also provides data of improved quality for the main filters. ■■

Example 6.5: The external model is the nonlinear DE in (2.1.36) for a sinusoid whose frequency is unknown, and the filter model is the (*1*-dimensional) DE for 2nd-degree polynomials. Again the $\boldsymbol{T}$ matrix would then be based on the transition matrix of the latter.

This situation could arise if we made the decision to approximate the sinusoid by a sequence of 2nd-degree polynomials.

This is done in the program ***13_Gauss_Newton_C*** so that high-data-rate observations can be prefiltered by the polynomial filters discussed in Chapters 12 and 13, again in order to conserve machine resources and only cycle the main filters – Gauss–Newton, Swerling or Kalman with their sinusoidal filter models – from time and to time, rather than at the high data-rate of the observations. ■■

This deliberate mismatch between the external and filter models is analysed in Project 6.5 where we show the following:

Theorem 6.3:

◊ Let $\boldsymbol{X}_{e,t_n}$ be the true state vector of the external trajectory.

◊ Let $\boldsymbol{T}_e$ be the total observation matrix based on the transition matrix for the external model.

◊ Let $\boldsymbol{W}_f$ be the filter matrix derived from the $\boldsymbol{T}$ matrix which is based on the transition matrix for the filter model.

Then the bias error vector in the filter estimate $\boldsymbol{X}^*_{n,n}$ will be

$$\boldsymbol{E}_{bias\ in\ \boldsymbol{X}^*_{n,n}} = (\boldsymbol{W}_f\boldsymbol{T}_e - \boldsymbol{I})\boldsymbol{X}_{e,t_n} \tag{6.8.3}$$

■■

If the filter model is identical to the external model, then $\boldsymbol{T}_e$ in (6.8.3) will be the same as the $\boldsymbol{T}$ matrix that was used to build $\boldsymbol{W}_f$. In that case the product $\boldsymbol{W}_f\boldsymbol{T}_e$ will be equal to $\boldsymbol{I}$ and $\boldsymbol{E}_{bias\ in\ \boldsymbol{X}^*_{n,n}}$ in (6.8.3) will be **Ø**.

However, if the filter model differs from the external model, then $\boldsymbol{T}_e$ in (6.8.3) will not be the same as the $\boldsymbol{T}$ matrix that was used to build $\boldsymbol{W}$. The product $\boldsymbol{W}_f\boldsymbol{T}_e$ will then not be equal to $\boldsymbol{I}$, and $\boldsymbol{E}_{bias\ in\ \boldsymbol{X}^*_{n,n}}$ in (6.8.3) will not be equal to **Ø**.

Theorem 6.3 is validated in a most convincing way in Problems 6.4 and 6.5 (together with the two computer programs referred to there).

Note: The matrix $\boldsymbol{W}_f$ may have more rows than $\boldsymbol{T}_e$ has columns, in which case we must neglect the extra rows in $\boldsymbol{W}_f$ so that the bias error in (6.8.3) is defined. (Neglecting rows in $\boldsymbol{W}_f$ means that we refrain from estimating some items).

As an example, suppose that we are approximating a sinusoid by a 2nd-degree polynomial. Then $\boldsymbol{W}_f$ will normally have three rows because we would normally use it to estimate x, $\dot{x}$ and $\ddot{x}$. However, the state vector in the external model DE for sinusoidal motion involves only x and $\dot{x}$ (see (2.4.22)) and so the matrix $\boldsymbol{T}_e$ would have only two columns.

Discarding the third row of $\boldsymbol{W}_f$ means that the filter is only estimating x and $\dot{x}$.

This is considered in greater detail in Projects 6.5 and 6.6.

6.9 Demonstration of bias errors

Three demonstrations of bias errors are provided in the supplementary material.

1. In Project 6.1 we use classical least squares as the estimation method.[3]

   By spending a few minutes to work through it, you will

   ◊ Familiarize yourself with the method of classical least squares (if you don't already know what it is all about).

   ◊ See a numerical demonstration of the following:

      ◊ There are no bias errors when the filter model is perfectly matched to the external model.

      ◊ Bias errors are clearly present when the match is not good.

2. In the program ***02_Bias_Errors*** we use polynomials of degrees *0* to *4* to estimate sections of sinusoidal waveforms. The program then displays the resulting bias errors that arise because the filter model and the external model are not the same.

   As the degree of the polynomial filter model is increased, the shrinking of the bias errors is clearly apparent.

3. In the programs ***03_Gauss_Aitken*** and ***13_Gauss_Newton_C*** we enable you to add bias errors of various magnitudes to the observation vectors. Because many of the items in these programs are only discussed in later chapters, we defer further discussion until Chapter 10.

6.10 Summary

For a filter to produce unbiased estimates, the following conditions must all be present:

◊ The external model must be a good representation of the physical process.

◊ The filter model must either be the same as the external model, or it must emulate it closely.

[3] We have included the project in full detail so that anybody not familiar with the method of least squares will be able to see how it works. We will refer to that project from time to time.

◊ The observation instrument must be properly calibrated (bore-sighted) so that the total observation vector $\boldsymbol{Y}_n$ is acceptably free of bias errors.

◊ Cycle by cycle, the filter matrix $\boldsymbol{W}_n$ must satisfy the exactness constraint relative to the matrix $\boldsymbol{T}_n$ that has been incorporated into the filter.

In the above discussion we elected to consider bias errors in only the simplest case, namely ***Case 1*** for which both the observation equation and the filter model are linear.

However, everything that we have derived applies with only minor changes to the other three cases in which nonlinearities are present.

The major difference is that in ***Case 1*** above we worked with whole vectors, whereas in ***Cases 2***, ***3*** and ***4*** we would be working with perturbation vectors.

The following items are in the supplementary material

Problems for Chapter 6

Solutions to Problems for Chapter 6

Projects for Chapter 6

Appendix 6.1

Carl Friedrich Gauss on random and bias errors[1]

However carefully one takes observations of the magnitudes of objects in nature, the results are always subject to larger or smaller errors. In general theses errors are not simple but arise from many sources acting together. Two of these must be carefully distinguished.

Certain causes of error are such that their effect on any one observation depends on varying circumstances that seem to have no essential connection with the observation itself. Errors arising in this way are called irregular or random, and they are no more subject to calculation than the circumstances on which they depend. Such errors come from the imperfections of our senses and random external causes, as when shimmering air disturbs our fine vision. Many defects in instruments, even the best, fall into this category, e.g. a roughness in the inner part of a level, a lack of absolute rigidity, etc.

On the other hand, other sources of error by their nature have a constant effect on all observations of the same class. Or if the effect is not absolutely constant, its size varies regularly with circumstances that are essentially connected with the observations. These errors are called constant or regular.

Now it is clear that this distinction is to some extent relative and depends on how broadly we take the notion of observations of the same class. For example, consider irregularities in the graduations of an instrument for measuring angles. If we need to measure a given angle again and again, then the irregularities produce constant errors, since the same defective graduation is used repeatedly. On the other hand, the same errors can be regarded as random when one is measuring unknown angles of arbitrary magnitude, since there is no table of the errors in the individual graduations.

We explicitly exclude the considerations of regular errors from this investigation. Of course, it is up to the observer to ferret out all sources of constant error and remove them. Failing that, he should at least scrutinize their origins and magnitudes, so that their effects on any given observation can be determined and removed, after which it will be the same as if the error had never occurred.

[1] Excerpted from Reference 4, pp. 3 and 4 and reproduced here with the kind permission of the Society for Industrial and Applied Mathematics (SIAM), Philadelphia, USA.

Irregular errors are essentially different, since by their nature they are not subject to calculation. For this reason we have to put up with them in the observations themselves; however, we should reduce their effects on derived quantities as far as possible by using *judicious combinations*[2] of the observations.

[2] Italics by the author. By 'judicious combinations', Gauss is referring to filter engineering techniques similar to those explored in this book.

Appendix 6.2
Proof of Theorem 6.2

In Chapter 6 we derived the following relationship:

$$\boldsymbol{W}\boldsymbol{T} = \boldsymbol{I} \tag{A6.2.1}$$

This equation is called the *exactness constraint* and it ensures that a filter's estimate vector will be unbiased if the observation vector is unbiased. Accordingly, when building filters we will use only $\boldsymbol{W}$ matrices that satisfy (A6.2.1).

There is an important consequence to (A6.2.1), and that is the following:

If two matrices $\boldsymbol{W}$ and $\boldsymbol{T}$ are capable of satisfying $\boldsymbol{W}\boldsymbol{T} = \boldsymbol{I}$ then we can immediately conclude that $\boldsymbol{W}$ has full row-rank and $\boldsymbol{T}$ has full column-rank.

We state this as follows:

***Theorem 6.2**:* Let the matrices $\boldsymbol{W}$ and $\boldsymbol{T}$ satisfy

$$\boldsymbol{W}\boldsymbol{T} = \boldsymbol{I} \tag{A6.2.2}$$

Then $\boldsymbol{W}$ must have full row-rank and $\boldsymbol{T}$ full column-rank. ■■

***Proof 1**:* As an example, let $\boldsymbol{W}$ be 3×6, let $\boldsymbol{T}$ be 6×3, and let $\boldsymbol{I}$ be 3×3.

Suppose, for contradiction, that (A6.2.1) is satisfied, but that $\boldsymbol{W}$ has less than full row-rank and so its rows are linearly dependent. Then there exists a *3*-dimensional non-zero row-vector $\boldsymbol{a}$ such that

$$\boldsymbol{a}\boldsymbol{W} = \boldsymbol{\emptyset}_6 \tag{A6.2.3}$$

in which $\boldsymbol{\emptyset}_6$ is the *6*-dimensional zero row-vector. We now premultiply the left side of (A6.2.1) by $\boldsymbol{a}$ obtaining

$$\boldsymbol{a}\boldsymbol{W}\boldsymbol{T} = \boldsymbol{\emptyset}_3 \tag{A6.2.4}$$

in which $\boldsymbol{\emptyset}_3$ is the *3*-dimensional zero row-vector.

On the other hand, if we premultiply the right side of (A6.2.1) by $\boldsymbol{a}$ we obtain

$$\boldsymbol{aI} = \boldsymbol{a} \tag{A6.2.5}$$

This is a contradiction, and so the only way in which (A6.2.3) can be trues is if $\boldsymbol{a}$ is the *3*-dimensional zero row-vector, which in turn means that the rows of $\boldsymbol{W}$ must be linearly independent, and so $\boldsymbol{W}$ must have full row-rank.

By a similar argument we post-multiply both sides of (A6.2.1) by a column vector $\boldsymbol{b}$ and show that $\boldsymbol{T}$ must have full column-rank. The proof is then complete. ■■

Corollary: If the matrices $\boldsymbol{W}$ and $\boldsymbol{T}$ satisfy $\boldsymbol{WT} = \boldsymbol{I}$, then one of the following must be true:

◊ $\boldsymbol{W}$ and $\boldsymbol{T}$ are both square.

◊ $\boldsymbol{W}$ is wide and $\boldsymbol{T}$ is high.

It is not possible for $\boldsymbol{W}$ to be high and $\boldsymbol{T}$ to be wide.

Proof follows from the fact that if $\boldsymbol{WT} = \boldsymbol{I}$ then the rows of $\boldsymbol{W}$ must be linearly independent, and that is not possible if $\boldsymbol{W}$ is high. Similarly, the columns of $\boldsymbol{T}$ must be linearly independent, and that cannot be possible if $\boldsymbol{T}$ is wide.

Proof 2: As an example, let $\boldsymbol{W}$ be 3×6, let $\boldsymbol{T}$ be 6×3 and let $\boldsymbol{I}$ be 3×3.

Let the row-vectors of $\boldsymbol{W}$ be called $\boldsymbol{w}_1$, $\boldsymbol{w}_2$ and $\boldsymbol{w}_3$, and let the column vectors of $\boldsymbol{T}$ be called $\boldsymbol{t}_1$, $\boldsymbol{t}_2$ and $\boldsymbol{t}_3$.

Then

$$\boldsymbol{W} = \begin{pmatrix} \boldsymbol{w}_1 \\ \boldsymbol{w}_2 \\ \boldsymbol{w}_3 \end{pmatrix} \quad \text{and} \quad \boldsymbol{T} = (\boldsymbol{t}_1 \ \boldsymbol{t}_2 \ \boldsymbol{t}_3) \tag{A6.2.6}$$

and the statement $\boldsymbol{WT} = \boldsymbol{I}$ becomes

$$\begin{pmatrix} \boldsymbol{w}_1\boldsymbol{t}_1 & \boldsymbol{w}_1\boldsymbol{t}_2 & \boldsymbol{w}_1\boldsymbol{t}_3 \\ \boldsymbol{w}_2\boldsymbol{t}_1 & \boldsymbol{w}_2\boldsymbol{t}_2 & \boldsymbol{w}_2\boldsymbol{t}_3 \\ \boldsymbol{w}_3\boldsymbol{t}_1 & \boldsymbol{w}_3\boldsymbol{t}_2 & \boldsymbol{w}_2\boldsymbol{t}_3 \end{pmatrix} = \begin{pmatrix} 1 & 0 & 0 \\ 0 & 1 & 0 \\ 0 & 0 & 1 \end{pmatrix} \tag{A6.2.7}$$

From the products on the diagonal it follows that none of $\boldsymbol{w}_1$, $\boldsymbol{w}_2$, $\boldsymbol{w}_3$ or $\boldsymbol{t}_1$, $\boldsymbol{t}_2$, $\boldsymbol{t}_3$ can be zero vectors.

Suppose now that the rows of $\boldsymbol{W}$ are linearly dependent, that is that there exist constants a, b, c, not all zero, such that

$$a\boldsymbol{w}_1 + b\boldsymbol{w}_2 + c\boldsymbol{w}_3 = \boldsymbol{\emptyset}_6 \tag{A6.2.8}$$

Let a be non-zero. Then from (A6.2.8)

$$\boldsymbol{w}_1 = -\left(\frac{b}{a}\right)\boldsymbol{w}_2 - \left(\frac{c}{a}\right)\boldsymbol{w}_3 \tag{A6.2.9}$$

and so the product $\boldsymbol{w}_1\boldsymbol{t}_2$ becomes

$$\boldsymbol{w}_1\boldsymbol{t}_2 = -\left(\frac{b}{a}\right)\boldsymbol{w}_2\boldsymbol{t}_2 - \left(\frac{c}{a}\right)\boldsymbol{w}_3\boldsymbol{t}_2 = -\left(\frac{b}{a}\right)1 - \left(\frac{c}{a}\right)0 = -\frac{b}{a} \tag{A6.2.10}$$

But from (A6.2.7) $\boldsymbol{w}_1\boldsymbol{t}_2 = 0$, and so (A6.2.10) means that $b = 0$, and by a similar argument using $\boldsymbol{w}_1\ \boldsymbol{t}_3$ we could show that $c = 0$.

But then by (A6.2.8), $\boldsymbol{w}_1$ must be the zero vector, which is a contradiction.

Thus the only way in which (A6.2.8) can be true is if all of a, b and c are zero, making the row-vectors $\boldsymbol{w}_1$, $\boldsymbol{w}_2$ and $\boldsymbol{w}_3$ linearly independent, and so $\boldsymbol{W}$ has full row-rank.

A similar argument shows that $\boldsymbol{T}$ has full column-rank, and the proof is complete.

■■

Chapter 7
Three tests for ECM consistency

There is little point in providing an estimate if we do not also give an indication of its accuracy. For this reason a complete filter always includes a filter covariance matrix.

However, it doesn't end there. Critical decisions are often based on that matrix, and so providing it is only the first step.

We must also be as certain as possible that the covariance matrix provided by the filter really *does* match the actual estimation-error covariance matrix – i.e. that the filter is ECM consistent.

In this chapter, we focus on three tests that determine if ECM consistency is present or absent.

◊ The first – called the *matrix-to-matrix ECM test* – is a Monte-Carlo type test.

◊ The second and third – called the *3-sigma* and the *Chi-squared ECM tests* – can be run both as Monte-Carlo and as single-shot tests.[1]

◊ All three tests can only be used in simulations.

In Chapter 9 we consider three tests for CR-consistency. Like the three ECM-consistency tests, those tests can also only be used in simulations.

In Chapter 10 we examine the *goodness-of-fit (GOF) test* that enables us to determine if the estimate vector is biased or not, and in Chapter 12 we examine the *sigma monitoring (SM) test* that enables us to keep an eye on the performance of the observation instrument.

These last two tests can be used in the field as well as in simulations.

7.1 End-to-end

When running a simulation it is a simple matter to derive the actual estimation error vector, defined as

$$\boldsymbol{N}^* = \boldsymbol{X}^* - \boldsymbol{X} \tag{7.1.1}$$

[1] See comments in Section 1.5 on *Monte-Carlo* and *single-shot.*

This definition of $\boldsymbol{N}^*$ is *end-to-end*, by which we mean that $\boldsymbol{X}^*$ is the final estimate vector and $\boldsymbol{X}$ is the true state vector of the observed trajectory.

An error-vector which is not end-to-end could arise when filtering with the cascaded structure shown in Figure 7.1 below.

Here Filter 1 is acting as a prefilter for Filter 2. If we considered only the errors between the input and output of Filter 2 we would write the error equation as

$$\boldsymbol{N}^* = \boldsymbol{X}^*_2 - \boldsymbol{X}^*_1 \tag{7.1.2}$$

which is *not* end-to-end. On the other hand, the equation

$$\boldsymbol{N}^* = \boldsymbol{X}^*_2 - \boldsymbol{X} \tag{7.1.3}$$

is end-to-end.

In Chapter 1 we pointed out that the basic objective of filtering is to derive an estimate of the true state vector $\boldsymbol{X}$. Accordingly, it is the end-to-end definition of the error vector $\boldsymbol{N}^*$ in (7.1.1) that we will always use.

Cascaded filtering of the type shown in Figure 7.1 will be considered in Chapters 11 and 12.

7.2 The matrix-to-matrix ECM test

This test makes use of equation (7.1.1) for estimating the actual estimation-error covariance matrix $\boldsymbol{S}_{actual}$. It operates as follows:

◊ We run a Monte-Carlo which generates M draws of the error vector $\boldsymbol{N}^*$ defined in (7.1.1).

◊ All M Monte-Carlo cycles must terminate at precisely the same point on the observed trajectory.

◊ Let the rth draw of $\boldsymbol{N}^*$ be called $\boldsymbol{N}^*_r$. As each $\boldsymbol{N}^*_r$ is obtained, it is used to form the matrix $\boldsymbol{N}^*_r\boldsymbol{N}^*_r{}^T$ and these matrices are accumulated while the runs are under way to give the sum

$$\sum_{r=1}^{M} \boldsymbol{N}^*_r\boldsymbol{N}^*_r{}^T$$

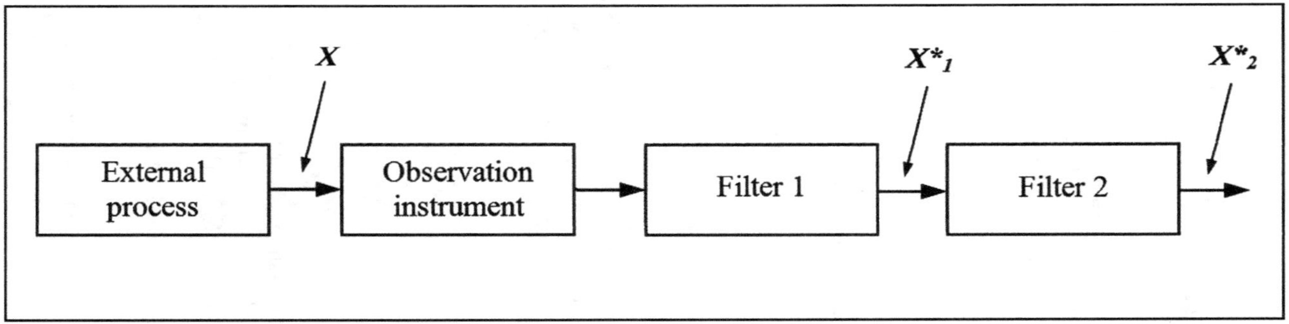

Figure 7.1 Cascaded filtering

◊ At some point, depending on the level of accuracy desired, the Monte-Carlo is terminated and the sum is divided by M to give

$$\boldsymbol{S}_{actual,\,estimated} = 1/M \sum_{r=1}^{M} \boldsymbol{N^*}_r \boldsymbol{N^*}_r^{\,T} \tag{7.2.1}$$

Then (7.2.1) is an estimate of the actual estimation-error covariance matrix $\boldsymbol{S}_{actual}$.

Testing for ECM consistency consists of comparing $\boldsymbol{S}_{actual,\,estimated}$ to the filter covariance matrix $\boldsymbol{S^*}$.

We said earlier that all of the runs must terminate at exactly the same point on the observed trajectory. Let the rth estimate vector at that point be $\boldsymbol{X^*}_r$ with filter covariance matrix $\boldsymbol{S^*}_r$ where $1 \leq r \leq M$.

These M covariance matrices at the termination point could differ slightly by virtue of small variations in the possible transformations (e.g. transforming from radar to Cartesian coordinates) that are used in each of the Monte-Carlo runs.

Such variations could arise because we have elected to use observed values in the transformation equations rather than the exact values that we wouldn't have in the field (see Item 1 in Section 5.1.3).

If such small variations do, in fact, exist, then in addition to accumulating the matrices $\boldsymbol{N^*}_r \boldsymbol{N^*}_r^{\,T}$ in (7.2.1), we must also accumulate the filter covariance matrices $\boldsymbol{S^*}_r$ at the termination point and finally divide that sum by M to give

$$\boldsymbol{S^*}_{ave} \equiv 1/M \sum_{r=1}^{M} \boldsymbol{S^*}_r \tag{7.2.2}$$

Then this is the matrix that must be compared to $\boldsymbol{S}_{actual,\,estimated}$.

However, if there are no variations in the filter covariance matrices $\boldsymbol{S^*}$ from run-to-run then this step can be skipped, and we would simply use any one of those matrices in place of $\boldsymbol{S^*}_{ave}$ for comparison with $\boldsymbol{S}_{actual,\,estimated}$.

The matrix-to-matrix test compares $\boldsymbol{S^*}_{ave}$ to $\boldsymbol{S}_{actual,\,estimated}$ at the selected point on the observed trajectory.

If the match is good, i.e. $\boldsymbol{S^*}_{ave}$ is a good representation of $\boldsymbol{S}_{actual,\,estimated}$ then the filter is ECM consistent at that point on the trajectory, and if the match is not good then it is ECM inconsistent at that point on the trajectory.

We now consider how we assess the match between two matrices.

7.2.1 Comparing two matrices

There are a number of ways in which two matrices can be compared to see if they are well matched, but the one that we have found to be the best is by examining what we call their *ratio matrix*. The following example makes clear what this means.

Example 7.1: Consider the following three matrices:

$$\begin{pmatrix} 2.503 & -2.993 \\ -2.993 & 5.011 \end{pmatrix} \quad \begin{pmatrix} 2.498 & -3.014 \\ -3.014 & 4.992 \end{pmatrix} \quad \begin{pmatrix} 1.002 & 0.993 \\ 0.993 & 1.004 \end{pmatrix}$$

The first two appear to be similar. If we divide each element of the first by the corresponding element of the second then we obtain the third. Thus:

$$2.503 \div 2.498 = 1.002, \quad -2.993 \div -3.014 = 0.993, \quad 5.011 \div 4.992 = 1.004$$

The third matrix is called the *ratio matrix* of the first two.

In this example all elements of the ratio matrix are acceptably close to *1*, and so we say that the match between the first two matrices is good. ■■

Definition 7.1: Let $\boldsymbol{A}$ and $\boldsymbol{B}$ be two matrices that have the same number of rows and columns.

◊ Their **ratio matrix** is the matrix $\boldsymbol{C}$ whose i, jth element is given by

$$[\boldsymbol{C}]_{i,j} = [\boldsymbol{B}]_{i,j} \div [\boldsymbol{A}]_{i,j} \tag{7.2.3}$$

◊ Their **square-root ratio matrix** is the matrix $\boldsymbol{C}'$ whose i, jth element is given by

$$[\boldsymbol{C}']_{i,j} = Sqr\left(Abs([\boldsymbol{B}]_{i,j} \div [\boldsymbol{A}]_{i,j})\right) \tag{7.2.4}$$

■■

Note: The square-root ratio matrix is of value because its diagonals contain the *accuracy ratios*.

There is nothing sacred about the order on the right of (7.2.3) and (7.2.4) and the order could just as well be reversed. If, for example, $\boldsymbol{A}$ is derived from a Monte-Carlo then it is unlikely to have any of its elements precisely equal to *0*, and so *it* would be the one whose elements are placed in the denominator. Either way, if any elements of the chosen denominator matrix happen to be zero then skip the division and display the result, for example, by a string of *1's*, namely *11111*.

7.2.2 *A few words of caution regarding ratio matrices*

Consider the following example:

Example 7.2: The first two matrices below appear to be similar and we wish to compare them.

$$\begin{pmatrix} 64.98 & 0.009 \\ 0.009 & 90.11 \end{pmatrix} \quad \begin{pmatrix} 65.03 & 0.001 \\ 0.001 & 89.92 \end{pmatrix} \quad \begin{pmatrix} 0.999 & 9.000 \\ 9.000 & 1.002 \end{pmatrix}$$

Dividing each element of the first by the corresponding element of the second we obtain the third – their ratio matrix.

The on-diagonal elements of the ratio matrix are acceptably close to *1,* but the off-diagonal elements differ significantly from *1*.

The reason for this discrepancy is that the off-diagonal values in the first two matrices were estimates of quantities whose exact values were *0*. ■■

◊ When estimating a quantity that is exactly *0* – for example a correlation coefficient – then the estimates will seldom be precisely *0* but will instead be small random values that differ from *0*, and this is true no matter how much time we spend deriving such estimates. The quotients of two such random quantities could then differ *significantly* from *1*.

  Accordingly, we cannot depend solely on the ratio matrix, but will sometimes also have to examine the elements of the two matrices from which it was derived. (This is especially true for off-diagonal elements. We are estimating covariance matrices that are always positive-definite and so their on-diagonal elements should never be zero.)

◊ A second item to keep in mind is that the on-diagonal terms of the two matrices from which the ratio matrix is formed are variances (because we are comparing covariance matrices), whereas what we are really after is standard deviations. To compare standard deviations would require that the on-diagonal elements in the ratio matrix be square-rooted, and if they are close to unity before their square roots are taken, then they will be even closer after.

◊ Finally, keep in mind that we are comparing variances and covariances, and not estimates of quantities like x or $\dot{x}$. If two estimates of a variance or a standard deviation differ by *10%* then there is little to argue about, whereas we might well have grounds for dispute if the two estimates were the coordinates of an impact point and they differed by *10%*.

7.2.3 An actual example

In Figure 7.2 below we show a ratio matrix obtained from a Gauss filter that was being evaluated for ECM consistency.

The ratios are the elements of $\boldsymbol{S}_{actual\,,\,estimated}$ divided by $\boldsymbol{S}^*_{ave}$ – both obtained from a *10,000*-cycle Monte Carlo in which each cycle used the same *3*-dimensional 2nd-degree flight path that terminated at the same end point.

The dashed lines in Figure 7.2 subdivide the matrix into nine 3×3 submatrices. Numbering the submatrices in the first row as *1*, *2*, *3*, those in the second row as *4*, *5*, *6*, and those in the third row as *7*, *8*, *9*, we have the following.

- ◊ Submatrix-*1* relates to $(x\,,\, \dot{x}\,,\, \ddot{x})$, submatrix-*5* to $(y\,,\, \dot{y}\,,\, \ddot{y})$ and submatrix-*9* to $(z\,,\, \dot{z}\,,\, \ddot{z})$.
- ◊ The diagonal elements of submatrix-*1* relate to the variances σ_x^2, $\sigma_{\dot{x}}^2$, $\sigma_{\ddot{x}}^2$, and the off-diagonal terms to the covariances. Similarly for submatrices *5* and *9*.
- ◊ Submatrices *2*, *3*, *4*, *6*, *7*, *8* contain the remaining covariances.

We observe that all submatrices involving $x\,,\, \dot{x}\,,\, \ddot{x}\,,\, y\,,\, \dot{y}\,,\, \ddot{y}$, namely *1, 2, 4* and *5* contain elements that are acceptably close to *1*. However all submatrices involving $z\,,\, \dot{z}\,,\, \ddot{z}$, namely *3, 6, 7, 8, 9*, contains elements that are close to *0.4*.

Accordingly, $\boldsymbol{S}^*_{ave}$ was not a good representation of $\boldsymbol{S}_{actual\,,\,estimated}$ and so the filter was declared to be ECM inconsistent.

Examination of the values in the figure suggested that the x and y channels of the filter were working correctly, but that something was wrong in the z channel. That in turn led us to the problem, which we were soon able to fix.

In Figure 7.3 we show the ratio matrix of the same filter after the problem was corrected. All elements of the ratio matrix are now acceptably close to *1* and so (together with testing in a number of other ways) the filter was declared to be ECM consistent.

Taking the square roots of the nine diagonal values in Figure 7.3 to obtain ratios of the standard deviations gives the following nine numbers, which are even closer to *1*:

1.003, 1.007, 1.008, 1.006, 1.008, 1.009, 0.960, 0.960, 0.960 ■■

Note: We were under considerable time pressure when this study was done, and each Monte-Carlo took about *24* hours to run. We therefore had to be satisfied with the numbers in Figure 7.3. However, we believe that we could have improved further on the problem in the z-channel, which would have resulted in values related to $z, \dot{z}, \ddot{z}$ being even closer to *1*.

1.0016294	1.0058577	1.0108596	1.0543981	1.0499461	1.0503303	0.37880406	0.41724740	0.44224248
1.0058577	1.0019240	1.0042917	1.0528435	1.0408186	1.0407008	0.38610087	0.38885537	0.38839239
1.0108596	1.0042917	1.0047085	1.0551564	1.0417122	1.0399196	0.38942061	0.37863689	0.37151441
1.0543981	1.0528435	1.0551564	1.0004268	1.0007551	1.0029269	0.36556828	0.35436135	0.35164473
1.0499461	1.0408186	1.0417122	1.0007551	0.9999075	1.0032090	0.36776640	0.36428640	0.37151912
1.0503303	1.0407008	1.0399196	1.0029269	1.0032090	1.0058210	0.37040674	0.37394475	0.38587870
0.3788040	0.3861008	0.3894206	0.3655682	0.3677664	0.3704067	0.37038939	0.37902665	0.38529082
0.4172474	0.3888553	0.3786368	0.3543613	0.3642864	0.3739447	0.37902665	0.37662407	0.37940492
0.4422424	0.3883923	0.3715144	0.3516447	0.3715191	0.3858787	0.38529082	0.37940492	0.38067827

Figure 7.2 Ratio matrix (before the problem was fixed)

1.0066045	1.0118055	1.0129139	1.0173365	1.0202732	1.0204651	0.8805430	0.89403505	0.90283178
1.0118055	1.0133689	1.0155868	1.0244045	1.0205750	1.0203042	0.8637400	0.90245099	0.93053056
1.0129139	1.0155868	1.0170566	1.0265498	1.0231537	1.0220800	0.8469556	0.88726359	0.91223754
1.0173365	1.0244045	1.0265498	1.0135435	1.0204296	1.0229615	0.9662483	0.96626771	0.96838309
1.0202732	1.0205750	1.0231537	1.0204296	1.0166609	1.0176924	1.0064925	0.95353785	0.93090772
1.0204651	1.0203042	1.0220800	1.0229615	1.0176924	1.0178600	1.0414824	0.98086749	0.95654096
0.8805430	0.8637400	0.8469556	0.9662483	1.0064925	1.0414824	0.9136686	0.91953106	0.92449484
0.8940350	0.9024509	0.8872635	0.9662677	0.9535378	0.9808674	0.9195310	0.91627471	0.92017935
0.9028317	0.9305305	0.9122375	0.9683830	0.9309077	0.9565409	0.9244948	0.92017935	0.92278658

Figure 7.3 Ratio matrix (after the problem was fixed)

7.3 Block diagram of the three tests

Figure 7.4 below is a block diagram of the three ECM-consistency tests that we are considering.

◊ In the upper right-hand box in the figure, multiple draws of $\boldsymbol{N}^*$ are being generated by a Monte-Carlo using (7.1.1).

◊ These draws are then used to derive $\boldsymbol{S}_{actual,\, estimated}$ according to (7.2.1).

◊ On the left, the filter covariance matrices $\boldsymbol{S}^*$ are being accumulated, from which $\boldsymbol{S}^*_{ave}$ is obtained according to (7.2.2).

◊ The matrix $\boldsymbol{S}_{actual,\, estimated}$ is then compared to $\boldsymbol{S}^*_{ave}$ by forming their ratio matrix in the lower bold black box labeled 'Matrix-to-matrix ECM test'.

The two ECM-consistency tests that we discuss further down appear in the upper two bold black boxes. Both are shown as single-shot tests for which the inputs are a single draw of $\boldsymbol{N}^*$ and a single filter covariance matrix $\boldsymbol{S}^*$. The tests then determine if $\boldsymbol{S}^*$ could be the covariance matrix of $\boldsymbol{N}^*$, and if the answer is *yes* then it means that $\boldsymbol{S}^*$ is well matched to $\boldsymbol{S}_{actual}$ – the actual estimation-error covariance matrix of the population from which $\boldsymbol{N}^*$ was drawn.

If run in single-shot mode, these two tests free us from the need to run a potentially time-consuming Monte-Carlo.[2] However, this freedom comes at a price:

◊ When running the matrix-to-matrix ECM test, if the test fails then the numbers in the ratio matrix could be of help in locating the problem, and as we saw above, that was indeed the case.

◊ When running the *3*-sigma and Chi-squared ECM tests in single-shot mode, there will be no such help. All that we will learn is that ECM-consistency is present or absent at the instants at which the tests are carried out, and when ECM consistency is absent there will be no clues suggesting where the problem might lie.

The *3*-sigma and Chi-squared ECM tests can also be run in Monte-Carlo mode. However, for them we cannot form a ratio matrix and must instead determine whether or not ECM consistency is present by another method. We will return to this further down.

[2] Monte-Carlo runs of fully configured filtering systems can sometimes take up to *24* hours or even more to complete.

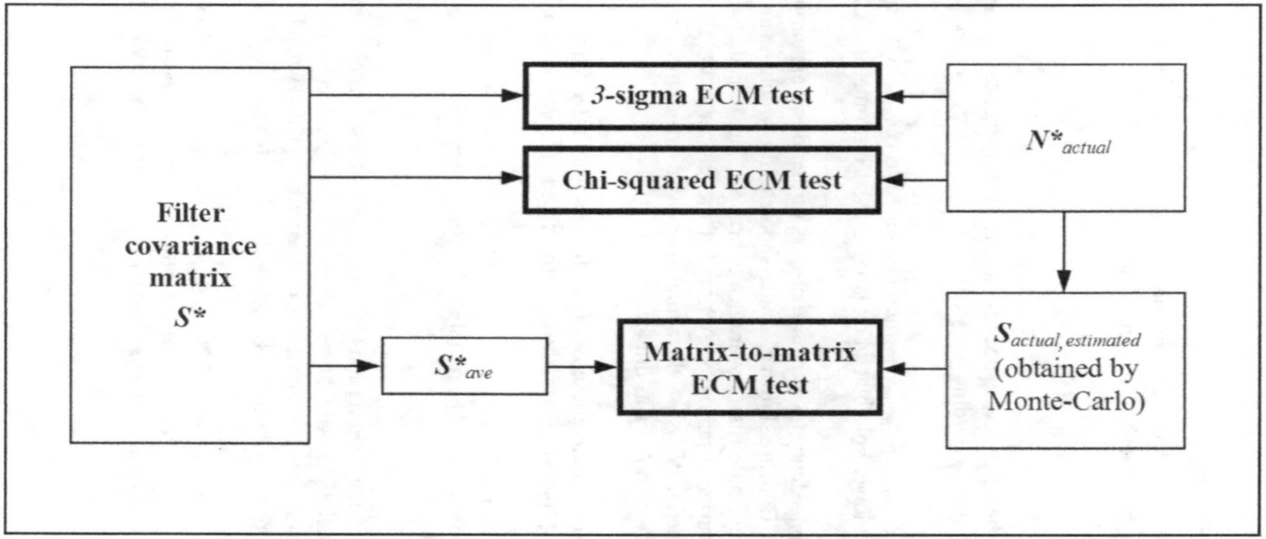

Figure 7.4 The three ECM-consistency tests

7.4 The multivariate-Gaussian and the Chi-squared PDFs

In most systems in which tracking filters are used we can assume that the errors in the observation vector $\boldsymbol{Y}$ are multivariate-Gaussian. Because our filters are all linear transformations of the form

$$\boldsymbol{X}^* = \boldsymbol{W}\boldsymbol{Y} \tag{7.4.1}$$

it follows that the random errors in $\boldsymbol{X}^*$ will then also be multivariate-Gaussian.

7.4.1 *Multivariate-Gaussian RVs*

(See References 141 or 142.)

***Definition** 7.2:* Let $\boldsymbol{X}^*$ be an m-vector and let its error vector be

$$\boldsymbol{N}^* = (\nu^*_1, \nu^*_2, \dots \nu^*_m)^T \tag{7.4.2}$$

Let $\boldsymbol{S}_{N^*}$ be the covariance matrix of $\boldsymbol{N}^*$ and let $\boldsymbol{Z} = (z_1, z_2, \dots z_m)^T$ be an m-vector of algebraic variables.

A. If $\boldsymbol{N}^*$ **is multivariate-Gaussian, zero-mean** (i.e. $\boldsymbol{X}^*$ is unbiased) then the PDF of $\boldsymbol{N}^*$ is

$$f_{N^*}(\boldsymbol{Z}) = (2\pi)^{-m/2}\,|\boldsymbol{S}_{N^*}|^{-1/2}\, exp\left(-{}^1\!/_2\boldsymbol{Z}^T\boldsymbol{S}_{N^*}{}^{-1}\boldsymbol{Z}\right) \tag{7.4.3}$$

B. If $\boldsymbol{N}^*$ is **multivariate-Gaussian with mean** $\boldsymbol{N}^*_b$ (i.e. $\boldsymbol{X}^*$ is biased) then the PDF of $\boldsymbol{N}^*$ is

$$f_{N^*}(\boldsymbol{Z}) = (2\pi)^{-m/2}\,|\boldsymbol{S}_{N^*}|^{-1/2}\, exp\left(-{}^1\!/_2(\boldsymbol{Z}-\boldsymbol{N}^*_b)^T\boldsymbol{S}_{N^*}{}^{-1}(\boldsymbol{Z}-\boldsymbol{N}^*_b)\right) \tag{7.4.4}$$

■■

In what follows we will assume initially that $\boldsymbol{N}^*_b = \boldsymbol{\emptyset}$ (i.e. $\boldsymbol{N}^*$ is zero-mean), but we will also examine what happens when that is not the case.

***Definition** 7.3:* Let $\boldsymbol{N}^*$ be an m-vector of Gaussian RVs that are zero-mean, independent and of variance $\sigma^2 = 1$. Then the covariance matrix of $\boldsymbol{N}^*$ is $\boldsymbol{S}_{N^*} = \boldsymbol{I}_m$ (an $m \times m$ identity matrix) and $\boldsymbol{N}^*$ is said to be ***m*-variate standard normal.** ■■

7.4.2 *Chi-squared RVs*

Let $\boldsymbol{N}^*$ be m-variate standard normal, and form the quantity c_m defined by

$$c_m \equiv \boldsymbol{N}^{*T}\boldsymbol{N}^* = \nu^{*2}_1 + \nu^{*2}_2 + \dots + \nu^{*2}_m \tag{7.4.5}$$

Then c_m is a scalar RV consisting of the sum of squares of m independent standard normal RVs.

Definition 7.4: The PDF of c_m in (7.4.5) is said to be **central Chi-squared with *m* degrees of freedom** for which the analytical expression is

$$f_{c_m}(x) = \begin{cases} \left(2^{m/2}\,\Gamma(m/2)\right)^{-1} x^{(m-2)/2} exp(-x/2) & (x \geq 0) \\ 0 & (x < 0) \end{cases} \tag{7.4.6}$$

■■

Note: See Projects 10.6 and 10.7 for further discussion of the Chi-squared PDF and the gamma function $\Gamma(x)$ that is present in its definition.[3]

Suppose next that the vector $\boldsymbol{N^*}$ is multivariate Gaussian zero-mean, but with covariance matrix $\boldsymbol{S}_{\boldsymbol{N^*}}$ that is any positive-definite matrix. Let the expression for c_m be

$$c_m = \boldsymbol{N^*}^T \boldsymbol{S}_{\boldsymbol{N^*}}{}^{-1} \boldsymbol{N^*} \tag{7.4.7}$$

Then it is still the case that c_m is central Chi-squared with *m* degrees of freedom. We state this as follows:

Theorem 7.1: Let the *m*-vector $\boldsymbol{N^*}$ be zero-mean, multivariate Gaussian with covariance matrix $\boldsymbol{S}_{\boldsymbol{N^*}}$. Then the RV

$$c_m = \boldsymbol{N^*}^T \boldsymbol{S}_{\boldsymbol{N^*}}{}^{-1} \boldsymbol{N^*} \tag{7.4.8}$$

is central Chi-squared with *m* degrees of freedom.

Proof is given in Problem 7.1.

■■

If $\boldsymbol{N^*}$ is not zero-mean then the PDF of the random variable c_m in (7.4.8) ceases to be *central* Chi-squared and becomes instead *non-central* Chi-squared which has an analytical definition that differs from (7.4.6) with substantially different properties.[4]

From now on, when we speak about the Chi-squared PDF we will drop the word *central* and use only the words *Chi-squared with m degrees of freedom* which is assumed to be central.

[3] Don't let the expression for the PDF in (7.4.6) intimidate you. In Project 10.7 we show you that only two simple facts are needed in order to turn the Gamma function $\Gamma(m/2)$ into numerical values, and from there the Chi-squared PDF is readily evaluated.

[4] See Reference 134.

In (7.4.8) the matrix $\boldsymbol{S}_{N*}$ is positive definite, and hence so also is $\boldsymbol{S}_{N*}{}^{-1}$. The error vector $\boldsymbol{N}^*$ cannot be the zero vector. Thus the quantity c_m can only be positive. Moreover, c_m will possess all of the properties inherent in a quadratic form on a positive definite matrix. This is discussed further in Project 7.6 where we examine the extreme sensitivity of the Chi-squared ECM test.

Chi-squared RVs possess an important property called *invariance under a transformation* that can save us time and trouble in filter engineering. This is discussed in Appendix 7.1 where we prove the following:

***Theorem** 7.2:* The Chi-squared random variable is invariant under a nonsingular linear or nonlinear transformation. ■■

The vector $\boldsymbol{N}^*$ must be zero-mean in order that c_m in (7.4.7) be central Chi-squared. When $\boldsymbol{N}^*$ is not zero-mean and we evaluate the quadratic form appearing in (7.4.7) – as we will be doing below – then the value of c_m could be significantly greater than what we might expect if its PDF *were* central Chi-squared.

When such inflation occurs, we can infer that bias errors might be present in $\boldsymbol{N}^*$. However, there are other factors that could cause c_m to become inflated. We will return to this topic further down.

◀▶

The values shown in Figure 7.5 below are called the ***99%* thresholds for Chi-squared RVs with *m* degrees of freedom.**

$m = 2$	$p(c_2 < 9.21) = 99\%$
3	$p(c_3 < 11.35) = 99\%$
4	$p(c_4 < 13.28) = 99\%$
5	$p(c_5 < 15.09) = 99\%$
6	$p(c_6 < 16.81) = 99\%$
7	$p(c_7 < 18.48) = 99\%$
8	$p(c_8 < 20.09) = 99\%$
9	$p(c_9 < 21.67) = 99\%$

Figure 7.5 Chi-squared 99% thresholds

As shown in equation (7.4.9), the thresholds are obtained by finding the value of x for which the integral of the function appearing in (7.4.6) is equal to 0.99, for various values of m.

$$p(c_m < x) = \int_0^x f_{c_m}(x')dx' = 0.99 \tag{7.4.9}$$

Note: The values in Figure 7.5 can be obtained by using a program called ***DistCalc*** that you can download from the Internet. (For down-loading details, see either the folder ***Miscellaneous\Thresholds***, or else Reference 80.) ***DistCalc.exe*** will enable you to extend Figure 7.5 as far as you like or else to modify the confidence level of 0.99. When creating the program ***13_Gauss_Newton_C*** it was necessary that the value of m should continue *indefinitely*, and ***DistCalc.exe*** provided the basis for the method that we developed (see Project 10.9).

Example 7.3: For $m = 3$, the Chi-squared RV c_3 is less than 11.35 with probability 0.99. Suppose we make repeated draws of c_3. Then we would expect that *99%* of them would be less than 11.35. However, if we frequently come up with values greater than 11.35 then clearly something would be wrong. ■■

7.5 The single-shot *3*-sigma ECM test

A simulator's estimate vector $\boldsymbol{X}^*$ has an error vector $\boldsymbol{N}^*$ whose elements are supposedly zero-mean. (We say *supposedly* because bias errors might be present.)

In the univariate case we know that a zero-mean Gaussian RV will lie in the range -3σ to $+3\sigma$ with probability *99.73%*. In the multivariate case the number *99.73%* changes slightly, but that is of little importance.

What *is* important is that the RVs in a multivariate zero-mean Gaussian population will lie inside their respective -3σ to $+3\sigma$ ranges a very high percentage of the time.

7.5.1 The single-shot 3-sigma ECM test

At some instant during a simulation:

◊ Let the true state vector of the observed trajectory be the m-vector $\boldsymbol{X}$.

◊ Let the filter's estimate vector be the m-vector $\boldsymbol{X}^*$.

◊ Let the actual error vector in the filter's estimate be the m-vector $\boldsymbol{N}^*$.

◊ Let the filter covariance matrix be $\boldsymbol{S}^*$ that is $m \times m$.

Step 1: Find the square roots of the diagonal elements of $\boldsymbol{S}^*$ which we call $s_1^*, s_2^*, \ldots s_m^*$. These are supposedly the standard deviations of the elements of $\boldsymbol{N}^*$.

Step 2: Multiply each of $s_1^*, s_2^*, \ldots s_m^*$ by *3*. The results are supposedly the *3*-sigma values of the elements of $\boldsymbol{N}^*$.

Step 3: Derive $\boldsymbol{N}^*$ using the end-to-end definition

$$\boldsymbol{N}^* = \boldsymbol{X}^* - \boldsymbol{X} \tag{7.5.1}$$

and let its elements be called $\nu_1^*, \nu_2^*, \ldots \nu_m^*$.

Step 4: If all of the ν^* lie inside their supposed $\pm 3\sigma$ range, i.e. if

$$|\nu_i^*| < 3s_i^* \qquad i = 1, 2, \ldots m \tag{7.5.2}$$

then the filter has passed the *3*-sigma ECM-consistency test at the instant at which the test was run.

On the other hand, if any of the ν^* lies significantly outside that range, then the filter has failed the test at that instant and is ECM inconsistent. ■■

Example 7.4: Figure 7.6 below is another example of Kalman/Swerling instability, similar to the one shown in Figure 1.10.

◊ The Kalman/Swerling error/sigma ratios all lie significantly outside the ranges specified in (7.5.2) and so those filters have failed the *3*-sigma ECM-consistency test.

◊ The Gauss–Newton error/sigma ratios all lie inside the specified ranges and so that filter has passed the test.

■■

The single-shot *3*-sigma ECM test – unlike the matrix-to-matrix ECM test – is based on a snapshot of the filter's state. The rationale behind the test is as follows:

◊ The actual covariance matrix of $\boldsymbol{N}^*$ is $\boldsymbol{S}_{actual}$ and if we used three times the square roots of the diagonal elements of $\boldsymbol{S}_{actual}$ as the *3*-sigma values in (7.5.2), then the test would almost always pass.

◊ In fact, we don't use the diagonal elements of $\boldsymbol{S}_{actual}$ and instead we use those of $\boldsymbol{S}^*$.

If the test then almost always passes, it means that the diagonal elements of $\boldsymbol{S}^*$ are good representations of the diagonal elements of $\boldsymbol{S}_{actual}$ and that, at least as far as the diagonal elements are concerned, the filter is ECM consistent.

```
                    Kalman Filter (Cycle #109)

              Error             Sigma          Error/Sigma

    x     -2.5468436e-2     1.7381474e-3     -14.652633
  x-dot   -1.6184613e-6     1.2569783e-7     -12.875809
  omega    1.1253123e-7     8.5437827e-9      13.171125

            Chi-squared = 2.2258217589271e+02 (11.35)


                   Swerling Filter (Cycle #109)

              Error             Sigma          Error/Sigma

    x     -2.5468436e-2     1.7381474e-3     -14.652633
  x-dot   -1.6184613e-6     1.2569783e-7     -12.875809
  omega    1.1253123e-7     8.5437827e-9      13.171125

            Chi-squared = 2.2258217589258e+02 (11.35)


                          Gauss-Newton

              Error             Sigma          Error/Sigma

    x     -6.6063557e-3     3.5859331e-3     -1.8422975
  x-dot    3.2044458e-7     1.4231528e-7      2.2516526
  omega    2.4949517e-8     1.1673397e-8      2.137297

                 Chi-squared = 5.5347756 (11.35)
```

Figure 7.6 Kalman/Swerling instability

The test has been implemented in the following seven programs:

***03_Gauss_Aitken*, *05_Pol_Filter_EMP*, *06_Pol_Filter_FMP*,**

***06b_FMP_Bug*, *11_Gauss_Newton_A*, *13_Gauss_Newton_C*,**

14_Orbit_Track

In most of these programs the filter is permitted to run continuously, and whenever ***D*** is pressed the run is halted and numerical values are displayed. Among them are the errors in the elements of $\boldsymbol{X}^*$ divided by their *3*-sigma values.

If these quotients are all absolutely less than *3* then ECM consistency is present, and if any of them exceeds *3* by a significant amount then ECM consistency is absent.

7.6 The single-shot Chi-squared ECM test

The above method of testing for ECM consistency is extremely easy to implement, but it has the drawback of considering only the the diagonal elements of $\boldsymbol{S}^*$.

The test that we now discuss considers the *entire* matrix, but has the drawback that $\boldsymbol{S}^*$ must be inverted, and that is not always as easy as one might think.

7.6.1 The single-shot Chi-squared ECM test

At some instant during a simulation:

◊ Let the state vector of the observed trajectory be the m-vector $\boldsymbol{X}$.

◊ Let the filter's estimate vector be the m-vector $\boldsymbol{X}^*$.

◊ Let the actual error vector in the filter's estimate be the m-vector $\boldsymbol{N}^*$.

◊ Let the filter covariance matrix of $\boldsymbol{N}^*$ be $\boldsymbol{S}^*$ which is $m \times m$.

Step 1: Invert $\boldsymbol{S}^*$ to give $\boldsymbol{S}^{*-1}$.

Step 2: Derive $\boldsymbol{N}^*$ using the end-to-end definition

$$\boldsymbol{N}^* = \boldsymbol{X}^* - \boldsymbol{X} \tag{7.6.1}$$

Step 3: Compute c_m from

$$c_m = \boldsymbol{N}^{*T}\boldsymbol{S}^{*-1}\boldsymbol{N}^* \tag{7.6.2}$$

Step 4: If c_m is less than the threshold value shown in Figure 7.5 for the given value of m, then the filter has passed the Chi-squared ECM-consistency test at the instant at which it was run.

On the other hand, if c_m significantly exceeds the threshold value, then the filter has failed the Chi-squared test at that instant and is ECM inconsistent. ■■

Example 7.4: Referring to Figure 7.6 above:

◊ Both the Kalman and Swerling Chi-squared values significantly exceed the threshold value shown in Figure 7.5 for the value of $m = 3$ and so those filters have failed the Chi-squared test.

◊ The Gauss–Newton filter's Chi-squared value is less than the threshold value of *11.35* and so that filter has passed the Chi-squared test. ■■

The single-shot Chi-squared test – like the single-shot *3*-sigma test – is also based on a single snapshot of the filter's state. The rationale behind this test is similar to that of the *3*-sigma test:

◊ The actual covariance matrix of $\boldsymbol{N}^*$ is $\boldsymbol{S}_{actual}$, and if we used $\boldsymbol{S}_{actual}$ in the test, then it would almost always pass.

◊ In fact, we don't use $\boldsymbol{S}_{actual}$ and instead we use $\boldsymbol{S}^*$.

If the test then almost always passes, it means that $\boldsymbol{S}^*$ is a good representation of $\boldsymbol{S}_{actual}$ and that the filter is ECM consistent.

However, if the test almost always fails, then it can only mean that $\boldsymbol{S}^*$ is not a good representation of $\boldsymbol{S}_{actual}$ and that the filter is ECM inconsistent.

The Chi-squared test has been implemented in the same seven programs as those listed above for the *3*-sigma test, and is operated in the same way.

7.7 Failure of the *3*-sigma and Chi-squared tests

Failure of either of the tests can be caused by two factors, either singly or in combination:

1. ECM consistency is absent.

2. Bias errors are present in $\boldsymbol{N}^*$.

   The tests are based on the assumption that the PDF from which $\boldsymbol{N}^*$ was drawn is zero-mean, and that the PDF of $\boldsymbol{N}^{*T}\boldsymbol{S}^{*-1}\boldsymbol{N}^*$ is then central Chi-squared. If bias errors are present then these assumptions are no longer valid and the test thresholds could easily be breached.

   Figure 7.7 shows the four possibilities that can occur.

ECM consistent Bias errors absent	ECM consistent Bias errors present
ECM inconsistent Bias errors absent	ECM inconsistent Bias errors present

Figure 7.7 ECM consistency and bias errors

We are testing for ECM consistency, and so

◊ In the top left box the tests will give a true positive.

◊ In the lower left box the tests will give a true negative.

◊ In the top right box the tests will give a false negative.

◊ In the lower right box the tests will give a true negative.

When the outcome of the test is negative we will find ourselves in one of the three negative boxes in the diagram and not know which one. However, this will tell us that either ECM consistency is absent or bias errors are present – both of which constitute a problem – and that we need to undertake an investigation in order to rectify the situation.

The GOF test that we discuss in Chapter 10 is only sensitive to bias, and so its results can provide help in determining the source of the problem and correcting it. We discuss this further in Project 10.3 entitled 'How bias errors affect the performance tests'.

3. There is a third factor that could cause the Chi-squared test to give a false negative, one that does not apply to the *3*-sigma test, namely that matrix inversion problems exist. If the elements of the error vector $\boldsymbol{N}^*$ are highly correlated, then the filter covariance matrix $\boldsymbol{S}^*$ could be difficult to invert (see further down).

The primary purpose of the *3*-sigma and Chi-squared tests is to act as sentinels that watch for ECM inconsistency.

We should not become too alarmed if the Chi-squared values occasionally exceed their *99%* thresholds.

◊ *1%* of the time we can expect them to do that anyway.

◊ The Chi-squared test is an extremely sensitive one, and when it fails you will see numbers that repeatedly exceed the *99%* thresholds by *significant* amounts, sometimes even by multiples of *1000* or more. We refer you to Project 7.6 where we discuss the sensitivity of the Chi-squared test, from which you will get a feel for why the test is so sensitive and why the Chi-squared values increase so quickly when there is a mismatch between an error vector and its supposed covariance matrix.

  Accordingly, if the Chi-squared values occasionally exceed their thresholds by only small amounts, judgment should be used as to whether or not this is significant.

◊ Another factor to bear in mind is that the *3*-sigma and Chi-squared tests complement each other, and when one of them suggests that something is amiss, then so also should the other. It is therefore wise to consider the results of the two tests as a composite indicator of what is taking place, rather than basing judgments on either of them singly.

7.7.1 Seeing the tests in action

In Project 7.2 we take you through the steps to run the program ***05_Pol_Filter_EMP*** which will enable you to see the single-shot *3*-sigma and Chi-squared tests in action.

7.8 Additional comments on the *3*-sigma and Chi-squared tests

We make the following additional comments regarding the two tests.

7.8.1 Bias errors are present

When the filter model is a bad fit to the external model then it is possible for both the *3*-sigma and the Chi-squared ECM-consistency tests to give false negatives – the cause of the threshold violations being bias errors rather than ECM inconsistency. The estimation errors are no longer zero mean, and so both methods could give readings that breach the thresholds, in the case of the Chi-squared test possibly by huge multiples (see Project 7.6).

The presence or absence of bias errors can be resolved by running the GOF test (Chapter 10), and so we suggest that it be included whenever possible in filter engineering projects. (We included the GOF test in three of our programs, namely ***03_Gauss_Aitken***, ***13_Gauss_Newton_C*** and ***14_Orbit_Track***.)

7.8.2 Matrix inversion problems exist

When the following state of affairs exists:

◊ The *3*-sigma test passes

◊ The thresholds of the Chi-squared test are repeatedly breached

then the underlying reason could be that a matrix inversion problem exists, caused by the fact that the elements of the vector $\boldsymbol{N}^*$ are highly correlated.

Referring back to (4.2.23), if the elements of $\boldsymbol{N}^*$ *are* highly correlated then the correlation matrix for $\boldsymbol{N}^*$ will have values that are close to ± 1 in its off-diagonal elements. This signals that $\boldsymbol{S}^*$ is ill-conditioned and difficult to invert, and that the computation in (7.6.2) could be producing inaccurate values.

There are two approaches to mitigating the problem.

◊ The first would be to use a computer language that employs higher numerical precision. Hopefully the matrix inversion can then be carried out with sufficient accuracy so that the results of (7.6.2) can be relied on. However, it may not be possible to do this, and even if it is, it may not always work.

◊ An alternate and much simpler approach is to reduce the size of the matrices that must be inverted so that they are 2×2, and in so doing to make it trivial for the existing computer language to invert them accurately.

As an example, assume that the error vector $\boldsymbol{N}^*$ is a 4-vector and so the filter covariance matrix $\boldsymbol{S}^*$ is 4×4. Thus

$$\boldsymbol{S}^* = \begin{pmatrix} s^*_{1,1} & s^*_{1,2} & s^*_{1,3} & s^*_{1,4} \\ s^*_{2,1} & s^*_{2,2} & s^*_{2,3} & s^*_{2,4} \\ s^*_{3,1} & s^*_{3,2} & s^*_{3,3} & s^*_{3,4} \\ s^*_{4,1} & s^*_{4,2} & s^*_{4,3} & s^*_{4,4} \end{pmatrix} \tag{7.8.1}$$

Then each of the principal 2×2 submatrices in $\boldsymbol{S}^*$ is the covariance matrix of an appropriate 2-vector from $\boldsymbol{N}^*$. Moreover, the RVs that are formed from those submatrices and subvectors according to (7.6.2) will also be Chi-squared, but now with only two degrees of freedom.

Instead of trying to invert the entire matrix $\boldsymbol{S}^*$ and then forming c_4 using the entire vector $\boldsymbol{N}^*$, rather extract a number of 2×2 principal submatrices from $\boldsymbol{S}^*$ and then perform the computations using 2-vectors taken from $\boldsymbol{N}^*$ as follows:

$$c_{2,1} = \begin{pmatrix} \nu^*_1 & \nu^*_2 \end{pmatrix} \begin{pmatrix} s^*_{1,1} & s^*_{1,2} \\ s^*_{2,1} & s^*_{2,2} \end{pmatrix}^{-1} \begin{pmatrix} \nu^*_1 \\ \nu^*_2 \end{pmatrix} \tag{7.8.2}$$

$$c_{2,2} = \begin{pmatrix} \nu^*_1 & \nu^*_3 \end{pmatrix} \begin{pmatrix} s^*_{1,1} & s^*_{1,3} \\ s^*_{3,1} & s^*_{3,3} \end{pmatrix}^{-1} \begin{pmatrix} \nu^*_1 \\ \nu^*_3 \end{pmatrix} \tag{7.8.3}$$

The matrix inversions are now significantly more accurate, and if either ECM inconsistency or bias errors are present, then they will be truly reflected in the values obtained from (7.8.2) and (7.8.3).

The problem of matrix inversion is less serious when $\boldsymbol{S}^*$ is 3×3 or smaller, but when it is of higher order then this approach works well.

It is not necessary to consider all of the possible principal 2×2 submatrices in order to detect ECM inconsistency, and only a few of them will be required. This is especially true when $\boldsymbol{S}^*$ is as large as 9×9, as would be the case if we were operating with second-degree polynomials in three dimensions (see (2.3.10)).

7.8.3 Demonstration

For a good demonstration of this technique of using submatrices taken from $\boldsymbol{S}^*$, see the discussion of the program ***06b_FMP_Bug*** in Project 7.1.

It was only after we had split the single 4×4 Chi-squared computation for the 3rd-degree FMP filter into smaller 2×2 Chi-squared computations that we were able to track down and fix an extremely troublesome problem.[5]

◆▶

Up to now we have only considered the *3*-sigma and Chi-squared tests in single-shot mode. In the next section we discuss how they can also be used in Monte-Carlo mode.

7.9 Four illustrative computer runs

The program ***04_Monte_Carlo*** runs the following four tests:

◊ The ECM-consistency matrix-to-matrix test

◊ The ECM-consistency Chi-squared test in Monte-Carlo mode

◊ The CR-consistency matrix-to-matrix test

◊ The CR-consistency Chi-squared test in Monte-Carlo mode

04_Monte_Carlo operates in seven modes and we examine four of them. However, rather than doing it here, we do it in Project 7.5.

Please take the time to read the project, and then run the program yourself. One of the benefits of doing so is that you will then understand how the Chi-squared test is run in Monte-Carlo mode.

[5] When debugging the program ***06_Pol_Filter_FMP***, it took us almost a month (on and off) to find the bug that was causing fleeting but serious difficulties. We were using the Chi-squared test and so we *absolutely knew* that a bug existed, and we were finally able to locate it. See Project 7.1 for how to reproduce the situation.

7.10 Final comments

The *3*-sigma and Chi-squared ECM-consistency tests are implemented in single-shot mode in the following seven programs:

03_Gauss_Aitken, ***05_Pol_Filter_EMP,*** ***06_Pol_Filter_FMP,***

06b_FMP_Bug, ***11_Gauss_Newton_A,*** ***13_Gauss_Newton_C,***

14_Orbit_Track

The matrix-to-matrix ECM test and the Chi-squared ECM test in Monte-Carlo mode are implemented in the following two Monte-Carlo programs:

04_Monte_Carlo, ***12_Gauss_Newton_B***

The GOF test is implemented in the following three programs:

03_Gauss_Aitken, ***13_Gauss_Newton_C,*** ***14_Orbit_Track***

It is always a source of comfort to know that something complex that one has engineered is operating correctly.

You will soon find that the four tests – matrix-to-matrix, *3*-sigma, Chi-squared and GOF – go a long way to providing a sense of confidence that all is well, and once you start using them, you won't want to leave home without them.

The following items are in the supplementary material

Problems for Chapter 7

Solutions to Problems for Chapter 7

Projects for Chapter 7

Appendix 7.1

Invariance of Chi-squared under non-singular transformations

Chi-squared RVs possess an important property called *invariance under non-singular transformations* that can be of great use to us in filter engineering.

Linear transformations

Suppose first that $\boldsymbol{N}^*$ undergoes a non-singular *linear* transformation by the matrix $\boldsymbol{A}$ as follows:

$$\boldsymbol{M}^* = \boldsymbol{A}\boldsymbol{N}^* \tag{A7.1.1}$$

in which $\boldsymbol{M}^*$ and $\boldsymbol{N}^*$ are both zero-mean m-vectors and $\boldsymbol{A}$ is non-singular. Then by Theorem 4.1 the covariance matrices of $\boldsymbol{N}^*$ and $\boldsymbol{M}^*$ will be related as follows:

$$\boldsymbol{S}_{\boldsymbol{M}^*} = \boldsymbol{A}\boldsymbol{S}_{\boldsymbol{N}^*}\boldsymbol{A}^T \tag{A7.1.2}$$

Now assume that $\boldsymbol{N}^*$ is multivariate Gaussian, and form the Chi-squared RV:

$$c_m = \boldsymbol{N}^{*T}\boldsymbol{S}_{\boldsymbol{N}^*}^{-1}\boldsymbol{N}^* \tag{A7.1.3}$$

Because $\boldsymbol{N}^*$ is multivariate Gaussian it follows from (A7.1.1) that $\boldsymbol{M}^*$ is also multivariate Gaussian, and by Theorem 7.1 that in turn means that the RV d_m defined by

$$d_m \equiv \boldsymbol{M}^{*T}\boldsymbol{S}_{\boldsymbol{M}^*}^{-1}\boldsymbol{M}^* \tag{A7.1.4}$$

will also be Chi-squared with m degrees of freedom.

We now show that d_m is equal to c_m.

Starting from (A7.1.4) we apply the transformation equations (A7.1.1) and (A7.1.2) and obtain the following:

$$\begin{aligned} d_m &= \boldsymbol{M}^{*T}\boldsymbol{S}_{\boldsymbol{M}^*}^{-1}\boldsymbol{M}^* \\ &= \left(\boldsymbol{A}\boldsymbol{N}^*\right)^T\left(\boldsymbol{A}\boldsymbol{S}_{\boldsymbol{N}^*}\boldsymbol{A}^T\right)^{-1}\left(\boldsymbol{A}\boldsymbol{N}^*\right) = \left(\boldsymbol{A}\boldsymbol{N}^*\right)^T\left((\boldsymbol{A}^T)^{-1}\boldsymbol{S}_{\boldsymbol{N}^*}^{-1}\boldsymbol{A}^{-1}\right)\left(\boldsymbol{A}\boldsymbol{N}^*\right) \\ &= \boldsymbol{N}^{*T}\boldsymbol{A}^T(\boldsymbol{A}^T)^{-1}\boldsymbol{S}_{\boldsymbol{N}^*}^{-1}\boldsymbol{A}^{-1}\boldsymbol{A}\boldsymbol{N}^* = \boldsymbol{N}^{*T}\boldsymbol{S}_{\boldsymbol{N}^*}^{-1}\boldsymbol{N}^* = c_m \end{aligned} \tag{A7.1.5}$$

Thus the Chi-squared variable formed from $\boldsymbol{N}^*$ with covariance matrix $\boldsymbol{S}_{N^*}$ remains unchanged when $\boldsymbol{N}^*$ and $\boldsymbol{S}_{N^*}$ are passed through a non-singular linear transformation.

Non-linear transformations

Suppose now that a nonsingular *non*-linear transformation is applied to a random vector and its covariance matrix (as we did in section 5.1). Then almost the same result as in the linear transformation applies, and can be stated as follows:

The Chi-squared variable formed from a Gaussian random vector and its covariance matrix is (to first order) invariant under a non-singular non-linear transformation.

Proof is straight-forward. We saw in (5.1.21) and (5.1.22) that under a non-singular non-linear transformation, error vectors and their covariance matrices transform as follows:

$$\boldsymbol{N}_C = \boldsymbol{G}\boldsymbol{N}_R \tag{A7.1.6}$$

$$\boldsymbol{R}_{N_C} = \boldsymbol{G}\boldsymbol{R}_{N_R}\boldsymbol{G}^T \tag{A7.1.7}$$

in which $\boldsymbol{G}$ is the Jacobi matrix of the non-linear transformation.

But this is a non-singular *linear* transformation of precisely the same structure as (A7.1.1) and (A7.1.2), and so what applied there regarding the invariance of the Chi-squared value must apply here as well.

We have thus proved the following:

***Theorem** 7.2:* The Chi-squared random variable that is formed from a multivariate Gaussian random vector and its covariance matrix is invariant under both a non-singular linear transformation and a non-singular non-linear transformation. ■■

Applications to filter engineering

The invariance of the Chi-squared RV under a non-singular transformation is an important result which can save us time and trouble in filter engineering, as the following example shows.

Example** A7.1.1:* In the program ***14_Orbit_Track we derive estimates in inertial Cartesian coordinates of a satellite's state vector:

$$\boldsymbol{X}^* = (x^*,\ \ \dot{x}^*,\ \ y^*,\ \ \dot{y}^*,\ \ z^*,\ \ \dot{z}^*)^T \tag{A7.1.8}$$

and we also derive the filter covariance matrix $\boldsymbol{S}^*$.

Starting from $\boldsymbol{X}^*$ we then use a fairly complicated non-singular, non-linear transformation to derive the Keplerians of the satellite's orbit, namely the six quantities

Ω, i, ω, e, A and τ.

(The transformation and the meaning of these symbols is discussed in the documentation to ***14_Orbit_Track***.)

It is relatively simple to obtain the Chi-squared values in inertial Cartesian coordinates. Knowing that they are invariant under a non-singular non-linear transformation we can then use that same Chi-squared value for the Keplerians without the need to re-derive it, and in this way we save ourselves a substantial amount of time and effort. ■■

Part 2

Non-recursive filtering

Chapter 8

Minimum variance and the Gauss–Aitken filters

8.1 Introduction

The Gauss filters discussed in this chapter and the next are implementations of what is known as the *minimum variance algorithm* (MVA). In this chapter we derive the MVA and we begin our exploration of how it is used in filter engineering.

Eight Gauss filters will emerge, all of them based on the MVA – two Gauss–Aitken filters in this chapter for use in the all-linear environment of ***Case 1***, and six Gauss–Newton filters in Chapter 9 for use in the non-linear environments of ***Cases 2, 3*** and ***4***.

Further down in this chapter we discuss the meanings of the words *non-recursive* and *recursive*, and it will become clear that the Gauss filters are all non-recursive implementations of the MVA.

In Chapter 11 we will see that the Swerling filter is a recursive implementation of the MVA. We will also see that an algebraic transformation applied to the Swerling filter takes us to the Kalman filter, and so it too is a recursive implementation of the MVA.

The polynomial filters that we discuss in Chapters 12 and 13 are recursive, and are based on a simplified version of the MVA.

Thus every filter that we consider in this book is, in one way or another, an implementation of the MVA.

8.1.1 Minimum variance

Minimum variance is to estimation theory something like what the speed of light is to physics – we can't exceed it. However, unlike the speed of light, minimum variance is something that we *can* achieve – the MVA that we derive in this chapter gives results that are consistent with the Cramér–Rao lower bound, the way in which the 'speed of light' is usually specified in estimation theory.

Not surprisingly, there are many ways in which to derive so profound a result. We will present five of them – one in this chapter, one in the next, a third in Appendix 9.3 and a fourth and fifth as Problems 9.3 and 9.4.

The MVA is used in basically two ways:

◊ Outside of filter engineering the models are almost always *algebraic* equations.

◊ When used in filter engineering the models are *differential* equations.

This book is concerned with filter engineering and so we will consider only the second.

8.1.2 The MVA and filter names

The method of least squares (which we are now referring to as the MVA) forms the basis for all the filters that we discuss in this book.

The method of least squares was developed by many people over many years. Starting in the final decade of the eighteenth century, Gauss, Laplace, Legendre and others made their contributions, and in 1934 Alexander C. Aitken contributed further. The story is complex and lengthy, and well outside the scope of this book.[1]

Our decision to give the non-recursive filters discussed in this book the generic name 'Gauss filters' came about because Gauss played a founding role in the development.

The name 'Gauss–Aitken' for the ***Case 1*** filters was selected because Aitken was the second person who made major contributions to the formulation of the MVA.

The name 'Gauss–Newton' for the ***Cases 2, 3*** and ***4*** filters was selected because that is the name that has been in use for many years for the algorithm on which they are based.

In what now follows we deliberately refrain from stating who was responsible for which development, and will instead simply present the ideas.

8.2 The residuals

The residuals are the starting point for deriving the MVA. We commence by writing the total observation equation

$$\boldsymbol{Y}_n = \boldsymbol{T}_n \boldsymbol{X}_n + \boldsymbol{N}_n \tag{8.2.1}$$

◊ $\boldsymbol{Y}_n$ is a long-vector of observations that is formed by concatenation of the short-vectors $\boldsymbol{Y}_n, \boldsymbol{Y}_{n-1}, \ldots, \boldsymbol{Y}_{n-L}$. For simplicity we will assume that each of the short-vectors $\boldsymbol{Y}_{n-k}$ is a scalar, and so we are observing only a single quantity at

[1] The interested reader is referred References 10, 17 and 143, to the afterword in Reference 4 and to the further list of references contained in Reference 4.

each time-stage. However, what we derive would apply equally well if we were observing more than one quantity, and the Y vectors were of higher dimension.

◊ $\boldsymbol{T}_n$ is the total observation matrix defined in (3.2.12).

◊ X_n is the true state vector of the observed trajectory at time t_n.

◊ $\boldsymbol{N}_n$ is the long-vector of zero-mean errors in $\boldsymbol{Y}_n$.

◊ The covariance matrix of $\boldsymbol{Y}_n$ in (8.2.1) is $\boldsymbol{R}_{\boldsymbol{Y}_n}$

The total observation equation in (8.2.1) assumes that we are observing X_n. If we now replace X_n in that equation by any estimate vector $X^*_{n,n}$ we obtain

$$\boldsymbol{Y}_n = \boldsymbol{T}_n X^*_{n,n} + \boldsymbol{N}^*_n \tag{8.2.2}$$

Note that $\boldsymbol{Y}_n$ is still present on the left, but the error vector has changed from $\boldsymbol{N}_n$ to a vector called $\boldsymbol{N}^*_n$ in order that the equation will balance. We now make $\boldsymbol{N}^*_n$ the subject of the equation, obtaining

$$\boldsymbol{N}^*_n = \boldsymbol{Y}_n - \boldsymbol{T}_n X^*_{n,n} \tag{8.2.3}$$

Definition 8.1: The vector $\boldsymbol{N}^*_n$ in (8.2.3) is called the **residual vector with respect to $X^*_{n,n}$** and $\boldsymbol{T}_n X^*_{n,n}$ is called the **fitted observation vector**. ■■

From (8.2.3) we see the following:

◊ The residual vector $\boldsymbol{N}^*_n$ is the total observation vector $\boldsymbol{Y}_n$ minus the fitted observation vector $\boldsymbol{T}_n X^*_{n,n}$.

◊ The residual vector depends on the estimate vector $X^*_{n,n}$, and if $X^*_{n,n}$ were to change then so would $\boldsymbol{N}^*_n$. Hence the words *with respect to* $X^*_{n,n}$ in Definition 8.1.

The residual vector is depicted in Figure 8.1 below.

◊ The time axis and five observation instants are shown at the bottom of the figure.

◊ The small circles are the observations. We are assuming that each of the short observation vectors Y_{n-k} is a scalar, and so each of them can be represented by a single circle.

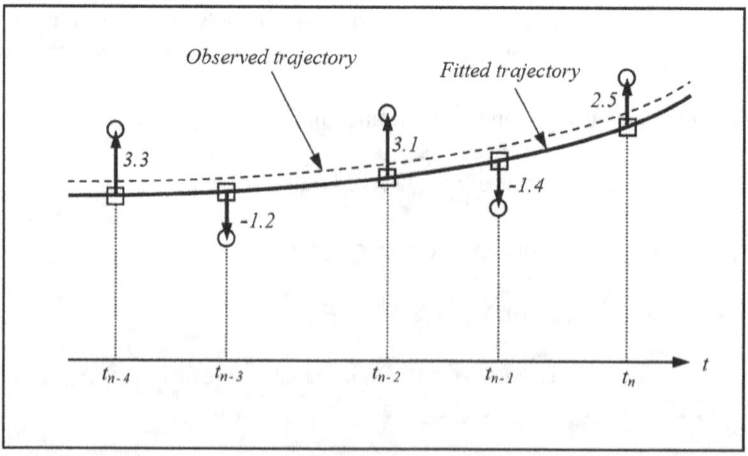

Figure 8.1 The residuals

◊ The dashed curve is the observed trajectory on which the observations were made. (This is discussed in Chapter 2, Subsection 2.3.4.) The observations seldom lie precisely on the dashed curve because of the observation errors, but they do nevertheless follow it.

◊ The solid black curve is the fitted trajectory based on $\boldsymbol{X}^*_{n,n}$.

◊ The small squares are the fitted observations that make up the vector $\boldsymbol{T}_n\boldsymbol{X}^*_{n,n}$.

◊ The arrows are the elements of the residual vector $\boldsymbol{N}^*_n$ – the actual observations minus the fitted observations. Up-arrows signify positive residuals and down-arrows negative ones.

We pause to make the following comment: The transition matrix $\boldsymbol{\Phi}$ is an essential component of the matrix $\boldsymbol{T}_n$. Transition matrices were discussed in Chapter 2 and we made extensive use of them in Chapter 3.

The transition matrix arises in filter engineering because we use differential equations to model the systems that we observe. By contrast, outside of filter engineering the models are frequently algebraic equations and for them the transition matrix $\boldsymbol{\Phi}$ does not arise.

Taken as a set, the residuals in Figure 8.1 form the 5-vector

$$\boldsymbol{N}^*_n = (\nu^*_1, \quad \nu^*_2, \quad \nu^*_3, \quad \nu^*_4, \quad \nu^*_5)^T_n \tag{8.2.4}$$

The numerical values shown in the figure are for the residuals with respect to the particular estimate $X^*_{n,n}$ appearing there, and so in this case the residual vector would evaluate to

$$\mathbf{N}^*_n = (2.5, \quad -1.4, \quad 3.1, \quad -1.2, \quad 3.3)^T \tag{8.2.5}$$

In Figure 8.1 we see how the residuals are anchored to the fitted trajectory $\mathbf{T}_n X^*_{n,n}$ and if $\mathbf{T}_n X^*_{n,n}$ were moved up or down by a change in $X^*_{n,n}$ then each of the values in (8.2.5) would change accordingly. The vector $\mathbf{N}^*_n$ is therefore clearly dependent on $X^*_{n,n}$.

8.2.1 *The sum of the weighted squared residuals*

As our first step in deriving the MVA we form the inner product

$$\mathbf{N}^{*T}_n \mathbf{N}^*_n = (\mathbf{Y}_n - \mathbf{T}_n X^*_{n,n})^T (\mathbf{Y}_n - \mathbf{T}_n X^*_{n,n}) \tag{8.2.6}$$

The scalar $\mathbf{N}^{*T}_n \mathbf{N}^*_n$ is then simply the sum of the squared residuals (*SSR*), namely

$$\mathbf{N}^*_n \mathbf{N}^*_n = SSR = (v^{*2}_1 + v^{*2}_2 + v^{*2}_3 + v^{*2}_4 + v^*_5)_n \tag{8.2.7}$$

For the particular set of values appearing in (8.2.5) *SSR* would evaluate as follows:

$$SSR = 2.5^2 + 1.4^2 + 3.1^2 + 1.2^2 + 3.3^2 = 30.15 \tag{8.2.8}$$

The next step in our development is to assume initially that the covariance matrix of $\mathbf{Y}_n$ is diagonal, namely that

$$\mathbf{R}_{\mathbf{Y}_n} = diag\,(\sigma^2_{y_1}, \quad \sigma^2_{y_2}, \quad \sigma^2_{y_3}, \quad \sigma^2_{y_4}, \quad \sigma^2_{y_5}) \tag{8.2.9}$$

We now rewrite (8.2.7) in the following way:

$$\mathbf{N}^{*T}_n \mathbf{R}_{\mathbf{Y}}^{-1} \mathbf{N}^*_n = SSR = \frac{v^{*2}_1}{\sigma^2_{y1}} + \frac{v^{*2}_2}{\sigma^2_{y2}} + \frac{v^{*2}_3}{\sigma^2_{y3}} + \frac{v^{*2}_4}{\sigma^2_{y4}} + \frac{v^{*2}_5}{\sigma^2_{y5}} \tag{8.2.10}$$

Observe how each squared residual in (8.2.10) is being *weighted* by the inverse of the variance of the error in its associated observation.

The rationale behind this weighting is as follows:

◊ The best $X^*_{n,n}$ will be the one that minimizes (8.2.10).

◊ Assume, for example, that y_1 and y_2 are by far the most accurate observations in the dataset, and so $\sigma^2_{y_1}$ and $\sigma^2_{y_2}$ are very much smaller than the other three variances.

◊ Dividing by the variances as shown in (8.2.10) would then result in the first two terms dominating the expression for *SSR* from which the best $X^*_{n,n}$ will be derived.

◊ Since the observations associated with the first two squared residuals are much more accurate than the others, this will lead to an $X^*_{n,n}$ that has better accuracy.

Another way to put it would be to say that dividing each squared residual by its associated observation-error variance will ensure that the final estimate will have a smaller variance.

We may not immediately grasp the fact that that the *smaller* variance will be the smallest *possible* variance – as in Cramér–Rao – but in Chapter 9 we will prove that to be the case.[2] The step taken in (8.2.10) therefore has the profoundest of implications.

The next step is to permit $\boldsymbol{R}_{\boldsymbol{Y}_n}$ to be *any* positive-definite matrix whatever, and not limited to the diagonal form that was used in (8.2.9). Then this gives us the following:

Definition 8.2: The **sum of the weighted squared residuals** is the scalar quantity $e(X^*_{n,n})$ defined by ■■

$$e(X^*_{n,n}) \equiv \boldsymbol{N}^{*T}_n \boldsymbol{R}^{-1}_{\boldsymbol{Y}_n} \boldsymbol{N}^*_n = (\boldsymbol{Y}_n - \boldsymbol{T}_n X^*_{n,n})^T \boldsymbol{R}^{-1}_{\boldsymbol{Y}_n} (\boldsymbol{Y}_n - \boldsymbol{T}_n X^*_{n,n}) \tag{8.2.11}$$

◊ Observe first that equation (8.2.11) includes the expression in (8.2.10) because if $\boldsymbol{R}_{\boldsymbol{Y}_n}$ has the simple form shown in (8.2.9) then (8.2.11) immediately gives us

$$\boldsymbol{N}^{*T}_n \boldsymbol{R}^{-1}_{\boldsymbol{Y}_n} \boldsymbol{N}^*_n = \frac{v^{*2}_1}{\sigma^2_{y_1}} + \frac{v^{*2}_2}{\sigma^2_{y_2}} + \frac{v^{*2}_3}{\sigma^2_{y_3}} + \frac{v^{*2}_4}{\sigma^2_{y_4}} + \frac{v^{*2}_5}{\sigma^2_{y_5}} \tag{8.2.12}$$

◊ Observe next in (8.2.11) that *SSR* is now called $e(X^*_{n,n})$ which is shown as a function of the vector $X^*_{n,n}$. This makes sense, because $X^*_{n,n}$ is the only unknown on the right of (8.2.11), and hence the only item that can affect the value of $e(X^*_{n,n})$.

◊ Finally, observe how (8.2.11) states the sum of the squared residuals in *matrix format*, an essential step in the development of the filters to come.

Regarding the quantity $e(X^*_{n,n})$ we draw the reader's attention to the following:

◊ The matrix $\boldsymbol{R}_{\boldsymbol{Y}_n}$ in (8.2.11) is always positive-definite, and so by Properties 4 and 5 in Section 4.4 its inverse exists and is also positive-definite.

[2] Keep in mind that *smallest possible variance* also means *most accurate estimate*.

◊ The vector $\boldsymbol{N}^*_n$ will never be the zero vector.

Thus $e(\boldsymbol{X}^*_{n,n}) = \boldsymbol{N}^{*T}_n \boldsymbol{R}^{-1}_{\boldsymbol{Y}_n} \boldsymbol{N}^*_n$ is a quadratic form on a positive-definite matrix and so it can only be positive. Therefore, it indeed must have a minimum.

8.3 Deriving the MVA: Method 1

We will derive the MVA in two ways, the first of which is to solve the following:

Problem Statement 1: Starting from the definition in (8.2.11), find the *best* $\boldsymbol{X}^*_{n,n}$, namely the one that minimizes $e(\boldsymbol{X}^*_{n,n})$. ■■

The second derivation is based on what we call *Problem Statement* 2, and will be considered in Chapter 9.

We noted above that the scalar $e(\boldsymbol{X}^*_{n,n})$ in (8.2.11) is a quadratic form on the positive-definite matrix $\boldsymbol{R}^{-1}_{\boldsymbol{Y}_n}$, and so it is always positive. In Figure 8.2 we show a paraboloidal surface that represents $e(\boldsymbol{X}^*_{n,n})$ for the case where $\boldsymbol{X}^*_{n,n}$ is a 2-vector, and we note its similarity to Figure 4.5.

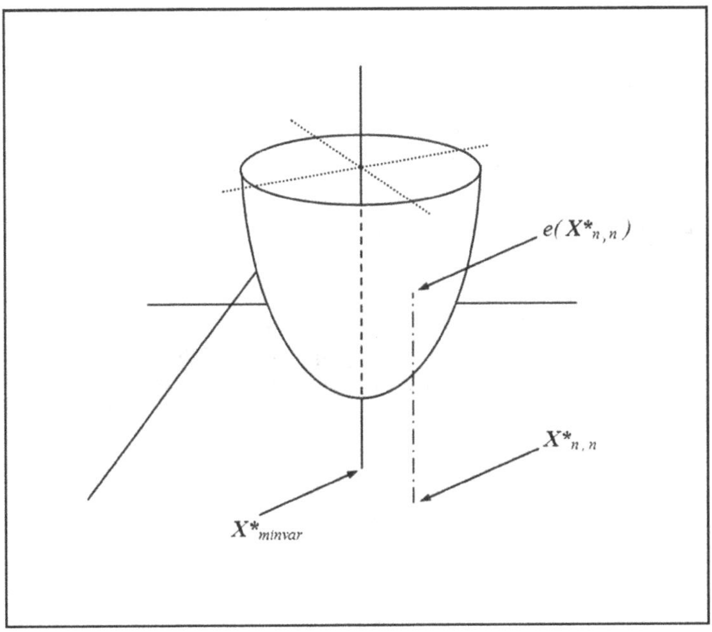

Figure 8.2 $e(\boldsymbol{X}^*_{n,n})$ *as a surface*

◊ The horizontal plane beneath the surface is conceptually the $\boldsymbol{X}^*_{n,n}$ plane.

◊ In Figure 4.5 the surface touched the origin, but here it is shifted up and away from it for two reasons:

  ◊ The vector $\boldsymbol{Y}_n - \boldsymbol{T}_n \boldsymbol{X}^*_{n,n}$, in (8.2.11) can never be the zero vector. This shifts the surface upwards.

  ◊ The vector $\boldsymbol{Y}_n$ is present in $\boldsymbol{Y}_n - \boldsymbol{T}_n \boldsymbol{X}^*_{n,n}$ and this shifts the surface sideways.

As we move around the horizontal plane beneath the surface we are conceptually ranging through all possible values of $\boldsymbol{X}^*_{n,n}$, each of which gives rise to a point on the surface above.

From the figure we see that there is a smallest value on the surface, and we are seeking the point in the plane called $\boldsymbol{X}^*_{minvar}$ that gives rise to that smallest value.

Note: The subscript *minvar* stands for *minimum variance*, but we are only showing here that $\boldsymbol{X}^*_{minvar}$ causes $e(\boldsymbol{X}^*_{n,n})$ to have its smallest value. In Chapter 9 we will prove that $\boldsymbol{X}^*_{minvar}$ is also the *minimum variance estimate*, and we will explain precisely what that means.

8.3.1 Finding $\boldsymbol{X}^*_{minvar}$

By analogy to what we would do in the scalar case, to find $\boldsymbol{X}^*_{minvar}$ we differentiate $e(\boldsymbol{X}^*_{n,n})$ in (8.2.11) with respect to $\boldsymbol{X}^*_{n,n}$ and set the result equal to $\boldsymbol{\emptyset}$.

Forming the derivative with respect to the vector $\boldsymbol{X}^*_{n,n}$ of the quantity

$$e(\boldsymbol{X}^*_{n,n}) = (\boldsymbol{Y}_n - \boldsymbol{T}_n\boldsymbol{X}^*_{n,n})^T \boldsymbol{R}_{\boldsymbol{Y}_n}^{-1}(\boldsymbol{Y}_n - \boldsymbol{T}_n\boldsymbol{X}^*_{n,n}) \tag{8.3.1}$$

is not difficult, but in order to keep the developments moving we carry it out in Appendix 8.1. The result obtained there is as follows:

$$\boldsymbol{X}^*_{minvar} = \boldsymbol{W}_{minvar}\boldsymbol{Y}_n \tag{8.3.2}$$

where

$$\boldsymbol{W}_{minvar} = (\boldsymbol{T}_n^T\boldsymbol{R}_{\boldsymbol{Y}_n}^{-1}\boldsymbol{T}_n)^{-1}\boldsymbol{T}_n^T\boldsymbol{R}_{\boldsymbol{Y}_n}^{-1} \tag{8.3.3}$$

Using (8.3.2) for $\boldsymbol{X}^*_{n,n}$ in (8.2.11) will then cause the scalar $e(\boldsymbol{X}^*_{n,n})$ to have its smallest possible value.

We need one more expression, namely the expression for the filter covariance matrix $\boldsymbol{S}^*_{minvar}$ that goes with $\boldsymbol{X}^*_{minvar}$. For this we note that (8.3.2) gives $\boldsymbol{X}^*_{minvar}$ as a linear transformation on $\mathbf{Y}_n$ and so we can apply Theorem 4.1, obtaining

$$\boldsymbol{S}^*_{minvar} = \boldsymbol{W}_{minvar}\mathbf{R}_{\mathbf{Y}_n}\boldsymbol{W}^T_{minvar} \tag{8.3.4}$$

Taking the above three equations together then gives us the following:

The MVA (Version 1)

$$\boldsymbol{W}_n = (\mathbf{T}_n^T\mathbf{R}_{\mathbf{Y}_n}^{-1}\mathbf{T}_n)^{-1}\mathbf{T}_n^T\mathbf{R}_{\mathbf{Y}_n}^{-1} \tag{8.3.5}$$

$$\boldsymbol{X}^*_{n,n} = \boldsymbol{W}_n\mathbf{Y}_n \tag{8.3.6}$$

$$\boldsymbol{S}^*_{n,n} = \boldsymbol{W}_n\mathbf{R}_{\mathbf{Y}_n}\boldsymbol{W}_n^T \tag{8.3.7}$$

We will return to Version 1 further down.

8.3.2 *Obtaining Version 2 of the algorithm*

In Appendix 8.2 we show that $\boldsymbol{S}^*_{n,n}$ in (8.3.7) reduces to

$$\boldsymbol{S}^*_{n,n} = (\mathbf{T}_n^T\mathbf{R}_{\mathbf{Y}_n}^{-1}\mathbf{T}_n)^{-1} \tag{8.3.8}$$

The right-hand side of this equation appears in (8.3.5), and so if we replace (8.3.7) by (8.3.8) then we can restate the MVA as follows:

The MVA (Version 2)

$$\boldsymbol{S}^*_{n,n} = (\mathbf{T}_n^T\mathbf{R}_{\mathbf{Y}_n}^{-1}\mathbf{T}_n)^{-1} \tag{8.3.9}$$

$$\boldsymbol{W}_n = \boldsymbol{S}^*_{n,n}\mathbf{T}_n^T\mathbf{R}_{\mathbf{Y}_n}^{-1} \tag{8.3.10}$$

$$\boldsymbol{X}^*_{n,n} = \boldsymbol{W}_n\mathbf{Y}_n \tag{8.3.11}$$

This is the way in which the MVA is usually presented. We note that Version 1 and Version 2 differ in their $\boldsymbol{S}^*$ equations, and also that Version 2 avoids two of the seven matrix multiplications that are present in Version 1.

The two versions yield identical results, and because Version 2 requires less computation than Version 1 – other things being equal – that would be the one to use. However, later we will see that Version 1 offers us a valuable engineering option that is not offered by Version 2.

Before continuing, we note the following:

The expression for the MVA filter matrix in both versions is

$$W_{minvar} = (T_n^T R_{Y_n}^{-1} T_n)^{-1} T_n^T R_{Y_n}^{-1} \tag{8.3.12}$$

where R_{Y_n} is always positive-definite. In order to compute W_{minvar} it is necessary that the matrix $T_n^T R_{Y_n}^{-1} T_n$ be non-singular so that it can be inverted, and by Theorem 4.4 $T_n^T R_{Y_n}^{-1} T_n$ *will* be non-singular if and only if T_n has full column-rank. From the discussion given in Project 5.3, T_n will have full column-rank if we are making observations in such a way that the required estimates can be obtained from them.

◊ If the matrix $T_n^T R_{Y_n}^{-1} T_n$ is then readily inverted, that will be a confirmation that the observation scheme which we have established is well constructed, and the expression for W_{minvar} will then be easily computed.

◊ However, if $T_n^T R_{Y_n}^{-1} T_n$ is difficult to invert, or possibly singular and non-invertible, then that will be a warning to us that the observation scheme is not well constructed, and that we need to re-examine its details.[3]

8.4 Attributes of the MVA

The MVA possesses an extensive list of attributes. We state a few of them here and others will emerge later.

◊ Equations (8.3.6) and (8.3.11) are both of the form

$$X^* = WY \tag{8.4.1}$$

The minimum variance estimate X^*_{minvar} is thus a linear transformation on the observation vector Y.

◊ In Chapter 6 we derived the exactness constraint, expressed in (6.5.6) as

$$WT = I \tag{8.4.2}$$

If we post-multiply the expression for W_{minvar} in either version of the MVA by T_n we obtain

$$W_{minvar} T_n = \left((T_n^T R_{Y_n}^{-1} T_n)^{-1} T_n^T R_{Y_n}^{-1} \right) T_n = I \tag{8.4.3}$$

[3] Whenever a matrix inversion is called for, it is wise to include in one's code a validation of the result by multiplying the derived inverse by the original matrix, and then checking to see if the product is close to an identity matrix. It has been our experience that computer languages cannot always be relied on to warn one that a matrix which is to be inverted is singular or nearly so, especially when the matrices involved are large.

The MVA therefore satisfies the exactness constraint, and so the estimates that it produces will be unbiased.

◊ From (8.4.3) we see that as $\boldsymbol{T}_n$ changes with n, so $\boldsymbol{W}_{minvar}$ will also change in just the right way for the exactness constraint always to be satisfied.

◊ We can use a different positive-definite matrix $\boldsymbol{R}_d$ in $\boldsymbol{W}$ (i.e. one that that differs from $\boldsymbol{R}_{\boldsymbol{Y}_n}$) and the exactness constraint will *still* be satisfied. Thus

$$\left((\boldsymbol{T}_n^T \boldsymbol{R}_d^{-1} \boldsymbol{T}_n)^{-1} \boldsymbol{T}_n^T \boldsymbol{R}_d^{-1}\right) \boldsymbol{T}_n = \boldsymbol{I} \tag{8.4.4}$$

Later we will see why we might wish to do this.

◊ By Theorem 6.2, if the matrices $\boldsymbol{W}$ and $\boldsymbol{T}$ satisfy $\boldsymbol{W}\boldsymbol{T} = \boldsymbol{I}$ then $\boldsymbol{W}$ will have full row-rank.

From (8.4.3) we see that $\boldsymbol{W}_{minvar}$ and $\boldsymbol{T}$ do indeed always satisfy $\boldsymbol{W}\boldsymbol{T} = \boldsymbol{I}$, and so $\boldsymbol{W}_{minvar}$ will always have full row-rank.

◊ Both versions of the MVA are ECM consistent. (We discuss this further down.)

◊ Both versions are of the MVA are also CR consistent. (This is also discussed further down.)

◊ In Section 8.7 below we will see what is perhaps the most fascinating property of the MVA – its ability to *mutate* between a non-recursive and a recursive form.

◊ Finally, we note the following: The only challenging operation that is required in order to obtain $\boldsymbol{W}_{minvar}$ in (8.3.5) or (8.3.10) is to build the matrix $\boldsymbol{T}$. After that the operations are simply matrix transposition, inversion and multiplication. For this reason we devoted all of Chapter 3 and its appendices to building the $\boldsymbol{T}$ matrix.

8.4.1 The ECM consistency of the MVA

The issue of ECM consistency in a filter arises because of the unfortunate proclivity of the extended Kalman and Swerling filters to become unstable, or what we call *spontaneously ECM inconsistent* (see Chapter 11), and were it not for that, then the ECM consistency of properly constructed and fully de-bugged algorithms could always be taken for granted.

In this section we review briefly the steps that have been taken to ensure that the MVA is ECM consistent.

The MVA is ECM consistent for the following reason. The actual estimation-error covariance matrix is defined by

$$\boldsymbol{S}_{actual} \equiv \underset{M\to\infty}{\text{Lim}} \frac{1}{M}\sum_{r=1}^{M} \boldsymbol{N}_r^* \boldsymbol{N}_r^{*T} \tag{8.4.5}$$

in which N_r^* is the rth draw from the actual estimation-error population defined by

$$\boldsymbol{N}_r^* = \boldsymbol{X}^*_{minvar} - \boldsymbol{X} \tag{8.4.6}$$

Theorem 4.1 states that if a vector $\boldsymbol{X}^*_{minvar}$ is produced from an observation vector $\boldsymbol{Y}$ according the equation

$$\boldsymbol{X}^*_{minvar} = \boldsymbol{W}_{minvar}\boldsymbol{Y} \tag{8.4.7}$$

then the correct mathematical formula for the quantity on the right of (8.4.5) will be

$$\boldsymbol{S}^*_{minvar} = \boldsymbol{W}_{minvar}\boldsymbol{R}_{\boldsymbol{Y}}\boldsymbol{W}_{minvar} \tag{8.4.8}$$

Since (8.4.8) is what Version 1 of the MVA uses, it follows that it is ECM consistent.

Putting it another way, of all the formulae for $\boldsymbol{S}^*$ that we might have included in Version 1 of the MVA, (8.4.8) is the best one to use if we expect that the numbers obtained from evaluating that formula for $\boldsymbol{S}^*$ will be equal to the numbers that we obtain from evaluating (8.4.5) for $\boldsymbol{S}_{actual}$.

The expression for $\boldsymbol{S}^*_{minvar}$ in Version 2 is algebraically equivalent to the one in Version 1, and so Version 2 is also ECM consistent.

From a practical standpoint, $\boldsymbol{S}_{actual}$ in (8.3.5) is estimated by

$$\boldsymbol{S}_{actual} \approx \frac{1}{M}\sum_{r=1}^{M} \boldsymbol{N}_r^* \boldsymbol{N}_r^{*T} \tag{8.4.9}$$

ECM consistency means that if the formula for $\boldsymbol{S}^*$ in (8.4.8) were to be evaluated at each instant t_n along a flight path, then the results would be the same as those that would be obtained if (8.4.9) were being used to estimate $\boldsymbol{S}_{actual}$ at each of those instants.

However, it is impossible in practice to evaluate (8.4.9) at each instant along a flight path because that requires that we run a Monte-Carlo for each of those instants, and it is for this reason that we use the formula for $\boldsymbol{S}^*$ instead. Because the algorithm is ECM consistent, we can then depend on $\boldsymbol{S}^*$ to give us accurate values for $\boldsymbol{S}_{actual}$.

For Version 1 we state this as follows:

$$1/M \sum_{r=1}^{M} \boldsymbol{N}^{*}_{n,n,r} \boldsymbol{N}^{*T}_{n,n,r} \approx \boldsymbol{W}_{minvar,\ n,n} \boldsymbol{R}_{\boldsymbol{Y}_n} \boldsymbol{W}^{T}_{minvar,\ n,n} \quad \text{(for all } t_n) \quad (8.4.10)$$

and for Version 2 we state it as

$$1/M \sum_{r=1}^{M} \boldsymbol{N}^{*}_{n,n,r} \boldsymbol{N}^{*T}_{n,n,r} \approx (\boldsymbol{T}_n^T \boldsymbol{R}_{\boldsymbol{Y}_n}^{-1} \boldsymbol{T}_n)^{-1} \quad \text{(for all } t_n) \quad (8.4.11)$$

8.4.2 Empirical verification

To verify empirically that a given configuration of the MVA is ECM consistent, we must program it and then verify its ECM consistency by running one or more of the three tests discussed in Chapter 7.

With this in mind we always run at least some of those tests in any filters that we create, and we do that testing at multiple points on a number of trajectories.

The following 8 programs make use of the MVA in either a Gauss–Aitken filter or a Gauss–Newton filter, some of them in very complex configurations:

03_Gauss_Aitken ***04_Monte_Carlo*** ***05_Pol_Filter_EMP***

06_Pol_Filter_FMP ***11_Gauss_Newton_A*** ***12_Gauss_Newton_B***

13_Gauss_Newton_C ***14_Orbit_Track***

The final two also include Kalman and Swerling filters that process the same data as the Gauss filters.

Each of these programs includes at least two of the three ECM-consistency tests discussed in Chapter 7, and over the years we have run most of these programs hundreds of times, applying the ECM-consistency tests over the allowable ranges of the parameters in each of the configurations.

We have never yet seen a case where the MVA, as implemented in a de-bugged Gauss–Aitken or Gauss–Newton filter, can be declared to be ECM inconsistent.

Clearly we cannot conduct verifications on *every possible* configuration of the MVA. However, the ones listed above are sufficiently diverse so that we can regard our verification as having been sufficient.

By contrast, we have often seen both the extended Kalman and Swerling filters becoming unstable, that is ECM inconsistent, and both of them are also based on the MVA.

However, the problem there is not with the MVA, but with the structures of the extended Kalman and Swerling filters. Using precisely the same data as processed by those filters, the Gauss–Newton filter running in the same programs always continues to be ECM consistent. This will be discussed further in Chapter 11.

8.4.3 The CR consistency of the MVA

Unlike ECM consistency discussed earlier, the CR consistency of the MVA is something that *can* be proved mathematically.

In Section 1.10 we discussed the multivariate Cramér–Rao lower bound and we mentioned the existence of the Cramér–Rao covariance matrix $\boldsymbol{S}_{CR}$.

In Chapter 9 we will show that, for a given set of observations, the diagonal elements of the matrices $\boldsymbol{S}^*_{minvar}$ in (8.3.7) or (8.3.9) and those of the Cramér–Rao covariance matrix $\boldsymbol{S}_{CR}$ corresponding to those observations, satisfy

$$[\boldsymbol{S}^*_{minvar}]_{i,\,i} = [\boldsymbol{S}_{CR}]_{i,\,i} \qquad 1 \le i \le m \tag{8.4.12}$$

and that this is true for observation errors $\boldsymbol{N}$ with *any* PDF.

We will also show that if the PDF of $\boldsymbol{N}$ is multivariate Gaussian, then the two *entire* matrices (i.e. both the on-diagonal and the off-diagonal elements) are equal to the same matrix, namely

$$\boldsymbol{S}^*_{minvar} = (\boldsymbol{T}^T \boldsymbol{R}_Y^{-1} \boldsymbol{T})^{-1} = \boldsymbol{S}_{CR} \tag{8.4.13}$$

Either way – Gaussian or not – the diagonal elements of $\boldsymbol{S}^*_{minvar}$ and those of $\boldsymbol{S}_{CR}$ will be shown to be equal.

Observe that we are talking here about the diagonal elements of the filter matrix $\boldsymbol{S}^*_{minvar}$ in relation to those of $\boldsymbol{S}_{CR}$, whereas (see Figure 1.8) the definition of CR consistency requires that the diagonal elements of $\boldsymbol{S}_{actual}$ be equal to the diagonal elements of $\boldsymbol{S}_{CR}$.

However, as discussed earlier, the MVA is ECM consistent, which means that $\boldsymbol{S}^*_{minvar}$ yields the same values as those contained in $\boldsymbol{S}_{actual}$. Accordingly, the diagonal elements of the three matrices must satisfy the following:

$$[\boldsymbol{S}^*_{minvar}]_{i,\,i} = [\boldsymbol{S}_{actual}]_{i,\,i} = [\boldsymbol{S}_{CR}]_{i,\,i} \qquad 1 \le i \le m \tag{8.4.14}$$

and from the last two of these and Definition 1.6 we can conclude that the MVA is CR consistent.

The matrix $\boldsymbol{R_Y}$ is normally used in both MVA versions – in the $\boldsymbol{W}$ equations (8.3.5) and (8.3.10) as well as in the $\boldsymbol{S}^*$ equations (8.3.7) and (8.3.9).

Later on we will explore why we might wish to use a matrix $\boldsymbol{R_d}$ that differs from $\boldsymbol{R_Y}$ in the $\boldsymbol{W}$ equations. However, for now everything that we have derived assumes that $\boldsymbol{R_Y}$ is used in all four places. We state this as follows:

***Theorem 8.1**:* If we use $\boldsymbol{R_Y}$ in both the $\boldsymbol{W}$ and $\boldsymbol{S}^*$ equations in either version of the MVA then the algorithm is both ECM and CR consistent. ■■

8.5 The Gauss–Aitken filters – Versions 1 and 2

Version 1 of the Gauss–Aitken filter is shown in Figure 8.3 – a direct implementation of Version 1 of the MVA.

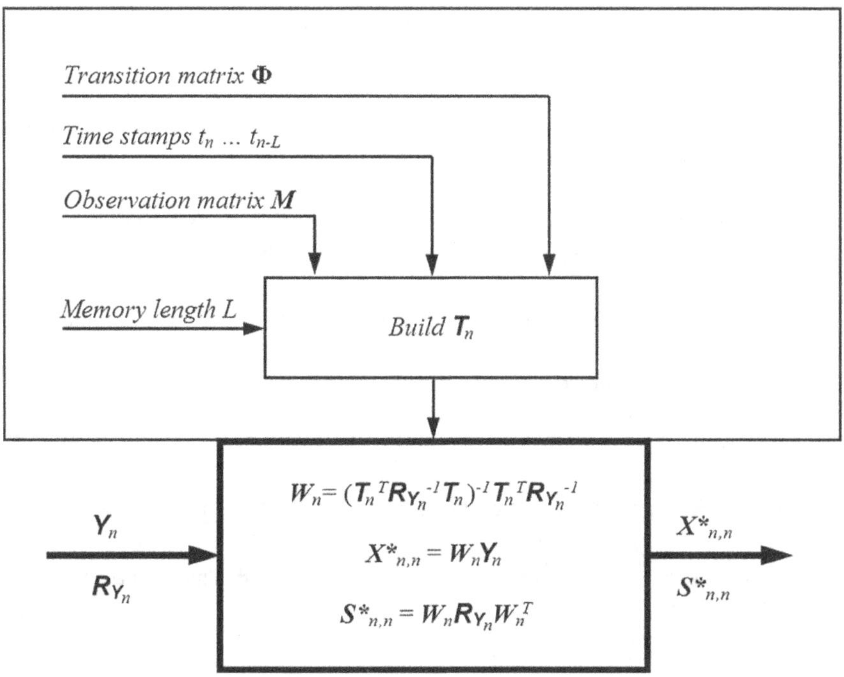

Figure 8.3 The Gauss–Aitken filter (Version 1)

Lower down we discuss how the filter is actually operated, but first we consider what is required to complete one of its cycles.

Task 1 (*the upper box*):

The first task on every cycle of the filter is to build the matrix $\boldsymbol{T}$. From (3.2.12) its full expression is

$$\boldsymbol{T}_n = \begin{pmatrix} \boldsymbol{M} \\ \hdashline \boldsymbol{M}\boldsymbol{\Phi}(t_{n-1} - t_n) \\ \hdashline \vdots \\ \hdashline \boldsymbol{M}\boldsymbol{\Phi}(t_{n-L} - t_n) \end{pmatrix} \tag{8.5.1}$$

Building $\boldsymbol{T}_n$ thus requires the following:

(1) Import the memory-length parameter L and the time stamps appearing in (8.5.1).

(2) Evaluate the transition matrices $\boldsymbol{\Phi}(t_n - t_n)$, $\boldsymbol{\Phi}(t_{n-1} - t_n)$,… , $\boldsymbol{\Phi}(t_{n-L} - t_n)$ using the expression for $\boldsymbol{\Phi}$ for the selected model.

(3) Pre-multiply each of the $\boldsymbol{\Phi}$ matrices in Step (2) by the matrix $\boldsymbol{M}$ for the selected observation equation.

(4) Assemble the products into $\boldsymbol{T}_n$ as shown in (8.5.1) and Task 1 is complete.

Once $\boldsymbol{T}_n$ has been assembled, continue as follows:

Task 2 (*the lower box*):

(1) Import $\boldsymbol{R}_{\boldsymbol{Y}_n}$ and find its inverse (this matrix is of the same order as $\boldsymbol{Y}_n$).

(2) Assemble the matrix $\boldsymbol{T}_n^T \boldsymbol{R}_{\boldsymbol{Y}_n}^{-1} \boldsymbol{T}_n$ and find its inverse (this matrix is of the same order as $\boldsymbol{X}^*_{n,n}$).

(3) Form the filter matrix $\boldsymbol{W}_n = (\boldsymbol{T}_n^T \boldsymbol{R}_{\boldsymbol{Y}_n}^{-1} \boldsymbol{T}_n)^{-1} \boldsymbol{T}_n^T \boldsymbol{R}_{\boldsymbol{Y}_n}^{-1}$.

(4) Import $\boldsymbol{Y}_n$ and pre-multiply it by $\boldsymbol{W}_n$ to obtain the estimate $\boldsymbol{X}^*_{n,n}$.

(5) Form $\boldsymbol{W}_n \boldsymbol{R}_{\boldsymbol{Y}} \boldsymbol{W}_n^T$ to obtain the filter covariance matrix $\boldsymbol{S}^*_{n,n}$, and one cycle of the filter is complete.

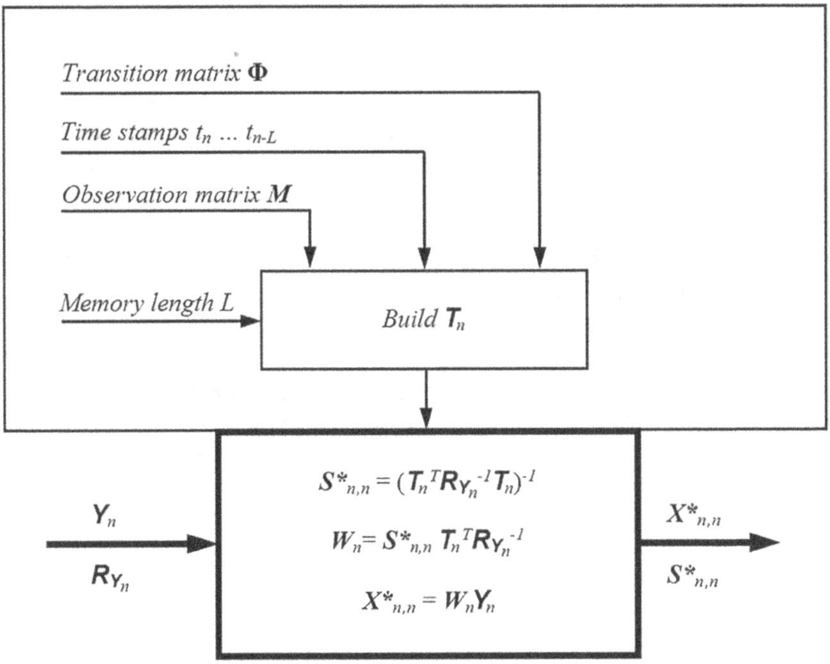

Figure 8.4 The Gauss–Aitken filter (Version 2)

Version 2 of the Gauss–Aitken filter shown in Figure 8.4 is similarly the direct implementation of Version 2 of the MVA.

Task 1 is the same as for Version 1, but Task 2 now changes slightly and we have the following:

Task 2 (*the lower box*):

(1) Import $\boldsymbol{R}_{\boldsymbol{Y}_n}$ and find its inverse.

(2) Assemble the matrix $\boldsymbol{T}_n^T \boldsymbol{R}_{\boldsymbol{Y}_n}^{-1}$ and save it.

(3) Post-multiply $\boldsymbol{T}_n^T \boldsymbol{R}_{\boldsymbol{Y}_n}^{-1}$ by $\boldsymbol{T}_n$ and find the inverse of the product. This gives the filter covariance matrix $\boldsymbol{S}_{n,n}^* = (\boldsymbol{T}_n^T \boldsymbol{R}_{\boldsymbol{Y}_n}^{-1} \boldsymbol{T}_n)^{-1}$.

(4) Post-multiply $\boldsymbol{S}_{n,n}^*$ by $\boldsymbol{T}_n^T \boldsymbol{R}_{\boldsymbol{Y}_n}^{-1}$ from Step (2) to form $\boldsymbol{W}_n = \boldsymbol{S}_{n,n}^* \boldsymbol{T}_n^T \boldsymbol{R}_{\boldsymbol{Y}_n}^{-1}$.

(5) Import $\boldsymbol{Y}_n$ and pre-multiply it by $\boldsymbol{W}_n$ to obtain the estimate $\boldsymbol{X}_{n,n}^*$ and one cycle of the filter is complete.

8.5.1 Operating the Gauss–Aitken filters

We consider now the two stages required to operate either of the Gauss–Aitken filters.

8.5.1.1 Stage 1: wait until sufficient observations have been received

The Gauss–Aitken filters require no initialization, and all that we have to do is to wait until sufficient observations have been received. We consider what this means.

We are in ***Case 1*** in which the filter model and the observation equation are both linear. In the following discussion we assume for simplicity that the filter model is based on a 1-dimensional 1st-degree polynomial.

If the filter model is instead based on 3-dimensional polynomials, as in (2.3.9) or (2.3.10), or on some other constant-coefficient linear DE, then the reader should have no difficulty in making the necessary extensions. Non-linear models and observation equations will be considered in Chapter 9.

For a 1-dimensional 1st-degree polynomial the observed trajectory's state vector is $\boldsymbol{X}_n = (x\,, \dot{x})_n^T$ for which the transition matrix is

$$\boldsymbol{\Phi}(\zeta) = \begin{pmatrix} 1 & \zeta \\ 0 & 1 \end{pmatrix} \tag{8.5.2}$$

We are in ***Case 1*** and so the observation equation is

$$\boldsymbol{Y}_n = \boldsymbol{M}\boldsymbol{X}_n + \boldsymbol{N}_n \tag{8.5.3}$$

and assuming that we are observing only x, this equation becomes

$$y_n = \begin{pmatrix} 1 & 0 \end{pmatrix} \begin{pmatrix} x \\ \dot{x} \end{pmatrix}_n + \nu_n \tag{8.5.4}$$

From (8.5.4) we see that

$$\boldsymbol{M} = \begin{pmatrix} 1 & 0 \end{pmatrix} \tag{8.5.5}$$

Let the time at which the first observation was received be called t_1.

We assume for simplicity that the observations $y_1, y_2, \ldots, y_n$ are uncorrelated and that the variances of their errors are respectively $r_1, r_2, \ldots, r_n$. Then the covariance matrix of the long-vector $\boldsymbol{Y}_n$ will be the diagonal matrix $\boldsymbol{R}_{\boldsymbol{Y}_n} = diag(r_1, r_2, \ldots, r_n)$. Given $\boldsymbol{M}$ and $\boldsymbol{\Phi}$ and the time stamps as specified above, the total observation matrix becomes

$$\boldsymbol{T}_n = \begin{pmatrix} \boldsymbol{M\Phi}(t_n - t_n) \\ \hdashline \boldsymbol{M\Phi}(t_{n-1} - t_n) \\ \hdashline \vdots \\ \hdashline \boldsymbol{M\Phi}(t_1 - t_n) \end{pmatrix} = \begin{pmatrix} 1 & 0 \\ 1 & t_{n-1} - t_n \\ \vdots & \vdots \\ 1 & t_1 - t_n \end{pmatrix} \tag{8.5.6}$$

Both Versions 1 and 2 of the Gauss–Aitken filter call for two matrices to be inverted on each cycle, namely $\boldsymbol{R}_{\boldsymbol{Y}_n}$ and $\boldsymbol{T}_n^T \boldsymbol{R}_{\boldsymbol{Y}_n}^{-1} \boldsymbol{T}_n$.

In our present example we assume that $\boldsymbol{R}_{\boldsymbol{Y}_n}$ is diagonal, and so its inverse is simply

$$\boldsymbol{R}_{\boldsymbol{Y}_n}^{-1} = diag(1/r_1, 1/r_2, \ldots, 1/r_n) \tag{8.5.7}$$

Note: In Section 8.8 we discuss what is involved in inverting $\boldsymbol{R}_{\boldsymbol{Y}_n}$ when it is not diagonal or not even block-diagonal.

By Theorem 4.4 the inversion of the second matrix $\boldsymbol{T}_n^T \boldsymbol{R}_{\boldsymbol{Y}_n}^{-1} \boldsymbol{T}_n$ will be possible if and only if $\boldsymbol{T}_n$ in (8.5.6) has full column-rank, and so our ability to cycle the Gauss–Aitken filters then depends on the column-rank of $\boldsymbol{T}_n$.

From (8.5.6) the matrix $\boldsymbol{T}$ at time t_1 will be

$$\boldsymbol{T}_1 = \boldsymbol{M} = (1 \quad 0) \tag{8.5.8}$$

This is a *2*-column matrix with rank *1*, and so its column-rank is less than full. As the following verifies, $\boldsymbol{T}_1^T \boldsymbol{R}_{\boldsymbol{Y}_1}^{-1} \boldsymbol{T}_1$ will not be invertible.

At time t_1 the variance of the errors in the observation y_1 is r_1 and so we have

$$\boldsymbol{T}_1^T \boldsymbol{R}_{\boldsymbol{Y}_1}^{-1} \boldsymbol{T}_1 = \begin{pmatrix} 1 \\ 0 \end{pmatrix} \frac{1}{r_1} (1 \quad 0) = \begin{pmatrix} 1/r_1 & 0 \\ 0 & 0 \end{pmatrix} \tag{8.5.9}$$

The final matrix in this equation cannot be inverted and we can hardly expect otherwise, for if it were invertible then we would be able to complete a cycle of the filter and obtain an estimate of both x and $\dot{x}$ from only the single observation y_1.

Waiting until time t_2 when the second observation is received, (8.5.6) becomes

$$\boldsymbol{T}_2 = \begin{pmatrix} \boldsymbol{M} \\ \text{-------------} \\ \boldsymbol{M\Phi}(t_1 - t_2) \end{pmatrix} = \begin{pmatrix} 1 & 0 \\ 1 & t_1 - t_2 \end{pmatrix} \tag{8.5.10}$$

This is a 2-column matrix with rank 2 (assuming that $t_1 \neq t_2$) and so its column-rank is now full. Using $\boldsymbol{T}_2$ we obtain

$$\boldsymbol{T}_2^T \boldsymbol{R}_{\boldsymbol{Y}_2}^{-1} \boldsymbol{T}_2 = \begin{pmatrix} 1 & 1 \\ 0 & t_1 - t_2 \end{pmatrix} \begin{pmatrix} 1/r_1 & 0 \\ 0 & 1/r_2 \end{pmatrix} \begin{pmatrix} 1 & 0 \\ 1 & t_1 - t_2 \end{pmatrix}$$

$$= \begin{pmatrix} 1/r_1 + 1/r_2 & (t_1 - t_2)/r_2 \\ (t_1 - t_2)/r_2 & (t_1 - t_2)^2/r_2 \end{pmatrix} \tag{8.5.11}$$

The determinant of the final matrix is $(t_1 - t_2)^2(r_1 r_2)$ which is nonzero (assuming again that $t_1 \neq t_2$). As correctly predicted by Theorem 4.4, the matrix $\boldsymbol{T}_2^T \boldsymbol{R}_{\boldsymbol{Y}_2}^{-1} \boldsymbol{T}_2$ is now invertible and operation of the filter can commence.

8.5.1.2 Stage 2: operating the filters

Once sufficient observations have been received, operation of either the Version 1 or the Version 2 filter proceeds as follows:

◊ At time t_2 the filter is cycled, using as the input $\boldsymbol{Y}_2 = (y_1,\ y_2)^T$ and the matrix $\boldsymbol{R}_{\boldsymbol{Y}_2} = diag(r_1,\ r_2)$. The outputs will be $\boldsymbol{X}^*_{2,2}$ and $\boldsymbol{S}^*_{2,2}$.

◊ At time t_3 the filter is cycled, using as the input $\boldsymbol{Y}_3 = (y_1,\ y_2,\ y_3)^T$ and the matrix $\boldsymbol{R}_{\boldsymbol{Y}_3} = diag(r_1,\ r_2,\ r_3)$. The outputs will be $\boldsymbol{X}^*_{3,3}$ and $\boldsymbol{S}^*_{3,3}$.

◊ At time t_4 the filter is cycled using $\boldsymbol{Y}_4 = (y_1,\ y_2,\ y_3,\ y_4)$ and $\boldsymbol{R}_{\boldsymbol{Y}_4} = diag(r_1,\ r_2,\ r_3,\ r_4)$. The outputs will be $\boldsymbol{X}^*_{4,4}$, $\boldsymbol{S}^*_{4,4}$ and so on.

The filter is thus being operated with an expanding memory, and so each of the successive estimates $\boldsymbol{X}^*_{n,n}$ should be more accurate than its predecessor.[4]

[4] In the final section of this chapter we discuss the improvement of accuracy that comes about as the number of observations increases.

We can either continue with an expanding memory in this way or at some instant we can change to a fixed-length memory. Supposing the latter, let the instant at which we change be called t_m. Then:

◊ At time t_m the filter is cycled using $\boldsymbol{Y}_m = (y_m, \ y_{m-1}, \ \ldots, y_2)^T$ and covariance matrix $\boldsymbol{R}_{\boldsymbol{Y}_m} = diag(r_m, \ r_{m-1}, \ \ldots, r_2)$, in which we have dropped y_1 and r_1.

◊ At time t_{m+1} we use $\boldsymbol{Y}_{m+1} = (y_{m+1}, y_m, \ \ldots, y_3)^T$ and $\boldsymbol{R}_{\boldsymbol{Y}_{m+1}} = diag(r_{m+1}, r_m, \ \ldots, r_3)$ in which we have dropped y_2, r_2 and so on.

The filter will now have a fixed-length memory, and so the accuracy of the estimate $\boldsymbol{X}^*$ will remain approximately fixed.

The memory length in the above discussion is depicted in Figure 8.5 for the case where the change from expanding to fixed memory takes place at time t_{51}.

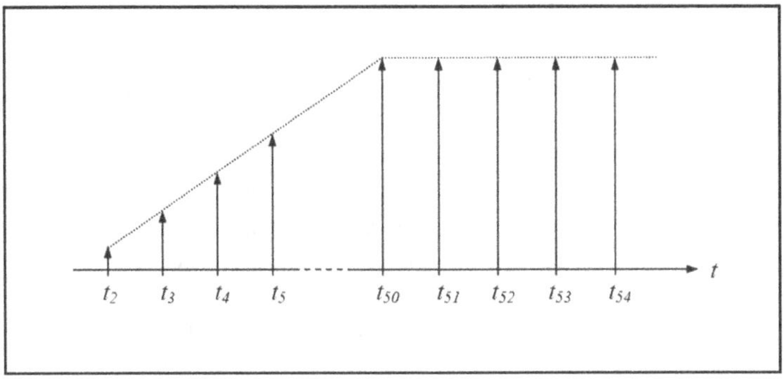

Figure 8.5 Memory length of the Gauss–Aitken filters

In the above simple example the model was a 1st-degree polynomial, and so we had to wait until two observations had been received.

If the model had been a 2nd-degree polynomial then $\boldsymbol{T}$ would have had three columns, and we would have had to wait until three observations had been received, and so on.

8.6 The flexibility of the Gauss–Aitken filters

From Figures 8.3 and 8.4 we can now see the extreme flexibility of the Gauss–Aitken filters. Two items are involved:

8.6.1 Filter memory

In Section 1.13 we spoke about filter memory and we identified the following three types:

◊ Fixed-length memory.

◊ Variable-length memory.

◊ Expanding memory.

We are required to build $\boldsymbol{T}$ each time that a Gauss filter is cycled, and in doing so we have three ways in which to proceed:

◊ Whenever a new observation is received the oldest observation can be dropped. This would result in the filter having a fixed-length memory.

◊ Whenever a new observation is received, none, one or more of the oldest observations can be dropped. This would result in the filter having a variable-length memory.

◊ Whenever a new observation is received, none of the oldest observations is dropped. This would result in the filter having an expanding memory.

We see that in this way the Gauss–Aitken filters can easily emulate filters with expanding memories like the Kalman and Swerling filters. However, the reverse – in which a Kalman or Swerling filter possesses a fixed or variable-length memory – can only be accomplished with difficulty using complex filter engineering.

We can also impose fading on any of these memory modes. The way in which that is done is discussed in Project 8.5.

These same facts also apply to the Gauss–Newton filters discussed in Chapter 9, and so it is clear that the Gauss filters all possess total flexibility regarding their memory.

In Chapter 10 we discuss the master control algorithms (MCAs) and we will see how they make use of this flexibility when manoeuvring targets are being tracked.

8.6.2 The filter model

The flexibility of the Gauss filters goes even further than memory, and also includes the filter model.

We are required to evaluate the transition matrix $\boldsymbol{\Phi}$ and the observation matrix $\boldsymbol{M}$ on every cycle when we build $\boldsymbol{T}$, and in doing so we can change both $\boldsymbol{\Phi}$ and $\boldsymbol{M}$ if we so desire.

For example, we may decide to switch from the $\boldsymbol{\Phi}$ for a polynomial model of 1st degree to the $\boldsymbol{\Phi}$ for 2nd degree, leaving the memory length L unchanged – or increase or reduce L while leaving the filter model unchanged – or change the degree of the model as well as L.

The MCAs in Chapter 10 also make use of this flexibility regarding the filter model.

8.7 Engineering two Gauss–Aitken filters

In Project 8.1 we take you through all the steps needed to engineer two Gauss–Aitken filters – one based on polynomials and the other on sinusoids.

8.8 The option offered by Version 1

As noted earlier, Version 2 of the MVA requires five matrix multiplications, whereas Version 1 requires seven, and so, all else being equal, we would use Version 2. However, Version 1 offers us a valuable engineering option that is not offered by Version 2.

8.8.1 The matrix $\mathbf{R_Y}$ *and its inverse*

$\mathbf{R_Y}$ is the covariance matrix of the errors in the vector $\mathbf{Y}$, and both Versions 1 and 2 require that we find its inverse. Consider its structure and what is required to invert it.

Let the number of elements in $\mathbf{Y}$ be $N_\mathbf{Y}$. Then $\mathbf{R_Y}$ will be $N_\mathbf{Y} \times N_\mathbf{Y}$ and because $N_\mathbf{Y}$ could easily be 300 or more[5] it seems that when we execute either version we will have to invert a very large matrix.

It thus appears as though the potentially large matrix $\mathbf{R_Y}$ poses a major drawback for the Gauss–Aitken filters. However, that is definitely not the case.

Keep in mind that the subvectors $\boldsymbol{Y}_n, \boldsymbol{Y}_{n-1}, \ldots, \boldsymbol{Y}_{n-L}$ in the long-vector $\mathbf{Y}$ are almost always stage-wise uncorrelated[6] which means that $\mathbf{R_Y}$ has the block-diagonal structure shown in (8.8.1):

$$\mathbf{R_Y} = \begin{pmatrix} \boldsymbol{R}_n & & & \\ & \boldsymbol{R}_{n-1} & & \\ & & \ddots & \\ & & & \boldsymbol{R}_{n-L} \end{pmatrix} \tag{8.8.1}$$

[5] We have worked on an application in which $N_\mathbf{Y}$ was of the order of *2,000*.

[6] See Definition 5.4 for the meaning of *stage-wise correlation.*

The sizes of the blocks are the same as the dimensions of the short-vectors $Y_n, Y_{n-1}, \ldots Y_{n-L}$ which is frequently *3*, and so the matrix $\boldsymbol{R}_{\boldsymbol{Y}}^{-1}$ then consists of the inverses of those *3* × *3* matrices, namely

$$\boldsymbol{R}_{\boldsymbol{Y}}^{-1} = \begin{pmatrix} \boldsymbol{R}_n^{-1} & & & \\ & \boldsymbol{R}_{n-1}^{-1} & & \\ & & \ddots & \\ & & & \boldsymbol{R}_{n-L}^{-1} \end{pmatrix} \tag{8.8.2}$$

Under these conditions the largest matrices that we will have to invert are of the order of the submatrices $\boldsymbol{R}_{n-k}$ and not of the order of $\boldsymbol{R}_{\boldsymbol{Y}}$.

Moreover, we won't have to invert all of them every time that the filter is cycled, because – with the exception of the most recent – the other inverses were needed during the previous cycle and so they are already available. Thus, at most only one of the typically *3* × *3* submatrices will have to be inverted on each cycle.

That same or a similar submatrix would also have to be inverted on each cycle of a Swerling or a Kalman filter, and so in this regard the presence of the large matrix $\boldsymbol{R}_{\boldsymbol{Y}}$ in the Gauss–Aitken filter equations poses no drawback whatsoever.

8.8.2 If the observations are stage-wise correlated

If the observations are stage-wise correlated then $\boldsymbol{R}_{\boldsymbol{Y}}$ is no longer block-diagonal, and its presence in the Gauss filter equations now constitutes a decided advantage – something that is not shared by the Kalman and Swerling filters.

The possibility of stage-wise correlation is not of merely academic interest. We have worked on a system in which stage-wise correlation was unavoidably present, and it was only by making use of the techniques that we are now discussing that we were able to engineer a solution for the problem.

The discussion is now continued in Project 8.2.

8.9 Non-recursive and recursive minimum variance

In this final section of this chapter, we discuss the meanings of the words *non-recursive* and *recursive*. These words occur often in filter engineering, and are at the heart of the differences between the Gauss filters in this and the next chapter which are non-recursive, and the Kalman and Swerling filters in Chapter 11 which are recursive.

We carry out the discussion by means of four worked examples, all of which centre on the MVA. The first two involve scalars, and in the second two we move to higher dimensions.

In the course of the discussion we will encounter what is perhaps the most fascinating property of the MVA – its ability to *mutate* from a non-recursive to a recursive form.

8.9.1 Rule for combining observations

Example 8.1: The following rule is well known to engineers and scientists who process data:

Let y_1 and y_2 be two independent – and hence uncorrelated – estimates of the same constant scalar quantity x, and let their variances be $\sigma_{y_1}^2$ and $\sigma_{y_2}^2$. Then the best estimate of x in the sense of most accurate is as follows:

$$x^* = \frac{y_1/\sigma_{y_1}^2 + y_2/\sigma_{y_2}^2}{1/\sigma_{y_1}^2 + 1/\sigma_{y_2}^2} \tag{8.9.1}$$

with variance

$$\sigma_{x^*}^2 = \frac{1}{1/\sigma_{y_1}^2 + 1/\sigma_{y_2}^2} \tag{8.9.2}$$

■■

We now show that these are exactly the same equations that would emerge if we used the MVA to combine y_1 and y_2.

Under the present circumstances, the process that we are observing is a constant function of time which means that the DE for the external model is

$$Dx(t) = 0 \tag{8.9.3}$$

Equation (8.9.3) is also the filter model. The observation equations are

$$y_1 = mx + v_1 \quad \text{and} \quad y_2 = mx + v_2 \tag{8.9.4}$$

in which

◊ y_1 and y_2 that we earlier called estimates of x will now be called observations on x.

◊ The quantity m is the observation 'matrix'. In this case m is a scalar and equal to 1.

◊ v_1 and v_2 are the observation errors in y_1 and y_2.

The transition matrix for the DE in (8.9.3) is the scalar $\boldsymbol{\Phi} = 1$. Because the observation matrix is the scalar $m = 1$, it follows from (8.5.1) that the $\boldsymbol{T}$ matrix will be

$$T = \begin{pmatrix} 1 \\ 1 \end{pmatrix} \tag{8.9.5}$$

The total observation equation $\mathbf{Y} = \mathbf{T}\mathbf{X} + \mathbf{N}$, thus, becomes

$$\begin{pmatrix} y_1 \\ y_2 \end{pmatrix} = \begin{pmatrix} 1 \\ 1 \end{pmatrix} x + \begin{pmatrix} \nu_1 \\ \nu_2 \end{pmatrix} \tag{8.9.6}$$

and the covariance matrix of the observation errors and its inverse become

$$\mathbf{R_Y} = \begin{pmatrix} \sigma_{y_1}^2 & \\ & \sigma_{y_2}^2 \end{pmatrix} \quad \mathbf{R_Y^{-1}} = \begin{pmatrix} 1/\sigma_{y_1}^2 & \\ & 1/\sigma_{y_2}^2 \end{pmatrix} \tag{8.9.7}$$

Notes: These two matrices are diagonal because of the assumption that y_1 and y_2 are uncorrelated.

We now apply Version 2 of the MVA. Equation (8.3.9) gives us

$$\sigma_{x^*}^2 = (\mathbf{T}^T\mathbf{R_Y}^{-1}\mathbf{T})^{-1} = \left(\begin{pmatrix} 1 & 1 \end{pmatrix} \begin{pmatrix} 1/\sigma_{y_1}^2 & \\ & 1/\sigma_{y_2}^2 \end{pmatrix} \begin{pmatrix} 1 \\ 1 \end{pmatrix} \right)^{-1} \tag{8.9.8}$$

which reduces to

$$\sigma_{x^*}^2 = \frac{1}{1/\sigma_{y_1}^2 + 1/\sigma_{y_2}^2} \tag{8.9.9}$$

Equations (8.3.10) and (8.3.11) then give

$$x^* = \sigma_{x^*}^2 \mathbf{T}^T\mathbf{R_Y}^{-1}\mathbf{Y} = \frac{1}{1/\sigma_{y_1}^2 + 1/\sigma_{y_2}^2} \begin{pmatrix} 1 & 1 \end{pmatrix} \begin{pmatrix} 1/\sigma_{y_1}^2 & \\ & 1/\sigma_{y_2}^2 \end{pmatrix} \begin{pmatrix} y_1 \\ y_2 \end{pmatrix} \tag{8.9.10}$$

which reduces to

$$x^* = \frac{y_1/\sigma_{y_1}^2 + y_2/\sigma_{y_2}^2}{1/\sigma_{y_1}^2 + 1/\sigma_{y_2}^2} \tag{8.9.11}$$

Comparing (8.9.11) and (8.9.9) to (8.9.1) and (8.9.2), we see that the combination rule stated at the beginning of this example is indeed the best way in which to combine y_1 and y_2, where *best* is in the sense of minimum variance and hence the most accurate.

■■

8.9.1.1 Summary

Let y_1 and y_2 be two uncorrelated observations on the same constant scalar quantity x, and let their variances be $\sigma^2_{y_1}$ and $\sigma^2_{y_2}$. Let them be combined according to the MVA to give x^*_2 and its variance $\sigma^2_{x^*_2}$. Then

$$x^*_2 = \frac{y_1/\sigma^2_{y_1} + y_2/\sigma^2_{y_2}}{1/\sigma^2_{y_1} + 1/\sigma^2_{y_2}} \tag{8.9.12}$$

$$\sigma^2_{x^*_2} = \frac{1}{1/\sigma^2_{y_1} + 1/\sigma^2_{y_2}} \tag{8.9.13}$$

in which the subscript 2 in x^*_2 signifies that estimation is based on the combination of *two* observations.

Observe that (8.9.12) and (8.9.13) constitute a complete filter because the equations for both an estimate and a covariance matrix have been provided.

We leave it as an exercise for the reader to show that for m observations these equations become

$$x^*_m = \frac{y_1/\sigma^2_{y_1} + y_2/\sigma^2_{y_2} + \cdots + y_m/\sigma^2_{y_m}}{1/\sigma^2_{y_1} + 1/\sigma^2_{y_2} + \cdots + 1/\sigma^2_{y_m}} \tag{8.9.14}$$

$$\sigma^2_{x^*_m} = \frac{1}{1/\sigma^2_{y_1} + 1/\sigma^2_{y_2} + \cdots + 1/\sigma^2_{y_m}} \tag{8.9.15}$$

8.9.2 Non-recursive and recursive

We now extend what we have derived above

Example 8.2: Suppose, as before, that we make two uncorrelated observations on the same constant scalar quantity and combine them by the use of (8.9.12) and (8.9.13) to obtain the minimum variance estimate.

Suppose that a third uncorrelated observation is then obtained.

Question: What is the best way to incorporate that third observation?

Answer: There are two ways to go about it. We call them *non-recursive* and *recursive*, and they are as follows:

Non-recursive: Start again, and combine all three observations simultaneously, using the MVA. From (8.9.14) and (8.9.15) the algorithm would be

$$x_3^* = \frac{y_1/\sigma_{y_1}^2 + y_2/\sigma_{y_2}^2 + y_3/\sigma_{y_3}^2}{1/\sigma_{y_1}^2 + 1/\sigma_{y_2}^2 + 1/\sigma_{y_3}^2} \tag{8.9.16}$$

$$\sigma_{x_3^*}^2 = \frac{1}{1/\sigma_{y_1}^2 + 1/\sigma_{y_2}^2 + 1/\sigma_{y_3}^2} \tag{8.9.17}$$

Recursive: Retain the results x_2^* with variance $\sigma_{x_2^*}^2$ that were obtained from combining the first two observations using the MVA, and then combine the third observation with that estimate, again using the MVA. The algorithm would be as follows:

$$x_3^* = \frac{x_2^*/\sigma_{x_2^*}^2 + y_3/\sigma_{y_3}^2}{1/\sigma_{x_2^*}^2 + 1/\sigma_{y_3}^2} \tag{8.9.18}$$

$$\sigma_{x_3^*}^2 = \frac{1}{1/\sigma_{x_2^*}^2 + 1/\sigma_{y_3}^2} \tag{8.9.19}$$

Observe, in these two equations, the presence of terms from the earlier result $(x_2^*, \sigma_{x_2^*}^2)$ as well as terms involving the most recent observation $(y_3, \sigma_{y_3}^2)$. ■■

We would be rightfully disappointed in the MVA if the two methods did not give precisely the same result.

◊ In the first, we combined three observations using a one-stage non-recursive method that is supposedly minimum variance.

◊ In the second we combined those same three observations using a two-stage recursive method that is also supposedly minimum variance.

We are thus entitled to expect that the two results will be identical. Examination of equation (8.9.20) below shows that (8.9.16) and (8.9.18) *are* indeed the same, and so our faith in the MVA can remain unshaken and possibly even somewhat enhanced.

Observe how the recursive version of the algorithm on the left has mutated into the non-recursive version on the right (and *vice versa*).

The reader can verify that (8.9.17) and (8.9.19) also give the same result for the variance, thereby showing that the two methods – non-recursive and recursive – in fact, *do* give precisely the same result.

8.9.3 The two forms of the MVA

We now see that the MVA can take on two possible forms, the first appearing in (8.9.16) and (8.9.17) and the second in (8.9.18) and (8.9.19). We will refer to them as

◊ The *non-recursive form* or *Form A*.

◊ The *recursive form* or *Form B*.

$$x_3^* = \frac{x_2^*/\sigma_{x_2^*}^2 + y_3/\sigma_{y_3}^2}{1/\sigma_{x_2^*}^2 + 1/\sigma_{y_3}^2} = \frac{\left(\dfrac{\dfrac{y_1/\sigma_{y_1}^2 + y_2/\sigma_{y_2}^2}{1/\sigma_{y_1}^2 + 1/\sigma_{y_2}^2}}{\dfrac{1}{1/\sigma_{y_1}^2 + 1/\sigma_{y_2}^2}}\right) + y_3/\sigma_{y_3}^2}{\left(\dfrac{1}{\dfrac{1}{1/\sigma_{y_1}^2 + 1/\sigma_{y_2}^2}}\right) + 1/\sigma_{y_3}^2} = \frac{y_1/\sigma_{y_1}^2 + y_2/\sigma_{y_2}^2 + y_3/\sigma_{y_3}^2}{1/\sigma_{y_1}^2 + 1/\sigma_{y_2}^2 + 1/\sigma_{y_3}^2} \tag{8.9.20}$$

8.9.3.1 Summary

Assume that we make the following $k-1$ uncorrelated observations on a constant scalar quantity x:

$$(y_1, \sigma^2_{y_1}), \ldots, (y_{k-1}, \sigma^2_{y_{k-1}})$$

and that we combine them by the use of the MVA to form the composite estimate $\left(x^*_{k-1}, \sigma^2_{x^*_{k-1}}\right)$.

Suppose that the further uncorrelated observation $(y_k, \sigma^2_{y_k})$ is now obtained. Then the latter can be incorporated by either of two ways that give the same result:

Non-recursive: The new estimate can be obtained by using *Form A* of the MVA as

$$x^*_k = \frac{y_1/\sigma^2_{y_1} + \cdots + y_{k-1}/\sigma^2_{y_{k-1}} + y_k/\sigma^2_{y_k}}{1/\sigma^2_{y_1} + \cdots + 1/\sigma^2_{y_{k-1}} + 1/\sigma^2_{y_k}} \tag{8.9.21}$$

$$\sigma^2_{x^*_k} = \frac{1}{1/\sigma^2_{y_1} + \cdots + 1/\sigma^2_{y_{k-1}} + 1/\sigma^2_{y_k}} \tag{8.9.22}$$

Recursive: The new estimate can be obtained by using *Form B* of the MVA as

$$x^*_k = \frac{x^*_{k-1}/\sigma^2_{x^*_{k-1}} + y_k/\sigma^2_{y_k}}{1/\sigma^2_{x^*_{k-1}} + 1/\sigma^2_{y_k}} \tag{8.9.23}$$

$$\sigma^2_{x^*_k} = \frac{1}{1/\sigma^2_{x^*_{k-1}} + 1/\sigma^2_{y_k}} \tag{8.9.24}$$

We now generalise to higher dimensions what we have learned from the above two scalar examples.

8.9.4 *Forming a higher dimension minimum-variance estimate*

Example 8.3: Let $\boldsymbol{X} = (x_1, x_2)^T$ be the state vector of a constant physical process (e.g. the coordinates of a fixed point in 2-space) on which we make two uncorrelated 2-vector observations. Let those observation vectors be $\boldsymbol{Y}_1 = (y_1, y_2)^T_1$ and $\boldsymbol{Y}_2 = (y_1, y_2)^T_2$ where

$$\begin{pmatrix} y_1 \\ y_2 \end{pmatrix}_1 = \begin{pmatrix} x_1 \\ x_2 \end{pmatrix} + \begin{pmatrix} v_1 \\ v_2 \end{pmatrix}_1 \tag{8.9.25}$$

and

$$\begin{pmatrix} y_1 \\ y_2 \end{pmatrix}_2 = \begin{pmatrix} x_1 \\ x_2 \end{pmatrix} + \begin{pmatrix} v_1 \\ v_2 \end{pmatrix}_2 \tag{8.9.26}$$

■■

Then (8.9.25) and (8.9.26) become

$$\boldsymbol{Y}_1 = \boldsymbol{X} + \boldsymbol{N}_1 \tag{8.9.27}$$

$$\boldsymbol{Y}_2 = \boldsymbol{X} + \boldsymbol{N}_2 \tag{8.9.28}$$

Let the covariance matrices of $\boldsymbol{Y}_1$ and $\boldsymbol{Y}_2$ be $\boldsymbol{R}_1$ and $\boldsymbol{R}_2$.

Question: What is the rule for combining $\boldsymbol{Y}_1, \boldsymbol{Y}_2$ and their covariance matrices so that the result is minimum variance?

Answer: We are assuming that the physical process under observation is unchanging, i.e. that $\boldsymbol{X}$ satisfies the DE

$$D\boldsymbol{X}(t) = \boldsymbol{\emptyset} \tag{8.9.29}$$

It therefore, follows that the transition matrix is

$$\boldsymbol{\Phi}(\zeta) = \boldsymbol{I} \text{ for all } \zeta \tag{8.9.30}$$

in which $\boldsymbol{I}$ is the 2×2 identity matrix. Then

$$\mathbf{Y} = \begin{pmatrix} \boldsymbol{Y}_1 \\ \text{----} \\ \boldsymbol{Y}_2 \end{pmatrix} \qquad \mathbf{T} = \begin{pmatrix} \boldsymbol{I} \\ \text{----} \\ \boldsymbol{I} \end{pmatrix}$$

$$\mathbf{R} = \left(\begin{array}{c|c} \boldsymbol{R}_1 & \\ \hline & \boldsymbol{R}_2 \end{array}\right) \qquad \boldsymbol{R}^{-1} = \left(\begin{array}{c|c} \boldsymbol{R}_1^{-1} & \\ \hline & \boldsymbol{R}_2^{-1} \end{array}\right) \tag{8.9.31}$$

The MVA now gives the estimate based on $\boldsymbol{Y}_1$ and $\boldsymbol{Y}_2$ as shown in (8.9.32):

$$\begin{aligned} \boldsymbol{X}_2^* &= (\boldsymbol{T}^T\boldsymbol{R}^{-1}\boldsymbol{T})^{-1}\boldsymbol{T}^T\boldsymbol{R}^{-1}\boldsymbol{Y} \\ &= \left(\left(\boldsymbol{I} \;\vdots\; \boldsymbol{I} \right) \left(\begin{array}{c:c} \boldsymbol{R}_1^{-1} & \\ \hdashline & \boldsymbol{R}_2^{-1} \end{array} \right) \left(\begin{array}{c} \boldsymbol{I} \\ \hdashline \boldsymbol{I} \end{array} \right) \right)^{-1} \left(\boldsymbol{I} \;\vdots\; \boldsymbol{I} \right) \left(\begin{array}{c:c} \boldsymbol{R}_1^{-1} & \\ \hdashline & \boldsymbol{R}_2^{-1} \end{array} \right) \left(\begin{array}{c} \boldsymbol{Y}_1 \\ \hdashline \boldsymbol{Y}_2 \end{array} \right) \end{aligned} \tag{8.9.32}$$

Simplifying (8.9.32), we obtain

$$X_2^* = S_2^*(R_1^{-1}Y_1 + R_2^{-1}Y_2) \tag{8.9.33}$$

in which

$$S_2^* = (R_1^{-1} + R_2^{-1})^{-1} \tag{8.9.34}$$

As we did in Example 8.2, we can also express (8.9.33) and (8.9.34) recursively, as follows:

$$X_k^* = S_k^*(S_{k-1}^{*-1}X_{k-1}^* + R_k^{-1}Y_k) \tag{8.9.35}$$

where

$$S_k^* = (S_{k-1}^{*-1} + R_k^{-1})^{-1} \tag{8.9.36}$$

In the first of these equations the term $R_1^{-1}Y_1$ in (8.9.33) has been replaced by the term $S_{k-1}^{*-1}X_{k-1}^*$ – both of which have the same structure – and in the second equation the term R_1^{-1} has been replaced by S_{k-1}^{*-1}.

8.9.4.1 Summary

Let X be the state vector of a constant physical process on which we make $k-1$ uncorrelated vectors of observations. Let them be $Y_1, Y_2, \ldots, Y_{k-1}$ with covariance matrices $R_1, R_2, \ldots, R_{k-1}$ and assume that we combine them by the use of the MVA to form the composite estimate X_{k-1}^* with covariance matrix S_{k-1}^*.

Suppose we now obtain the further uncorrelated observation vector Y_k with covariance matrix R_k. Then these can be incorporated in either of two ways that give the same result:

◊ ***Non-recursively:*** The composite estimate X_k^* with covariance matrix S_k^* can be obtained from the observation matrices and their covariance matrices using *Form A* of the MVA as follows:

$$X_k^* = S_k^*(R_1^{-1}Y_1 + \cdots + R_k^{-1}Y_k) \tag{8.9.37}$$

where

$$S_k^* = (R_1^{-1} + \cdots + R_k^{-1})^{-1} \tag{8.9.38}$$

◊ ***Recursively:*** The composite estimate X_k^* with covariance matrix S_k^* can be obtained from X_{k-1}^*, S_{k-1}^*, Y_k and R_k using *Form B* of the MVA as follows:

$$X_k^* = S_k^{*-1}(S_{k-1}^* X_{k-1}^* + R_k^{-1} Y_k) \tag{8.9.39}$$

where

$$S_k^* = (S_{k-1}^{*-1} + R_k^{-1})^{-1} \tag{8.9.40}$$

Observe once again from (8.9.37) and (8.9.39) how the MVA mutates from non-recursive to recursive.

Definition 8.3: A smoothing algorithm is said to be **non-recursive** if, during its execution, its estimate and covariance matrix are produced solely from the observations and their covariance matrix, without any of the preceding estimates and their covariance matrices being used. ■■

Definition 8.4: A smoothing algorithm is said to be **recursive** if, during its execution, its estimate and covariance matrix are produced from the observations and their covariance matrix *and* one or more of the preceding estimates and their covariance matrices. ■■

Taking another look at Figures 8.3 and 8.4 where we displayed the two versions of the Gauss–Aitken filters, we see, by Definition 8.3, that both are non-recursive filters – nowhere do preceding estimates and covariance matrices appear.

8.9.5 Further extensions

Earlier we were assuming that the physical process under observation did not change with time, and so the filter model had the very simple DE appearing in (8.9.3) and (8.9.29).

When the models are higher order DEs which generate state vectors that *do* change with time, then the transition matrices of those DEs come into play, and with them the full $\boldsymbol{T}$ matrix, rather than the simplified one that appeared in (8.9.31).

◊ The non-recursive *Method A* for combining observations then becomes the Gauss–Aitken filters in their two versions shown in Figures 8.3 and 8.4.

◊ In Chapter 9 we will see that *Method A* also gives rise to the Gauss–Newton filters in six versions, all of them non-recursive.

◊ In Chapter 11 we will see that *Method B* takes us to the Swerling filter which is recursive, and that an algebraic transformation of the Swerling filter gives us the Kalman filter.

◊ In Chapters 12 and 13 the polynomial filters will be seen to be based on a simplified implementation of the MVA. Those filters are of two types and both of them are recursive.

8.9.6 Data fusion

Example 8.4: In Figure 8.6 we show an array of k minimum variance filters $\boldsymbol{MV}_1$ through $\boldsymbol{MV}_k$. ■■

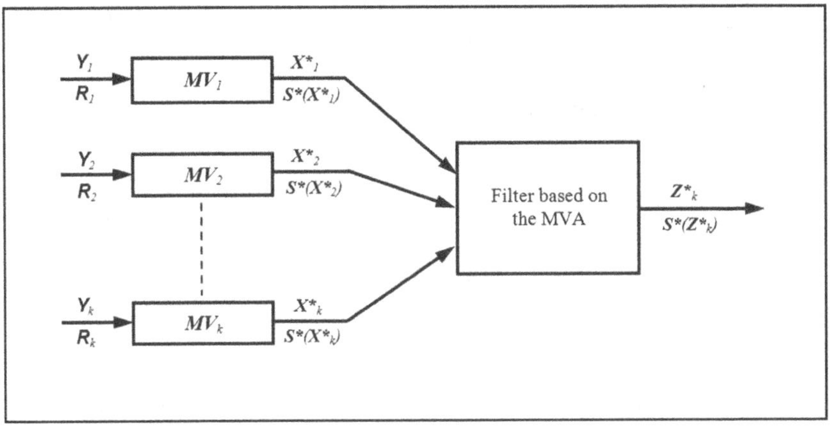

Figure 8.6 Minimum variance filter array in data fusion

◊ Each of the filters receives observations that are made on the same physical process.

◊ The observations are uncorrelated from filter to filter.

◊ The observation equations and filter models used in the filters can be either linear or non-linear, and need not be the same from filter to filter.

◊ The outputs of the filters are the minimum variance estimates and their covariance matrices $(\boldsymbol{X}_1^*,\ \boldsymbol{S}^*(\boldsymbol{X}_1^*)), \dots, (\boldsymbol{X}_k^*,\ \boldsymbol{S}^*(\boldsymbol{X}_k^*))$.

◊ These k estimates can then be combined in either of two ways to produce the more accurate minimum variance estimate $(\boldsymbol{Z}_k^*,\ \boldsymbol{S}^*(\boldsymbol{Z}_k^*))$ as follows:

◊ ***Non-recursively***:

The estimate $\boldsymbol{Z}_k^*$ and its covariance matrix $\boldsymbol{S}^*(\boldsymbol{Z}_k^*)$ can be obtained from $(\boldsymbol{X}_1^*,\ \boldsymbol{S}^*(\boldsymbol{X}_1^*)), \dots, (\boldsymbol{X}_k^*,\ \boldsymbol{S}^*(\boldsymbol{X}_k^*))$ using *Form A* of the MVA as follows:

$$Z^*_k = \left(S^*(Z^*_k)\right)\left(\left((S^*(X^*_1)\right)^{-1}X^*_1 + \cdots + \left(S^*(X^*_k)\right)^{-1}X^*_k\right) \tag{8.9.41}$$

in which

$$S^*(Z^*_k) = \left(\left(S^*(X^*_1)\right)^{-1} + \cdots + \left(S^*(X^*_k)\right)^{-1}\right)^{-1} \tag{8.9.42}$$

◊ *Recursively*:

Z^*_k and $S^*(Z^*_k)$ can be obtained from $(Z^*_{k-1},\ S^*(Z^*_{k-1}))$ and $(X^*_k,\ S^*(X^*_k))$ using *Form B* of the MVA as follows:

$$Z^*_k = \left(S^*(Z^*_k)\right)\left(\left(S^*(Z^*_{k-1})\right)^{-1}Z^*_{k-1} + \left(S^*(X^*_k)\right)^{-1}X^*_k\right) \tag{8.9.43}$$

in which

$$S^*(Z^*_k) = \left(\left(S^*(Z^*_{k-1})\right)^{-1} + \left(S^*(X^*_k)\right)^{-1}\right)^{-1} \tag{8.9.44}$$

8.9.7 Recursive algorithms must be initialized[7]

In closing we note the following: When we first cycle Equations (8.9.43) and (8.9.44) we start with $k = 1$, and so we must provide values for Z^*_0 and $S^*(Z^*_0)$. In this simple case we can use

$$Z^*_0 = \text{Ø} \quad \text{and} \quad S^*(Z^*_0)^{-1} = \text{Ø} \tag{8.9.45}$$

The algorithm will then start up correctly as follows:

$$S^*(Z^*_1) = \left(\left(S^*(Z^*_0)\right)^{-1} + \left(S^*(X^*_1)\right)^{-1}\right)^{-1} \tag{8.9.46}$$

$$= \left(\text{Ø} + (S^*(X^*_1)^{-1}\right)^{-1} = S^*(X^*_1) \tag{8.9.47}$$

$$Z^*_1 = \left(S^*(Z^*_1)\right)\left(\left(S^*(Z^*_0)\right)^{-1}Z^*_0 + \left(S^*(X^*_1)\right)^{-1}X^*_1\right)$$

$$= \left(S^*(X^*_1)\right)\left(\text{Ø} + \left(S^*(X^*_1)\right)^{-1}X^*_1\right) = X^*_1 \tag{8.9.48}$$

[7] Recursive algorithms can occasionally be self-initializing, but not very often. In Chapters 12 and 13 we consider a set of extremely useful recursive algorithms called the EMP filters that *are* self-initializing.

Using $\boldsymbol{S}^*(\boldsymbol{Z}^*_0)^{-1} = \boldsymbol{\emptyset}$ is equivalent to saying that the variances of the errors in the elements of $\boldsymbol{Z}^*_0$ are infinitely large, and so the MVA correctly weights those elements with infinitely large numbers in their denominators, resulting in their being entirely ignored.

8.9.8 Non-recursive algorithms do not require initialization

Equations (8.9.41) and (8.9.42) are complete, and no additional values are required for their execution – neither at start-up nor at any other time.

8.9.9 Fundamental assumption when combining observations

Take another look at the sentence three lines above (8.9.41): 'These k estimates can then be combined in either of two ways to produce the *more accurate* minimum variance estimate ...'

This is the fundamental assumption on which the combination of observations is based, namely

The more data we use to form an estimate, the more accurate will be the result.

Intuitively we know that this is true, but it is subject to mathematical proof. Thus we must be able to show that the most recent estimate $\boldsymbol{Z}^*_k$ on the left of (8.9.43) is more accurate than the previous estimate $\boldsymbol{Z}^*_{k-1}$ on the right.

Proof is provided in Problem 8.19.

The following items are in the supplementary material

Problems for Chapter 8

Solutions to Problems for Chapter 8

Projects for Chapter 8

Appendix 8.1

Minimization of $e(X^*_{n,n})$

We start from (8.2.11):

$$e(X^*_{n,n}) = (Y_n - \boldsymbol{T}_n X^*_{n,n})^T \boldsymbol{R}_{\mathbf{Y}_n}^{-1} (Y_n - \boldsymbol{T}_n X^*_{n,n}) \tag{A8.1.1}$$

To minimize $e(X^*_{n,n})$ we differentiate with respect to $(X^*_{n,n})$ and set the result equal to $\boldsymbol{\emptyset}$.

We note that (A8.1.1) is a quadratic form on a symmetric coefficient matrix. Accordingly, (see the file ***Miscellaneous\Vector_and_Matrix_Differentiation***) we obtain

$$D_{X^*}\Big((\mathbf{Y}_n - \boldsymbol{T}_n X^*_{n,n})^T \boldsymbol{R}_{\mathbf{Y}_n}^{-1} (\mathbf{Y}_n - \boldsymbol{T}_n X^*_{n,n})\Big)$$

$$= -2\boldsymbol{T}_n^T \boldsymbol{R}_{\mathbf{Y}_n}^{-1} (\mathbf{Y}_n - \boldsymbol{T}_n X^*_{n,n}) \tag{A8.1.2}$$

Setting this result equal to $\boldsymbol{\emptyset}$ and dropping the *-2* we obtain

$$(\boldsymbol{T}_n^T \boldsymbol{R}_{\mathbf{Y}_n}^{-1} \boldsymbol{T}_n) X^*_{n,n} = \boldsymbol{T}_n^T \boldsymbol{R}_{\mathbf{Y}_n}^{-1} \mathbf{Y}_n \tag{A8.1.3}$$

In order that this equation be solvable for $X^*_{n,n}$ it is necessary that the matrix $\boldsymbol{T}_n^T \boldsymbol{R}_{\mathbf{Y}_n}^{-1} \boldsymbol{T}_n$ be non-singular, and that is something to which we devoted a great deal of attention in Chapter 4.

◊ By Theorem 4.4, $\boldsymbol{T}_n^T \boldsymbol{R}_{\mathbf{Y}_n}^{-1} \boldsymbol{T}_n$ will be non-singular if and only if the high-matrix $\boldsymbol{T}_n$ has full column-rank (assuming, of course, that $\boldsymbol{R}_{\mathbf{Y}_n}^{-1}$ is positive-definite).

◊ In Project 5.3 we examined the column-rank of the matrix $\boldsymbol{T}_n$, and we concluded that if the observation scheme is properly constructed then $\boldsymbol{T}_n$ *will* have full column-rank.

Under these circumstances, by Theorem 4.4 $\boldsymbol{T}_n^T \boldsymbol{R}_{\mathbf{Y}_n}^{-1} \boldsymbol{T}_n$ will be non-singular, and so we can solve for $X^*_{n,n}$ from (A8.1.3), obtaining

$$X^*_{n,n} = (T_n^T R_{Y_n}^{-1} T_n)^{-1} T_n^T R_{Y_n}^{-1} Y_n \tag{A8.1.4}$$

We now restate (A8.1.4) as follows:

$$X^*_{minvar} = W_{minvar} Y_n \tag{A8.1.5}$$

where

$$W_{minvar} = (T_n^T R_{Y_n}^{-1} T_n)^{-1} T_n^T R_{Y_n}^{-1} \tag{A8.1.6}$$

The rank of W_{minvar}

In Appendix 5.2 we pointed out that any wide-matrices that we intend to use as filter matrices must have full row-rank, and so we now investigate the row-rank of the wide-matrix W_{minvar} in (A8.1.6).

In Chapter 6 we derived Theorem 6.2, which we restate here:

Theorem 6.2: Let the matrices W and T satisfy

$$WT = I \tag{A8.1.7}$$

Then W must have full row-rank and T full column-rank. ■■

Applying the theorem to W_{minvar} in (A8.1.6), we post-multiply by the matrix T_n obtaining

$$W_{minvar} T_n = (T_n^T R_{Y_n}^{-1} T_n)^{-1} T_n^T R_{Y_n}^{-1} T_n = I \tag{A8.1.8}$$

Equation (A8.1.7) is therefore satisfied, and so by Theorem 6.2 W_{minvar} has full row-rank and can be used as a filter matrix.

Appendix 8.2

The MVA filter covariance matrix S^*_{minvar}

From (A8.1.5) and (A8.1.6) in Appendix A8.1:

$$X^*_{minvar} = W_{minvar}\mathbf{Y} \tag{A8.2.1}$$

where

$$W_{minvar} = (\mathbf{T}^T\mathbf{R}_{\mathbf{Y}}^{-1}\mathbf{T})^{-1}\mathbf{T}^T\mathbf{R}_{\mathbf{Y}}^{-1} \tag{A8.2.2}$$

■■

From (A8.2.1) and Theorem 4.1 it then follows that the formula for the MVA filter covariance matrix will be

$$S^*_{minvar} = W_{minvar}\mathbf{R}_{\mathbf{Y}}W^T_{minvar}$$

$$\begin{aligned} S^*_{minvar} &= ((\mathbf{T}^T\mathbf{R}_{\mathbf{Y}}^{-1}\mathbf{T})^{-1}\mathbf{T}^T\mathbf{R}_{\mathbf{Y}}^{-1})\mathbf{R}_{\mathbf{Y}}(\mathbf{R}_{\mathbf{Y}}^{-1}\mathbf{T}(\mathbf{T}^T\mathbf{R}_{\mathbf{Y}_n}^{-1}\mathbf{T})^{-1}) \\ &= (\mathbf{T}^T\mathbf{R}_{\mathbf{Y}}^{-1}\mathbf{T})^{-1}(\mathbf{T}^T\mathbf{R}_{\mathbf{Y}}^{-1}\mathbf{T})(\mathbf{T}^T\mathbf{R}_{\mathbf{Y}}^{-1}\mathbf{T})^{-1} = (\mathbf{T}^T\mathbf{R}_{\mathbf{Y}}^{-1}\mathbf{T})^{-1} \end{aligned} \tag{A8.2.3}$$

We have thus shown that

$$S^*_{minvar} = (\mathbf{T}^T\mathbf{R}_{\mathbf{Y}}^{-1}\mathbf{T})^{-1} \tag{A8.2.4}$$

■■

Note that if a matrix $\mathbf{R}_d$ that differs from $\mathbf{R}_{\mathbf{Y}}$ is used to form the filter matrix W, then the above simplification no longer takes place. Instead, here's what happens:

We start from the two equations

$$X^*_d = W_d\mathbf{Y} \tag{A8.2.5}$$

and

$$W_d = (\mathbf{T}^T\mathbf{R}_d^{-1}\mathbf{T})^{-1}\mathbf{T}^T\mathbf{R}_d^{-1} \tag{A8.2.6}$$

and we note that W_d still satisfies the exactness constraint when post-multiplied by $\mathbf{T}$.

However, from (A8.2.5) and Theorem 4.1

$$\begin{aligned} S^*_d &= W_d\mathbf{R}_{\mathbf{Y}}W_d^T \\ &= \left((\mathbf{T}^T\mathbf{R}_d^{-1}\mathbf{T})^{-1}\mathbf{T}^T\mathbf{R}_d^{-1}\right)\mathbf{R}_{\mathbf{Y}}\left(\mathbf{R}_d^{-1}\mathbf{T}(\mathbf{T}^T\mathbf{R}_d^{-1}\mathbf{T})^{-1}\right) \end{aligned} \tag{A8.2.7}$$

and we can go no further.

Appendix 8.3

Obtaining the covariance matrix $\boldsymbol{R_Y}$

In Figure 1.6 we showed the basic elements of the filtering procedure, and a number of matrices appeared there, among them $\boldsymbol{R_Y}$, the covariance matrix of the errors in the observations.

The matrix $\boldsymbol{R_Y}$ differs from the actual estimation-error covariance matrix $\boldsymbol{S}_{actual}$ in the sense that the latter can be estimated by running a Monte-Carlo, whereas the former cannot.

Obtaining an estimate of $\boldsymbol{R_Y}$ is typically done in one of two ways:

◊ Using calibration and measurement equipment on the observation instrument.

◊ By the use of analysis carried out by the engineers who built the instrument, this being the way in which an estimate of $\boldsymbol{R_Y}$ is usually obtained in radar systems.

Either way, an estimate of the matrix $\boldsymbol{R_Y}$ comes to us from an external source, and it is for this reason that in Figure 1.6 we show Item 5 as '$\boldsymbol{R}_{\boldsymbol{Y},engrs}$, the supposed covariance matrix of the errors in $\boldsymbol{Y}$'.

Chapter 9

Minimum variance and the Gauss–Newton filters

In Chapter 8 we derived the minimum variance algorithm (MVA) by solving what we called *Problem Statement 1*, namely by finding the vector $\boldsymbol{X}^*$ that minimizes the quantity $e(\boldsymbol{X}^*)$.

This led us to two versions of the Gauss–Aitken filter for use in ***Case 1***, in which both the filter model and the observation equation are linear.

In this chapter we re-derive the MVA by solving what we call *Problem Statement 2*. This will accomplish three objectives:

◊ Explain what the words *minimum variance* mean.

◊ Provide the link between the MVA and Cramér–Rao.

◊ Enable us to create three tests by which to determine whether or not a filter is CR consistent.

In the final sections of the chapter we show how the MVA gives rise to the six versions of the Gauss–Newton filter for use when nonlinearities are present and we are operating in ***Cases 2***, ***3*** or ***4***.

◄►

In Chapter 1 we mentioned that Gauss gave two justifications for his method of least squares:

◊ In 1809, he showed that if the errors are normal (i.e. Gaussian), then weighted least squares is equivalent to maximum likelihood estimation. We provide a proof for this statement in Appendix 9.3.

◊ In the 1820s he proved his celebrated *minimum variance theorem*, namely: Of all linear combinations of measurements for estimating an unknown, the weighted least squares estimate has the greatest precision (i.e. the smallest variance), regardless of the distribution of the observation errors.

We provide a proof for this statement lower down in this chapter, where we refer to it as Theorem 9.2.

9.1 Deriving the MVA a second way

9.1.1 Stating the problem in general terms

In general terms, the problem that we are about to consider is as follows:

Figure 9.1 is a depiction of the total observation equation $\boldsymbol{Y} = \boldsymbol{TX} + \boldsymbol{N}$.

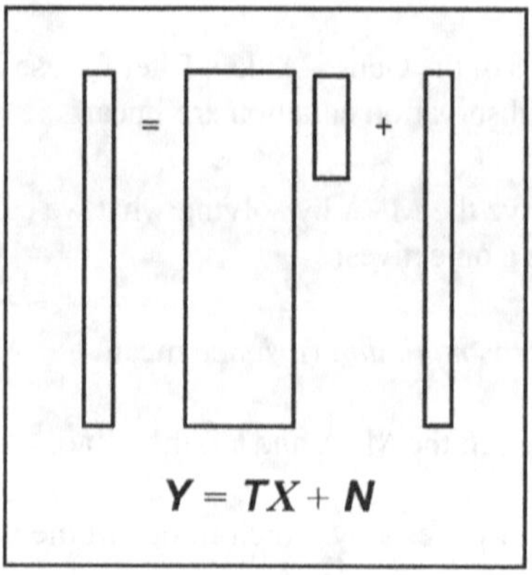

Figure 9.1 $\boldsymbol{Y} = \boldsymbol{TX} + \boldsymbol{N}$

◊ $\boldsymbol{Y}$ and $\boldsymbol{N}$ are long vectors of dimension k.

◊ $\boldsymbol{T}$ is $k \times m$.

◊ $\boldsymbol{X}$ is a short vector of dimension m.

◊ $k > m$ and so $\boldsymbol{T}$ is high.

◊ $\boldsymbol{T}$ has full column-rank.

◊ The covariance matrix of $\boldsymbol{Y}$ is $\boldsymbol{R}_{\boldsymbol{Y}}$ which is $k \times k$.

Figures 9.2(a) and 9.2(b) on the next page are depictions of the vector $\boldsymbol{Y}$ and its covariance matrix $\boldsymbol{R}_{\boldsymbol{Y}}$ being transformed by one of infinitely many filter matrices $\boldsymbol{W}_j$ according to the equations

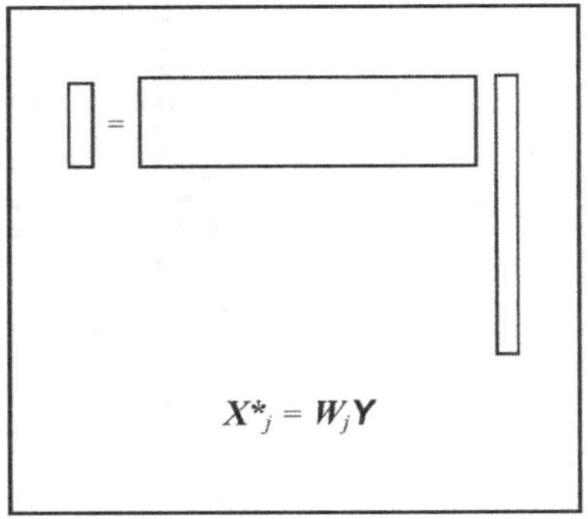

Figure 9.2 (a) $\boldsymbol{X}^*_j = \boldsymbol{W}_j\mathbf{Y}$

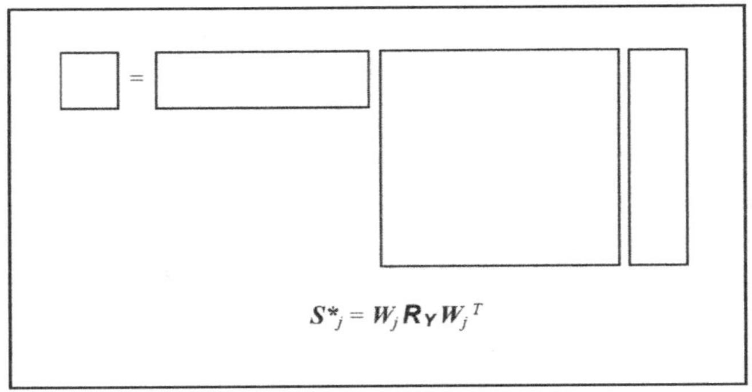

Figure 9.2(b) $\boldsymbol{S}^*_j = \boldsymbol{W}_j\mathbf{R}_\mathbf{Y}\boldsymbol{W}_j^T$

$$X_j^* = W_j \mathbf{Y} \qquad j = 1, 2, \ldots \qquad S_j^* = W_j \mathbf{R_Y} W_j^T \tag{9.1.1}$$

◊ The X^*_j are short m-vectors and their covariance matrices S^*_j are $m \times m$.

◊ The filter matrices W_j are $m \times k$.

◊ $k > m$ and so the W_j matrices are wide.

The filter matrices W_j are all assumed to satisfy the exactness constraint, namely

$$W_j \mathbf{T} = I \qquad j = 1, 2, \ldots \tag{9.1.2}$$

in which $\mathbf{T}$ is the matrix appearing in Figure 9.1, and the identity matrix I is $m \times m$. Because the W_j all satisfy the exactness constraint, it follows from Theorem 6.2 that they all have full row-rank.

Consider now the infinitely many covariance matrices S_j^* that are produced by this procedure, and for now, let them all be 3×3.

Their diagonal elements are all positive and so, for a given diagonal position, e.g. the *1, 1* position, one of those infinitely-many matrices must contain the smallest value.

Call that matrix the *min-1,1 matrix*.

The same must also be true for the remaining two diagonal positions, and so we can identify two further matrices which we call the *min-2,2* and *min-3,3 matrices*.

◊ It is a remarkable fact that these three matrices are all *one and the same*, and are precisely the matrix S^*_{minvar} which we identified in Chapter 8.

Thus among the infinitely many covariance matrices S_j^* there exists a special one called S^*_{minvar} which possesses what we shall refer to as the *minimum variance property,* namely:

All of the diagonal elements of each and every S_j^* and those of S^*_{minvar} satisfy

$$[S_j^*]_{i,i} \geq [S^*_{minvar}]_{i,i} \qquad 1 \leq i \leq m \qquad j = 1, 2, \ldots \tag{9.1.3}$$

Thus the lower bounds of *all* of the diagonal elements of *every* matrix S_j^* are *simultaneously* present in the single matrix S^*_{minvar}.

In what now follows, our objective will be to find the expression for W_{minvar} that gives rise to the estimate vector X^*_{minvar} whose covariance matrix is that one special matrix called S^*_{minvar}.

9.1.2 Stating the problem formally

Problem Statement 2

Given all of the following:

◊ The fixed total observation vector $\mathbf{Y}$ and the fixed total observation matrix $\mathbf{T}$ in the equation

$$\mathbf{Y} = \mathbf{T}X + \mathbf{N} \tag{9.1.4}$$

in which $\mathbf{T}$ has full column-rank and $\mathcal{E}\{\mathbf{N}\} = \emptyset$

◊ That $\mathbf{R}$ is the positive-definite covariance matrix of $\mathbf{Y}$, i.e. $\mathbf{R} = \mathcal{E}\{\mathbf{N}\mathbf{N}^T\}$

◊ That $\mathbf{Y}$ is transformed by infinitely many filter matrices W_j to give

$$X^*_j = W_j \mathbf{Y} \tag{9.1.5}$$

$$j = 1,\ 2,\ \ldots$$

$$S^*_j = W_j \mathbf{R}_{\mathbf{Y}} W^T_j \tag{9.1.6}$$

◊ That the filter matrices W_j all satisfy

$$W_j \mathbf{T} = I \qquad j = 1,\ 2,\ \ldots \tag{9.1.7}$$

Problem:

Find the expression for the matrix W_{minvar} which gives rise to X^*_{minvar} whose filter covariance matrix S^*_{minvar} and all other covariance matrices S^*_j satisfy

$$[S^*_j]_{i,i} \geq [S^*_{minvar}]_{i,i} \qquad 1 \leq i \leq m \qquad j = 1,\ 2,\ \ldots \tag{9.1.8}$$

■■

Thus, given the full column-rank matrix $\mathbf{T}$ and the positive-definite matrix $\mathbf{R}$ we wish to find the matrix W for which the diagonal elements of $S^* = W\mathbf{R}W^T$ are simultaneously smallest out of all the S^*_j, subject to the constraint that $W\mathbf{T} = I$.

From the problem statement we see that $\boldsymbol{S}^*_{minvar}$ has the smallest possible diagonal elements. Those diagonal elements are variances, and so $\boldsymbol{S}^*_{minvar}$ can now be called the *minimum variance* covariance matrix.

We also see the following:

◊ By virtue of (9.1.5) the estimates $\boldsymbol{X}^*_j$ are all obtained from linear transformations on the observation vector $\boldsymbol{Y}$.

◊ By virtue of (9.1.7) the estimates $\boldsymbol{X}^*_j$ are all unbiased.

Accordingly, as we did in Chapter 1, we will refer to the vectors $\boldsymbol{X}^*_j$ as *linear unbiased* (*LU*) estimates, and to $\boldsymbol{X}^*_{minvar}$ as the *best linear unbiased* (*BLU*) estimate.

9.1.3 Solving the problem

Problem Statement 2 is a minimization problem (the diagonal elements of $\boldsymbol{S}^*$) subject to a constraint ($\boldsymbol{W}_j\boldsymbol{T} = \boldsymbol{I}$). It is therefore perfectly suited to an application of Lagrange's method of undetermined multipliers.[1]

In Problem 9.2 and its solution we take you through the application of Lagrange's method of undetermined multipliers to Problem Statement 2, ending up with the required expression for the filter matrix $\boldsymbol{W}_{minvar}$.

As might be expected, the matrix $\boldsymbol{W}_{minvar}$ that gives rise to the *BLU* estimate $\boldsymbol{X}^*_{minvar}$ is no different from the $\boldsymbol{W}_{minvar}$ appearing in (8.3.5) or (8.3.10) namely

$$\boldsymbol{W}_{minvar} = (\boldsymbol{T}^T\boldsymbol{R}_{\boldsymbol{Y}}^{-1}\boldsymbol{T})^{-1}\boldsymbol{T}^T\boldsymbol{R}_{\boldsymbol{Y}}^{-1} \tag{9.1.9}$$

and so the algorithms that we obtain by solving Problem Statement 2 will be no different from Versions 1 and 2 of the MVA in Chapter 8.

However, we are now fully entitled to speak about the MVA as the *minimum variance* algorithm.

We note that $\boldsymbol{X}^*_{minvar}$ is obtained from $\boldsymbol{Y}$ using the linear transformation

$$\boldsymbol{X}^*_{minvar} = \boldsymbol{W}_{minvar}\boldsymbol{Y} \tag{9.1.10}$$

and so by Theorem 4.1 $\boldsymbol{X}^*_{minvar}$ will have a covariance matrix that is given by

$$\boldsymbol{S}^*_{minvar} = \boldsymbol{W}_{minvar}\boldsymbol{R}_{\boldsymbol{Y}}\boldsymbol{W}^T_{minvar} \tag{9.1.11}$$

[1] See, for example, Reference. 114.

This expression was shown in Appendix 8.2 to reduce to

$$\boldsymbol{S}^*_{minvar} = (\boldsymbol{T}^T \boldsymbol{R}_{\boldsymbol{Y}}^{-1} \boldsymbol{T})^{-1} \tag{9.1.12}$$

9.1.4 What minimum variance means

In Section 9.1 we gave a descriptive explanation of what minimum variance means, and we made the statement there that the lower bounds of all of the diagonal elements of each and every covariance matrix $\boldsymbol{S}_j^*$ are simultaneously present in the special covariance matrix $\boldsymbol{S}^*_{minvar}$. We called this the minimum variance property of the matrix $\boldsymbol{S}^*_{minvar}$.

In Section 9.1.3 we used Lagrange's method of undetermined multipliers to arrive at the expression for $\boldsymbol{S}^*_{minvar}$, and what we proved was that for the matrix $\boldsymbol{S}^*_{minvar}$ to possess the minimum variance property it must be as stated in (9.1.12).

In this section we prove the converse, namely that if the matrix $\boldsymbol{S}^*_{minvar}$ has the definition given in (9.1.2) then it possesses the minimum variance property. Proof is based on the following theorem:

***Theorem 9.1**:* Let $\boldsymbol{S}^*_{minvar}$ and $\boldsymbol{S}_j^*$ be defined as follows:

$$\boldsymbol{S}^*_{minvar} = (\boldsymbol{T}^T \boldsymbol{R}_{\boldsymbol{Y}}^{-1} \boldsymbol{T})^{-1} \quad \text{and} \quad \boldsymbol{S}_j^* = \boldsymbol{W}_j \boldsymbol{R}_{\boldsymbol{Y}} \boldsymbol{W}_j^T \tag{9.1.13}$$

where $\boldsymbol{W}_j$ is different from $\boldsymbol{W}_{minvar}$ defined in (9.1.9.) but satisfies the exactness constraint $\boldsymbol{W}_j \boldsymbol{T} = \boldsymbol{I}$.

Then the matrix

$$\boldsymbol{G}_j = \boldsymbol{S}_j^* - \boldsymbol{S}^*_{minvar} \tag{9.1.14}$$

is positive-semidefinite. ■■

Proof is given in Appendix 9.1. We now examine what the theorem tells us.

For simplicity, let the matrices in (9.1.14) be 3×3. Then by Property 11 for positive-semidefinite matrices, the diagonal elements of $\boldsymbol{G}_j$ must satisfy

$$[\boldsymbol{G}_j]_{1,1} \geq 0, \quad [\boldsymbol{G}_j]_{2,2} \geq 0, \quad [\boldsymbol{G}_j]_{3,3} \geq 0 \tag{9.1.15}$$

by which we mean that they can be positive or zero, but they cannot be negative.

Suppose that in all three cases they are zero, i.e. that (9.1.15) is

$$[\boldsymbol{G}_j]_{1,1} = 0, \quad [\boldsymbol{G}_j]_{2,2} = 0, \quad [\boldsymbol{G}_j]_{3,3} = 0 \tag{9.1.16}$$

Then we have a positive-semidefinite matrix whose diagonal elements are all zero, and by Property 13, $\boldsymbol{G}_j$ must be the null matrix. That being the case, by (9.1.14)

$$S_j^* = S_{minvar}^* \tag{9.1.17}$$

However, the theorem specifically states that $\boldsymbol{W}_j$ is different from $\boldsymbol{W}_{minvar}$ which means that $\boldsymbol{S}_j^*$ cannot be equal to $\boldsymbol{S}_{minvar}^*$.

This in turn means that $\boldsymbol{G}_j$ cannot have all of its diagonal elements equal to zero, and so at least one, and possibly more than one, must be strictly positive.

Suppose that only the first diagonal element of $\boldsymbol{G}_j$ is strictly positive. (The reader can easily make the extensions to the case where more than one is positive.) Then (9.1.15) becomes

$$[\boldsymbol{G}_j]_{1,1} > 0, \quad [\boldsymbol{G}_j]_{2,2} \geq 0, \quad [\boldsymbol{G}_j]_{3,3} \geq 0 \tag{9.1.18}$$

Equation (9.1.18) now means that

$$[\boldsymbol{S}_{minvar}^*]_{1,1} < [\boldsymbol{S}_j^*]_{1,1}, \qquad [\boldsymbol{S}_{minvar}^*]_{2,2} \leq [\boldsymbol{S}_j^*]_{2,2}, \qquad [\boldsymbol{S}_{minvar}^*]_{3,3} \leq [\boldsymbol{S}_j^*]_{3,3} \tag{9.1.19}$$

This tells us the following: At least one of the diagonal elements of $\boldsymbol{S}_{minvar}^*$ is strictly less than the corresponding diagonal element of any of the $\boldsymbol{S}_j^*$, while the remaining diagonal elements of $\boldsymbol{S}_{minvar}^*$ are either equal to or less than their counterparts in $\boldsymbol{S}_j^*$.

We can generate infinitely many matrices $\boldsymbol{S}_j^*$ which satisfy Theorem 9.1 relative to a given $\boldsymbol{S}_{minvar}^*$ as follows:

Starting with a given $\boldsymbol{T}$ and a given $\boldsymbol{R}_{\boldsymbol{Y}}$ we have

$$\boldsymbol{W}_{minvar} = (\boldsymbol{T}^T \boldsymbol{R}_{\boldsymbol{Y}}^{-1} \boldsymbol{T})^{-1} \boldsymbol{T}^T \boldsymbol{R}_{\boldsymbol{Y}}^{-1} \tag{9.1.20}$$

from which

$$\boldsymbol{S}_{minvar}^* = (\boldsymbol{T}^T \boldsymbol{R}_{\boldsymbol{Y}}^{-1} \boldsymbol{T})^{-1} \tag{9.1.21}$$

Now let $\boldsymbol{R}_{d_j}$ be any positive definite matrix different from $\boldsymbol{R}_{\boldsymbol{Y}}$, and when we say *different* we are excluding matrices of the form $k\boldsymbol{R}_{\boldsymbol{Y}}$ in which k is a positive scalar.

Using such matrices $\boldsymbol{R}_{d_j}$ we now form the infinitely many matrices

$$\boldsymbol{W}_j = (\boldsymbol{T}^T \boldsymbol{R}_{d_j}^{-1} \boldsymbol{T})^{-1} \boldsymbol{T}^T \boldsymbol{R}_{d_j}^{-1} \tag{9.1.22}$$

Then all such $\boldsymbol{W}_j$ satisfy the exactness constraint relative to the given $\boldsymbol{T}$. (These matrices form a subset of the infinitely many $\boldsymbol{W}$ matrices that were discussed in Section 6.6.)

If such $\boldsymbol{W}_j$ were to be used as the filter matrix in the equation

$$\boldsymbol{X}_j^* = \boldsymbol{W}_j \boldsymbol{Y} \tag{9.1.23}$$

then by Theorem 4.1 the resulting filter covariance matrix would be

$$\boldsymbol{S}_j^* = \boldsymbol{W}_j \boldsymbol{R}_{\boldsymbol{Y}} \boldsymbol{W}_j^T \tag{9.1.24}$$

According to Theorem 9.1, if we now form the infinitely many matrices $\boldsymbol{G}_j$ as follows:

$$\boldsymbol{G}_j = \boldsymbol{S}_j^* - \boldsymbol{S}_{minvar}^* = \boldsymbol{W}_j \boldsymbol{R}_{\boldsymbol{Y}} \boldsymbol{W}_j^T - (\boldsymbol{T}^T \boldsymbol{R}_{\boldsymbol{Y}}^{-1} \boldsymbol{T})^{-1} \tag{9.1.25}$$

then all such $\boldsymbol{G}_j$ will be positive-semidefinite and so the following will be true:

All matrices $\boldsymbol{S}_j^*$ will have at least one diagonal element that is strictly greater than the corresponding element in $\boldsymbol{S}_{minvar}^*$ and the remaining diagonal elements of $\boldsymbol{S}_j^*$ will be as large, or greater than, their counterparts in $\boldsymbol{S}_{minvar}^*$.

We have therefore shown that if $\boldsymbol{S}_{minvar}^*$ is as defined in (9.1.13) then it possesses the minimum variance property.

***Example 9.1**:* Let $\boldsymbol{S}_{minvar}^*$ be the positive-definite matrix

$$\boldsymbol{S}_{minvar}^* = \begin{pmatrix} 3 & 1 \\ 1 & 2 \end{pmatrix} \tag{9.1.26}$$

Then the following three positive-definite matrices are possibilities for $\boldsymbol{S}_j^*$:

(1) $\boldsymbol{S}_1^* = \begin{pmatrix} 4 & 3 \\ 3 & 6 \end{pmatrix}$

because

$$\boldsymbol{S}_1^* - \boldsymbol{S}_{minvar}^* = \begin{pmatrix} 1 & 2 \\ 2 & 4 \end{pmatrix}$$

is positive-semidefinite. In this instance, both diagonal elements of $\boldsymbol{S}_1^*$ are greater than their counterparts in $\boldsymbol{S}_{minvar}^*$.

(2) $\boldsymbol{S}_2^* = \begin{pmatrix} 3.1 & 1 \\ 1 & 2 \end{pmatrix}$

because

$$S_2^* - S_{minvar}^* = \begin{pmatrix} 0.1 & 0 \\ 0 & 0 \end{pmatrix}$$

is positive-semidefinite. In this instance, the first diagonal element of S_2^* is greater than its counterpart in S_{minvar}^* while the second diagonal element of S_2^* is equal to its counterpart in S_{minvar}^*.

(3) $S_3^* = \begin{pmatrix} 3 & 1 \\ 1 & 2.05 \end{pmatrix}$

because

$$S_3^* - S_{minvar}^* = \begin{pmatrix} 0 & 0 \\ 0 & 0.05 \end{pmatrix}$$

is positive-semidefinite. In this instance, the second diagonal element of S_3^* is greater than its counterpart in S_{minvar}^* while the first diagonal element of S_3^* is equal to its counterpart in S_{minvar}^*.

The following two positive-definite matrices cannot be possibilities for S_j^*:

(4) $S_4^* = \begin{pmatrix} 3 & 1/2 \\ 1/2 & 2 \end{pmatrix}$

because

$$S_4^* - S_{minvar}^* = \begin{pmatrix} 0 & -1/2 \\ -1/2 & 0 \end{pmatrix}$$

is not positive-semidefinite.

(5) $S_5^* = \begin{pmatrix} 3 & 1.1 \\ 1.1 & 2 \end{pmatrix}$

because

$$S_5^* - S_{minvar}^* = \begin{pmatrix} 0 & 0.1 \\ 0.1 & 0 \end{pmatrix}$$

is not positive-semidefinite.

In (4) and (5) both diagonal elements of S_4^* and S_5^* are equal to their counterparts in S_{minvar}^* but their differences from S_{minvar}^* are not positive-semidefinite.

■■

9.1.5 *Necessary and sufficient*

We can now pull together Problem Statement 2 and Theorem 9.1 in the follo wing way.

First, we define what we call the *lower bound matrix*.

***Definition 9.1**:* The **lower-bound matrix** is the matrix whose diagonal elements are the lower bounds on all of the the diagonal elements of the covariance matrix of every *LU* estimate vector. ■■

◊ By solving Problem Statement 2 we proved that if S^* is the lower-bound matrix then it is equal to $(T^T R_Y^{-1} T)^{-1}$.

◊ On the other hand, Theorem 9.1 proves the reverse, namely that if S^* is equal to $(T^T R_Y^{-1} T)^{-1}$ then it is the lower-bound matrix.

Taken together, this means that S^* is the lower-bound matrix if and only if it is equal to $(T^T R_Y^{-1} T)^{-1}$.

9.1.6 *Minimum variance and Cramér–Rao*

The lower-bound matrix was derived by solving Problem Statement 2 using Lagrange's method of undetermined multipliers, and that led us to the expression for S_{minvar}^*.

Observe that when we did that, we made no reference whatsoever to the PDF of the population from which the error vector N was drawn. We have thus provided a proof of *Gauss' minimum variance theorem*, namely:

***Theorem 9.2**:* Of all linear combinations of measurements estimating an unknown, the weighted least squares estimate has the greatest precision, regardless of the probability density function of the observation errors.[2] ■■

[2] See Appendix 9.4 which is an excerpt from Reference 117.

Quite independently of the above that was based on Lagrange's method, there is a second method for deriving the lower-bound matrix, namely by making use of Cramér–Rao theory to derive the expression for $\boldsymbol{S}_{CR}$.

The following is proved in Appendix 9.2:

Theorem 9.3:

1. Regardless of the precise form of the PDF of the errors in $\boldsymbol{Y}$, the diagonal elements of $\boldsymbol{S}^*_{minvar}$ are equal to their counterparts in $\boldsymbol{S}_{CR}$, namely

$$[\boldsymbol{S}^*_{minvar}]_{i,i} = [\boldsymbol{S}_{CR}]_{i,i} \qquad 1 \le i \le m \tag{9.1.27}$$

2. If the PDF of the errors in $\boldsymbol{Y}$ is multivariate Gaussian, then the *entire matrices* $\boldsymbol{S}^*_{minvar}$ and $\boldsymbol{S}_{CR}$ are equal, and satisfy

$$\boldsymbol{S}^*_{minvar} = (\boldsymbol{T}^T \boldsymbol{R}_Y^{-1} \boldsymbol{T})^{-1} = \boldsymbol{S}_{CR} \tag{9.1.28}$$

■■

In Part 1 of this theorem (any PDF) it is only the diagonal elements of $\boldsymbol{S}^*_{minvar}$ and $\boldsymbol{S}_{CR}$ that are necessarily equal, whereas in Part 2 of the theorem (Gaussian PDF) *all* of the elements – both on and off the diagonal – are equal.

Either way – Gaussian PDF or not – the diagonal elements of $\boldsymbol{S}^*_{minvar}$ and those of $\boldsymbol{S}_{CR}$ are equal, and those diagonal elements constitute the diagonal elements of the lower-bound matrix. In Chapter 8 we made use of this result to show that the MVA is CR consistent.

9.1.7 Minimum variance under prediction or retrodiction[3]

The following is an important proposition that is proved in Reference 6, starting on p.197:

Theorem 9.4: The minimum variance property is preserved under prediction or retrodiction. ■■

In mathematical terms this means the following: Suppose we apply the transition matrix $\boldsymbol{\Phi}$ to $\boldsymbol{X}^*_{minvar}$ as follows:

$$\boldsymbol{X}^*_{t_n+\zeta,\,t_n} = \boldsymbol{\Phi}(\zeta)\boldsymbol{X}^*_{minvar,\,t_n,\,t_n} \tag{9.1.29}$$

[3] Merriam-Webster: A retrodiction is the opposite of a prediction, i.e. it is an inference of a past state of affairs made from the present ones.

The vector $\boldsymbol{X}^*_{minvar,\, t_n,\, t_n}$ on the right (whose validity instant is t_n) is the *BLU* estimate produced from $\boldsymbol{Y}$, and being minimum variance means that its covariance matrix $\boldsymbol{S}^*_{minvar,\, t_n,\, t_n}$ has the smallest possible diagonal elements out of the covariance matrices $\boldsymbol{S}^*_{t_n,\, t_n}$ of all *LU* estimates that might also be produced from the same $\boldsymbol{Y}$.

The theorem then tells us that $\boldsymbol{X}^*_{t_n+\zeta,\, t_n}$ appearing on the left (whose validity instant is $t_n+\zeta$) will *also* be a *BLU* estimate, meaning that *its* covariance matrix will have the smallest possible diagonal elements out of the covariance matrices $\boldsymbol{S}^*_{t_n+\zeta,\, t_n}$ of all *LU* estimates that might be produced from $\boldsymbol{Y}$.

Thus the theorem tells us that equation (9.1.29) should be written as

$$\boldsymbol{X}^*_{minvar,\, t_n+\zeta,\, t_n} = \boldsymbol{\Phi}(\zeta)\boldsymbol{X}^*_{minvar,\, t_n,\, t_n} \tag{9.1.30}$$

in which the vector $\boldsymbol{X}^*$ on the left can now also be subscripted with *minvar*, and that its covariance matrix should be written as $\boldsymbol{S}^*_{minvar,\ t_n+\zeta,\, t_n}$ that satisfies

$$\boldsymbol{S}^*_{minvar,\, t_n+\zeta,\, t_n} = \boldsymbol{\Phi}(\zeta)\boldsymbol{S}^*_{minvar,\, t_n,\, t_n}\boldsymbol{\Phi}(\zeta)^T \tag{9.1.31}$$

9.2 The three CR consistency tests

In Chapter 8 we made use of the following result from Theorem 9.3:

◊ If the PDF of $\boldsymbol{N}$ is multivariate Gaussian, then the two matrices $\boldsymbol{S}^*_{minvar}$ and $\boldsymbol{S}_{CR}$ are equal to the same matrix, namely

$$\boldsymbol{S}^*_{minvar} = (\boldsymbol{T}^T\boldsymbol{R}_{\boldsymbol{Y}}^{-1}\boldsymbol{T})^{-1} = \boldsymbol{S}_{CR} \tag{9.2.1}$$

In most filtering environments in which we operate, the observation errors *are* multivariate Gaussian, and so (9.2.1) enables us to create the three CR consistency tests shown in Figure 9.3 below.

Comparing Figure 9.3 to Figure 7.4 shows that the two figures differ only in that every occurrence of the term ECM in the earlier figure has now been replaced by the term CR. Specifically:

◊ In Chapter 7 the three ECM consistency tests compared a filter's $\boldsymbol{S}_{actual}$ to its filter covariance matrix $\boldsymbol{S}^*$.

◊ In Figure 9.3 the three CR consistency tests are comparing $\boldsymbol{S}_{actual}$ to the Cramér–Rao covariance matrix $\boldsymbol{S}_{CR}$.

In order to run the three CR consistency tests, simply follow the steps for the corresponding ECM consistency tests discussed in Chapter 7, except substitute the covariance matrix $\boldsymbol{S}_{CR}$ for the filter covariance matrix $\boldsymbol{S}^*$ using the expression

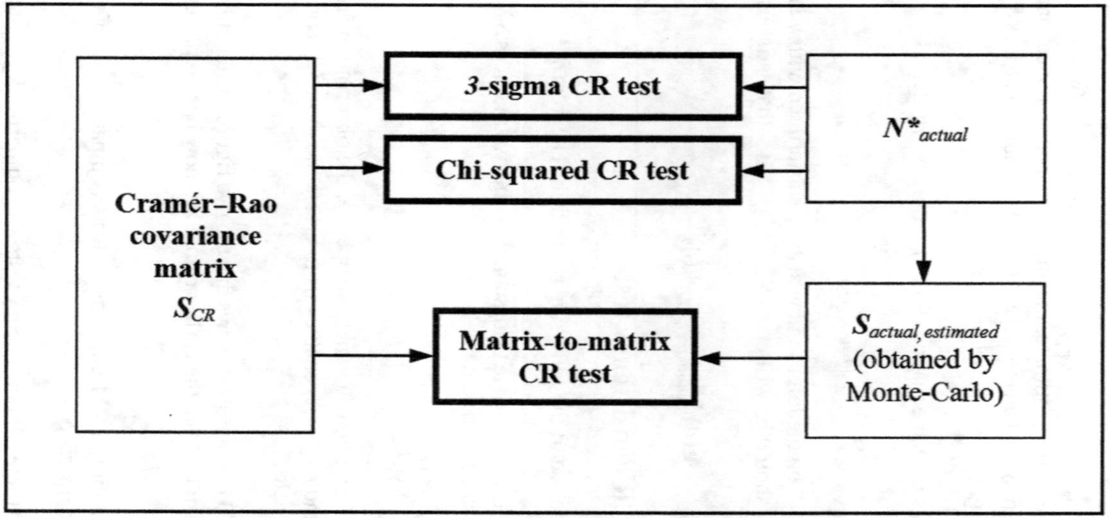

Figure 9.3 The three CR consistency tests

$$\boldsymbol{S}_{CR} = (\boldsymbol{T}^T \boldsymbol{R}_Y^{-1} \boldsymbol{T})^{-1} \tag{9.2.2}$$

We point out the following:

1. When the observations are stage-wise uncorrelated and so $\boldsymbol{R}_Y$ is block-diagonal, we use Version 2 of the MVA for which the $\boldsymbol{S}^*$ and $\boldsymbol{W}$ equations are

$$\boldsymbol{S}^* = (\boldsymbol{T}^T \boldsymbol{R}_Y^{-1} \boldsymbol{T})^{-1} \tag{9.2.3}$$

$$\boldsymbol{W} = \boldsymbol{S}^* \boldsymbol{T}^T \boldsymbol{R}_Y^{-1} \tag{9.2.4}$$

   The Gauss filters are then both ECM and CR consistent.

   If we then run any of the CR consistency tests in Figure 9.3 to confirm CR consistency, we would use the expression in (9.2.2) for $\boldsymbol{S}_{CR}$. However, that is precisely what we would use for $\boldsymbol{S}^*$ in the ECM consistency tests in Figure 7.4, and so under these circumstances the CR tests will simply be repeats of the ECM tests, and no useful information will be obtained from them.

2. If the observations are stage-wise correlated and so $\boldsymbol{R}_Y$ is not block-diagonal, we use Version 1 for which

   ◊ The $\boldsymbol{W}$ equation will be

$$\boldsymbol{W} = (\boldsymbol{T}^T \boldsymbol{R}_d^{-1} \boldsymbol{T})^{-1} \boldsymbol{T}^T \boldsymbol{R}_d^{-1} \tag{9.2.5}$$

   in which $\boldsymbol{R}_d$ differs from $\boldsymbol{R}_Y$ because the former is block-diagonal and easy to invert (see the discussion in Section 8.8 and in Projects 8.2, 8.3 and 8.4). The filter is now theoretically CR inconsistent and the CR tests enable us to see how far it has departed from CR consistency, thereby yielding useful information.

   ◊ The $\boldsymbol{S}^*$ equation will be

$$\boldsymbol{S}^* = \boldsymbol{W} \boldsymbol{R}_Y \boldsymbol{W}^T \tag{9.2.6}$$

   The filter should be ECM consistent, and running the ECM tests should confirm that fact.

   The two sets of tests will now yield differing answers and both will provide valuable information.

The program ***04_Monte_Carlo*** is a full implementation of everything that we have been discussing here. The ECM-consistency and CR-consistency matrix-to-matrix tests are implemented, as well as the ECM and CR-consistency Chi-squared tests in Monte-Carlo mode.

Reading Projects 8.2, 8.3 and 8.4 and running the program will give you a complete understanding of all of the above, as well as of other issues relating to the ECM and CR consistency tests.

9.3 Other methods for deriving the MVA

We have derived the MVA by two methods:

◊ Solving Problem Statement 1.

◊ Solving Problem Statement 2.

The MVA can be derived by at least three other methods that we are aware of.

1. By direct algebraic manipulation of the expression

$$e(X^*) = (\boldsymbol{Y} - \boldsymbol{T}X^*)^T \boldsymbol{R}_{\boldsymbol{Y}}^{-1} (\boldsymbol{Y} - \boldsymbol{T}X^*) \tag{9.3.1}$$

in a way that is similar to completing the squares. This method forms the subject of Problem 9.3.

2. If the observation errors are multivariate Gaussian, then the maximum likelihood estimate becomes the minimum variance estimate (see Appendix 9.3). This was the first of two the justifications that Gauss gave for his method of weighted least squares.

3. By orthogonal projection. This method forms the subject of Problem 9.4 and is fully covered there.

9.4 Minimum variance and least squares

Restating Version 1 of the MVA we have:

$$W = (\boldsymbol{T}^T \boldsymbol{R}_{\boldsymbol{Y}}^{-1} \boldsymbol{T})^{-1} \boldsymbol{T}^T \boldsymbol{R}_{\boldsymbol{Y}}^{-1} \tag{9.4.1}$$

$$X^* = W\boldsymbol{Y} \tag{9.4.2}$$

$$S^* = W\boldsymbol{R}_{\boldsymbol{Y}} W^T \tag{9.4.3}$$

The estimate vector X^* in (9.4.2) minimizes the sum of the weighted squared residuals

$$e(X^*) \equiv \boldsymbol{N}^{*T}\boldsymbol{R}_{\boldsymbol{Y}}^{-1}\boldsymbol{N}^* = (\boldsymbol{Y} - \boldsymbol{T}X^*)^T\boldsymbol{R}_{\boldsymbol{Y}}^{-1}(\boldsymbol{Y} - \boldsymbol{T}X^*) \tag{9.4.4}$$

This procedure is usually referred to as *weighted least squares*, in which it is the inverse of $\boldsymbol{R}_{\boldsymbol{Y}}$ that is providing the weights in (9.4.4).

If the matrix $\boldsymbol{R}_{\boldsymbol{Y}}$ in (9.4.4) is omitted, then the name of the procedure changes from *weighted least squares* to simply *least squares* and (9.4.1) becomes

$$W = (\boldsymbol{T}^T\boldsymbol{T})^{-1}\boldsymbol{T}^T \tag{9.4.5}$$

In this case (9.4.2) reduces to

$$X^* = (\boldsymbol{T}^T\boldsymbol{T})^{-1}\boldsymbol{T}^T\boldsymbol{Y} \tag{9.4.6}$$

The use of least squares should be restricted to situations in which the observations in the long-vector $\boldsymbol{Y}$ all have the *same* physical dimensions (i.e. they are all observations of *position*, or all observations of *velocity*, etc.).

If that is not the case and $\boldsymbol{Y}$ is made up of observations that have mixed dimensions, then least squares should not be used. That is because the resulting estimate X^* is then not unique, but depends on the choice of units that were selected for the mixed dimensions.

However, this restriction does not apply to the MVA (i.e. with $\boldsymbol{R}_{\boldsymbol{Y}}$ present), for which the dimensions of the observations in $\boldsymbol{Y}$ *can* be mixed.

The estimate vector X^* is then definitely unique, regardless of whether or not the observations have mixed dimensions, and regardless of the choice of units in which they are expressed, provided only that we are consistent. By this we mean that whatever dimension we use for a given observation in $\boldsymbol{Y}$, we must also use the square of that dimension for its variance in $\boldsymbol{R}$.

Proof of these statements is given in Project 9.4.

9.5 The Gauss–Newton filters

Nonlinearities are frequently present in filter models or observation equations. In this section we discuss how the MVA gives rise to the Gauss–Newton filters that are required for such situations.

In Chapter 3 we saw that there are the four possible cases shown in Figure 9.4.

The most desirable situation, if it exists or can be brought about, is the one in the top left box where the filter model and the observation equation are both linear, thereby

Case 1 Filter model linear Observation equation linear	***Case 3*** Filter model nonlinear Observation equation linear
Case 2 Filter model linear Observation equation nonlinear	***Case 4*** Filter model nonlinear Observation equation nonlinear

Figure 9.4 The four cases

putting us in ***Case 1***. We can then use either of the Gauss–Aitken filters exactly as shown in Figures 8.3 or 8.4.

However, that may not always be possible, and so we now examine what must be done when we are located in any of the other three boxes of Figure 9.4.

9.5.1 Cycle and iteration

Referring to Figure 9.5 we consider first the meanings of the words *cycle* and *iteration*.

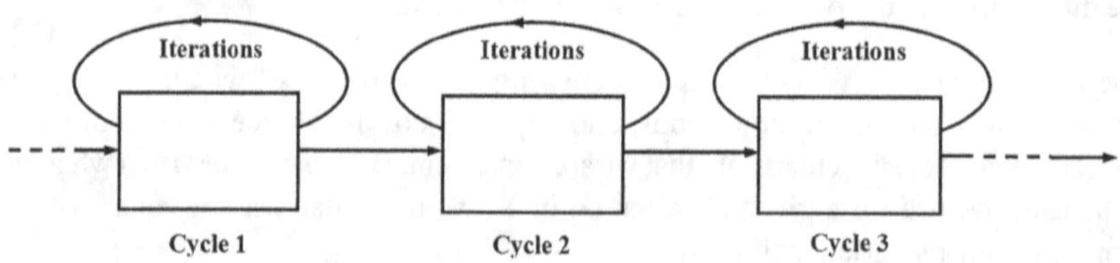

Figure 9.5 Cycle and iteration

When we say that 'one cycle of a Gauss–Aitken filter has been completed' we mean that the observation vector $\boldsymbol{Y}$ and its covariance matrix $\boldsymbol{R_Y}$ have been processed to produce the *BLU* estimate $\boldsymbol{X}^*$ and its covariance matrix $\boldsymbol{S}^*$.

The Gauss–Aitken filters are non-iterative, and so each time that they are operated, one of their cycles is completed.

By contrast, the Gauss–Newton filters that we are now discussing are iterative procedures, and in order to complete one of their cycles they must iterate multiple times. As an example, three iterations might be required to derive the estimate vector $\boldsymbol{X}^*$ and its covariance matrix $\boldsymbol{S}^*$.

◊ During each iteration within that cycle, the MVA is executed using the *same* $\boldsymbol{Y}$ and $\boldsymbol{R}$.

◊ Each iteration produces a tentative X^* and S^*, and only when the stopping rule is satisfied do the tentative X^* and S^* become final.

At that point the current cycle of the filter has been completed and we can move on to the next cycle.

9.5.2 The six Gauss–Newton filters

The six Gauss–Newton filters make use of

◊ The MVA (this is the Gauss part).

◊ Perturbation vectors and the method of local linearization (this is the Newton part).

The filters operate as follows:

Let X^* be the *BLU* estimate of the true state vector X that we are seeking and let $\bar{X}$ be the state vector of a known trajectory – called a *nominal trajectory* – that is close to, and satisfies the same DE as the vector X. Let δX^* be the perturbation vector defined as follows:

$$X^* = \bar{X} + \delta X^* \tag{9.5.1}$$

If we could find δX^* directly from the observations then we could add it to the known $\bar{X}$ in (9.5.1), thereby obtaining the desired estimate X^*. The Gauss–Newton filters do precisely that – but in a sequence of iterative steps.

For ***Cases 2*** and ***4*** (the observation equation is nonlinear) we start from

$$\boldsymbol{Y} = \boldsymbol{G}(X) + \boldsymbol{N} \tag{9.5.2}$$

to which we apply local linearization, obtaining the equation

$$\delta\boldsymbol{Y} = \boldsymbol{T}(\bar{X})\delta X + \boldsymbol{N} \tag{9.5.3}$$

(See (3.5.11) for ***Case 2*** or (3.3.36) for ***Case 4***.)

For ***Case 3*** the observation equation is linear, but even then we must use the nominal trajectory $\bar{X}$ when building $\boldsymbol{T}$ because the filter model is nonlinear (see (3.4.14)). The observation equation then becomes (see (3.4.15))

$$\delta \boldsymbol{Y} = \boldsymbol{T}(\bar{X})\delta X + \boldsymbol{N} \tag{9.5.4}$$

and so for all three ***Cases 2***, ***3*** and ***4*** we have an equation of the form appearing in (9.5.4).

We now replace δX in (9.5.4) with δX^*, obtaining the new equation

$$\delta \boldsymbol{Y} = \boldsymbol{T}(\bar{X})\delta X^* + \boldsymbol{N}^* \tag{9.5.5}$$

in which $\boldsymbol{N}$ has been replaced by the vector $\boldsymbol{N}^*$ so that (9.5.5) will balance.

We note that what we did to obtain (9.5.5) is similar to what we did to obtain (8.2.2), except that here we are dealing with perturbation vectors δY and δX^*, whereas there we were dealing with whole vectors $\boldsymbol{Y}$ and X^*.

We also note from (9.5.4) that the covariance matrix of δY is the covariance matrix of the vector $\boldsymbol{N}$, namely $\boldsymbol{R}_{\boldsymbol{Y}}$.

For the all-linear ***Case 1*** we minimized the quantity

$$e(X^*) = (\boldsymbol{Y} - \boldsymbol{T}X^*)^T \boldsymbol{R}_{\boldsymbol{Y}}^{-1} (\boldsymbol{Y} - \boldsymbol{T}X^*) \tag{9.5.6}$$

and this resulted in the expression for the required estimate as

$$X^*_{minvar} = (\boldsymbol{T}^T \boldsymbol{R}_{\boldsymbol{Y}}^{-1} \boldsymbol{T})^{-1} \boldsymbol{T}^T \boldsymbol{R}_{\boldsymbol{Y}}^{-1} \boldsymbol{Y} \tag{9.5.7}$$

For ***Cases 2, 3*** and ***4*** we must instead minimize the quantity

$$e(\delta X^*) = \left(\delta \boldsymbol{Y} - \boldsymbol{T}(\bar{X})\delta X^*\right)^T \boldsymbol{R}_{\boldsymbol{Y}}^{-1} \left(\delta \boldsymbol{Y} - \boldsymbol{T}(\bar{X})\delta X^*\right) \tag{9.5.8}$$

which leads to the expression

$$\delta X^*_{minvar} = \left(\boldsymbol{T}(\bar{X})^T \boldsymbol{R}_{\boldsymbol{Y}}^{-1} \boldsymbol{T}(\bar{X})\right)^{-1} \boldsymbol{T}(\bar{X})^T \boldsymbol{R}_{\boldsymbol{Y}}^{-1} \delta \boldsymbol{Y} \tag{9.5.9}$$

Having found δX^* we then add it to $\bar{X}$ in (9.5.1), hoping to obtain the required estimate X^*_{minvar}. However, that will not quite be the case for the following reason:

Recall, in Chapter 3, that in deriving the total observation equations for ***Cases 2***, ***3*** and ***4*** we used the method of local linearization, during which we neglected the higher order terms in the Taylor series in order to remove the nonlinearities. Had we not done that we would not have arrived at (9.5.5), but the price must now be paid.

When we add δX^* to $\bar{X}$ in (9.5.1) we move *linearly* in the direction of the *BLU* estimate X^*_{minvar}. However, we don't quite reach it, because by assumption our problem involves nonlinearities. We have made an improvement over the starting trajectory defined by $\bar{X}$, but we have not yet reached the best estimate towards which the observations are pulling us.

However, what we have derived above can now be used to build the six iterative procedures that are known as the Gauss–Newton filters.

◊ In Project 9.5 we discuss the details of the filters for each of ***Cases 2***, ***3*** and ***4***. (The fact that there are two for each case comes about because there are two versions of the MVA.)

◊ In Project 9.6 we discuss possible ways in which a stopping rule for the filters might be created.

9.6 Initializing the Gauss–Newton filters

Initialization of the Gauss–Newton algorithms centres on finding a nominal trajectory $\bar{X}$ with which to get them started. The full discussion appears in Projects 9.7 and 9.8.

9.7 Enhancements to the Gauss–Newton filters

On occasions the Gauss–Newton filters run into difficulties – they converge slowly or are unable to converge at all.

A considerable amount of investigation has been carried out in this regard and a number of highly effective enhancements have been devised, among them the Levenberg-Marquardt algorithm.

These are beyond the scope of our discussion, but if such difficulties are encountered then the reader has a variety of places to turn for help.[4]

In Project 9.9 we discuss a further way in which Gauss–Newton convergence can be assisted.

The following items are in the supplementary material

Problems for Chapter 9

Solutions to Problems for Chapter 9

Projects for Chapter 9

[4] See References 19, 60, 61, 62, 85, 86, 87, 88, 89, 175, 176, 177, 199, 200, 201.

Appendix 9.1
Proof of Theorem 9.1

Theorem 9.1: Let $\boldsymbol{S}^*_{minvar}$ and $\boldsymbol{S}^*_j$ be defined as follows:

$$\boldsymbol{S}^*_{minvar} = (\boldsymbol{T}^T\boldsymbol{R}_{\boldsymbol{Y}}^{-1}\boldsymbol{T})^{-1} \quad \text{and} \quad \boldsymbol{S}^*_j = \boldsymbol{W}_j\boldsymbol{R}_{\boldsymbol{Y}}\boldsymbol{W}_j^T \tag{A9.1.1}$$

where $\boldsymbol{W}_j$ is different from $\boldsymbol{W}_{minvar}$ defined in (9.1.9.) but satisfies the exactness constraint $\boldsymbol{W}_j\boldsymbol{T} = \boldsymbol{I}$.

Then the matrix

$$\boldsymbol{G}_j = \boldsymbol{S}^*_j - \boldsymbol{S}^*_{minvar} \tag{A9.1.2}$$

is positive-semidefinite. ■■

Writing out the expressions for $\boldsymbol{S}^*_j$ and $\boldsymbol{S}^*_{minvar}$ we have

$$\boldsymbol{G}_j = \boldsymbol{W}_j\boldsymbol{R}_{\boldsymbol{Y}}\boldsymbol{W}_j^T - (\boldsymbol{T}^T\boldsymbol{R}_{\boldsymbol{Y}}^{-1}\boldsymbol{T})^{-1} \tag{A9.1.3}$$

in which $\boldsymbol{W}_j$ is any of the infinitely many filter matrices in Problem Statement 2 that satisfy

$$\boldsymbol{W}_j\boldsymbol{T} = \boldsymbol{I} \qquad j = 1, 2, \ldots \tag{A9.1.4}$$

and where $\boldsymbol{R}_{\boldsymbol{Y}}$ is $k \times k$, $\boldsymbol{W}_j$ is $m \times k$ and wide, $\boldsymbol{T}$ is $k \times m$ and high, and so $k > m$.

Equation (A9.1.4) together with Theorem 6.2 means that $\boldsymbol{W}_j$ has full row-rank and $\boldsymbol{T}$ has full column-rank.

We are required to prove that $\boldsymbol{G}_j$ in (A9.1.3) is positive- semidefinite. Proof is based on the following intermediate result.

Lemma A9.1.1: Let $\boldsymbol{B}$ be a full row-rank $m \times k$ matrix where $k > m$ (i.e. $\boldsymbol{B}$ is wide), and let $\boldsymbol{H}$ be a full column-rank $k \times m$ matrix (i.e. $\boldsymbol{H}$ is high). Then the matrix

$$K = BB^T - BH(H^TH)^{-1}H^TB^T \tag{A9.1.5}$$

is positive-semidefinite. ■■

Proof of the lemma is given in Problem 9.9.

Using this result, we now prove Theorem 9.1 as follows:

R_Y is positive-definite and so by Property 7 in Section 4.4 its square root $R_Y^{1/2}$ exists and is positive-definite. By Properties 4 and 5 $R_Y^{-1/2}$ exists and is also positive-definite. Moreover, R_Y is $k \times k$ and so $R_Y^{1/2}$ and $R_Y^{-1/2}$ are also $k \times k$.

Let $B = W_j R_Y^{1/2}$. Because W_j has full row-rank and $R_Y^{1/2}$ is positive-definite it follows that B has full row-rank (see Problem 9.10).

Similarly, let $H = R_Y^{-1/2}T$. Because T has full column-rank and $R_Y^{-1/2}$ is positive-definite it follows that H has full column-rank (see Problem 9.10).

We note that B is $m \times k$ and wide with full row-rank, and H is $k \times m$ and high with full column-rank, and so all conditions for Lemma A9.1.1 are now met. Starting from (A9.1.5) we then have

$$\begin{aligned} K &= BB^T - BH(H^TH)^{-1}H^TB^T = W_jR_YW_j^T - W_jT(T^TR_Y^{-1}T)^{-1}T^TW_j^T \\ &= W_jR_YW_j^T - I(T^TR_Y^{-1}T)^{-1}I = G_j \text{ in Theorem 9.1.} \end{aligned} \tag{A9.1.6}$$

But from Lemma A9.1.1 we know that K in (A9.1.6) is positive-semidefinite and so the matrix G_j is also positive-semidefinite and the proof of Theorem 9.1 is complete.

■■

Theorem 9.1 tells us that every filter covariance matrix S_j^* is 'larger' than the matrix S_{minvar}^* in the sense that the former minus the latter is positive-semidefinite.

This result is often written symbolically as follows:

$$S_j^* - S_{minvar}^* \geq 0 \tag{A9.1.7}$$

by which we mean that the matrix formed by the difference of the two matrices on the left is positive-semidefinite.

Equation (A9.1.7) is also written symbolically as

$$S_j^* \geq S_{minvar}^* \tag{A9.1.8}$$

Consider now the diagonal terms of the matrices $\boldsymbol{G}_j$, $\boldsymbol{S}_j^*$ and $\boldsymbol{S}_{minvar}^*$. We have shown that $\boldsymbol{G}_j$ is positive-semidefinite, and so we have

$$[\boldsymbol{G}_j]_{i,i} = [\boldsymbol{S}_j^*]_{i,i} - [\boldsymbol{S}_{minvar}^*]_{i,i} \geq 0 \qquad 1 \leq i \leq m \tag{A9.1.9}$$

Then

$$[\boldsymbol{S}_j^*]_{i,i} \geq [\mathbf{S}_{minvar}^*]_{i,i} \qquad 1 \leq i \leq m \tag{A9.1.10}$$

showing that the diagonal elements of $\boldsymbol{S}_{minvar}^*$ are lower bounds on those of all possible covariance matrices $\boldsymbol{S}_j^*$.

We have therefore shown that if $\boldsymbol{S}_{minvar}^*$ is as defined in (A9.1.1) then it possesses the minimum variance property.

Appendix 9.2

Minimum variance and Cramér–Rao

We assume the following:

◊ An observation vector $\boldsymbol{Y}$ has been obtained according to the equation

$$\boldsymbol{Y} = \boldsymbol{TX} + \boldsymbol{N} \tag{A9.2.1}$$

in which $\boldsymbol{Y}$ and $\boldsymbol{N}$ are k-vectors, $\boldsymbol{T}$ is $k \times m$, $\boldsymbol{X}$ is an m-vector and $k > m$. The covariance matrix of $\boldsymbol{Y}$ is $\boldsymbol{R_Y}$ which is $k \times k$.

◊ $\boldsymbol{Y}$ is linearly transformed by infinitely many $m \times k$ filter matrices $\boldsymbol{W}_j$ according to the equations

$$\left.\begin{aligned} \boldsymbol{X}_j^* &= \boldsymbol{W}_j \boldsymbol{Y} \\ \boldsymbol{S}_j^* &= \boldsymbol{W}_j \boldsymbol{R_Y} \boldsymbol{W}_j^T \end{aligned}\right\} \quad j = 1, 2, \ldots \tag{A9.2.2}$$

in which

- ◊ The $\boldsymbol{X}_j^*$ are m-vectors and their covariance matrices $\boldsymbol{S}_j^*$ are $m \times m$.
- ◊ The filter matrices $\boldsymbol{W}_j$ are all assumed to satisfy the exactness constraint, namely

  $$\boldsymbol{W}_j \boldsymbol{T} = \boldsymbol{I} \quad j = 1, 2, \ldots \tag{A9.2.3}$$

  in which $\boldsymbol{T}$ is the same as in (A9.2.1) and where $\boldsymbol{I}$ is an $m \times m$ identity matrix.

A9.2.1 Minimum variance

Given the above set of assumptions, solving Problem Statement 2 by the use of Lagrange's method of undetermined multipliers shows that among the infinity of covariance matrices $\boldsymbol{S}_j^*$ in (A9.2.2), there exists one called $\boldsymbol{S}_{minvar}^*$ with the following property:

The diagonal elements of $\boldsymbol{S}^*_{minvar}$ are the lower bounds on the diagonal elements of each and every other covariance matrix $\boldsymbol{S}^*_j$, namely

$$[\boldsymbol{S}^*_j]_{i,i} \geq [\boldsymbol{S}^*_{minvar}]_{i,i} \qquad 1 \leq i \leq m \qquad j = 1, 2, \ldots \infty \tag{A9.2.4}$$

This gives us the following:

The minimum variance algorithm (Version 1)

$$\boldsymbol{W}_{minvar} = (\boldsymbol{T}^T \mathbf{R}_{\mathbf{Y}}^{-1} \boldsymbol{T})^{-1} \boldsymbol{T}^T \mathbf{R}_{\mathbf{Y}}^{-1} \tag{A9.2.5}$$

$$\boldsymbol{X}^*_{minvar} = \boldsymbol{W}_{minvar} \mathbf{Y} \tag{A9.2.6}$$

$$\boldsymbol{S}^*_{minvar} = \boldsymbol{W}_{minvar} \mathbf{R}_{\mathbf{Y}} \boldsymbol{W}^T_{minvar} \tag{A9.2.7}$$

It was shown in Appendix 8.2 that (A9.2.7) reduces as follows:

$$\boldsymbol{S}^*_{minvar} = (\boldsymbol{T}^T \mathbf{R}_{\mathbf{Y}}^{-1} \boldsymbol{T})^{-1} \tag{A9.2.8}$$

Observe that at this stage that no restrictions whatever have been placed on the PDF of the error vector $\mathbf{N}$ in (A9.2.1).

A9.2.2 Cramér–Rao

Given the same set of assumptions, Cramér–Rao theory proves the following:

There exists an $m \times m$ matrix $\boldsymbol{S}_{CR}$ (called the Cramér–Rao covariance matrix) whose diagonal elements are the lower bounds on the diagonal elements of each and every other covariance matrix $\boldsymbol{S}^*_j$, namely

$$[\boldsymbol{S}^*_j]_{i,i} \geq [\boldsymbol{S}_{CR}]_{i,i} \quad 1 \leq i \leq m \quad j = 1, 2, \ldots \infty \tag{A9.2.9}$$

From Cramér–Rao theory the matrix $\boldsymbol{S}_{CR}$ is derived as follows:

$$\boldsymbol{S}_{CR} = \left(-\mathcal{E}\{\partial^2 ln(\mathsf{L})/\partial \boldsymbol{X}^T \partial \boldsymbol{X}\}\right)^{-1} \tag{A9.2.10}$$

in which L is the *likelihood function* of $\boldsymbol{X}$ given $\mathbf{Y}$ in (A9.2.1), and the notation $\partial^2 ln(\mathsf{L})/\partial \boldsymbol{X}^T \partial \boldsymbol{X}$ means that $ln(\mathsf{L})$ is to be differentiated twice with respect to $\boldsymbol{X}$ (see the document ***Miscellaneous\Vector_and_Matrix_Differentiation***).

Note: Readers unfamiliar with the concept of the likelihood function should consult References 96 and 97. Those unfamiliar with Cramér–Rao theory should consult References 11, 12 and 26, the first two of which deal with the univariate case and the third with the multivariate case.

We now compare equations (A9.2.4) and (A9.2.9) and we see the following:

From (A9.2.4):

$$[\boldsymbol{S}_j^*]_{i,i} \geq [\boldsymbol{S}_{minvar}^*]_{i,i} \quad 1 \leq i \leq m \tag{A9.2.11}$$

From (A9.2.9):

$$[\boldsymbol{S}_j^*]_{i,i} \geq [\boldsymbol{S}_{CR}]_{i,i} \quad 1 \leq i \leq m \tag{A9.2.12}$$

Thus $\boldsymbol{S}_{minvar}^*$ and $\boldsymbol{S}_{CR}$ are both the lower-bound matrix (see Definition 9.1) and so we must have

$$[\boldsymbol{S}_{minvar}^*]_{i,i} = [\boldsymbol{S}_{CR}]_{i,i} \quad 1 \leq i \leq m \tag{A9.2.13}$$

In applying Lagrange's method of undetermined multipliers which led to $[\boldsymbol{S}_{minvar}^*]_{i,i}$ in (A9.2.13), no mention was made of the PDF of the error vector $\boldsymbol{N}$ in (A9.2.1), and so far we have said nothing about that PDF when discussing Cramér–Rao theory.

Accordingly, we have the following:

For observation errors $\boldsymbol{N}$ with any PDF whatsoever, the diagonal elements of $\boldsymbol{S}_{minvar}^*$ and those of $\boldsymbol{S}_{CR}$ satisfy

$$[\boldsymbol{S}_{minvar}^*]_{i,i} = [\boldsymbol{S}_{CR}]_{i,i} \quad 1 \leq i \leq m \tag{A9.2.14}$$

A9.2.3 Multivariate Gaussian errors

If the PDF of the error vector $\boldsymbol{N}$ is multivariate Gaussian then we also have the following:

From (A9.2.1) the likelihood function of $\boldsymbol{X}$ given the observation vector $\boldsymbol{Y}$ becomes

$$L = (2\pi)^{-m/2} |\,\boldsymbol{R}\,|^{-1/2} exp\left(-\tfrac{1}{2}(\boldsymbol{Y}-\boldsymbol{TX})^T \boldsymbol{R_Y}^{-1}(\boldsymbol{Y}-\boldsymbol{TX})\right) \tag{A9.2.15}$$

from which

$$ln(L) = -\tfrac{1}{2}\left(m\, ln(2\pi) + ln|\,\boldsymbol{R}\,| + (\boldsymbol{Y}-\boldsymbol{TX})^T \boldsymbol{R_Y}^{-1}(\boldsymbol{Y}-\boldsymbol{TX})\right) \tag{A9.2.16}$$

In ***Miscellaneous\Matrix_Differentiation*** we show that double differentiation of $ln(L)$ in this equation with respect to $\boldsymbol{X}$ results in

$$\partial^2 ln(L)/\partial \boldsymbol{X}^T \partial \boldsymbol{X} = -\boldsymbol{T}^T \boldsymbol{R_Y}^{-1} \boldsymbol{T} \tag{A9.2.17}$$

Using (A9.2.17) in (A9.2.10) then gives the expression for $\boldsymbol{S}_{CR}$ for the multivariate Gaussian case as follows:

$$\boldsymbol{S}_{CR} = (-\mathcal{E}\{-\boldsymbol{T}^T\boldsymbol{R}_{\boldsymbol{Y}}^{-1}\boldsymbol{T}\})^{-1} = (\boldsymbol{T}^T\boldsymbol{R}_{\boldsymbol{Y}}^{-1}\boldsymbol{T})^{-1} \tag{A9.2.18}$$

Comparison with (A9.2.8) shows that when the observation errors are Gaussian, the *entire* matrices $\boldsymbol{S}^*_{minvar}$ and $\boldsymbol{S}_{CR}$ are equal, that is

$$\boldsymbol{S}^*_{minvar} = \boldsymbol{S}_{CR} \tag{A9.2.19}$$

We accordingly have the following:

***Theorem 9.3**:*

1. Regardless of the precise form of the PDF of the errors in $\boldsymbol{Y}$, the diagonal elements of $\boldsymbol{S}^*_{minvar}$ are equal to their counterparts in $\boldsymbol{S}_{CR}$, namely

$$[\boldsymbol{S}^*_{minvar}]_{i,i} = [\boldsymbol{S}_{CR}]_{i,i} \quad 1 \le i \le m \tag{A9.2.20}$$

2. If the PDF of the errors in $\boldsymbol{Y}$ is multivariate Gaussian, then the *entire matrices* $\boldsymbol{S}^*_{minvar}$ and $\boldsymbol{S}_{CR}$ are equal, and satisfy

$$\boldsymbol{S}^*_{minvar} = (\boldsymbol{T}^T\boldsymbol{R}_{\boldsymbol{Y}}^{-1}\boldsymbol{T})^{-1} = \boldsymbol{S}_{CR} \tag{A9.2.21}$$

■■

Appendix 9.3

Minimum variance and maximum likelihood

Given the linear observation equation

$$\mathbf{Y} = \mathbf{TX} + \mathbf{N} \tag{A9.3.1}$$

in which $\mathbf{Y}$ is the total observation vector with error vector $\mathbf{N}$, $\mathbf{X}$ is the true state vector of the observed trajectory and $\mathbf{T}$ is the total observation matrix, we assume that $\mathbf{N}$ is zero-mean multivariate Gaussian with covariance matrix $\mathbf{R_Y}$.

Let $\mathbf{X}^*$ be an estimate of $\mathbf{X}$. Then the likelihood function of $\mathbf{X}^*$ given the observation vector $\mathbf{Y}$ is

$$L = (2\pi)^{-k/2}|\,\mathbf{R}\,|^{-1/2} exp(-\tfrac{1}{2}(\mathbf{Y} - \mathbf{TX}^*)^T \mathbf{R_Y^{-1}}(\mathbf{Y} - \mathbf{TX}^*)) \tag{A9.3.2}$$

The *method of maximum likelihood* selects $\mathbf{X}^*$ so that L in (A9.3.2) is maximized. That in turn means that $\mathbf{X}^*$ must be selected so that the scalar exponent in that equation, namely

$$(\mathbf{Y} - \mathbf{TX}^*)^T \mathbf{R_Y^{-1}}(\mathbf{Y} - \mathbf{TX}^*) \tag{A9.3.3}$$

is minimized.

Because this is identical to the starting point in Appendix 8.1 for the derivation of the minimum variance algorithm, it follows that maximization of L in (A9.3.2) over $\mathbf{X}^*$ must lead to the same conclusion, namely

$$\mathbf{X}^*_{minvar} = (\mathbf{T}^T \mathbf{R_Y^{-1}} \mathbf{T})^{-1} \mathbf{T}^T \mathbf{R_Y^{-1}} \mathbf{Y} \tag{A9.3.4}$$

Thus, when the errors are multivariate Gaussian, the minimum variance algorithm is consistent with the maximum likelihood estimate. ■■

Appendix 9.4

Excerpt from 'Theory of the Combination of Observations Least Subject to Errors' [1]

Although Gauss had discovered the method of least squares during the last decade of the eighteenth century and used it regularly after 1801 in astronomical calculations, it was Legendre who introduced it to the world in an appendix to an astronomical memoir. Legendre stated the principle of least squares for combining observations and derived the normal equations from which the least squares estimate may be calculated. However, he provided no justification for the method, other than noting that it prevented extreme errors from prevailing by establishing a sort of equilibrium among all errors, and he was content to refer the calculator to the methods of the day for solving linear systems.

In 1809, towards the end of the treatise on *The Theory of the Motion of Heavenly Bodies*, Gauss gave a probabilistic justification of the method, in which he essentially showed that if the errors are normal[2] then least squares gives the maximum likelihood estimate. However, his reasons for assuming normality were tenuous, and Gauss himself later rejected the approach. In other respects the treatment was more successful. It contains the first mention of Gaussian elimination (worked out in detail in a later publication), which was also used to derive expressions for the precision of the estimates. He also described the Gauss–Newton method for solving non-linear least squares problems and gave a characterisation of what we would now call approximation in terms of ℓ_1 norms.

Shortly thereafter, Laplace turned to the subject and derived the method of least squares from the principle that the best estimate should have the smallest mean error, by which he meant the mean of the absolute value of the error. Since the mean absolute error does not lead directly to the least squares principle, Laplace gave an asymptotic argument based on his central limit theorem.

In the 1820s, Gauss returned to least squares in two memoirs, the first in two parts, published by the Royal Society of Gottingen under the common title *Theoria*

[1] Excerpted from Reference 117 and reproduced here by the kind permission of the Society for Industrial and Applied Mathematics (SIAM), Philadelphia, USA.
[2] That is, Gaussian.

Combinationis Observationum Erroribus Minimus Obnoxiae (*Theory of the Combination of Observations Least Subject to Errors*). In the *Pars Prior* of the first memoir, Gauss substituted the root mean square error for Laplace's mean absolute error. This enabled him to prove his minimum variance theorem: Of all linear combinations of measurements estimating an unknown, the least squares estimate has the greatest precision. The remarkable thing about this theorem is that it does not depend on the distributions of the errors, and, unlike Laplace's result, is not asymptotic.

Chapter 10

The master control algorithms and goodness-of-fit

10.1 Introduction

In this chapter we discuss two versions of what we call the *master control algorithm* (MCA) which is used to control the Gauss filters when manoeuvring targets are being tracked. We refer to them as MCA-1 and MCA-2.

The goodness-of-fit (GOF) test is used in both versions of the MCA. The GOF test can also be implemented stand-alone, i.e. independently of the MCAs.

The ECM and CR consistency tests that were discussed in Chapters 7 and 9 all require that the true state vector of the observed trajectory be known, which means that they can only be used in simulations.

By contrast, the GOF test does not involve the observed trajectory, and so it – and hence also the two versions of the MCA – can be operated both in the field and in simulations. This makes them particularly powerful constructs for use in filter engineering.

10.2 Background

The MCAs are based on the following six assumptions.

10.2.1 Kinds of manoeuvres

The manoeuvres that we envisage are of two basic kinds.

Type 1: The target's heading changes while its speed remains constant. This is illustrated in Figure 10.1 below.

This is the type of manoeuvre that would be performed by an aircraft flying at constant speed which then abruptly changes its heading.

Type 2: The target's speed changes while its heading remains constant. This is illustrated in Figure 10.2.

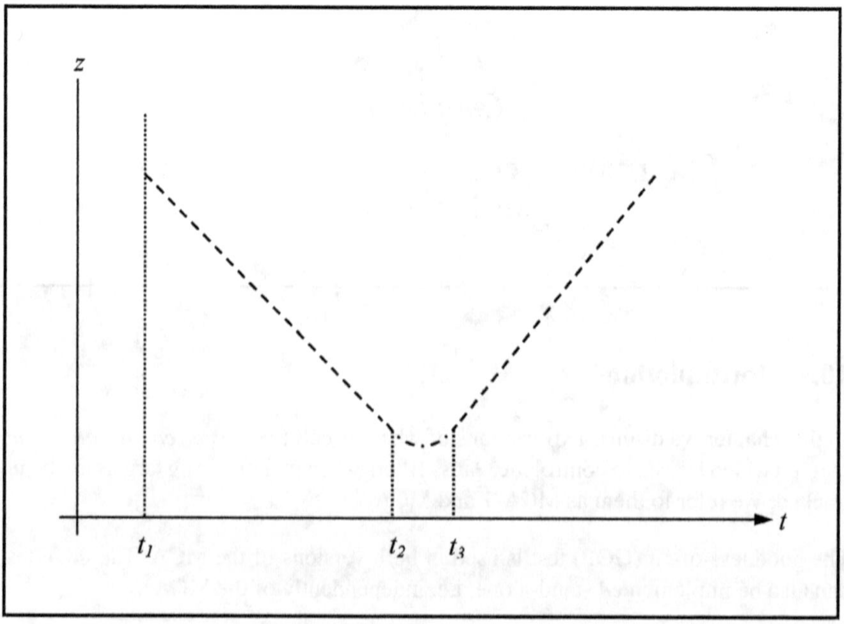

Figure 10.1 The heading of a constant-speed manoeuvring target

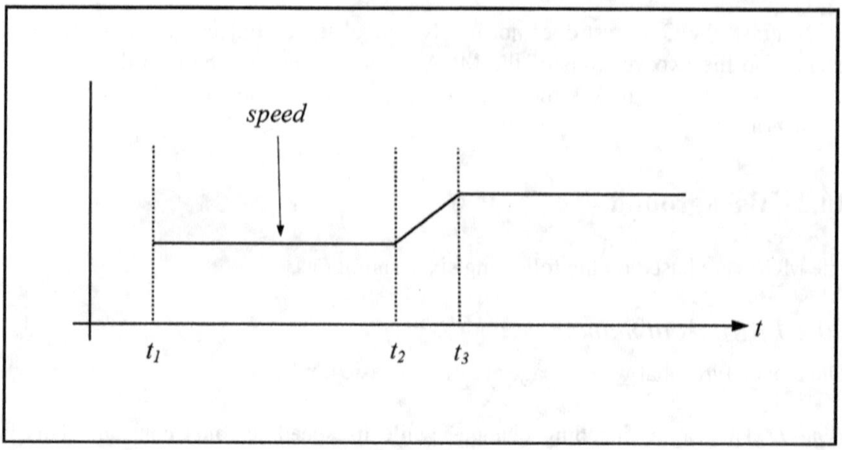

Figure 10.2 The speed of a constant-heading manoeuvring target

This is the type of manoeuvre that would be performed by a cruise missile flying on a straight-line path that abruptly changes its speed by changing the thrust from its engine.

Manoeuvres of virtually any type can be created from these two building blocks.

10.2.2 Filter models

The types of manoeuvring objects that we have in mind are fixed-wing aircraft, cruise missiles and helicopters, and the only feasible models that can be used for them are the DEs for polynomials. Accordingly, we assume that

◊ The filter model is the *3*-dimensional DE for 1st-degree polynomials (i.e. straight-line flight) when the target is not manoeuvring.

◊ The filter model is the *3*-dimensional DE for 2nd or higher degree polynomials when the target is manoeuvring.

10.2.3 The radar

Targets such as the ones being considered can be observed by two types of radar:

◊ Tracking radars: These have steerable antennas by which they observe and simultaneously track a *single target*, and if track is lost then further observations cease until track is re-established by the radar.

◊ Track-while-scan (TWS) radars: These have constantly rotating antennas by which they observe *multiple targets* without tracking them, and if tracking takes place then it is done by processing the observations in a computer that is separate from the radar. The observations made by TWS radars continue to be available regardless of whether or not successful tracking in the computer is taking place.

In the discussion that follows it is TWS radar that we have in mind.

10.2.4 Number of targets

We have used both MCA-1 and MCA-2 in simulations where multiple targets are present, as for example in Figure 1.12 in which three targets are being tracked by Gauss–Aitken filters under MCA control. We have also run simulations where there were many more than three targets.

However, the multiple-target problem is a complex one that is beyond the scope of this book.[1] Accordingly, we will assume that

[1] The problem of multiple-target tracking and the associated problems of initial track formation and track deletion are considered in References 30, 31, 50 and 100. See also References 106 to 110 and 157 for information on the auction algorithm that is used when multiple targets are involved.

◊ Only a single target is present.

◊ There is no radar clutter.

That being the case, every return received by the radar is a valid one that can be allocated to that single target.

10.2.5 The *T* matrix

We are assuming that the flight-path is a sequence of piecewise-continuous polynomials, and so the filter models will always be linear. Each section of the flight-path will have its own polynomial degree and hence its own transition matrix.

◊ If the observation equation is a linear one, i.e. of the form

$$\boldsymbol{Y} = \boldsymbol{T}\boldsymbol{X} + \boldsymbol{N} \tag{10.2.1}$$

then, together with the linear filter model we would be in ***Case 1*** for which the ***T*** matrix is (see (3.2.12)):

$$\boldsymbol{T}_n = \begin{pmatrix} \boldsymbol{M} \\ \hdashline \boldsymbol{M}\boldsymbol{\Phi}(t_{n-1} - t_n) \\ \hdashline \vdots \\ \hdashline \boldsymbol{M}\boldsymbol{\Phi}(t_{n-L} - t_n) \end{pmatrix} \tag{10.2.2}$$

◊ It is also possible that the observation equation is of the form

$$\boldsymbol{Y} = \boldsymbol{G}(\boldsymbol{X}) + \boldsymbol{N} \tag{10.2.3}$$

in which the observations are related to the observed trajectory ***X*** by a nonlinear transformation (e.g. radar coordinates).

We shall assume that the radar values are transformed to Cartesian coordinates immediately upon receipt, and that it is the resulting Cartesian values which are sent to the filter. The ***T*** matrix will therefore also be of the ***Case 1*** form shown in (10.2.2).

Examining $\boldsymbol{T}$ in (10.2.2), we see that it is contains two items:

- ◊ The transition matrix $\boldsymbol{\Phi}$ (in this case for polynomials) whose general form appears in (2.3.9) or (2.3.10), if the polynomial degrees are *1* or *2*, with similar extensions if the polynomial degree is higher than *2*.

- ◊ The parameter L (the filter's memory length).

Both versions of the MCA control the Gauss filters by specifying the degree of the polynomial (and hence the form of $\boldsymbol{\Phi}$), and by specifying the memory length (and hence the value of L).

10.2.6 The four manoeuvre sequences

In Figure 10.3 we show the four manoeuvre/non-manoeuvre sequences that must be considered.

Was in straight-line flight Now in straight-line flight	Was in straight-line flight Now manoeuvring
Was manoeuvring Now in straight-line flight	Was manoeuvring Now manoeuvring

Figure 10.3 The four manoeuvre sequences

- ◊ In the top-left box the target was in straight-line flight and is now in straight-line flight, and so *a non-manoeuvre is continuing.*

- ◊ In the top-right box the target was in straight-line flight and is now manoeuvring, and so *a manoeuvre has commenced.*

- ◊ In the bottom-left box the target was manoeuvring and is now in straight-line flight, and so *a manoeuvre has ended.*

- ◊ In the bottom-right box the target was manoeuvring and is now manoeuvring, and so *a manoeuvre is continuing.*

10.3 The MCAs for the Gauss–Aitken filters

10.3.1 Overview

Figure 10.4 below is a simplified diagram that shows either MCA-1 or MCA-2 being used to control a Gauss–Aitken filter.

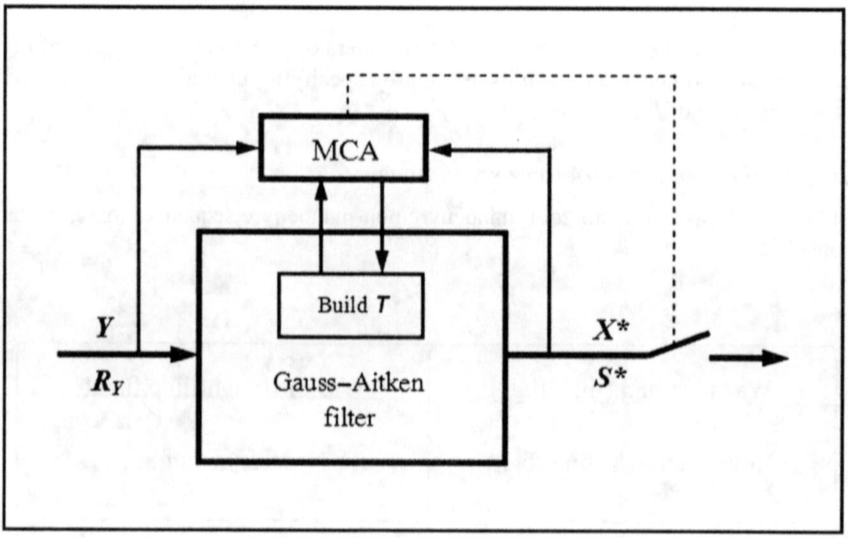

Figure 10.4 MCA overview

The sequence of operations is as follows:

1. The switch is opened.

2. An observation vector Y and its covariance matrix R_Y are received. These are appended to those already in memory to form the long-vector $\mathbf{Y}$ and its covariance matrix $\mathbf{R}_{\mathbf{Y}}$. Nothing is dropped and so the memory length has increased by *1* time stamp.

For MCA-1:

3. The filter is set to 1st-degree and then cycled, creating a tentative estimate X^* that is not yet accepted (i.e. the switch remains open).

4. The GOF test is run.

5. If the fit is good then X^* is accepted, the switch is closed and we return to *1*.

6. If the fit is not good then the filter is set to 2nd-degree and cycled, creating a new tentative estimate.

7. The GOF test is run a second time.

8. If the fit is good then $\boldsymbol{X}^*$ is accepted, the switch is closed and we return to *1*.

9. If the fit is not good for all allowable degrees then the memory length is shortened and we return to 3.

Thus with MCA-1, the estimate with the lowest degree and longest memory length that passes the GOF test is the one that is accepted.

For MCA-2:

3. The filter is set successively to each of the allowable degrees and cycled, creating a set of tentative estimates $\boldsymbol{X}^*$ that are not yet accepted (i.e. the switch remains open).

4. The GOF test is run after each $\boldsymbol{X}^*$ is obtained.

5. The best $\boldsymbol{X}^*$ is selected from among the estimates that pass their GOF tests.[2]

6. If none of the estimates passes its GOF test then the memory length is shortened and we return to 3.

10.3.2 MCA-1 in detail

In Figure 10.5 below we show the detailed flowchart for MCA-1. The flowchart for MCA-2 is discussed in Appendix 1.

Past observations and their covariance matrices are housed in a stack and the switch appearing in Figure 10.4 is initially open. The sequence of operations is as follows:

◊ **New observation**: An observation vector $\boldsymbol{Y}$ is received, together with its covariance matrix $\boldsymbol{R_Y}$. As a pair, these will be referred to as *an observation.*

◊ $\mathbf{L = L + 1}$ **and** $\mathbf{L_{max}}$? The stack-depth L can be restricted to a maximum value L_{max} (a user option).

  ◊ If L_{max} has not yet been reached: The stack is pushed down, the latest observation is placed at the top of the stack and no observations are dropped. The value of L therefore increases by *1*.

[2] In Appendix 10.1 we specify what *best* means.

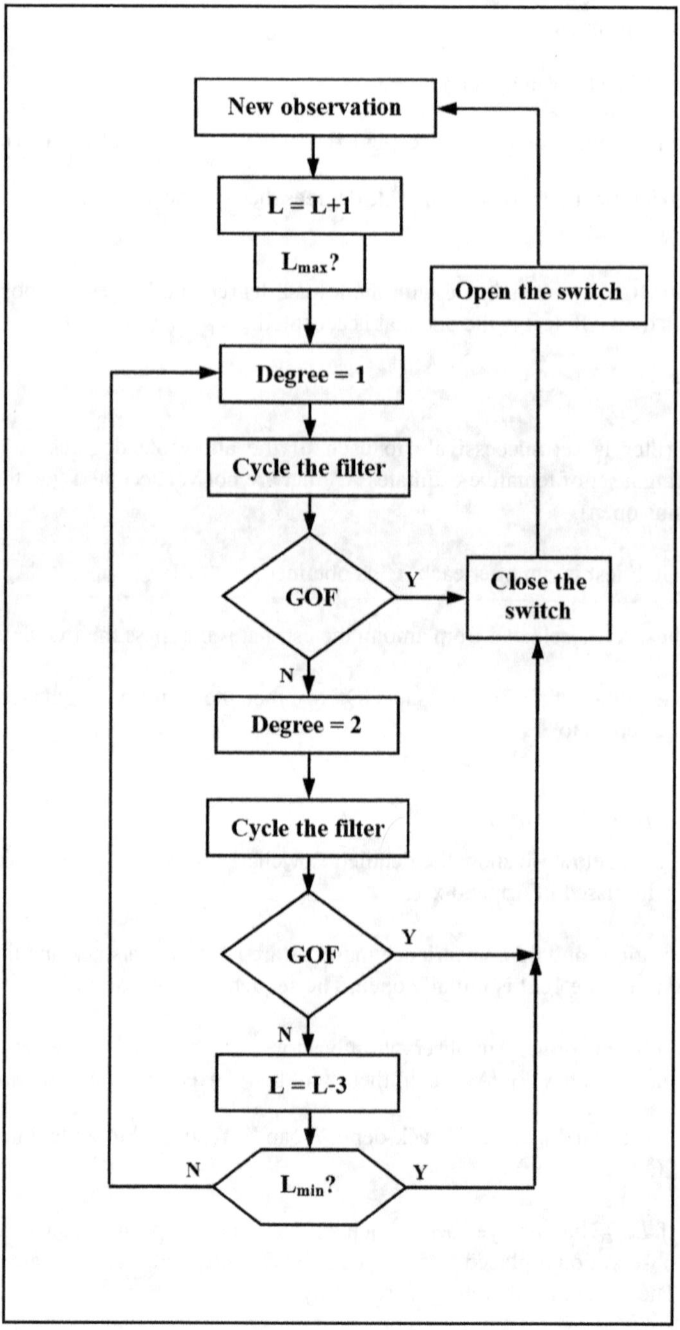

Figure 10.5 Detailed flowchart of MCA-1

- If L_{max} has been reached: The stack is pushed down, the oldest observation is dropped and the latest observation is placed at the top. The value of L therefore remains constant.

◊ **Degree 1**: $\boldsymbol{T}$ is built with memory length L and a 1st-degree transition matrix. The expression for $\boldsymbol{T}$ is shown in (10.2.2), and for $\boldsymbol{\Phi}$ in (2.3.9).

◊ **Cycle the filter**: The Gauss–Aitken filter is cycled, processing the data in the stack. This produces a 1st-degree estimate vector $\boldsymbol{X}^*$ and its covariance matrix $\boldsymbol{S}^*$.

◊ **GOF**: The GOF test is run.

- **Yes**: The fit is good, and so the value of L and the 1st-degree filter model are accepted. The switch is closed and $\boldsymbol{X}^*$ and $\boldsymbol{S}^*$ are made available to the subsequent stages of the system. The switch is then opened and the MCA awaits the next observation ($\boldsymbol{Y}$ and $\boldsymbol{R_Y}$).

- **No**: The fit is not good. The value of L and the 1st-degree model are not acceptable.

◊ **Degree 2**: $\boldsymbol{T}$ is built again with the same memory length L but using a 2nd-degree transition matrix. The expression for $\boldsymbol{T}$ is shown in (10.2.2), and for $\boldsymbol{\Phi}$ in (2.3.10).

◊ **Cycle the filter**: The Gauss filter is cycled, processing the same data in the stack. This produces a 2nd-degree estimate vector $\boldsymbol{X}^*$ and its covariance matrix $\boldsymbol{S}^*$.

◊ **GOF**: The GOF test is run.

- **Yes**: The fit is good, and so the value of L and the 2nd-degree model are accepted. The switch is closed and $\boldsymbol{X}^*$ and $\boldsymbol{S}^*$ are made available to the subsequent stages of the system. The switch is then opened and the MCA awaits the next observation ($\boldsymbol{Y}$ and $\boldsymbol{R_Y}$).

- **No**: The fit is not good. Neither the 1st-degree nor the 2nd-degree filter models using the current value of L gave a satisfactory fit to the data.

The user now has two options:

◊ Option 1: Include higher degree filters.

◊ Option 2: Limit MCA-1 to degrees *1* and *2*.

We have tried running MCA-1 using higher degree filters than *2*, but to us it appeared as though the overall system performance degraded if we went beyond 2nd-degree.

However, the reader's experience may differ from ours, especially if his/her system configuration differs markedly from ours.

Note: If a decision is made to include a 3rd-degree stage, then in Figure 10.6 we show what would be required.

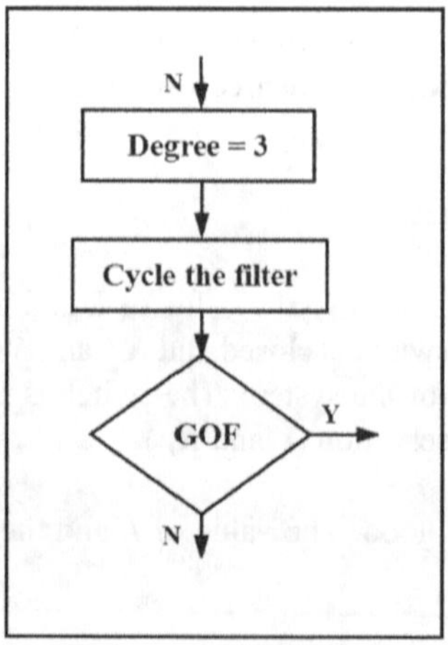

Figure 10.6 3rd-degree stage

◊ **L = L - 3**: The stack's depth is shortened by *3* and the observations ($\boldsymbol{Y}$ and $\boldsymbol{R_Y}$) with the three oldest time stamps are dropped. The stack's depth must be shortened by at least *1* but it can be shortened by more than *1*. (See comments below.)

◊ $\mathbf{L}_{\mathrm{min}}$**?** The user must set a minimum value for L. (See comments below.)

- ◊ If L_{min} has been reached then an unconditional exit takes place. The switch is closed and the estimate just obtained is made available to the system. The switch is then opened and the MCA awaits the next observation ($\boldsymbol{Y}$ and $\boldsymbol{R_Y}$).

- ◊ If L_{min} has not yet been reached then the MCA proceeds to the left in the flowchart and returns to the box labelled **Degree = 1**.

◊ The filter reprocesses the data in the stack using the reduced value of L and the same polynomial-model logic as before, starting with 1st-degree.

Eventually one of two outcomes takes place:

1. One of the GOF tests passes.

2. L_{min} is reached.

Either way, the switch is closed and the estimate is made available to subsequent stages of the system. The switch is then opened and the MCA awaits the next observation.

Comments:

◊ An exit from the MCA *will* eventually take place and so the possibility of endless cycling is excluded.

◊ To incorporate the latest observation, the filter might have to be cycled anywhere from *one* to *many* times to produce an acceptable estimate.

  ◊ Suppose that a target has previously been flying in a straight line and now continues to do so. When the newest observation arrives the fit could be immediately acceptable, and so in this case the filter would only be cycled once.

  ◊ Suppose instead that a target has been flying in a straight line for some time and that L has expanded to L_{max}. A manoeuvre then commences. The fit could now remain bad for more than one filter cycle while L is being reduced until an acceptable fit is obtained. The filter must be cycled either once or twice (or more, i.e. once for each degree) after each reduction, and so in this case it will be cycled many times to incorporate the latest observation.

◊ Allowing L to increase steadily when the target is in straight-line flight leads to estimates with diminishing variances and hence increasing accuracy, which is desirable. However, allowing L to increase too far has the following drawback:

If the target commences a manoeuvre with L at a large value then memory-shortening must take place, and with each shortening step the filter must be cycled possibly twice or more (i.e. once for each degree). It could then take many filter cycles to reach the value of L for which the GOF test passes and an acceptable fit is obtained, and this would add to the machine load (especially if we are considering the possibility of multiple targets that can all manoeuvre).

◊ We have tried reducing L by steps greater than *3*. This has the advantage of quickly reducing its value when a manoeuvre either begins or ends, thereby keeping down the number of filter cycles. However, it has the disadvantage of giving poorer resolution on L (i.e. the best value of L could be jumped over by the larger shortening steps). We also tried reducing L by steps less than *3* (i.e. by *1* or *2*) which gave improved resolution on L at the cost of more filter cycles. In the end we felt that *3* was a suitable compromise between these conflicting conditions.

◊ A typical value for L_{min} is *2* (i.e. *3* observations).

10.3.3 Two examples of the MCA in action

The first example is shown in Figure 10.7 below which contains *8* small sub-graphics. These were artificially created to form the illustration, and were not taken from an actual run. The flight-path is similar to the one that appeared in Figure 10.1.

- ◊ The first three graphics in the top row show the fitted curve expanding under MCA control during the straight-line leg. The fit is good each time that the filter is cycled using a 1st-degree polynomial, and so the memory length L expands as successively more observations are incorporated.
- ◊ In the fourth figure we see that the memory length has reached the user-selected maximum.
- ◊ The first figure in the second row relates only to the manoeuvre.
  - ◊ At the start of the manoeuvre the filter is run in 1st-degree with the L that existed in the fourth figure, and the GOF test fails.
  - ◊ The filter is then run with that same L and a 2nd-degree polynomial, but the GOF test fails again.
  - ◊ The two filters (1st and 2nd-degree) are run repeatedly, but the GOF tests all fail. After each 2nd-degre failure, L is reduced by *3*.
  - ◊ Eventually L is reduced to the point where a short 2nd-degree polynomial is successfully fitted to the observations (shorter than the one shown in the figure).

  All of the above take place before the next observation arrives.

  - ◊ The MCA then permits successive new observations to be absorbed into a 2nd-degree polynomial that grows in length while the manoeuvre is continuing. The resulting 2nd-degree polynomial at the end of the manoeuvre is depicted in the fifth box in the figure.
- ◊ The final three figures show what takes place after the manoeuvre ends and straight-line flight resumes.
  - ◊ Shortly after the end of the manoeuvre, after the memory length has been increased by *1*, the filter is cycled with a 1st-degree polynomial for which the GOF test fails. The filter is then cycled with a 2nd-degree polynomial and that same memory length, but again the GOF test fails.
  - ◊ The MCA now reduces the memory length by *3's*, trying first to fit a 1st-degree and then a 2nd-degree polynomial (without any new observations being incorporated). Eventually the memory length is sufficiently reduced so that a 1st-degree fit succeeds.

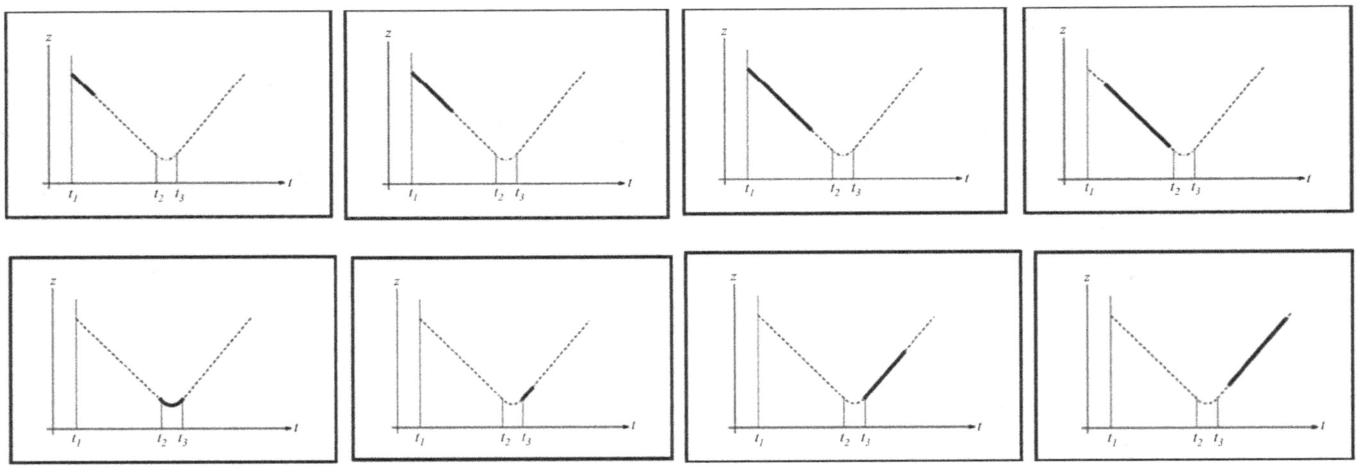

Figure 10.7 A simple demonstration of the MCA in action

◊ The MCA then permits successive new observations to be added to a 1st-degree polynomial that grows in length. This situation continues as long as the final straight leg of the flight-path continues, or the value of L_{max} is reached.

◂▸

The second set of illustrations came from the same simulator as the one that was used in Figure 1.12 in which we displayed three jets flying parallel courses. In this case there is only a single target – a simulated high-performance military jet.

In Figure 10.8 we see the target making a *7*-g turn. The actual flight-path lies directly beneath the filter output and so it is not visible.

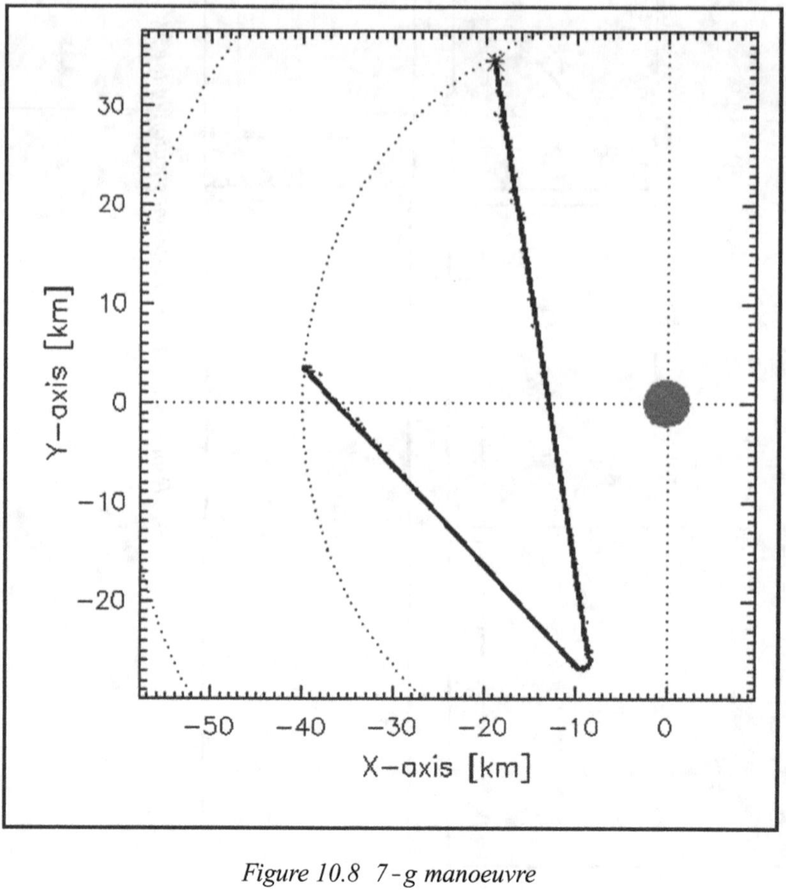

Figure 10.8 7 - g manoeuvre

Figure 10.9 is a plot of the memory length L under MCA control.

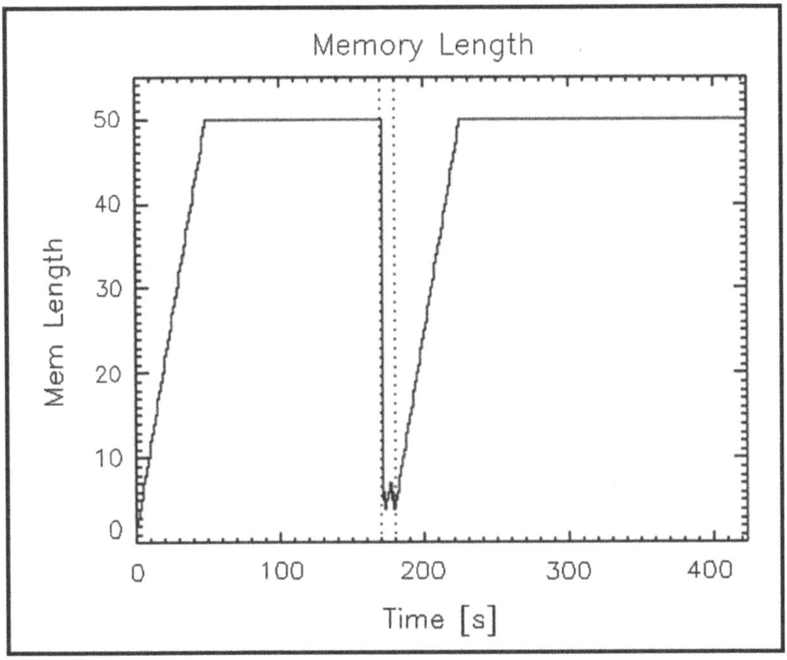

Figure 10.9 Memory length

◊ As seen in Figure 10.9, the aircraft is initially in non-manoeuvring straight-line flight and so L expands from L_{min} to *50,* which was the user-selected value of L_{max}.

◊ The MCA then holds the value of L at *50* for the remainder of that leg, until the start of the manoeuvre.

◊ Shortly after the manoeuvre commences the GOF test fails for both 1st and 2nd degrees. L is then decreased from *50* to *5* in steps of *3,* and after each decrease the filter is cycled again – once with 1st-degree and once with 2nd. (This reduction of memory length takes place before the next observation is received, and appears as the vertical line in Figure 10.9) After each cycle the GOF test is run but fails. Finally L reaches *5* and the GOF test passes for 2nd-degree.

◊ The MCA then maintains L at about *5* for the duration of the manoeuvre.

◊ When the manoeuvre ends, L once again increases until it reaches L_{max} where it remains.

In Figure 10.10 is a plot of the polynomial degree under MCA control.

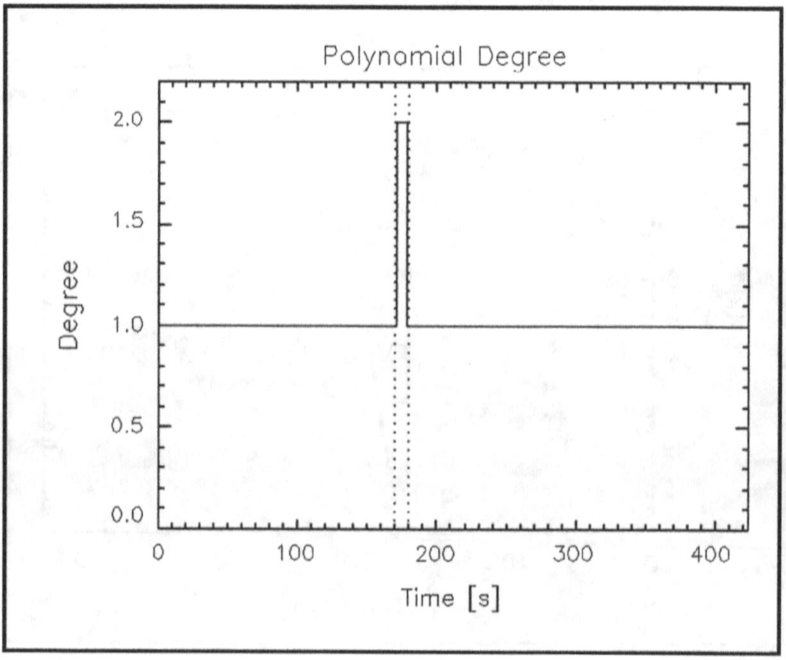

Figure 10.10 Polynomial degree

◊ The MCA sets the filter degree to *1* during the first straight-line leg.

◊ During the manoeuvre it sets the degree to *2* where it remains until the manoeuvre ends.

◊ The degree is then reduced to *1* for the final straight-line leg.

10.3.4 Summary

MCA-1 finds a value for the memory length L (anywhere from one more than the previous value down to L_{min}) and a polynomial degree (either *1* or *2* or possibly higher), such that the curve implied by the estimate $\boldsymbol{X}^*$ is a good fit to the total observation vector $\boldsymbol{Y}$.

The end result is an estimate $\boldsymbol{X}^*$ of the lowest possible degree whose implied curve has been fitted acceptably over the longest possible memory length.

When operation of the MCA has been completed and $\boldsymbol{X}^*$ is accepted:

◊ If the degree of $\boldsymbol{X}^*_{n-1,\,n-1}$ was *1* and the degree of $\boldsymbol{X}^*_{n,\,n}$ is now *1*, then a non-manoeuvre is continuing.

◊ If the degree of $\boldsymbol{X}^*_{n-1,\,n-1}$ was *1* and the degree of $\boldsymbol{X}^*_{n,\,n}$ is now *2*, then a manoeuvre has started.

◊ If the degree of $\boldsymbol{X}^*_{n-1,\,n-1}$ was *2* and the degree of $\boldsymbol{X}^*_{n,\,n}$ is now *1*, then a manoeuvre has ended.

◊ If the degree of $\boldsymbol{X}^*_{n-1,\,n-1}$ was *2* and the degree of $\boldsymbol{X}^*_{n,\,n}$ is now *2*, then a manoeuvre is continuing.

Thus all four sequences in Figure 10.3 have been allowed for.

We suggest that you now view the movie clips in the folder ***Video_Clips\TWS\Flights***. However, in order to understand fully what you are viewing, first please read the document entitled ***Video_Clips\TWS\Documents\Readme***.

10.4 Testing for goodness-of-fit – a brief description

In both MCA-1 and MCA-2 the estimate produced by the Gauss–Aitken filter is tested for goodness of fit (GOF). We now review briefly how that is done.

Definition 10.1: The total number of elements in the vector Y is symbolized as N_{Y} and the total number of elements in $\boldsymbol{X}^*$ as N_{X^*}. ■■

Example 10.1: Let the total observation vector Y_n consist of eleven short vectors Y_{n-k} as shown in (10.4.1).

$$\mathsf{Y}_n = \begin{pmatrix} Y_n \\ \text{------} \\ Y_{n-1} \\ \text{------} \\ \vdots \\ \text{------} \\ Y_{n-10} \end{pmatrix} \tag{10.4.1}$$

Let each of those Y vectors be a *3*-vector. Then there are thirty-three elements in Y_n and so $N_{\mathsf{Y}} = 33$.

Assume that the filter's estimate vector is

$$\boldsymbol{X}^* = (x^*, \dot{x}^*, \ddot{x}^*, y^*, \dot{y}^*, \ddot{y}^*, z^*, \dot{z}^*, \ddot{z}^*)^T \tag{10.4.2}$$

Then there are nine elements in $\boldsymbol{X}^*$, and so $N_{\boldsymbol{X}^*} = 9$. ■■

10.4.1 Overview of the GOF test

The GOF test appears in two places in the MCA-1 flowchart in Figure 10.5 and also in MCA-2 in Appendix 10.1. (It would appear in more than two places if polynomials of degree higher than 2nd are included.)

For a complete understanding of the GOF test it is essential that all of the material in the appendices to this chapter be absorbed, in particular Appendix 10.2 in which we state Theorem 10.1 on which the GOF test is based. (Proof of the theorem is considered in Problem 10.18.)

Very briefly the test operates as follows:

The filter has just produced an estimate vector $\boldsymbol{X}^*$ which must now be tested for GOF:

◊ Let $\boldsymbol{Y}$ be the total observation vector that was used when the filter produced $\boldsymbol{X}^*$.

◊ Let $\boldsymbol{T}$ be the total observation matrix that was used by the filter when it produced $\boldsymbol{X}^*$.

◊ Let $\boldsymbol{R_Y}^{-1}$ be the inverse of the covariance matrix of $\boldsymbol{Y}$ that was used when the filter produced $\boldsymbol{X}^*$.

◊ Let k be defined as follows:

$$k \equiv N_{\boldsymbol{Y}} - N_{\boldsymbol{X}^*} \tag{10.4.3}$$

Thus k is the number of elements in the filter's input minus the number of elements in its output.

The following three steps are then taken:

1. *SSR* (the sum of the squared residuals) is computed using

$$SSR = (\boldsymbol{Y} - \boldsymbol{T}\boldsymbol{X}^*)^T \boldsymbol{R_Y}^{-1} (\boldsymbol{Y} - \boldsymbol{T}\boldsymbol{X}^*) \tag{10.4.4}$$

2. *NSSR* (the normalized sum of the squared residuals) is computed using

$$NSSR_k = SSR/k \tag{10.4.5}$$

3. $NSSR_k$ is compared to the kth threshold δ_k as shown in Figure 10.11.

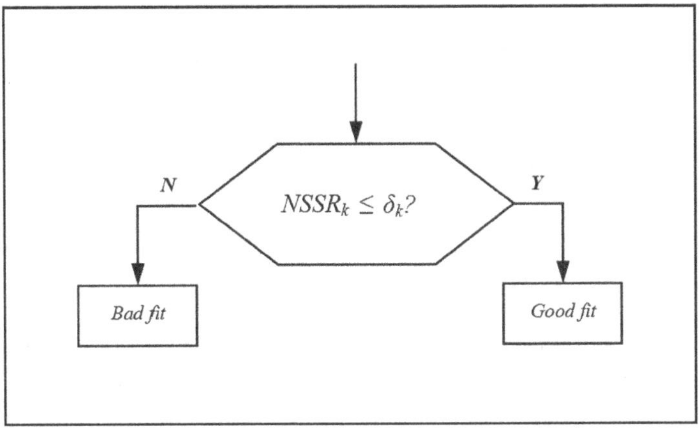

Figure 10.11 Testing for goodness-of-fit

If $NSSR_k \leq \delta_k$ then the fit is good and if $NSSR_k > \delta_k$ then the fit is not good.

The threshold δ_k depends on the value of k that was defined in (10.4.3).

The values of δ_k for $1 \leq k \leq 1260$, in a form in which they can be incorporated into a lookup table, are contained in the document entitled

Miscellaneous\Thresholds\NSSR_Thresholds.txt

For values of k greater than *1260* the thresholds δ_k can be generated at run-time using the formula that is contained in that same document.

10.4.2 Comments

◊ $\boldsymbol{Y}$, $\boldsymbol{T}$ and $\boldsymbol{R}^{-1}$ were required when the filter was cycled just before running the GOF test, and $\boldsymbol{X}^*$ is the resulting filter output. Thus all quantities needed to evaluate *SSR* in (10.4.4) already exist and so very little additional computation is required to implement or run the GOF test.

Note: The above comment is true only for the Gauss filters. By contrast, if a Kalman or Swerling filter were being used, then a considerable amount of programming would have to be carried out in order to implement the GOF test.

◊ The quantity *SSR* in (10.4.4) increases rapidly when the fit is bad, i.e. when a bias vector is present. In Project 10.5 we prove the following:

Theorem 10.2: Let E_b be the magnitude of a possible bias vector in the estimate $\boldsymbol{X}^*$. Then $\mathcal{E}\{SSR\}$ in (10.4.4) increases like E_b^2, i.e. it follows a square law. ■■

Square laws are fast, and so the GOF test is a sensitive one.

10.5 Stand-alone implementation of the GOF test

The tests for ECM consistency discussed in Chapter 7 and those for CR consistency in Chapter 9 all require values of the observed trajectory's true state vector $\boldsymbol{X}$, and so they can only be used in simulations.

By contrast the GOF test does not make use of $\boldsymbol{X}$ and is based instead on the residuals. Because these are available in the field as well as in simulations, the GOF test (and hence also the MCA) can be used in both of those environments.

Tests that can be used in the field are hard to come by[3] and so we now take a few minutes to consider what might be accomplished if a stand-alone GOF test (i.e. not as part of the MCA) were to be included in a filter in the field.

There are two stages of deployment to the field:

◊ Initial debugging.

◊ Final operation.

During initial debugging, if any of the modelling assumptions on which the filter is based are incorrect, then bias errors between the filter output and the observations could arise.

Such bias errors are precisely what the GOF test detects, and so it can be used to make us aware that something is amiss in the modelling assumptions.

If that occurs, then steps must be taken to track down the problem and eliminate it, and for this purpose alone the GOF test could be extremely valuable.

[3] We will discuss a second test, called the sigma monitoring test, when we reach the end of Chapter 12. That test is also based on the residuals and so it can also be used in the field.

Once initial in-field debugging has been completed, it could still be of value to keep the GOF test running as part of the filtering system. Bias errors may not have been present during initial debugging, but they could arise in actual field operations after the system has been fully commissioned. The failure of the GOF test would then signal that the filter outputs cannot be fully trusted.

10.5.1 Demonstrations

The three programs ***03_Gauss–Aitken***, ***13_Gauss_Newton_C*** and ***14_Orbit_Track*** demonstrate how the GOF test might be used in the field in ways that are admittedly simplistic but nevertheless illustrative of what can be accomplished.

1. The program 03_Gauss–Aitken

This program operates in a ***Case-1*** environment as follows:

Samples of a noisy sine wave whose frequency is known are fed to a Gauss–Aitken filter.

A menu button called *Bias* enables you to add a constant error of selectable size to the observations. This bias in the observations then results in a biased estimate from the filter (see equation (6.7.4)).

The following four performance monitors are included, and their results are displayed:

◊ The *3*-sigma test (Section 7.5).

◊ The Chi-squared test (Section 7.6).

◊ The GOF test.

◊ The SM test (see near the end of Chapter 12).

The first two tests cannot be used in the field but the last two can.

The program demonstrates how both the GOF and SM tests fail when bias errors are present. It also demonstrates how the first two of the above four tests pass when bias is absent, but fail when bias is present.

In Section 7.7 we pointed out that the failure of the *3*-sigma and Chi-squared tests could be caused either by ECM inconsistency or by bias errors or both, and we displayed the diagram shown as Figure 7.7 below.

Running the GOF test in addition to the *3*-sigma and Chi-squared tests provides critical information for resolving the ambiguities in the figure when these latter two tests fail.

ECM consistent Bias errors absent	ECM consistent Bias errors present
ECM inconsistent Bias errors absent	ECM inconsistent Bias errors present

Figure 7.7 ECM consistency and bias errors

2. *The program 13_Gauss_Newton_C*

This program operates in a ***Case-3*** environment (linear observation equation, nonlinear filter model). Samples of a noisy sine wave whose frequency is unknown are pre-filtered, and periodic samples of the pre-filter's output are then fed to three filters – Kalman, Swerling and Gauss–Newton. The intention is for these filters to estimate the unknown frequency.

A menu button called *Bias* is provided, and the following three performance monitors are included:

◊ The *3*-sigma test.

◊ The Chi-squared test.

◊ The GOF test.

The program demonstrates how the GOF test fails when bias errors are present, and how the first two tests pass when bias is absent, but fail when bias is present.

The GOF test was easily implemented in the Gauss–Newton filter and very few lines of code are needed for that purpose. By contrast, to implement the GOF test in the Kalman or Swerling filters would have required an extensive amount of code, and so it was not done. (Perhaps in a later release we will find the time to undertake that task.)

3. *The program 14_Orbit_Track*

This program operates in a ***Case-3*** environment (linear observation equation, nonlinear filter model). Radar observations are made on a user-selectable satellite. These are then transformed to Cartesian coordinate and fed to three filters – Kalman, Swerling and Gauss–Newton, the intention being to estimate the satellite's orbital parameters.

The following three performance monitors are included:

- ◊ The *3*-sigma test.
- ◊ The Chi-squared test.
- ◊ The GOF test.

We elected not to include a *Bias* button.

As with the previous program, the GOF test was implemented only in the Gauss–Newton filter because that is a very simple undertaking.

For a full discussion of how bias errors affect the performance tests we refer you to Project 10.3. For a detailed discussion on the Chi-squared PDF, see Projects 10.6 and 10.7.

The following items are in the supplementary material

Problems for Chapter 10

Solutions to Problems for Chapter 10

Projects for Chapter 10

Appendix 10.1

MCA-2

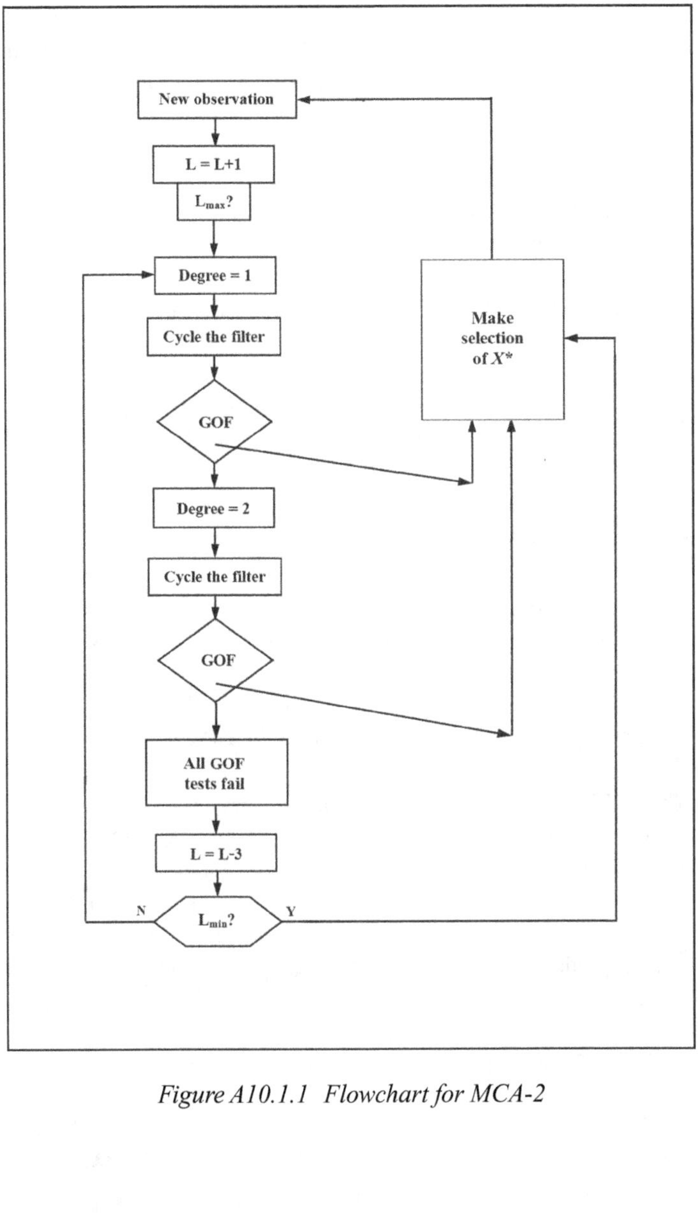

Figure A10.1.1 Flowchart for MCA-2

A10.1.1 Operation of MCA-2

Past observations and their covariance matrices are housed in a stack and the switch appearing in Figure 10.4 is initially open. The sequence of operations is as follows:

◊ **New observation**: An observation vector $\boldsymbol{Y}$ is received, together with its covariance matrix $\boldsymbol{R_Y}$. As a pair, these are referred to as *an observation*.

◊ **L = L+1 and $\mathbf{L_{max}}$?** The stack-depth L can be restricted to a maximum value L_{max} (a user option).

- ◊ *If L_{max} has not yet been reached*: The stack is pushed down, the latest observation is placed at the top and no observations are dropped, and so the value of L increases by *1*.

- ◊ *If L_{max} has been reached*: The stack is pushed down, the latest observation is placed at the top and the oldest observation is dropped, and so the value of L remains constant.

◊ **Degree 1**: $\boldsymbol{T}$ is built with memory length L and a first-degree transition matrix. The expression for $\boldsymbol{T}$ is shown in (10.2.2) and for $\boldsymbol{\Phi}$ in (2.3.9).

◊ **Cycle the filter:** The Gauss–Aitken filter is cycled, processing the data in the stack. This produces a first-degree estimate vector and its covariance matrix called $\boldsymbol{X}_1^*$ and $\boldsymbol{S}_1^*$.

◊ **GOF**: The GOF test is run.

- ◊ *Yes*: The fit is good, and so the values of L and the first-degree filter model are acceptable. $\boldsymbol{X}_1^*$, $\boldsymbol{S}_1^*$ and the value of SSR_1 (see Section A10.1.2) are placed in storage.

- ◊ *No*: The fit is not good. Nothing is placed in storage.

◊ **Degree 2**: $\boldsymbol{T}$ is built again with the same memory length L but now is using a second-degree transition matrix. The expression for $\boldsymbol{T}$ is shown in (10.2.2) and for $\boldsymbol{\Phi}$ in (2.3.10).

◊ **Cycle the filter**: The Gauss–Aitken filter is cycled, processing the same data stack. This produces a 2nd-degree estimate vector and its covariance matrix called $\boldsymbol{X}_2^*$ and $\boldsymbol{S}_2^*$.

◊ **GOF**: The GOF test is run.

- **Yes**: The fit is good, and so the value of L and the second-degree filter model are acceptable. X_2^*, S_2^* and the value of SSR$_2$ (see Section A10.1.2) are placed in storage.

- **No**: The fit is not good. Nothing is placed in storage.

Operation continues in this way using the same value for L until the estimate based on the filter model with maximum allowable degree has been obtained, and the values X_{max}^*, S_{max}^* and SSR_{max}^* have been placed in storage.

Note: In one system that we have worked on we found it beneficial to use models of degrees *1* to *5*.

A10.1.2 Computing the value of SSR

In Subsection 10.4.1 we gave an overview of how the GOF test is run for an estimate of particular degree, and we stated there that *SSR* is defined as follows:

$$SSR = (Y - TX^*)^T R_Y^{-1}(Y - TX^*) \tag{A10.1.1}$$

We noted that Y, T and R_Y^{-1} on the right of this equation are required to run the filter, and that X^* is produced by the filter, and so all quantities needed to evaluate *SSR* for each degree are already available. Thus *SSR* for each allowable degree is readily computed, and can be saved along with the estimate X^* and its covariance matrix S^*.

A10.1.3 Selecting the degree which gives the best fit

We must now select the X^* and S^* which gave the best fit. Selection is based on the following rationale:

> Select the lowest-degree estimate for which GOF was acceptable, unless there is a higher degree estimate for which the fit was *significantly* better.

We demonstrate what *significant* means by the following.

Example A10.1:

- Assume that filter degrees *1* to *5* are implemented.

- Assume that only degrees *3, 4* and *5* pass their respective GOF tests at the current memory length.

- Let their respective *SSR* values be called SSR_3, SSR_4 and SSR_5.

a. Compute the following *SSR* differences (lower-degree minus higher-degree):

$$\Delta_{3,4} = SSR_3 - SSR_4 \tag{A10.1.2}$$

$$\Delta_{3,5} = SSR_3 - SSR_5$$

$$\Delta_{4,5} = SSR_4 - SSR_5$$

Delete any Δ which is negative, and delete also the corresponding $\boldsymbol{X}^*$, $\boldsymbol{S}^*$ and *SSR* with the higher degree.

As an example, if $SSR_3 = 107$ and $SSR_4 = 132$ then $\Delta_{3,4} = 107 - 132 = -25$. We would then delete $\Delta_{3,4}$ as well as $\boldsymbol{X}_4^*$, $\boldsymbol{S}_4^*$ and SSR_4.

b. Assume that all Δ' s in (A 10.1.2) are positive. This means that the three *SSR* values satisfy

$$SSR_3 > SSR_4 > SSR_5 \tag{A10.1.3}$$

In going to a higher degree the fit therefore always improves.

We now determine if such improvement in fit was *significant*, for if not, then it would be better to select the lower degree.

c. The differences $\Delta_{3,4}$, $\Delta_{3,5}$ and $\Delta_{4,5}$ in (A10.1.2) are Chi-squared RVs, and their degrees of freedom (*dof*) are given by the difference in the number of elements in the two estimates involved. We recall that a *3*-dimensional 3rd-degree estimate vector has twelve elements, with corresponding numbers for 4th and 5th degrees $\boldsymbol{X}_4^*$ and $\boldsymbol{X}_5^*$, and so we have the following number of elements in $\boldsymbol{X}_3^*$, $\boldsymbol{X}_4^*$ and $\boldsymbol{X}_5^*$:

$$N_{X^*_3} = 12$$

$$N_{X^*_4} = 15 \tag{A10.1.4}$$

$$N_{X^*_5} = 18$$

Then the degrees of freedom of $\Delta_{3,4}$, $\Delta_{3,5}$ and $\Delta_{4,5}$ are as follows:

$$\begin{aligned} dof_{\Delta_{3,4}} &= N_{X^*_4} - N_{X^*_3} = 15 - 12 = 3 \\ dof_{\Delta_{3,5}} &= N_{X^*_5} - N_{X^*_3} = 18 - 12 = 6 \\ dof_{\Delta_{4,5}} &= N_{X^*_5} - N_{X^*_4} = 18 - 15 = 3 \end{aligned} \tag{A10.1.5}$$

d. The *SSRs* of the three estimates are Chi-squared, with degrees of freedom as follows:

Let dof_{SSR_3} be the degrees of freedom of SSR_3, dof_{SSR_4} be the degrees of freedom of SSR_4 and dof_{SSR_5} be the degrees of freedom of SSR_5. Then (see Section 10.4).

$$\begin{aligned} dof_{SSR_3} &= N_{\mathbf{Y}} - N_{\mathbf{X}_3^*} \\ dof_{SSR_4} &= N_{\mathbf{Y}} - N_{\mathbf{X}_4^*} \\ dof_{SSR_5} &= N_{\mathbf{Y}} - N_{\mathbf{X}_5^*} \end{aligned} \tag{A10.1.6}$$

in which $N_{\mathbf{Y}}$ is the total number of elements in the vector $\mathbf{Y}$ and $N_{\mathbf{X}^*}$ is the total number of elements in $\mathbf{X}^*$.

Assume that $\mathbf{Y}$ consists of *11* short-vectors Y, each of which is an observation on x, y and z. Then

$$N_{\mathbf{Y}} = 33 \tag{A10.1.7}$$

In (A10.1.5) we showed the values for $N_{\mathbf{X}_3*}$, $N_{\mathbf{X}_4*}$ and $N_{\mathbf{X}_5^*}$. Then the degrees of freedom of the Chi-squared RVs SSR_3, SSR_4 and SSR_5 are as follows:

$$\begin{aligned} dof_{SSR_3} &= N_{\mathbf{Y}} - N_{\mathbf{X}_3^*} = 33 - 12 = 21 \\ dof_{SSR_4} &= N_{\mathbf{Y}} - N_{\mathbf{X}_4^*} = 33 - 15 = 18 \\ dof_{SSR_5} &= N_{\mathbf{Y}} - N_{\mathbf{X}_5^*} = 33 - 18 = 15 \end{aligned} \tag{A10.1.8}$$

e. Compute the following three quantities:

$$\begin{aligned} F_{3,4} &\equiv (\Delta_{3,4}/dof_{\Delta_{3,4}})/(SSR_4/dof_{SSR_4}) \\ F_{3,5} &\equiv (\Delta_{3,5}/dof_{\Delta_{3,5}})/(SSR_5/dof_{SSR_5}) \\ F_{4,5} &\equiv (\Delta_{4,5}/dof_{\Delta_{4,5}})/(SSR_5/dof_{SSR_5}) \end{aligned} \tag{A10.1.9}$$

Then $F_{3,4}$, $F_{3,5}$ and $F_{4,5}$ are RVs with the F-distribution.[1]

[1] See References 194 or 195.

Note 1: The F-distribution is the *ratio* of two Chi-squared RVs c_a and c_b with degrees of freedom a and b respectively, where, before forming the ratio, each Chi-squared value is divided by its number of degrees of freedom.

Note 2: The *entire* procedure for selecting the best estimate requires that we form the ratios described in ***Note 1***.

f. We now make use of the table of *5%* critical values of the F-distribution (see ***Miscellaneous\Thresholds\F_Thresholds***) as follows:

Let the SSR values be as follows:

$$SSR_3 = 108, \qquad SSR_4 = 69.5$$

Then in (A10.1.9)

$$\Delta_{3,4}/dof_{\Delta_{3,4}} = (108 - 69.5)/3 = 12.83$$

and

$$SSR_4/dof_{SSR_4} = 69.5/18 = 3.86$$

and so

$$F_{3,4} = (\Delta_{3,4}/dof_{\Delta_{3,4}})/(SSR_4/dof_{SSR_4}) = 12.83/3.86 = 3.32$$

Using the table of *5%* critical values of the F-distribution, we have the following:

1. $dof_{\Delta_{3,4}} = 3$, which is the column number of the table
2. $dof_{SSR_4} = 18$, which is the row number of the table

The table value is *3.16*, and so $F_{3,4} = 3.32$ obtained above exceeds the table value.

This tells us that the fit of the fourth-degree estimate is *significantly* better than the fit of the third-degree estimate, and so we would use the fourth-degree estimate rather than the third-degree estimate.

We now determine if the fit of the fifth-degree estimate is *significantly* better than the fit of the fourth-degree estimate.

Let the SSR values be as follows:

$$SSR_4 = 69.5, \quad SSR_5 = 58.4$$

Then

$$\Delta_{4,5} / dof_{\Delta_{4,5}} = (69.5 - 58.4)/3 = 3.7$$

and

$$SSR_5 / dof_{SSR_5} = 58.4/15 = 3.89$$

and so

$$F_{4,5} = (\Delta_{3,4} / dof_{\Delta_{3,4}}) / (SSR_4 / dof_{SSR_4}) = 3.7/3.89 = 0.95$$

Using the table of *5%* critical values of the F-distribution, we have the following:

$dof_{\Delta_{4,5}} = 3$, which is the column number of the table

$dof_{SSR_5} = 15$, which is the row number of the table

The table value is then seen to be *3.29*, and so $F_{4,5} = 0.95$ obtained above is less than the table value.

This means that the fit of the fifth-degree estimate is *not* significantly better than the fit of the fourth-degree estimate, and so we would use the fourth-degree estimate in preference to the fifth-degree estimate.

The final decision in this case would be to use the fourth-degree estimate.

Appendix 10.2

Theorem 10.1 on which the GOF test is based

The complete expressions for the sum of the squared residuals (*SSR*) and for the minimum variance estimate X^*_{minvar} are as follows:

$$\text{SSR} = (\boldsymbol{Y} - \boldsymbol{T}X^*)^T \boldsymbol{R_Y}^{-1}(\boldsymbol{Y} - \boldsymbol{T}X^*) \tag{A10.2.1}$$

$$X^*_{minvar} = (\boldsymbol{T}^T \boldsymbol{R_Y}^{-1}\boldsymbol{T})^{-1}\boldsymbol{T}^T \boldsymbol{R_Y}^{-1}\boldsymbol{Y} \tag{A10.2.2}$$

Combining these two equations gives

$$\begin{aligned} SSR &= (\boldsymbol{Y} - \boldsymbol{T}(\boldsymbol{T}^T\boldsymbol{R_Y}^{-1}\boldsymbol{T})^{-1}\boldsymbol{T}^T\boldsymbol{R_Y}^{-1}\boldsymbol{Y})^T\boldsymbol{R_Y}^{-1}(\boldsymbol{Y} - \boldsymbol{T}(\boldsymbol{T}^T\boldsymbol{R_Y}^{-1}\boldsymbol{T})^{-1}\boldsymbol{T}^T\boldsymbol{R_Y}^{-1}\boldsymbol{Y}) \\ &= \boldsymbol{Y}^T(\boldsymbol{I_Y} - \boldsymbol{T}(\boldsymbol{T}^T\boldsymbol{R_Y}^{-1}\boldsymbol{T})^{-1}\boldsymbol{T}^T\boldsymbol{R_Y}^{-1})^T\boldsymbol{R_Y}^{-1}(\boldsymbol{I_Y} - \boldsymbol{T}(\boldsymbol{T}^T\boldsymbol{R_Y}^{-1}\boldsymbol{T})^{-1}\boldsymbol{T}^T\boldsymbol{R_Y}^{-1})\boldsymbol{Y} \end{aligned} \tag{A10.2.3}$$

in which $\boldsymbol{I_Y}$ is the identity matrix of the same order as the long-vector $\boldsymbol{Y}$.

We write (A10.2.3) as follows:

$$SSR = \boldsymbol{Y}^T \boldsymbol{Q} \boldsymbol{Y} \tag{A10.2.4}$$

where

$$\boldsymbol{Q} \equiv \boldsymbol{B}^T \boldsymbol{R_Y}^{-1} \boldsymbol{B} \quad \text{and} \quad \boldsymbol{B} \equiv \boldsymbol{I_Y} - \boldsymbol{T}(\boldsymbol{T}^T \boldsymbol{R_Y}^{-1} \boldsymbol{T})^{-1} \boldsymbol{T}^T \boldsymbol{R_Y}^{-1} \tag{A10.2.5}$$

Theorem 10.1:

◊ Let

$$\boldsymbol{Y} = \boldsymbol{T}\boldsymbol{X} + \boldsymbol{N} \tag{A10.2.6}$$

where $\boldsymbol{N}$ is zero-mean multivariate Gaussian with $\mathcal{E}\{\boldsymbol{N}\boldsymbol{N}^T\} = \boldsymbol{R_Y}$.

◊ Let $\boldsymbol{X}^*_{minvar}$ be the minimum variance estimate of $\boldsymbol{X}$ given by

$$\boldsymbol{X}^*_{minvar} = (\boldsymbol{T}^T \mathbf{R}_{\mathbf{Y}}^{-1} \boldsymbol{T})^{-1} \boldsymbol{T}^T \mathbf{R}_{\mathbf{Y}}^{-1} \mathbf{Y} \tag{A10.2.7}$$

◊ Let the number of elements in $\mathbf{Y}$ (i.e. its dimension) be $N_{\mathbf{Y}}$ and the number of elements in $\boldsymbol{X}^*_{minvar}$ be $N_{\boldsymbol{X}^*}$.

◊ Let $\boldsymbol{I}_{\mathbf{Y}}$ and $\boldsymbol{I}_{\boldsymbol{X}^*}$ be identity matrices of order $N_{\mathbf{Y}}$ and $N_{\boldsymbol{X}^*}$.

◊ Let

$$\boldsymbol{B} = \boldsymbol{I}_{\mathbf{Y}} - \boldsymbol{T}(\boldsymbol{T}^T \mathbf{R}_{\mathbf{Y}}^{-1} \boldsymbol{T})^{-1} \boldsymbol{T}^T \mathbf{R}_{\mathbf{Y}}^{-1} \quad \text{and} \quad \boldsymbol{Q} = \boldsymbol{B}^T \mathbf{R}_{\mathbf{Y}}^{-1} \boldsymbol{B} \tag{A10.2.8}$$

◊ Let c be the value of SSR that is obtained when the minimum variance estimate $\boldsymbol{X}^*_{minvar}$ is used in the expression

$$SSR = (\mathbf{Y} - \boldsymbol{T}\boldsymbol{X}^*)^T \mathbf{R}_{\mathbf{Y}}^{-1} (Y - \boldsymbol{T}\boldsymbol{X}^*) \tag{A10.2.9}$$

namely

$$c = \mathbf{Y}^T \boldsymbol{Q} \mathbf{Y} \tag{A10.2.10}$$

Then

a. c is central Chi-squared.

b. $\mathcal{E}\{c\} = N_{\mathbf{Y}} - N_{\boldsymbol{X}^*}$ which we call k.

c. The Chi-squared PDF of c has k degrees of freedom.

Proof of the theorem can be found in Problem 10.18.

Part 3

Recursive filtering

Chapter 11

The Kalman and Swerling filters

11.1 Introduction

This is a book about Gauss–Newton and polynomial filtering and not about the Kalman filter. However, no book on filter engineering would be complete without at least a brief look at both the Kalman filter and its forerunner the Swerling filter.

It is not our intention to list advantages and disadvantages of various filters. An undertaking of that sort would be both futile and senseless, since such advantages and disadvantages must of necessity relate to particular applications.

Nevertheless, there is one item that we will discuss, and that is the following:

◊ In theory the extended Kalman and Swerling filters[1] are CR consistent, which means that – for a given set of inputs – they should produce results that have the same accuracy as those of the (CR-consistent) Gauss–Newton filters.

◊ In practice however, the extended Kalman and Swerling filters must be operated in a way that makes them CR inconsistent, and so the results that they produce are in fact *less* accurate than those of the Gauss–Newton filters, and sometimes significantly so.

We will discuss this issue fully in Section 11.5.

11.1.1 What exactly is the Kalman filter?

The words *Kalman filter* mean different things to different people.

◊ First, there were the equations that appeared in the original papers.[2]

[1] The meaning of the word *extended* will be explained further down.
[2] References 1 and 2.

These were then reworked[3] to give what we call the *mainstream version*.[4] It is the mainstream version that has been the one most used over the past more than 50 years.

◊ Other versions abound, going all the way back to 1967 – and perhaps even earlier. Some were known as the *square root Kalman filters*[5] and some as the *UD Kalman filters*.[6]

◊ In more recent times there are the *central difference Kalman filters*,[7] the *sigma-point Kalman filters*,[8] the *unscented Kalman filters*[9] and the *interacting multiple model (IMM) Kalman filters* used for tracking manoeuvring targets.[10]

◊ In addition to all of these, some of our colleagues have told us that they use their own variations, each of whom swears by them.

Clearly we cannot consider them all. Accordingly, when we discuss the *Kalman filter*, it will be the mainstream version that appears in the seventeen references cited in Footnote 4.

11.1.2 The nonrecursive/recursive duality

In section 8.9 we examined the terms *nonrecursive* and *recursive* in some detail, and we showed that the MVA can be configured in two ways – nonrecursively (Form A) and recursively (Form B) – with the results being fully equivalent.

This nonrecursive/recursive duality accompanies most algorithms in filter engineering, and for each non-recursive implementation there is usually a recursive one that, in theory, produces the same results.

The Gauss–Aitken and Gauss–Newton filters in Chapters 8 and 9 are the nonrecursive implementations of the MVA when the filter model and the observation equation are taken into account.

In this chapter we derive the recursive counterparts to the Gauss filters. These are called the *Swerling filters*, the first of which is for the all-linear ***Case 1***, and the remaining three – known as the *extended* Swerling filters – for use in the nonlinear ***Cases 2***, ***3*** and ***4***.

[3] Reference 6, Chapter 15.
[4] References 3, 6, 22, 23, 24, 25, 30, 31, 48, 50, 53, 54, 55, 56, 57, 173 and 174.
[5] References 22, 43, 44, 45, 46, 47 and 59.
[6] References 22, 43, 45, 46 and 47.
[7] Reference 155.
[8] Reference 36.
[9] References 34, 37 and 39.
[10] References 185–189.

We also derive the equations for the Kalman filter using an approach that differs from the one used by Kalman and Bucy, and it will become clear that the Kalman filter and its three extensions constitute a second set of recursive counterparts to the Gauss filters.

These relationships are depicted in Figure 11.1 below.

Starting on the left we see the details of the filter model and the observation equation being incorporated into the MVA, leading to the eight Gauss filters.

In the centre the four Swerling filters are shown, with the restriction that for them the observation errors *must* be stage-wise uncorrelated. These are then the first set of recursive duals of the Gauss filters.

Finally, on the right we see four Kalman filters emerging from the four Swerling filters when the equations of the latter are rearranged by the use of what is known as the *inversion lemma*. These then form a second set of recursive duals of the Gauss filters.

11.2 The Swerling filters

In July 1955 the USA announced that during the International Geophysical Year which was to run from July 1957 to December 1958, an attempt would be made to place mankind's first artificial satellite in orbit.

Upon hearing this, the Soviet Union immediately made the decision to beat the Americans, and on 4 October 1957 they succeeded in doing so. The satellite that they placed in orbit was named Sputnik 1.[11]

Peter Swerling was a scientist who worked at the RAND Corporation in California in the 1950s.[12] His interests lay in radar and its applications, and his name is today well known in that field for his work on radar cross sections of airborne targets.

Swerling began examining the problem of satellite tracking by radar some time before Sputnik 1 was launched, and in January 1958, just 3 months after Sputnik was placed in orbit, he published the first of three ground-breaking papers on that topic.[13]

His starting point was Version 2 of the MVA that we derived in Chapter 8, namely

[11] Reference 40.
[12] Reference 41.
[13] References 7, 8 and 9.

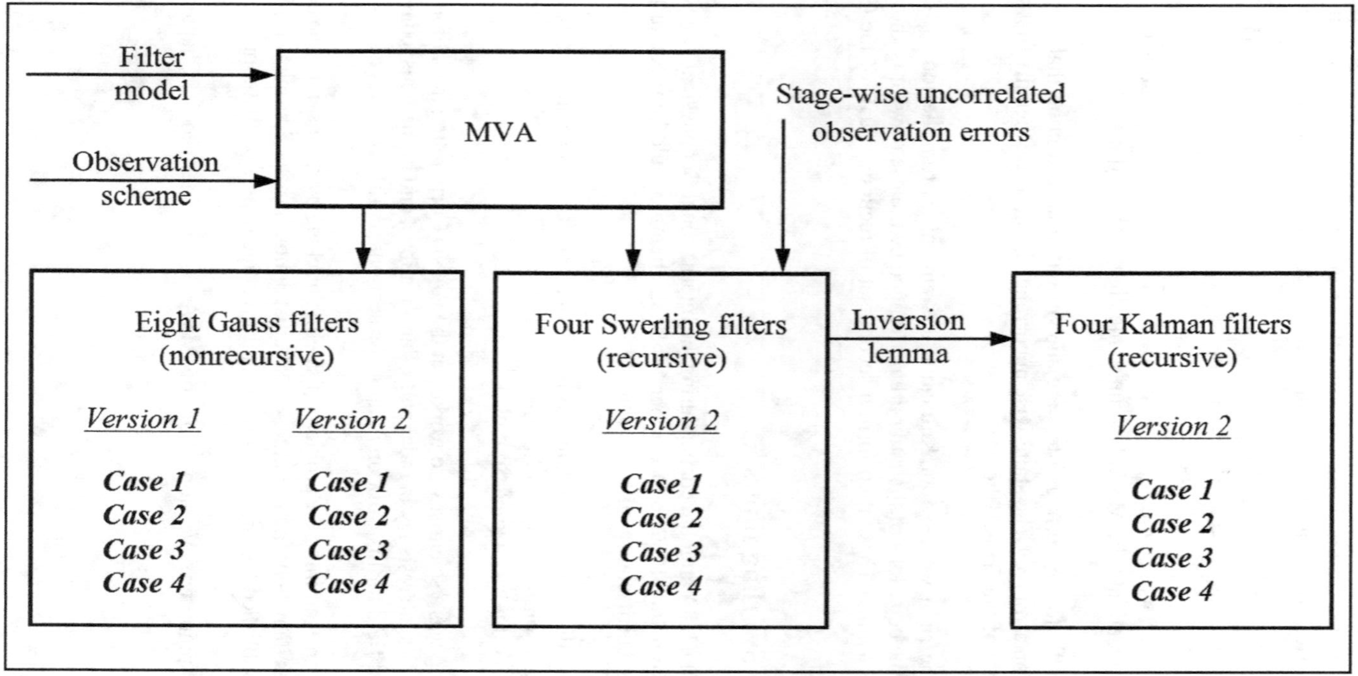

Figure 11.1 The Gauss, Swerling and Kalman filters

$$X^*_{n,n} = (T_n^T R_{Y_n}^{-1} T_n)^{-1} T_n^T R_{Y_n}^{-1} Y_n \tag{11.2.1}$$

$$S^*_{n,n} = (T_n^T R_{Y_n}^{-1} T_n)^{-1} \tag{11.2.2}$$

From there he arrived at the recursive filter shown in Figure 11.2 below.

The Swerling filter for Case 1

Inputs: Y_n, R_{Y_n}

$$X^*_{n,n-1} = \Phi(n,n-1) X^*_{n-1,n-1} \tag{11.2.3}$$

$$S^*_{n,n-1} = \Phi(n,n-1) S^*_{n-1,n-1} \Phi(n,n-1)^T \tag{11.2.4}$$

$$S^*_{n,n} = (S^{*\,-1}_{n,n-1} + M_n^T R_n^{-1} M_n)^{-1} \tag{11.2.5}$$

$$H_n = S^*_{n,n} M_n^T R_n^{-1} \tag{11.2.6}$$

$$X^*_{n,n} = X^*_{n,n-1} + H_n (Y_n - M_n X^*_{n,n-1}) \tag{11.2.7}$$

Outputs: $X^*_{n,n}$, $S^*_{n,n}$

Figure 11.2 The Swerling filter for ***Case-1***

Swerling's derivation is not a difficult one, but rather than carrying it out here we do it in Project 11.1. Once understood, it will be clear that his filter is simply a *recursive implementation* of the Gauss–Aitken Version 2 filter.

In Figure 11.2 we see the following:

- ◊ The Swerling filter is recursive, as evidenced by the presence of the preceding estimate $X^*_{n-1,n-1}$ and the preceding filter covariance matrix $S^*_{n-1,n-1}$ in its first two lines.

- ◊ The presence of the transition matrix Φ and observation matrix M tells us that the filter model and observation equations have been taken into account.

- ◊ The filter possesses what is known as the *predictor/corrector* structure:

  - ◊ In (11.2.3) and (11.2.4) the preceding estimate $X^*_{n-1,n-1}$ and the filter covariance matrix $S^*_{n-1,n-1}$ are projected forward from t_{n-1} to t_n using the filter-model's transition matrix to give $X^*_{n,n-1}$ and $S^*_{n,n-1}$. These two equations constitute the predictor portion.

  - ◊ In (11.2.5) the observation-error covariance matrix R_n is combined with $S^*_{n,n-1}$ to give the current-estimate covariance matrix $S^*_{n,n}$.

- ◊ In (11.2.6) the weight matrix $\boldsymbol{H}_n$ is derived.
- ◊ $\boldsymbol{H}_n$ is then used in (11.2.7) where the observation vector $\boldsymbol{Y}_n$ is incorporated to give the current estimate $\boldsymbol{X}^*_{n,n}$.

Equations (11.2.5) to (11.2.7) constitute the corrector portion of the algorithm.

Central to the derivation of the Swerling filter is the assumption that the observations are stage-wise uncorrelated. This means that there are no Swerling counterparts to the Gauss Version 1 filters, and so the options offered by Version 1 of the MVA regarding stage-wise correlated observation errors are not available when running any of the Swerling or Kalman filters.

The end result is that both the Swerling and Kalman filters are unable to process data that are stage-wise correlated, and if an attempt is made to do so by ignoring the stage-wise correlation terms, then they will be both ECM and CR inconsistent – the former of which is fatal.

The three extended Swerling filters (***Cases 2***, ***3*** and ***4***) are derived in Project 11.2.

11.3 The Kalman filters

Before discussing the Kalman filters it is necessary that we elaborate on two items. This may take a little time, but there is no alternative and we ask for the reader's patience.

11.3.1 Process noise and spontaneous ECM inconsistency

The Kalman filters can be configured in one of two possible ways – either with or without what is known as a $\boldsymbol{Q}$ matrix.[14]

When included, $\boldsymbol{Q}$ is a positive-definite matrix that is added to the second line of the filter (see (11.3.6)) further down).

There are two possibilities:

1. ***The external model contains a random forcing function known as process noise***

In the original papers describing their filter,[15] consideration was given by Kalman and Bucy to external models that include random forcing functions known as *process noise*.[16]

[14] Also true for the Swerling filters.
[15] References 1 and 2.
[16] Sometimes also known as *plant noise* or *driving noise*.

When the DE of the external model contains such a random forcing function, then it is entirely appropriate that process noise should also be included in the Kalman filter model, and mathematically this results in a positive-definite matrix (usually called $\boldsymbol{Q}$) appearing in the second equation of the filter.[17]

When this is done, we say that $\boldsymbol{Q}$ has been included *by necessity*.

However, the DE for the external model now contains a forcing function, and so (see Subsection 2.2.1.1) this situation will not be considered further in our discussion.

2. *The external model does not contain a random forcing function*

Even when the DE of the external model contains no random forcing function, it is almost always the case that a $\boldsymbol{Q}$ matrix is *still* included in the Kalman filter equations.

When this is done, we say that $\boldsymbol{Q}$ has been included *by choice*.

Some of the reasons offered for doing so will be listed further down, but perhaps the first of them is the most important, namely:

> If run without an appropriately selected $\boldsymbol{Q}$ matrix, the extended Kalman filters are afflicted with the proclivity to become unstable, meaning that they become (fatally) ECM inconsistent. In our discussion we will refer to this behaviour as *the problem of spontaneous ECM inconsistency*.

We consider briefly what spontaneous ECM inconsistency means.

It is possible to create a filter – Gauss–Newton, Kalman or Swerling – in which an implementation error is present, resulting in the filter *always* being ECM inconsistent throughout every run, and when that error is corrected the filter becomes ECM consistent and remains so throughout every run thereafter.

For the extended Kalman and Swerling filters a different type of ECM inconsistency is also possible.

◊ If a fully de-bugged extended Kalman filter without a $\boldsymbol{Q}$ matrix is repeatedly run from start-up, it will be fully ECM consistent throughout most of the runs, showing that there is nothing wrong with the computer code, the way in which the filter has been initialized, or any other aspect of the implementation.

◊ However, on *some* runs the filter can manifest what we call *spontaneous ECM inconsistency* in a number of ways:

[17] $\boldsymbol{Q}_n$ is the covariance matrix of the process-noise vector at time t_n.

- ◊ It can immediately start out being ECM inconsistent and remain so.
- ◊ It can start out being ECM consistent but then become ECM inconsistent after an unpredictable number of cycles.
- ◊ Once inconsistent, it can in some cases remain so thereafter, and in others it drifts into and out of ECM inconsistency.

This phenomenon is well known to users of the Kalman filter, and it is also well known that the way in which to prevent such behaviour from occurring is to include a $\boldsymbol{Q}$ matrix – even when no random forcing function is present in the DE of the filter model.

This procedure is known as *tuning* and is discussed in most books on the Kalman filter, the following being an example[18]:

> 'The procedure for tuning is to make the filter consistent, that is, the three criteria (a) – (c) from subsection 5.4.2 should be satisfied. While this is easier said than done, in practice one has to strive to make the filter as close to being consistent as feasible, while at the same time achieving small RMS estimation errors.'

It is clear from this statement that the authors are aware of the possibility of spontaneous Kalman ECM inconsistency – otherwise why speak about 'tuning' and why the statement 'one has to strive to make the filter as close to being consistent as feasible'?

They are evidently also aware of the fact that the presence of the $\boldsymbol{Q}$ matrix increases the RMS estimation errors – hence making the filter CR inconsistent – otherwise why speak about 'achieving small RMS errors'?

We will return to the CR consistency of the Kalman filter later on.

11.3.2 Examples of spontaneous Kalman/Swerling ECM inconsistency (instability)

In Section 1.12 we presented an example of spontaneous ECM inconsistency in the Kalman and Swerling filters that was produced by the program ***13_Gauss–Newton_C.***

In Projects 11.3 and 11.4 we discuss further examples of Kalman/Swerling instability. Some of these are produced by the program ***13_Gauss_newton_C*** and others by ***14_Orbit_Track***.

[18] See Reference 3.

They contain what we believe to be useful information regarding the Kalman/Swerling instability phenomenon.

11.3.3 Derivation of the Kalman filter equations

Our derivation of the Kalman filter equations starts from the assumption that there is no forcing function present in the external model – either related to process noise or for any other reason – and so our derivation differs from that used by Kalman and Bucy.

However, once we have derived the filter equations, a $\boldsymbol{Q}$ matrix can be included in the same way as is indicated in Reference 3, thereby arriving at exactly the same equations that the seventeen references cited in Footnote 4 call the *Kalman filter*.

The derivation starts from the Swerling equations to which we apply a mathematical relationship known as the *inversion lemma,* but rather than carrying it out here, we do it in Appendix 11.1. The end result for ***Case 1*** appears in Figure 11.3.

The Kalman filter for Case 1

Inputs: Y_n, $\boldsymbol{R}_n$

$$\boldsymbol{X}^*_{n,n-1} = \boldsymbol{\Phi}(n,n-1)\boldsymbol{X}^*_{n-1,n-1} \tag{11.3.1}$$

$$\boldsymbol{S}^*_{n,n-1} = \boldsymbol{\Phi}(n,n-1)\boldsymbol{S}^*_{n-1,n-1}\boldsymbol{\Phi}(n,n-1)^T \tag{11.3.2}$$

$$\boldsymbol{H}_n = \boldsymbol{S}^*_{n,n-1}\boldsymbol{M}_n^T(\boldsymbol{R}_n + \boldsymbol{M}_n\boldsymbol{S}^*_{n,n-1}\boldsymbol{M}_n^T)^{-1} \tag{11.3.3}$$

$$\boldsymbol{S}^*_{n,n} = (\boldsymbol{I} - \boldsymbol{H}_n\boldsymbol{M}_n)\boldsymbol{S}^*_{n,n-1} \tag{11.3.4}$$

$$\boldsymbol{X}^*_{n,n} = \boldsymbol{X}^*_{n,n-1} + \boldsymbol{H}_n(\boldsymbol{Y}_n - \boldsymbol{M}_n\boldsymbol{X}^*_{n,n-1}) \tag{11.3.5}$$

Outputs: $\boldsymbol{X}^*_{n,n}$, $\boldsymbol{S}^*_{n,n}$

*Figure 11.3 The Kalman Filter for **Case 1***

These equations, with the necessary modifications, form the basis for the extended Kalman filters for ***Cases 2***, ***3*** and ***4*** (see Project 11.2). If the user then wishes to include a $\boldsymbol{Q}$ matrix, it would be added into the second equation of the filter and so (11.3.2) would read as follows:

$$\boldsymbol{S}^*_{n,n-1} = \boldsymbol{\Phi}(n, n-1)\, \boldsymbol{S}^*_{n-1,n-1} + \boldsymbol{Q}_n \tag{11.3.6}$$

If it is further desired to add what is known as a $\boldsymbol{V}$ vector (as some authors do) then (11.3.1) would become

$$\boldsymbol{X}^*_{n,n-1} = \boldsymbol{\Phi}(n, n-1)\, \boldsymbol{X}^*_{n-1,n-1} + \boldsymbol{V}_n \tag{11.3.7}$$

Note: $\boldsymbol{V}_n$ is a vector of random numbers generated from a zero-mean multivariate Gaussian PDF with covariance matrix $\boldsymbol{Q}_n$.[19]

◀▶

In Figure 11.3 we see the following:

◊ The Kalman filter, like the Swerling filter, is recursive, as evidenced by the presence of the preceding estimate $\boldsymbol{X}^*_{n-1,n-1}$ and filter covariance matrix $\boldsymbol{S}^*_{n-1,n-1}$ in its first two lines.

◊ The presence of the transition matrix $\boldsymbol{\Phi}$ and observation matrix $\boldsymbol{M}$ tells us that the filter model and observation equations have been taken into account.

◊ The Kalman filter has the same predictor/corrector structure as the Swerling filter, but its third and fourth equations are in reverse order to those of Swerling.

  ◊ In (11.3.1) and (11.3.2) the preceding estimate $\boldsymbol{X}^*_{n-1,n-1}$ and preceding filter covariance matrix $\boldsymbol{S}^*_{n-1,n-1}$ are projected forward from t_{n-1} to t_n using the filter-model's transition matrix to give $\boldsymbol{X}^*_{n,n-1}$ and $\boldsymbol{S}^*_{n,n-1}$. These two equations constitute the predictor portion.

  ◊ In (11.3.3) the observation covariance matrix $\boldsymbol{R}_n$ is incorporated, giving the weight matrix $\boldsymbol{H}_n$.

  ◊ In (11.3.4) the updated filter covariance matrix $\boldsymbol{S}^*_{n,n}$ is obtained (which now includes $\boldsymbol{R}_n$).

  ◊ Finally in (11.3.5) the observation vector $\boldsymbol{Y}_n$ is incorporated, giving the updated estimate $\boldsymbol{X}^*_{n,n}$.

  Equations (11.3.3) to (11.3.5) constitute the corrector portion.

The derivation of the Kalman equations in Appendix 11.1 commences from those of the Swerling filter which is based on Version 2 of the MVA, together with the assumption that the observations are stage-wise uncorrelated. Thus – as with the Swerling filter – the options offered by Version 1 of the MVA are not available to the Kalman filter, and so it too cannot be used when the observations are stagewise correlated.

The three extended Kalman filters (***Cases 2***, ***3*** and ***4***) are derived in Project 11.2.

[19] See Project 4.7 for the details on how such a vector is generated.

11.4 Equivalence of the Gauss, Swerling and Kalman filters

We have seen that the Kalman and Swerling filters (without $\boldsymbol{Q}$ matrices) are simply the recursive duals of the Version-2 Gauss filters, and that all of them are based on the MVA. This is depicted in Figure 11.4.

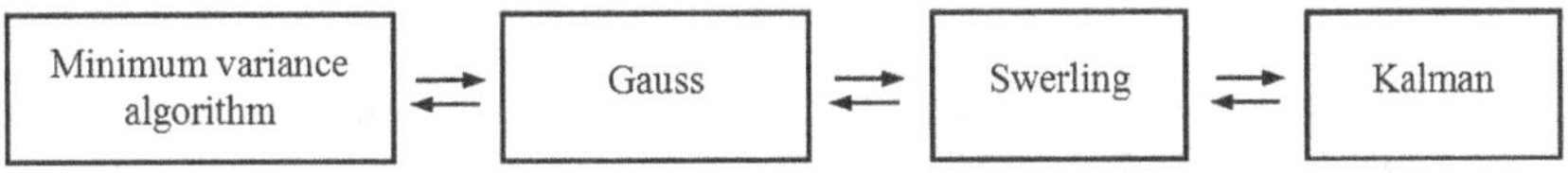

Figure 11.4 Minimum variance, Gauss, Swerling and Kalman filters

Given this equivalence, we can make the following statements:

1. If we exclude runs in which the Kalman/Swerling filters have become unstable, then they should both be ECM consistent.
2. If we exclude runs in which the Kalman/Swerling filters have become unstable, then they should both be CR consistent.

In Appendix 11.3 we present numerical results that support these statements.

11.4.1 Comments

1. We noted in Section 8.4 that the Gauss filters are *LU* (linear unbiased) estimators in the sense that $\boldsymbol{X}^* = \boldsymbol{WY}$, and $\boldsymbol{WT} = \boldsymbol{I}$. Because the Kalman and Swerling filters are the recursive duals of the Version-2 Gauss filters, it then follows that they are also *LU* estimators.
2. Given the equivalences in Figure 11.4 one might then ask why the extended Swerling and Kalman filters can, on occasions, both become unstable, resulting in spontaneous ECM inconsistency – yet we have never seen that occurring in the Gauss filters.

   This is a question for which we don't have an answer, and all that we can offer is the following:

   ◊ Algorithms can be mathematically equivalent and yet differ in their computational behaviour.

   ◊ The potential for spontaneous ECM inconsistency in the extended Swerling and Kalman filters appears to come about because of their recursive structures. We base this on the fact that in both of our programs in which these filters are implemented[20] if one of them becomes unstable then the

[20] *13_Gauss–Newton_C* and *14_Orbit_Track.*

other always does as well, while the nonrecursive Gauss–Newton filter that is processing exactly the same data remains fully ECM consistent.

3. The author has heard highly placed academics make the statement that the Swerling filter is inferior to the Kalman filter.

   This comes as something of a surprise for the following reasons:

   ◊ The two filters are implemented in the programs ***13_Gauss–Newton_C*** and ***14_Orbit_Track*** and are given identical inputs. Their outputs are always in agreement to eleven decimal digits.

   ◊ In hundreds of runs of those two programs, we have never yet seen an instance in which one of the filters becomes unstable without the other one doing likewise.

   ◊ The ***Case 1*** Swerling filter is self-initializing while the ***Case 1*** Kalman filter is not. (See Project 11.5.)

   It is difficult then to understand why anybody would regard the Swerling filter as inferior to the Kalman filter.

11.5 The Kalman/Swerling inconsistency dilemma

The Kalman filter has experienced explosive growth in popularity since first it was announced in 1960.

Because it is rich in concepts from linear systems theory, statistics and other fields, it has made outstanding material for university courses, and generations of engineering students and others have been introduced to it in this way. Large numbers of theses and many textbooks have been written about it, and the technical papers that have been published are by now almost too many to enumerate.

It is almost always the case that a $\boldsymbol{Q}$ matrix is included in the Kalman filter, some of the reasons offered for doing so being the following:

◊ To ensure that stability, i.e. ECM consistency, is always present.

◊ To allow for possible knowledge-gaps or inaccuracies in the filter model.

◊ To endow the filter with a fading memory.

◊ To prevent the filter from going into 'sleep mode'.

◊ To give the filter a 'higher bandwidth'.

◊ To keep the filter numerically healthy i.e. to avoid negative numbers or zeros on the diagonal of its covariance matrix and/or to avoid singular covariance matrices.

As stated in Section 11.1, we do not intend to list advantages and disadvantages of any of the filters. However, there is one feature of the extended Kalman and Swerling filters that does require comment, and which we now discuss.

11.5.1 The Kalman/Swerling inconsistency dilemma

Because of their potential instability, users of the extended Kalman and Swerling filters face the following dilemma:

◊ If run without a $\boldsymbol{Q}$ matrix then those filters are normally CR consistent, but on unpredictable occasions, and for reasons not well understood, they may become spontaneously ECM inconsistent (i.e. unstable).

◊ On the other hand, if an appropriately selected $\boldsymbol{Q}$ matrix is included then they will always be ECM consistent, but they then become CR inconsistent.

Simply put, the extended Kalman/Swerling filters can either be always ECM consistent or they can be CR consistent, but they cannot be both.

This means that whenever an extended Kalman/Swerling filter *with* a $\boldsymbol{Q}$ matrix is used in an application in which the external model DE has no forcing function (random or otherwise), then the person doing so should be aware that a Gauss–Newton filter is available that provides estimates with smaller sigma values, meaning that they are more accurate.

And if an extended Kalman/Swerling filter *without* a $\boldsymbol{Q}$ matrix is selected for such an application, then its results will normally be of the same accuracy as those of the Gauss–Newton filter, but the possibility exists that those filters could become spontaneously ECM inconsistent, which is fatal.

There can be little doubt that the dilemma exists.

◊ Numerical evidence is presented in Appendix 11.4, and in Projects 11.3 and 11.4.

◊ An analytical proof that the Kalman and Swerling filters with a $\boldsymbol{Q}$ matrix cannot be CR consistent is provided in Appendix 11.2.

11.6 Conclusion

Our treatment of the Kalman and Swerling filters has, of necessity, been brief. The following is a list of our references that the interested reader may wish to pursue:

1, 2, 3, 6, 22, 23, 24, 25, 30, 31, 32, 33, 36, 37, 38, 39, 43, 47, 53, 54, 55, 56, 57, 59, 81, 83, 84, 152, 153, 154, 155, 156, 158, 159, 160, 161, 162, 163, 164, 165, 166, 167, 168, 169, 170, 171, 173, 174.

The following items are in the supplementary material

Problems for Chapter 11

Solutions to Problems for Chapter 11

Projects for Chapter 11

Appendix 11.1

Derivation of the Case-1 Kalman filter

We use what is known as the *inversion lemma* to rearrange the equations of the Swerling filter in Figure A11.2, thereby obtaining the Kalman filter.

The inversion lemma: Let $\boldsymbol{S}$ and $\boldsymbol{R}$ be positive-definite matrices in which $\boldsymbol{S}$ is $m \times m$ and $\boldsymbol{R}$ is $k \times k$. Let $\boldsymbol{M}$ be a $k \times m$ matrix. Then

$$(\boldsymbol{S}^{-1} + \boldsymbol{M}^T \boldsymbol{R}^{-1} \boldsymbol{M})^{-1} = \boldsymbol{S} - \boldsymbol{S}\boldsymbol{M}^T (\boldsymbol{R} + \boldsymbol{M}\boldsymbol{S}\boldsymbol{M}^T)^{-1} \boldsymbol{M}\boldsymbol{S} \tag{A11.1.1}$$

Proof is considered in Problem 11.1. ■■

Post-multiplying both sides of (A11.1.1) by $\boldsymbol{M}^T \boldsymbol{R}^{-1}$ gives

$$(\boldsymbol{S}^{-1} + \boldsymbol{M}^T \boldsymbol{R}^{-1} \boldsymbol{M})^{-1} \boldsymbol{M}^T \boldsymbol{R}^{-1} = (\boldsymbol{S} - \boldsymbol{S}\boldsymbol{M}^T (\boldsymbol{R} + \boldsymbol{M}\boldsymbol{S}\boldsymbol{M}^T)^{-1} \boldsymbol{M}\boldsymbol{S}) \boldsymbol{M}^T \boldsymbol{R}^{-1} \tag{A11.1.2}$$

It is also proved in Problem 11.1 that (A11.1.2) reduces to

$$(\boldsymbol{S}^{-1} + \boldsymbol{M}^T \boldsymbol{R}^{-1} \boldsymbol{M})^{-1} \boldsymbol{M}^T \boldsymbol{R}^{-1} = \boldsymbol{S}\boldsymbol{M}^T (\boldsymbol{R} + \boldsymbol{M}\boldsymbol{S}\boldsymbol{M}^T)^{-1} \tag{A11.1.3}$$

We now introduce stars and subscripts into (A11.1.1) and (A11.1.3) obtaining

$$\begin{aligned}(\boldsymbol{S}^*_{n,n-1}{}^{-1} + \boldsymbol{M}_n^T \boldsymbol{R}_n^{-1} \boldsymbol{M}_n)^{-1} &= \\ \boldsymbol{S}^*_{n,n-1} - \boldsymbol{S}^*_{n,n-1} \boldsymbol{M}_n^T (\boldsymbol{R}_n &+ \boldsymbol{M}_n \boldsymbol{S}^*_{n,n-1} \boldsymbol{M}_n^T)^{-1} \boldsymbol{M}_n \boldsymbol{S}^*_{n,n-1}\end{aligned} \tag{A11.1.4}$$

and

$$\begin{aligned}(\boldsymbol{S}^*_{n,n-1}{}^{-1} + \boldsymbol{M}_n^T \boldsymbol{R}_n^{-1} \boldsymbol{M}_n)^{-1} \boldsymbol{M}_n^T \boldsymbol{R}_n^{-1} &= \\ \boldsymbol{S}^*_{n,n-1} \boldsymbol{M}_n^T (\boldsymbol{R}_n &+ \boldsymbol{M}_n \boldsymbol{S}^*_{n,n-1} \boldsymbol{M}_n^T)^{-1}\end{aligned} \tag{A11.1.5}$$

Using the expression for $\boldsymbol{S}^*_{n,n}$ from (11.2.5) in (A11.1.5) we obtain

$$\boldsymbol{S}^*_{n,n} \boldsymbol{M}_n^T \boldsymbol{R}_n^{-1} = \boldsymbol{S}^*_{n,n-1} \boldsymbol{M}_n^T (\boldsymbol{R}_n + \boldsymbol{M}_n \boldsymbol{S}^*_{n,n-1} \boldsymbol{M}_n^T)^{-1} \tag{A11.1.6}$$

From (11.2.6) we recall that the left side of (A11.1.6) is the weight matrix $\boldsymbol{H}_n$, and so we now have the following equivalent expression for the weight matrix:

$$\boldsymbol{H}_n = \boldsymbol{S}^*_{n,n-1}\boldsymbol{M}_n^T(\boldsymbol{R}_n + \boldsymbol{M}_n\boldsymbol{S}^*_{n,n-1}\boldsymbol{M}_n^T)^{-1} \tag{A11.1.7}$$

Equation (11.2.5) together with (A11.1.4) also gives

$$\boldsymbol{S}^*_{n,n} = \boldsymbol{S}^*_{n,n-1} - \boldsymbol{S}^*_{n,n-1}\boldsymbol{M}_n^T(\boldsymbol{R}_n + \boldsymbol{M}_n\boldsymbol{S}^*_{n,n-1}\boldsymbol{M}_n^T)^{-1}\boldsymbol{M}_n\boldsymbol{S}^*_{n,n-1} \tag{A11.1.8}$$

which together with (A11.1.7) means that

$$\boldsymbol{S}^*_{n,n} = \boldsymbol{S}^*_{n,n-1} - \boldsymbol{H}_n\boldsymbol{M}_n\boldsymbol{S}^*_{n,n-1} = (\boldsymbol{I} - \boldsymbol{H}_n\boldsymbol{M}_n)\,\boldsymbol{S}^*_{n,n-1} \tag{A11.1.9}$$

This gives us the following alternate expression for $\boldsymbol{S}^*_{n,n}$:

$$\boldsymbol{S}^*_{n,n} = (\boldsymbol{I} - \boldsymbol{H}_n\boldsymbol{M}_n)\,\boldsymbol{S}^*_{n,n-1} \tag{A11.1.10}$$

and the derivation is complete.

The two expressions (A11.1.7) and (A11.1.10) now give us a second way of writing the Swerling filter. The outcome is known as the *Kalman filter for* ***Case 1*** and is shown in Figure A11.1.1.

The Kalman filter for Case 1

Inputs: $\boldsymbol{Y}_n$, $\boldsymbol{R}_n$

$$\boldsymbol{X}^*_{n,n-1} = \boldsymbol{\Phi}(n,n-1)\boldsymbol{X}^*_{n-1,n-1} \tag{A11.1.11}$$

$$\boldsymbol{S}^*_{n,n-1} = \boldsymbol{\Phi}(n,n-1)\boldsymbol{S}^*_{n-1,n-1}\,\boldsymbol{\Phi}(n,n-1)^T \tag{A11.1.12}$$

$$\boldsymbol{H}_n = \boldsymbol{S}^*_{n,n-1}\boldsymbol{M}_n^T(\boldsymbol{R}_n + \boldsymbol{M}_n\boldsymbol{S}^*_{n,n-1}\boldsymbol{M}_n^T)^{-1} \tag{A11.1.13}$$

$$\boldsymbol{S}^*_{n,n} = (\boldsymbol{I} - \boldsymbol{H}_n\boldsymbol{M}_n)\boldsymbol{S}^*_{n,n-1} \tag{A11.1.14}$$

$$\boldsymbol{X}^*_{n,n} = \boldsymbol{X}^*_{n,n-1} + \boldsymbol{H}_n(\boldsymbol{Y}_n - \boldsymbol{M}_n\boldsymbol{X}^*_{n,n-1}) \tag{A11.1.15}$$

Outputs: $\boldsymbol{X}^*_{n,n}$, $\boldsymbol{S}^*_{n,n}$

Figure A11.1.1 The Kalman Filter for Case 1

Cases 2, ***3*** and ***4*** must still be accounted for, and that is done in Project 11.2.

Appendix 11.2

The Kalman/Swerling inconsistency-dilemma analytical proof

We show that the Kalman/Swerling filters with a positive-definite $\boldsymbol{Q}$ matrix cannot be CR consistent. We start from the Kalman covariance matrix equations, but we soon invoke the inversion lemma which moves us to the Swerling covariance matrix equations. Thus the proof covers both filters.

Without a $\boldsymbol{Q}$ matrix, the Kalman equations for updating its covariance matrix are as follows:

$$\boldsymbol{S}^*_{n,n-1} = \boldsymbol{\Phi}(n, n-1)\, \boldsymbol{S}^*_{n-1,n-1} \boldsymbol{\Phi}(n, n-1)^T \tag{A11.2.1}$$

$$\boldsymbol{H}_n = \boldsymbol{S}^*_{n,n-1} \boldsymbol{M}_n^T (\boldsymbol{R}_n + \boldsymbol{M}_n \boldsymbol{S}^*_{n,n-1} \boldsymbol{M}_n^T)^{-1} \tag{A11.2.2}$$

$$\boldsymbol{S}^*_{n,n} = (\boldsymbol{I} - \boldsymbol{H}_n \boldsymbol{M}_n)\, \boldsymbol{S}^*_{n,n-1} \tag{A11.2.3}$$

(These three equations constitute what is known as a *discrete Riccati equation*.[1])

In Appendix 11.1 we showed that the Kalman filter without a $\boldsymbol{Q}$ matrix is an algebraic reorganization of the Swerling filter which is a recursive version of the MVA. We know that the latter is CR consistent and so this means that the Kalman filter without a $\boldsymbol{Q}$ matrix is also CR consistent.

Accordingly, suppose that a Kalman filter has been running without a $\boldsymbol{Q}$ matrix up to time t_{n-1}, and so $\boldsymbol{S}^*_{n-1,n-1}$ in (A11.2.1) is CR consistent.

Then assume that at time t_n a positive-definite $\boldsymbol{Q}$ matrix is added into (A11.2.1).

That being the case, $\boldsymbol{S}^*_{n,n}$ in (A11.2.3) cannot be CR consistent.

Proof: Let the symbols $\boldsymbol{S}^*_{n,n-1}$, $\boldsymbol{H}_n$ and $\boldsymbol{S}^*_{n,n}$ be used when a $\boldsymbol{Q}$ matrix is *not* included and let them be written as $\boldsymbol{S}^*_{n,n-1,Q}$, $\boldsymbol{H}_{n,Q}$ and $\boldsymbol{S}^*_{n,n,Q}$ when a $\boldsymbol{Q}$ matrix *is* included.

[1] See Reference 53 or 54.

With a $\boldsymbol{Q}$ matrix added into (A11.2.1), we have

$$\boldsymbol{S}^*_{n,n-1,Q} = \boldsymbol{\Phi}(n,n-1)\boldsymbol{S}^*_{n-1,n-1}\boldsymbol{\Phi}(n,n-1)^T + \boldsymbol{Q} = \boldsymbol{S}^*_{n,n-1} + \boldsymbol{Q} \tag{A11.2.4}$$

and so by (A11.2.3)

$$\begin{aligned}\boldsymbol{S}^*_{n,n,Q} &= (\boldsymbol{I} - \boldsymbol{H}_{n,Q}\boldsymbol{M}_n)\,\boldsymbol{S}^*_{n,n-1,Q} \\ &= (\boldsymbol{I} - \boldsymbol{S}^*_{n,n-1,Q}\boldsymbol{M}_n^T(\boldsymbol{R}_n + \boldsymbol{M}_n\boldsymbol{S}^*_{n,n-1,Q}\boldsymbol{M}_n^T)^{-1}\boldsymbol{M}_n)\,\boldsymbol{S}^*_{n,n-1,Q} \\ &= \boldsymbol{S}^*_{n,n-1,Q} - \boldsymbol{S}^*_{n,n-1,Q}\boldsymbol{M}_n^T(\boldsymbol{R}_n + \boldsymbol{M}_n\boldsymbol{S}^*_{n,n-1,Q}\boldsymbol{M}_n^T)^{-1}\boldsymbol{M}_n\boldsymbol{S}^*_{n,n-1,Q}\end{aligned} \tag{A11.2.5}$$

Invoking the inversion lemma (A11.1.1), this now continues as

$$\begin{aligned} &= (\boldsymbol{S}^{*\,-1}_{n,n-1,Q} + \boldsymbol{M}_n^T\boldsymbol{R}_n^{-1}\boldsymbol{M}_n)^{-1} \\ &= \left((\boldsymbol{S}^*_{n,n-1} + \boldsymbol{Q})^{-1} + \boldsymbol{M}_n^T\boldsymbol{R}_n^{-1}\boldsymbol{M}_n\right)^{-1}\end{aligned} \tag{A11.2.6}$$

We have thus shown that when a $\boldsymbol{Q}$ matrix is added into (A11.2.1), then

$$\boldsymbol{S}^*_{n,n,Q} = \left((\boldsymbol{S}^*_{n,n-1} + \boldsymbol{Q})^{-1} + \boldsymbol{M}_n^T\boldsymbol{R}_n^{-1}\boldsymbol{M}_n\right)^{-1} \tag{A11.2.7}$$

We also know from (11.2.5) that if a $\boldsymbol{Q}$ matrix had *not* been added then

$$\boldsymbol{S}^*_{n,n} = (\boldsymbol{S}^{*\,-1}_{n,n-1} + \boldsymbol{M}_n^T\boldsymbol{R}_n^{-1}\boldsymbol{M}_n)^{-1} \tag{A11.2.8}$$

Note: If we had started from the Swerling equations rather than the Kalman equations, then we could have written (A11.2.7) by inspection. However, we elected to start from the Kalman equations so that this proof would cover both filters.

We now use Lemma A11.2.1 (see below) to prove that if $\boldsymbol{Q}$ is positive-definite, then each of the diagonal elements of $\boldsymbol{S}^*_{n,n,Q}$ in (A11.2.7) will be larger than the corresponding diagonal element of $\boldsymbol{S}^*_{n,n}$ in (A11.2.8).

Because $\boldsymbol{S}^*_{n,n}$ in (A11.2.8) was CR consistent, this proves that the Kalman/Swerling filters with a $\boldsymbol{Q}$ matrix cannot be CR consistent. ■■

Lemma A11.2.1: If $\boldsymbol{A}$ and $\boldsymbol{B}$ are positive-definite then $\boldsymbol{A} - \boldsymbol{B}$ is positive-definite *iff* $\boldsymbol{B}^{-1} - \boldsymbol{A}^{-1}$ is positive definite.

Proof is given in Problem 11.2. ■■

Note: In the scalar case the lemma would read: If a and b are positive numbers, then a-b is positive if and only if $1/b - 1/a$ is positive. As an example, let $a = 10$ and $b = 5$. Then $10 - 5$ is positive if and only if $1/5 - 1/10$ is positive.

Using the lemma, we subtract (A11.2.8) from (A11.2.7) obtaining

$$\begin{aligned} \boldsymbol{S}^*_{n,n,Q} - \boldsymbol{S}^*_{n,n} &= \left((\boldsymbol{S}^*_{n,n-1} + \boldsymbol{Q})^{-1} + \boldsymbol{M}_n^T \boldsymbol{R}_n^{-1} \boldsymbol{M}_n\right)^{-1} \\ &\quad - (\boldsymbol{S}^{*\,-1}_{n,n-1} + \boldsymbol{M}_n^T \boldsymbol{R}_n^{-1} \boldsymbol{M}_n)^{-1} \end{aligned} \tag{A11.2.9}$$

Consider now the first term on the right, namely $\left((\boldsymbol{S}^*_{n,n-1} + \boldsymbol{Q})^{-1} + \boldsymbol{M}_n^T \boldsymbol{R}_n^{-1} \boldsymbol{M}_n\right)^{-1}$:

◊ The matrix $\boldsymbol{S}^*_{n,n-1}$ is positive definite because it was formed in (A11.2.1) by a congruence transformation on the (assumed to be) positive-definite matrix $\boldsymbol{S}^*_{n-1,n-1}$. We are also assuming that $\boldsymbol{Q}$ is positive definite. Thus the sum $\boldsymbol{S}^*_{n,n-1} + \boldsymbol{Q}$ is positive-definite and hence so also is $(\boldsymbol{S}^*_{n,n-1} + \boldsymbol{Q})^{-1}$.

◊ $\boldsymbol{R}_n$ is positive-definite, and so, depending on the structure of $\boldsymbol{M}$ the matrix $\boldsymbol{M}_n^T \boldsymbol{R}_n^{-1} \boldsymbol{M}_n$ is either positive-definite or positive-semidefinite. (See Problem 11.5.)

◊ The matrix $(\boldsymbol{S}^*_{n,n-1} + \boldsymbol{Q})^{-1} + \boldsymbol{M}_n^T \boldsymbol{R}_n^{-1} \boldsymbol{M}_n$ is therefore the sum of $(\boldsymbol{S}^*_{n,n-1} + \boldsymbol{Q})^{-1}$ which is positive-definite, and $\boldsymbol{M}_n^T \boldsymbol{R}_n^{-1} \boldsymbol{M}_n$, which is either positive-definite or positive-semidefinite matrix. Either way $(\boldsymbol{S}^*_{n,n-1} + \boldsymbol{Q})^{-1} + \boldsymbol{M}_n^T \boldsymbol{R}_n^{-1} \boldsymbol{M}_n$ is positive-definite. (See Property 1 in Chapter 4 and Problem 11.5b.)

◊ The inverse matrix $\left((\boldsymbol{S}^*_{n,n-1} + \boldsymbol{Q})^{-1} + \boldsymbol{M}_n^T \boldsymbol{R}_n^{-1} \boldsymbol{M}_n\right)^{-1}$ is therefore positive definite.

By similar arguments the second term on the right of (A11.2.9), namely $(\boldsymbol{S}^{*\,-1}_{n,n-1} + \boldsymbol{M}_n^T \boldsymbol{R}_n^{-1} \boldsymbol{M}_n)^{-1}$ is also positive-definite.

We now start from (A11.2.9) and invoke Lemma A11.2.1:

$\boldsymbol{S}^*_{n,n,Q} - \boldsymbol{S}^*_{n,n} = \left((\boldsymbol{S}^*_{n,n-1} + \boldsymbol{Q})^{-1} + \boldsymbol{M}_n^T \boldsymbol{R}_n^{-1} \boldsymbol{M}_n\right)^{-1} - (\boldsymbol{S}^{*\,-1}_{n,n-1} + \boldsymbol{M}_n^T \boldsymbol{R}_n^{-1} \boldsymbol{M}_n)^{-1}$ is positive-definite

iff $(\boldsymbol{S}^{*\,-1}_{n,n-1} + \boldsymbol{M}_n^T \boldsymbol{R}_n^{-1} \boldsymbol{M}_n) - \left((\boldsymbol{S}^*_{n,n-1} + \boldsymbol{Q})^{-1} + \boldsymbol{M}_n^T \boldsymbol{R}_n^{-1} \boldsymbol{M}_n\right)$ is positive-definite (by Lemma A11.2.1)

iff $\boldsymbol{S}^{*\,-1}_{n,n-1} - (\boldsymbol{S}^*_{n,n-1} + \boldsymbol{Q})^{-1}$ is positive-definite (because $\boldsymbol{M}_n^T \boldsymbol{R}_n^{-1} \boldsymbol{M}_n$ cancels) *iff* $(\boldsymbol{S}^*_{n,n-1} + \boldsymbol{Q}) - \boldsymbol{S}^*_{n,n-1}$ is positive-definite (by Lemma A11.2.1)

iff $\boldsymbol{Q}$ is positive-definite (because $\boldsymbol{S}^*_{n,n-1}$ cancels)

which by assumption it is.

We have thus shown that $\boldsymbol{S}^*_{n,n,Q} - \boldsymbol{S}^*_{n,n}$ is positive definite. We now let

$$\boldsymbol{S}^*_{n,n,Q} - \boldsymbol{S}^*_{n,n} = \boldsymbol{G} \tag{A11.2.10}$$

Then $\boldsymbol{G}$ is a positive definite matrix. Rearranging (A11.2.10) we have

$$\boldsymbol{S}^*_{n,n,Q} = \boldsymbol{S}^*_{n,n} + \boldsymbol{G} \tag{A11.2.11}$$

The diagonal elements of $\boldsymbol{S}^*_{n,n}$ and $\boldsymbol{G}$ are all positive, and so it follows that the diagonal elements of $\boldsymbol{S}^*_{n,n,Q}$ must all be greater than the diagonal elements of $\boldsymbol{S}^*_{n,n}$ which was assumed to be CR consistent. Hence $\boldsymbol{S}^*_{n,n,Q}$ cannot be CR consistent, and the proof is complete. The Kalman/Swerling filters with a $\boldsymbol{Q}$ matrix cannot be CR consistent.

■■

In the same way, (A11.2.11) shows that the diagonal elements of $\boldsymbol{S}^*_{n,n,Q}$ on the left will be larger that those of $\boldsymbol{S}^*_{n,n}$ on the right *regardless of whether or not* the latter is CR consistent. The covariance matrix of a Kalman filter with a $\boldsymbol{Q}$ matrix must therefore grow steadily when compared to one without a $\boldsymbol{Q}$ matrix, and so the filter with a $\boldsymbol{Q}$ matrix must move steadily away from CR consistency when cycled repeatedly.

Appendix 11.3

Numerical evidence that the three filters are equivalent

In Section 11.4 we made the following two statements:

1. If we exclude runs in which the Kalman/Swerling filters are spontaneously ECM inconsistent, then they are both ECM consistent.

2. If we exclude runs in which the Kalman/Swerling filters are spontaneously ECM inconsistent, then they are both CR consistent.

In this appendix we present numerical evidence which supports these statements.

13_Gauss_Newton_C

In this program, high data-rate samples of a sinusoidal waveform whose frequency is unknown are pre-filtered, and periodic samples of the pre-filter's output are then fed to three filters – Kalman, Swerling and Gauss–Newton. The objective is for these filters to estimate the unknown frequency. The filter models are the autonomous nonlinear DE that appears in (2.1.36):

In Figure A11.3.1 below we show values from ***13_Gauss_Newton_C***.

1. The Kalman and Swerling filters have been cycled *2502* times, and so by now they are running in steady state, i.e. free of possible effects caused by initialization.

   Their error/sigma ratios are all within the range of ± 3, and their Chi-squared values are both less than 11.35. They are thus both ECM consistent.

   This confirms Statement 1 above.

2. We are considering a run in which Kalman and Swerling are not spontaneously inconsistent. Their sigma values are very close to those of Gauss–Newton, which is known to be always CR consistent. They do not contain $\boldsymbol{Q}$ matrices. Both Kalman and Swerling are CR consistent.

   This confirms Statement 2 above.

Kalman Filter (Cycle #2502)

	Error	Sigma	Error/Sigma
x	4.4329482e-5	1.2123561e-4	.36564738
x-dot	-1.4965868e-9	5.4546688e-9	-.27436805
omega	-5.5909261e-12	1.3782116e-11	-.40566528

Chi-squared = 1.8935706162522

Swerling Filter (Cycle #2502)

	Error	Sigma	Error/Sigma
x	4.4329482e-5	1.2123561e-4	.36564738
x-dot	-1.4965868e-9	5.4546688e-9	-.27436805
omega	-5.5909261e-12	1.3782116e-11	-.40566528

Chi-squared = 1.8935706162031

Gauss-Newton

	Error	Sigma	Error/Sigma
x	7.7454213e-5	1.2274131e-4	.63103623
x-dot	-5.312372e-9	5.4294942e-9	-.97842851
omega	-5.6928851e-12	1.3781705e-11	-.41307554

Chi-squared = 1.9958918 (11.35)

*Figure A11.3.1 Kalman and Swerling filters without a **Q** matrix*

14_Orbit_Track

We examine a set of results from the program ***14_Orbit_Track.*** This program simulates satellite tracking by radar and operates as follows[2]:

Observations on any one of four satellites arrive at a high data rate from a tracking radar, in this case at *2000/sec*. These are first processed by three prefilter (*x*, *y* and *z*) using the polynomial filters discussed in Chapters 12 and 13. Once every *1000* prefilter cycles their 0th-derivative output values are sent to both the Kalman and Swerling filters to be processed immediately, and are also saved for later processing by the Gauss–Newton filter.

In the run appearing in Figure A11.3.2 below, the observations came from the Russian satellite *Molniya*. Tracking continued for *300* seconds, by the end of which the Kalman and Swerling filters (both without ***Q*** matrices) had processed *600* observations. Those two filters were then halted and the same *600* observations were processed by the Gauss–Newton filter, after which the results for all three filters were displayed for comparison.

[2] See the program documentation for the details on how to run the program.

Molniya

Gauss-Newton K E P L E R I A N S

	Exact	Estimate	Error	Sigma	Error/Sigma	Rel error (%)
OMEGA (deg)	180	179.97135	-2.8648892e-2	2.4251738e-2	-1.1813130e+0	-1.5916051e-2
Incl (deg)	63.4	63.388858	-1.1141720e-2	9.4072490e-3	-1.1843760e+0	-1.7573691e-2
omega (deg)	270	269.99825	-1.7513529e-3	2.3060407e-2	-7.5946314e-2	-6.4864921e-4
Eccent	.7222	.72250568	3.0567692e-4	2.2462131e-4	1.3608545e+0	4.2325799e-2
Period(hrs)	11.959994	11.957818	-2.1764087e-3	4.2811507e-3	-5.0837003e-1	-1.8197406e-2
A (km)	26555	26551.778	-3.2216452e+0	6.3373974e+0	-5.0835461e-1	-1.2131972e-2
tau (sec)	10	13.747019	3.7470189e+0	2.7364676e+0	1.3692904e+0	3.7470189e+1

Chi-squared = 5.7228641 (16.81)

TraceQ = 0

Kalman K E P L E R I A N S

	Exact	Estimate	Error	Sigma	Error/Sigma	Rel error (%)
OMEGA (deg)	180	179.97137	-2.8632815e-2	2.4250651e-2	-1.1807030e+0	-1.5907120e-2
Incl (deg)	63.4	63.388865	-1.1135482e-2	9.4068307e-3	-1.1837655e+0	-1.7563852e-2
omega (deg)	270	269.99736	-2.6382111e-3	2.3059376e-2	-1.1440948e-1	-9.7711524e-4
Eccent	.7222	.72250055	3.0055272e-4	2.2461117e-4	1.3381023e+0	4.1616272e-2
Period(hrs)	11.959994	11.95756	-2.4338475e-3	4.2809296e-3	-5.6853248e-1	-2.0349906e-2
A (km)	26555	26551.397	-3.6027338e+0	6.3371156e+0	-5.6851319e-1	-1.3567064e-2
tau (sec)	10	14.063274	4.0632744e+0	2.7363436e+0	1.4849284e+0	4.0632744e+1

Chi-squared = 5.8892389 (16.81)

*Figure A11.3.2 Keplerian values for Molniya from **14_Orbit_Track***

Note: To save space we do not include the Swerling values. In every case they are almost identical to the Kalman values.

1. We are considering a run in which Kalman and Swerling are not spontaneously inconsistent (unstable). Their error/sigma ratios are all within the range of ± 3, and their Chi-squared values are both less than 16.81. They are thus both ECM consistent.

   This confirms Statement 1 above.

2. We are considering a run in which Kalman and Swerling are not spontaneously inconsistent. Their sigma values are very close to those of Gauss–Newton, which is known to be always CR consistent. Thus both Kalman and Swerling are CR consistent.

   This confirms Statement 2 above.

Appendix 11.4

Growth in the CR inconsistency of the extended Kalman and Swerling filters that include Q matrices

When Kalman/Swerling filters that include a $\boldsymbol{Q}$ matrix are cycled repeatedly, their accuracy falls behind that of a Gauss–Newton filter that is processing the same data.

The Gauss–Newton filter is known to be CR consistent, and so this relative loss of accuracy by the Kalman/Swerling filters means that they are becoming progressively more and more CR inconsistent.

Figure A11.4.1 below was produced by the program ***13_Gauss_Newton_C.*** The objective of the program is to estimate the value of the (unkown) frequency of a sine wave.

As examples, the figure shows that when $N = 20{,}000$, the Gauss–Newton estimate of the unknown frequency is about *56* times more accurate than the Kalman estimate, and when $N = 70{,}000$, the Gauss–Newton estimate is about *115* times more accurate than the Kalman estimate.

The $\boldsymbol{Q}$ matrix that was used for this run is the default that appears in the program, and is as follows:

$$\boldsymbol{Q} = diag(5e\text{-}11, \quad 1e\text{-}14, \quad 1e\text{-}100) \tag{A11.2.12}$$

This matrix was arrived at by a lengthy trial and error process, and it appeared to have the smallest diagonal elements that could prevent instability from occuring.

The Swerling filter gave almost exactly the same results as the Kalman filter, and so the above graph applies also to the Swerling filter.

For a complete discussion on how Figure A11.4.1 was obtained, see Project 11.9.

We also used the program ***14-Orbit_Track*** to investigate how the Kalman and Swerling filters with a $\boldsymbol{Q}$ matrix lose their CR consistency.

The results are shown in Figures A11.4.2 and A11.4.3 below.

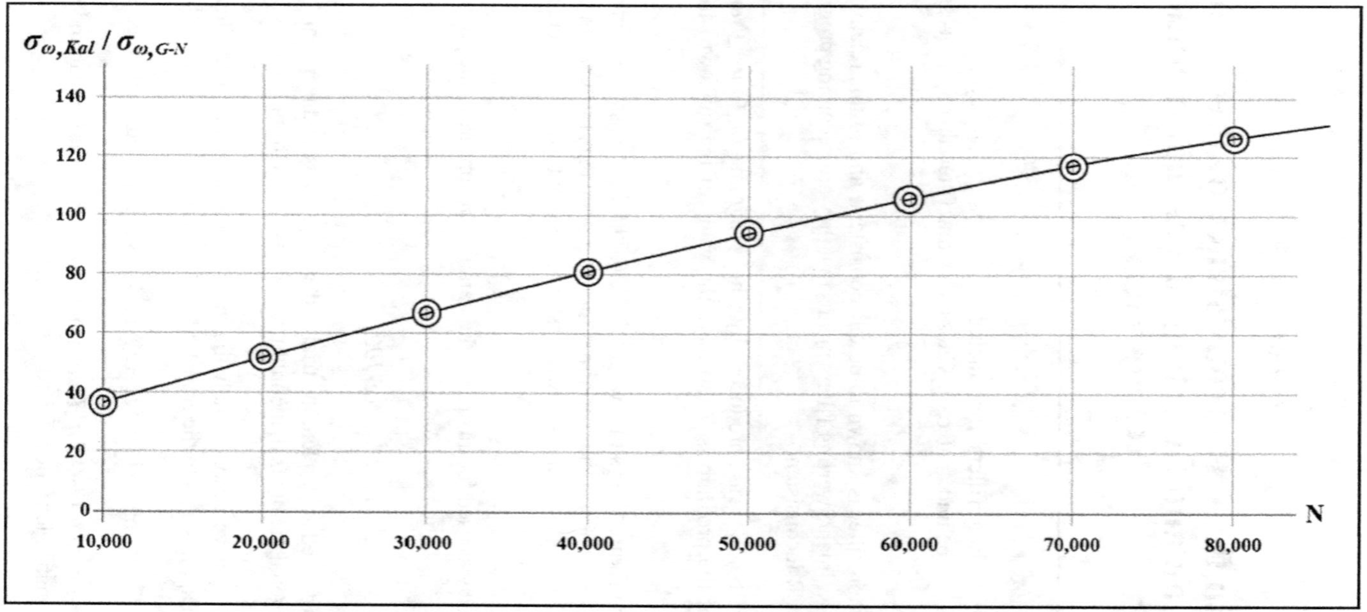

Figure A11.4.1 Number of times that Gauss–Newton's accuracy is better than Kalman's

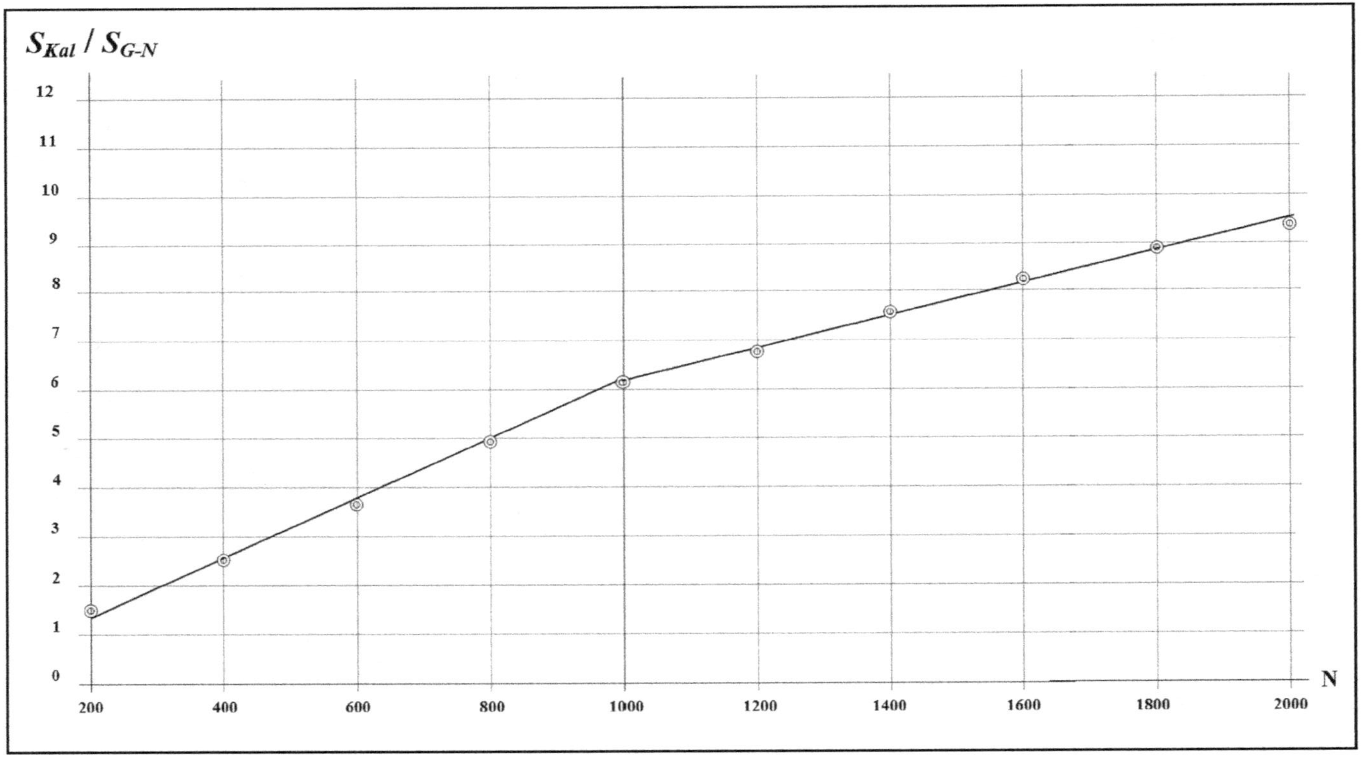

Figure A11.4.2 Number of times that Gauss–Newton's average x, y and z accuracy is better than Kalman's

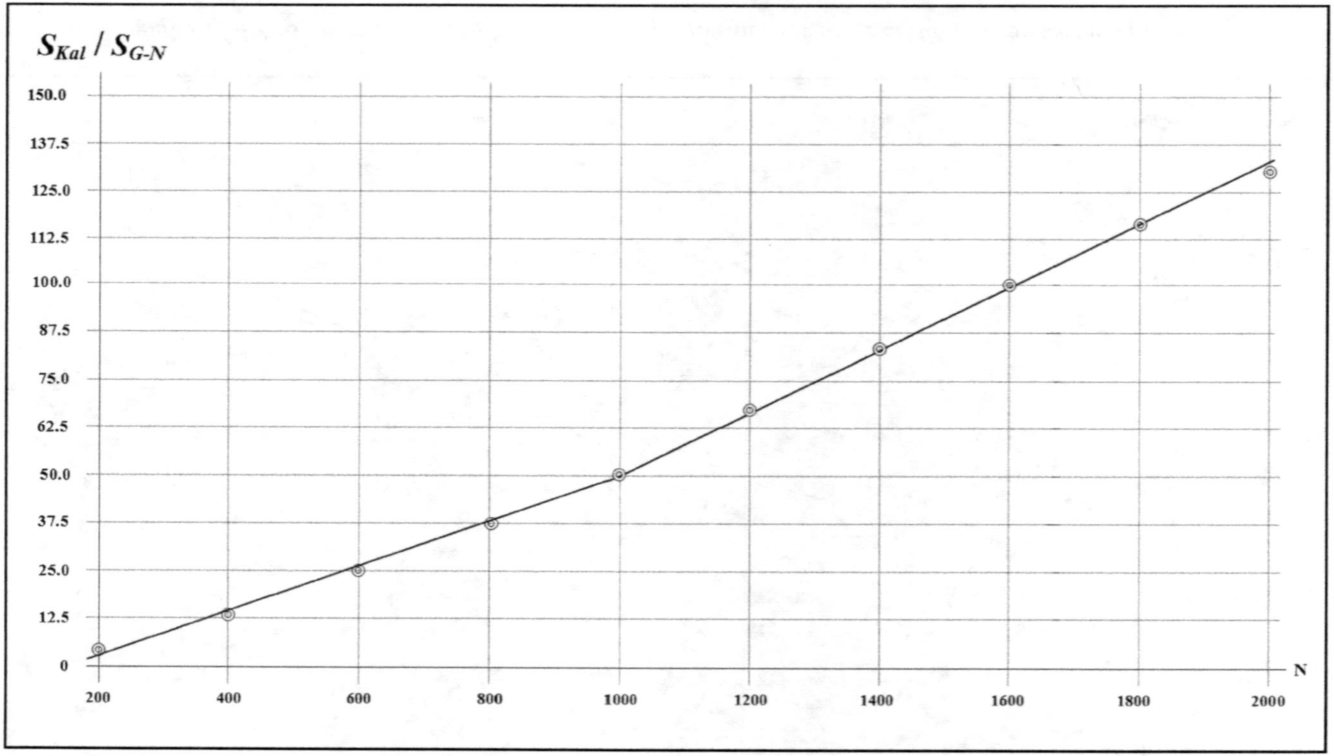

Figure A11.4.3 Number of times that Gauss–Newton's average $\dot{x}$, $\dot{y}$ and $\dot{z}$ accuracy is better than Kalman's

As an example, we see from Figure A11.16.3 that when $N = 1600$, on average the Gauss–Newton $\dot{x}$, $\dot{y}$ and $\dot{z}$ estimates are *100* times more accurate than those of Kalman.

For a complete discussion on how Figures A11.4.2 and A11.4.3 were obtained, see Project 11.9.

The $\boldsymbol{Q}$ matrix that was used for the run that gave these two plots is the default that appears in the program, and is as follows:

$$\boldsymbol{Q} = diag(0.25, \quad 0.25, \quad 0.25, \quad 0.025, \quad 0.025, \quad 0.025)$$

This matrix was arrived at by a lengthy trial and error process, and it appeared to have the smallest diagonal elements that could prevent instability from occuring.

I do not have an explanation for the change in slope that takes place at $N = 1000$ in the above two figures.

Chapter 12

Polynomial filtering – 1

12.1 Overview

The polynomial filters are based on the orthogonal polynomials of Legendre and Laguerre.

Orthogonal polynomials are widely used in applied mathematics, physics and engineering, and the Legendre and Laguerre polynomials are only two of infinitely many sets, each of which has its own *weight function.*

◊ The Legendre polynomials have a *uniform* weight function that leads naturally to filters with expanding memories. These are called the EMP (expanding memory polynomial) filters.

◊ The Laguerre polynomials have an *exponentially decaying* weight function that leads naturally to filters with fading memories. These are called the FMP (fading memory polynomial) filters.

A considerable amount of material on the polynomial filters can be found in Reference 6 which is where we first wrote them up.[1] However, over the years we have learned a great deal more about them, and in this chapter and the next we provide a self-contained discussion that includes most of that new material.

Note: In Project 9.8 we present a brief introduction to the EMP filters and how they can be used to initialize the Gauss–Newton filters. If you ever have to engineer a Gauss–Newton filter then that material will be of use.

If you are reading this chapter for the first time then we suggest that you follow the instructions in the box below entitled ***Polynomial filter quick start***. Running the two programs mentioned there will make it much easier to understand the material contained in this chapter and the next.

[1] See also References 24, 64 to 69, 112, 118 and 202.

Polynomial filter quick-start

Run the two programs

05_Pol_Filter_EMP.exe and ***06_Pol_Filter_FMP.exe***

Quick-start information is contained in the files

05_Quickstart.pdf and ***06_Quickstart.pdf***

Full documentation on the programs is contained in

05_Readme.pdf and ***06_Readme.pdf***

12.2 The polynomial filters and their properties

12.2.1 The five assumptions

The EMP and FMP filters are based on the following five assumptions:

1. The input is a sequence of *scalar values*.

2. The input values are spaced by a *constant sampling interval*.

3. The errors in the input values have *constant variance*.

4. The errors are *stage-wise uncorrelated*.

5. Smoothing is done by fitting a polynomial using *least squares*.

We will use the following five symbols:

1. The input values will be called $\ldots, y_{n-2}, y_{n-1}, y_n$.

2. The constant sampling interval will be called τ.

3. The errors in the input values will be called $\ldots, \nu_{n-2}, \nu_{n-1}, \nu_n$.

4. The constant variance of the input errors will be called σ_ν^2.

5. The degree of the fitted polynomial will be called m.

12.2.2 Versions

There are four versions of the polynomial filters. These are shown in Figure 12.1.

	Expanding memory	Fading memory
1-step predictor	*Version 1*	*Version 2*
Current estimate	*Version 3*	*Version 4*

Figure 12.1 The four versions

The four versions will be referred to as follows:

Version 1: 1-step predictor EMP *Version 2: 1-step predictor FMP*

Version 3: Current estimate EMP *Version 4: Current estimate FMP*

12.2.3 The algorithms

◊ The algorithms for the *1*-step predictor EMP and FMP filters are listed in Appendices 12.1 and 12.2 up to 4th degree.

◊ The algorithms for the current-estimate EMP and FMP filters are listed in Appendices 12.3 and 12.4 up to 4th degree.

◊ The *1*-step predictor EMP and FMP filters are also listed in the file entitled ***Miscellaneous\Polynomial_filters\1-step_predictors_up_to_8th_degree***.

◊ The current estimate EMP and FMP filters are also listed in the file entitled ***Miscellaneous\Polynomial_filters\Current_estimate_up_to_8th_degree***.

12.2.4 Applications

The EMP and FMP filters can be used as follows:

◊ As a pair, the *1*-step predictor EMP and FMP filters can form a high-data-rate tracking filter, both for data smoothing and for radar pedestal control (see Figure 12.2 below).

◊ As a pair, the current-estimate EMP and FMP filters can form a pre-filter for the Gauss–Newton, Kalman or Swerling filters (see Figure 12.3).

◊ The current-estimate EMP filter can be used alone as a self-initializing algorithm to initialize the Gauss–Newton, Kalman or Swerling filters (Figure 12.4).

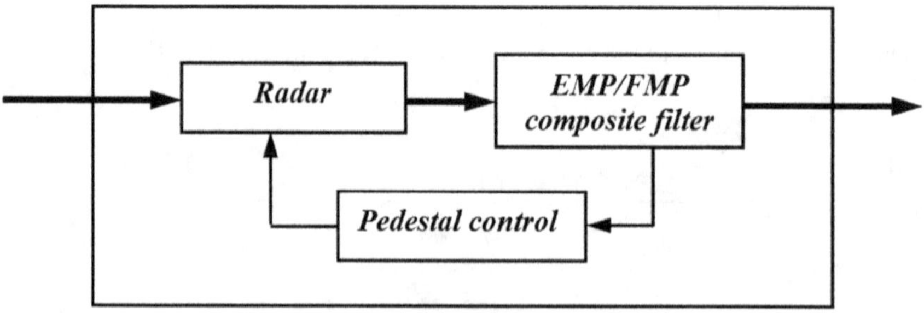

Figure 12.2 High-data-rate tracking

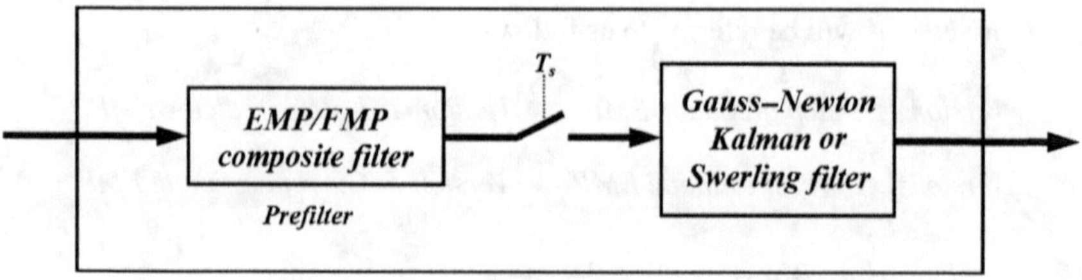

Figure 12.3 Prefiltering

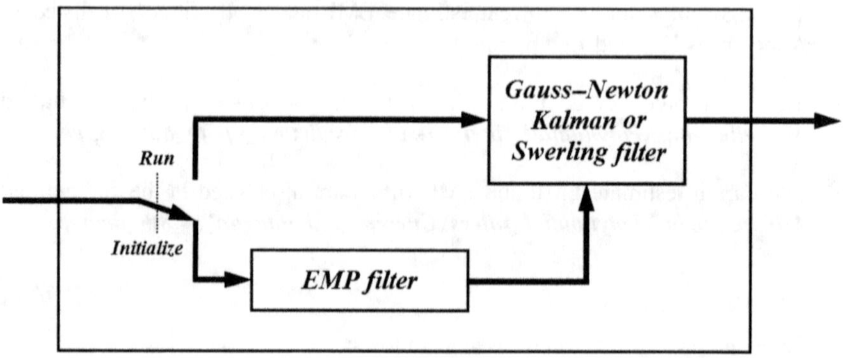

Figure 12.4 Initialization

◊ Together with a memory stack, the EMP filters can be operated as polynomial filters with fixed, variable or expanding memory lengths (see Figure 12.5 below).

◊ The EMP filters can be used to fit a polynomial of degrees *1, 2, …* by least squares to a sequence of noisy data. Derivatives can then be obtained from the fitted polynomial (see Figure 12.6).

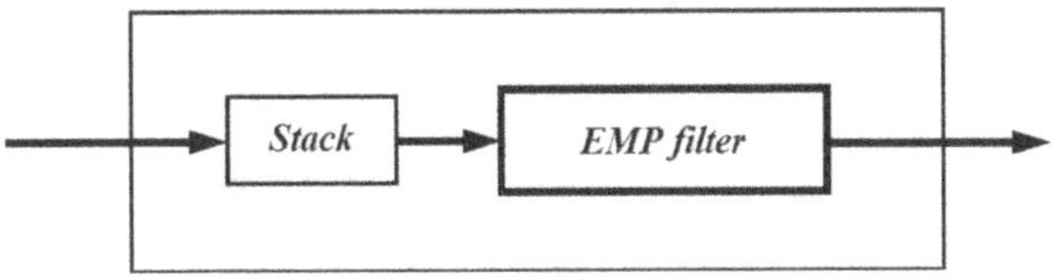

Figure 12.5 Fixed or variable memory length polynomial filtering

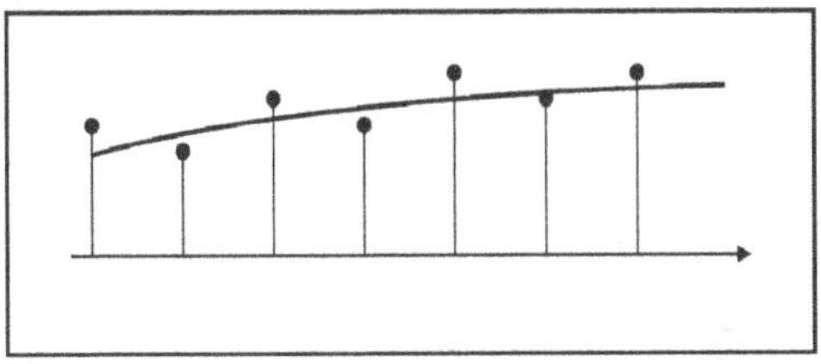

Figure 12.6 Fitting a polynomial to noisy data

12.2.5 Cycle times

Figure 12.7 shows the time in microseconds to cycle the *1*-step predictor EMP and FMP filters on the author's lap-top computer.[2,3]

Figure 12.8 shows the time in microseconds to cycle the current-estimate EMP and FMP filters on the same machine.

The cycle times appearing in Figures 12.8 and 12.9 are for compiled True Basic. They would be substantially less if a more powerful programming language were used.

If the number of times that the EMP filters will be cycled is limited to a value that is not too large, then their cycle times can be reduced to the same values as the FMP filters by pre-computing their *weights* (see lower down) and storing them in look-up tables.

Clearly, the EMP and FMP algorithms are extremely fast-running, and if need be, many hundreds or even thousands of them could be cycled without over-extending a modern desk-top or lap-top computer.

Note 1: The cycle times of the EMP algorithms can double if they are not properly coded. In Project 12.13 we consider how this comes about and how to avoid it.

Note 2: You can find the times shown in Figures 12.7 and 12.8 for your own computer by following the instructions given in Project 12.1.

[2] *Dell Studio XPS 1640* with dual Intel processors, rating *Core 2 CP P8700 @ 2.53 GHz, Base Score 5.9.*
[3] Computer language used was compiled True BASIC.

1-step predictor cycle time (μsec)		
Degree	*EMP*	*FMP*
0	*0.9*	*0.9*
1	*1.5*	*1.3*
2	*2.3*	*1.8*
3	*3.3*	*2.4*
4	*4.3*	*3.1*
5	*5.7*	*3.9*

Figure 12.7 1-step predictor EMP and FMP cycle times in microseconds

Current-estimate cycle time (μsec)		
Degree	*EMP*	*FMP*
0	*1.1*	*1.1*
1	*2.2*	*1.9*
2	*3.4*	*2.6*
3	*4.7*	*3.4*
4	*6.9*	*4.4*
5	*8.1*	*5.5*

Figure 12.8 Current-estimate EMP and FMP cycle times in microseconds

12.2.6 The filter inputs

In Figure 12.9 below we show the input to a polynomial filter as the scalar whose generic name is y_n.

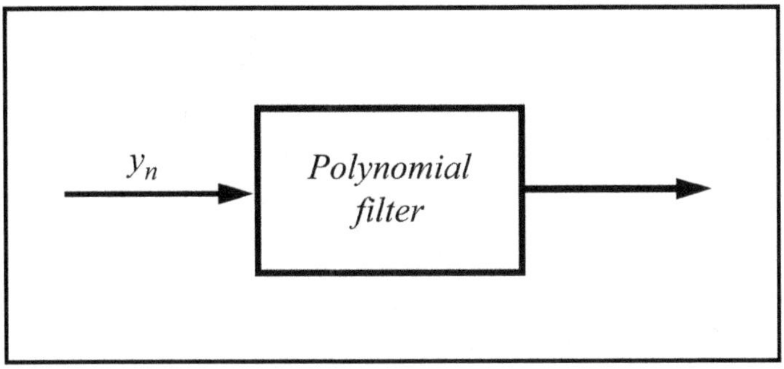

Figure 12.9 Filter input

For example:

◊ y_n could be the x coordinate of an object.

◊ y_n could be the y coordinate of an object.

◊ y_n could be the azimuth angle of an object.

◊ y_n could be the Doppler shift caused by a moving object.

Each item that we wish to smooth requires its own filter.

12.2.7 The sampling interval

The derivation of the EMP and FMP filters is based on the assumption that the observations are equally spaced on the time axis.

If there is a slight departure from that assumption – known as time-jitter – then they can still be used as they stand. However, see Ref. 112 if you want to take time-jitter fully into account.

12.2.8 Block diagram

In Figure 12.10 below we show the polynomial filtering block-diagram.

◊ On the far left we see $\boldsymbol{X}(t)$, the true state vector of an observed process. It is always assumed that $\boldsymbol{X}(t)$ has the following structure:

$$\boldsymbol{X}(t) = \left(x(t),\ \dot{x}(t),\ \ddot{x}(t), \ldots D^m x(t)\right)^T \tag{12.2.1}$$

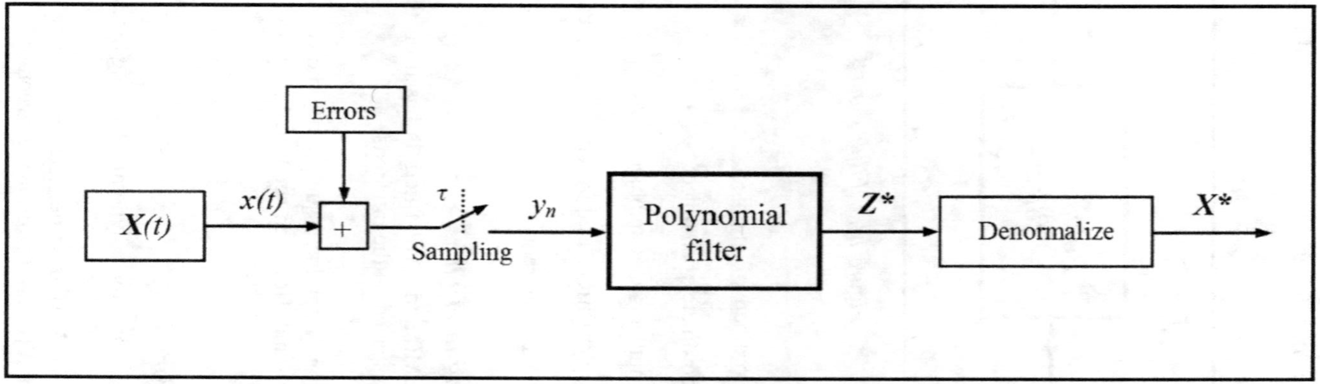

Figure 12.10 Polynomial filtering block diagram

◊ At time t_n this vector $\boldsymbol{X}$ is written as follows:

$$\boldsymbol{X}(t_n) = (x_0,\ x_1,\ x_2, \dots x_m)_n^T \tag{12.2.2}$$

◊ We wish to obtain smoothed estimates of $\boldsymbol{X}(t)$, which will be represented as

$$\boldsymbol{X}^* = (x_0^*,\ x_1^*,\ x_2^*, \dots x_m^*)^T \tag{12.2.3}$$

The elements of $\boldsymbol{X}^*$ therefore have the following meanings: x_0^* is the smoothed estimate of $x(t)$, x_1^* the smoothed estimate of $\dot{x}(t)$ and x_m^* is the smoothed estimate of $D^m x(t)$.

◊ We are using a polynomial of degree m as our filter model, and estimation is based on least squares.

◊ The observed trajectory is the 0th derivative of $\boldsymbol{X}(t)$ and is shown in Figure 12.10 as $x(t)$.

◊ $x(t)$ is sampled every τ seconds resulting in the observations $\dots y_{n-2}, y_{n-1}, y_n \dots$ where y_n is related to $x_{0,n}$ (the first element of $\boldsymbol{X}(t_n)$ in (12.2.2)) by the observation equation

$$y_n = x_{0,n} + \nu_n \tag{12.2.4}$$

◊ The quantity ν_n in (12.2.4) is the observation error with constant variance σ_ν^2.

◊ The sequence $\dots y_{n-2}, y_{n-1}, y_n \dots$ is the input to the filter.

◊ The output of the polynomial filter is either the vector $\boldsymbol{Z}_{n+1,n}^*$ (*1*-step predictor) or the vector $\boldsymbol{Z}_{n,n}^*$ (current-estimate). The elements of the *1*-step predictor are

$$\boldsymbol{Z}_{n+1,n}^* = \left(x_0^*,\ \tau x_1^*,\ (\tau^2/2!)x_2^*, \dots (\tau^m/m!)x_m^*\right)_{n+1,n}^T \tag{12.2.5}$$

and of the current-estimate

$$\boldsymbol{Z}_{n,n}^* = \left(x_0^*,\ \tau x_1^*,\ (\tau^2/2!)x_2^*, \dots (\tau^m/m!)x_m^*\right)_{n,n}^T \tag{12.2.6}$$

The vectors $\boldsymbol{Z}^*$ that have this structure were first considered in (2.3.12), and are called *normalized* vectors because of the way in which the sampling interval τ has been incorporated into them. As a result of that normalization the algorithms operate internally in an extremely efficient way that is completely independent of τ.

◊ When the EMP and FMP filters are used in conjunction with radar and we cycle them at the same rate as the radar's pulse repetition frequency (prf), then

$$\tau = 1/prf \tag{12.2.7}$$

As an example: Let $prf = 2000 \;\; pps$. Then

$$\tau = 1/2000 = 0.5\, msec \tag{12.2.8}$$

◊ The final block in Figure 12.10 shows that the vector $\boldsymbol{Z}^*$ must be *denormalized* in order to obtain the desired estimate vector $\boldsymbol{X}^*$. Denormalization is explained in the following example.

Example 12.1: In the algorithms appearing in Appendices 12.1 to 12.4, the elements of $\boldsymbol{Z}^*$ have the following names (assuming the 2nd-degree case):

◊ z_0^* is the normalized 0th derivative.

◊ z_1^* is the normalized 1st derivative.

◊ z_2^* is the normalized 2nd derivative, and so on.

Referring to (12.2.5) or (12.2.6) we see that for a 2nd-degree filter $(m = 2)$, the denormalized estimate vector $\boldsymbol{X}^*$ would be obtained from the filter output vector $\boldsymbol{Z}^*$ using the following denormalization equation:

$$\boldsymbol{X}^* = \begin{pmatrix} x_0^* \\ x_1^* \\ x_2^* \end{pmatrix} = \begin{pmatrix} z_0^* \\ (1/\tau)z_1^* \\ (2/\tau^2)z_2^* \end{pmatrix} \tag{12.2.9}$$

From (12.2.9) we see that the denormalization equation for both the *1*-step predictor and current-estimate filters is

$$\boldsymbol{X}^* = \boldsymbol{D}(\tau)\,\boldsymbol{Z}^* \tag{12.2.10}$$

in which $\boldsymbol{D}(\tau)$ is the diagonal matrix whose elements are

$$\left[\boldsymbol{D}(\tau)\right]_{i,i} = i!/\tau^i \quad 0 \le i \le m \tag{12.2.11}$$

Thus

$$D(\tau) = diag(0!/\tau^0,\ \ 1!/\tau^1,\ \ 2!/\tau^2, \ldots m!/\tau^m) \tag{12.2.12}$$

■■

Note: Extreme caution must be exercised with regard to denormalization, both for the estimate vectors as well as for their covariance matrices. (See Project 12.12 in the supplementary material for comments regarding denormalization in the computer programs.)

12.2.9 Two ways to obtain $\boldsymbol{Z}^*$ and $\boldsymbol{X}^*$

The output vector $\boldsymbol{Z}^*$ can be obtained in two ways:

1. By direct implementation of the algorithm of selected degree – either the *1*-step predictor or the current-estimate equations as given in the appendices.

2. By use of the transition equation to shift the validity instant. Thus:

   ◊ The *1*-step prediction can be obtained from the current estimate output using

   $$\boldsymbol{Z}^*_{n+1,\,n} = \boldsymbol{\Phi}(1)\boldsymbol{Z}^*_{n,\,n} \tag{12.2.13}$$

   ◊ The current estimate can be obtained from the *1*-step predictor output using

   $$\boldsymbol{Z}^*_{n,\,n} = \boldsymbol{\Phi}(-1)\boldsymbol{Z}^*_{n+1,\,n} \tag{12.2.14}$$

Method 2 will reduce the computation time if both $\boldsymbol{Z}^*_{n,\,n}$ and $\boldsymbol{Z}^*_{n+1,\,n}$ are required.

The general expression for the matrices $\boldsymbol{\Phi}(1)$ and $\boldsymbol{\Phi}(-1)$ appearing in (12.2.13) and (12.2.14) is given in (2.3.14) for any degree. Assuming that the filters are of 2nd degree, the matrices are shown in (2.3.17) and (2.3.18).

Once the required $\boldsymbol{Z}^*$ has been obtained, the related $\boldsymbol{X}^*$ is then obtained by denormalization as shown in (12.2.10).

12.2.10 Basic structure of the algorithms

The basic structure of the *1*-step predictor EMP is shown in the following set of equations from Appendix 12.1:

2nd-degree 1-step predictor EMP

$$
\begin{aligned}
e_n &= y_n - z_0{}^*_{n,n-1} \\
z_2{}^*_{n+1,n} &= z_2{}^*_{n,n-1} + \gamma e_n \\
z_1{}^*_{n+1,n} &= z_1{}^*_{n,n-1} + 2z_2{}^*_{n+1,n} + \beta e_n \\
z_0{}^*_{n+1,n} &= z_0{}^*_{n,n-1} + z_1{}^*_{n+1,n} - z_2{}^*_{n+1,n} + \alpha e_n
\end{aligned}
\tag{12.2.15}
$$

◊ The *1*-step predictor FMP has precisely the same basic structure, and so there is no need to repeat its equations. The only differences between the EMP and FMP algorithms are the expressions for α, β, γ appearing in (12.2.15). These are called the *filter weights*.

◊ From Appendices 12.1 and 12.3 we see that the weights for the current-estimate EMP filter are the same as the weights for the *1*-step predictor EMP filters for all degrees. However, the basic structures of the current estimate and *1*-step predictor EMP algorithms differ.

◊ Similarly, the weights for the current-estimate FMP filters are the same as the weights for the *1*-step predictor FMP filters for all degrees, but the basic structures of the current estimate and *1*-step predictor FMP algorithms differ.

◊ In Problem 12.13 we consider how to convert a filter (EMP or FMP) from its *1*-step predictor version to its current estimate version. By working that problem you will come to understand how the basic structures of the two versions of the algorithms come about.

The fact that the basic structures of the *1*-step predictor EMP and FMP filters are the same means that we can switch seamlessly from EMP to FMP simply by changing the weights. We will see further down how this is used to create the composite EMP/FMP filter shown Figure 12.11 below. The meaning of N_s in the figure will be explained at that time.

12.2.11 Combing

Returning to the 2nd-degree *1*-step predictor algorithm shown in (12.2.15):

◊ The first equation shows that the quantity e_n is computed as the difference between the most recent observation and the prediction of what that observation will be. The quantity e_n is then used in the next three equations, each time multiplied by the appropriate *filter weight*.

◊ In the second equation we see the previous value $z_2{}^*_{n,n-1}$ on the right being used to produce the new value $z_2{}^*_{n+1,n}$ on the left. The quantity $z_2{}^*_{n,n-1}$ will not be needed again and so it can be overwritten by $z_2{}^*_{n+1,n}$.

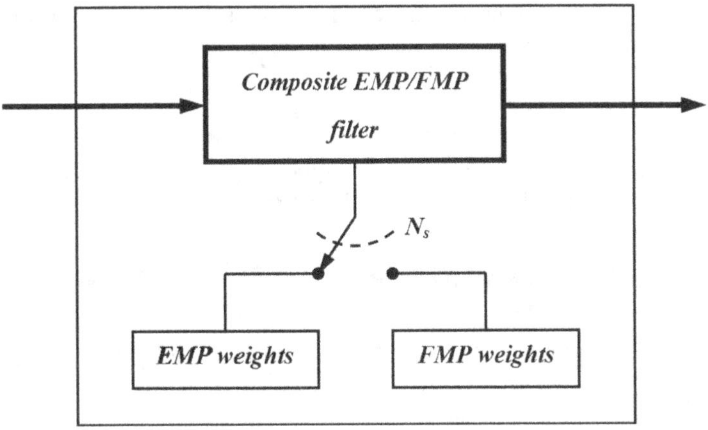

Figure 12.11 Composite EMP/FMP filter

◊ In the third equation we see the previous value $z_1{}^*_{n,n-1}$ on the right being used to produce the new value $z_1{}^*_{n+1,n}$ on the left. The quantity $z_1{}^*_{n,n-1}$ will also not be needed again and so it too can be overwritten by $z_1{}^*_{n+1,n}$.

◊ Finally in the fourth equation we see the previous value $z_0{}^*_{n,n-1}$ being used to produce the latest value $z_0{}^*_{n+1,n}$. The quantity $z_0{}^*_{n,n-1}$ can also be overwritten by $z_0{}^*_{n+1,n}$.

Taking all of the above into account, the algorithm would be implemented in computer code as follows:

2nd-degree 1-step predictor EMP

$$e = y - z_0^*$$

$$z_2^* = z_2^* + \gamma e$$

$$z_1^* = z_1^* + 2z_2^* + \beta e \tag{12.2.16}$$

$$z_0^* = z_0^* + z_1^* - z_2^* + \alpha e$$

This structure, in which each new estimate overwrites its previous value, applies to both the *1*-step predictor EMP and the *1*-step predictor FMP filters. We refer to it as *combing.*

Combing makes for maximum speed. It also makes for saving of RAM, but that is less important these days than it was in 1966, when these algorithms were first derived.

◆

In the current-estimate versions given in Appendices 12.3 and 12.4 we see that combing is not present. Thus, from Appendix 12.3:

2nd-degree current estimate EMP

$$e_n = y_n - (z_0{}^*_{n-1,n-1} + z_1{}^*_{n-1,n-1} + z_2{}^*_{n-1,n-1})$$

$$z_2{}^*_{n,n} = z_2{}^*_{n-1,n-1} + \gamma e_n$$

$$z_1{}^*_{n,n} = z_1{}^*_{n-1,n-1} + 2z_2{}^*_{n-1,n-1} + \beta e_n \tag{12.2.17}$$

$$z_0{}^*_{n,n} = z_0{}^*_{n-1,n-1} + z_1{}^*_{n-1,n-1} + z_2{}^*_{n-1,n-1} + \alpha e_n$$

The current-estimate FMP has precisely the same basic structure, and so there is no need to repeat its equations. The only difference between the current-estimate EMP and FMP algorithms is in the expressions for their weights.

◊ The first equation in (12.2.17) shows that the quantity e_n is now computed as the difference between the most recent observation and the sum of $z_0{}^*_{n-1,n-1}$, $z_1{}^*_{n-1,n-1}$ and $z_2{}^*_{n-1,n-1}$. (This sum is in fact the prediction of what that observation will be.) The quantity e_n is then used in the next three equations, each time multiplied by the appropriate filter *weight.*

◊ The filter weights for the current estimate EMP algorithms are the same as the filter weights for the *1*-step predictor EMP.

◊ The filter weights for the current estimate FMP algorithms are the same as the filter weights for the *1*-step predictor FMP.

◊ In (12.2.17) we have the previous values on the right again being used to produce new values on the left. However, here the previous values are required in more than one equation, and so they cannot be overwritten until the entire algorithm has been completed.

Taking all of the above into account, the algorithm would be implemented in computer code as shown in (12.2.18).

2nd-degree current estimate EMP

$$
\begin{aligned}
e &= y - (z_0^* + z_1^* + z_2^*) \\
Temp_2 &= z_2^* + \gamma e \\
Temp_1 &= z_1^* + 2z_2^* + \beta e \\
Temp_0 &= z_0^* + z_1^* + z_2^* + \alpha e \qquad (12.2.18) \\
z_2^* &= Temp_2 \\
z_1^* &= Temp_1 \\
z_0^* &= Temp_0
\end{aligned}
$$

The extended first line and the final three lines in these equations result in slightly longer execution times for the current-estimate filters than for the *1*-step predictors (see Figures 12.7 and 12.8).

This could be of significance if many thousands of current-estimate filters were being cycled on the same machine.

12.2.12 The EMP filters are self-initializing

Even though they are recursive, the EMP filters are completely self-initializing. This is true for both the *1*-step predictor and the current-estimate versions.

This property is highly unusual for recursive algorithms which are typically not self-initializing, and it makes the EMP filters extremely valuable in many ways. We now examine this property, but first we require the following:

Definition 12.1: The quantity $n^{(m)}$ is the product of m factors starting with n followed by $(n - 1), (n - 2)\ldots$ as follows:

$$n^{(m)} = \overbrace{n(n-1)(n-2)\ldots(n-m+1)}^{m\text{ factors}} \qquad (12.2.19)$$

Each factor appearing in (12.2.19) is *1* less than its predecessor. ■■

Example 12.2:

$(n + 4)^{(2)} = (n + 4)(n + 3)$ and is read as '$(n + 4)$ *count-down 2*'.

Compare this to

$(n+4)^2 = (n+4)(n+4)$ that is read as '$(n+4)$ *to-the-power-of 2*'.

In both cases two factors are being multiplied together.

◊ $(n+2)^{(3)} = (n+2)(n+1)n$ is seen to be the product of three factors, each of which is 1 less than its predecessor.

◊ Observe that $n^{(m)}$ is always equal to zero if n is a positive integer and $m > n$. As examples for $n = 4$:

- ◊ If $m = 4$ then $4^{(4)} = 4 \times 3 \times 2 \times 1 = 4!$
- ◊ If $m = 5$ then $4^{(5)} = 4 \times 3 \times 2 \times 1 \times 0 = 0$

◊ Observe also that

$$\binom{j}{v} = \frac{j!}{v!(j-v)!} = j^{(v)}/v! \tag{12.2.20}$$

We will make use of (12.2.20) in Chapter 13. ■■

Consider now the following filter shown in Appendix 12.1:

1st-degree 1-step predictor EMP

$$\begin{aligned} z_1{}^*{}_{n+1,n} &= z_1{}^*{}_{n,n-1} + \left(6/(n+2)^{(2)}\right)\left(y_n - z_0{}^*{}_{n,n-1}\right) \\ z_0{}^*{}_{n+1,n} &= z_0{}^*{}_{n,n-1} + z_1{}^*{}_{n+1,n} \\ &\quad + \left(2(2n+1)/(n+2)^{(2)}\right)\left(y_n - z_0{}^*{}_{n,n-1}\right) \end{aligned} \tag{12.2.21}$$

Important: *n always starts at zero for the EMP filters.*

Setting $n = 0$ for the first cycle gives us

$$\begin{aligned} z_1{}^*{}_{1,0} &= z_1{}^*{}_{0,-1} + 3\left(y_0 - z_0{}^*{}_{0,-1}\right) \\ z_0{}^*{}_{1,0} &= z_0{}^*{}_{0,-1} + z_1{}^*{}_{1,0} + 1\left(y_0 - z_0{}^*{}_{0,-1}\right) \end{aligned} \tag{12.2.22}$$

We must now provide the initial values $z_0{}^*_{0,\,-1}$ and $z_1{}^*_{0,\,-1}$. However, the EMP filters are self-initializing and so it is completely immaterial what initial values are used.

◊ The initial values used in the *m*th-degree EMP filter will be completely discarded after precisely $m + 1$ cycles. The algorithm's output will then be correctly based on the *m*th-degree polynomial that passes through the $m + 1$ data values that have been incorporated thus far.

◊ This self-initialization property is present in both the *1*-step predictor EMP as well as the current-estimate EMP versions.

A general proof that the EMP filters are self-initializing is provided in Reference 6 starting on p. 367.

Example 12.3: We show that the *1*-step predictor 0th-degree EMP filter is self-initializing. Its algorithm is given in Appendix 12.1 as follows:

$$z_0{}^*_{n+1,\,n} = z_0{}^*_{n,\,n-1} + \big(1/(n+1)\big)(y_n - z_0{}^*_{n,\,n-1}) \tag{12.2.23}$$

Setting $n = 0$ for the first cycle we obtain

$$z_0{}^*_{1,\,0} = z_0{}^*_{0,\,-1} + 1(y_0 - z_0{}^*_{0,\,-1}) = y_0 \tag{12.2.24}$$

This shows that whatever was used for $z_0{}^*_{0,\,-1}$ has been discarded after the first cycle.

Setting $n = 1$ for the second cycle gives

$$\begin{aligned} z_0{}^*_{2,\,1} &= z_0{}^*_{1,\,0} + \tfrac{1}{2}(y_1 - z_0{}^*_{1,0}) = y_0 + \tfrac{1}{2}(y_1 - y_0) \\ &= \tfrac{1}{2}(y_0 + y_1) \end{aligned} \tag{12.2.25}$$

This is the correct prediction of $z_0{}^*$ based on the least-squares fit of a 0th-degree polynomial to two data points. The reader can verify that on the third cycle the filter output will be

$$z_0{}^*_{3,\,2} = \tfrac{1}{3}(y_0 + y_1 + y_2) \tag{12.2.26}$$

which is correct for three data points. The filter has self-initialized correctly. ■■

In Problem 12.14 we ask you to verify that the 1st-degree *1*-step predictor EMP filter shown in (12.2.21) is self-initializing, and in Problem 12.15 we ask you to show that the 1st-degree current-estimate EMP filter is also self-initializing.

12.2.13 The FMP filters are not self-initializing

The FMP filters are not self-initializing and so they must be properly initialized if we wish to avoid start-up transients that would make their outputs completely useless. The screen captures appearing in Figures 12.12 and 12.13 show what those transients can look like.

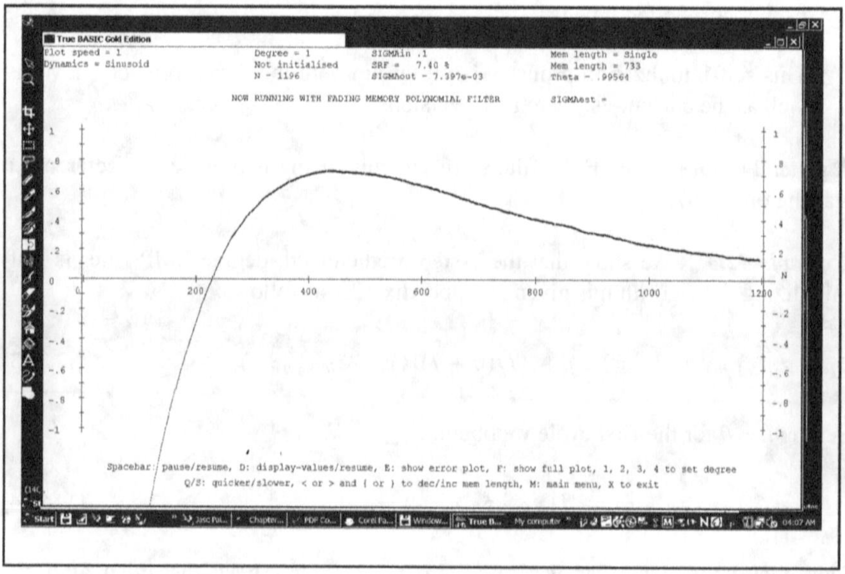

Figure 12.12 Start-up transient, 1st-degree 1-step predictor FMP filter

The inputs to the filters were samples of a low-frequency sinusoid, whose peak values were ± 1.

◊ The vertical scales in the figures run from *- 1* to *+1*.

◊ The figures show the errors in the filter outputs, i.e. the start-up transients.

For degrees *1* and *2* the transients are comparable in magnitude to the amplitude of the input sine wave, and for degrees *3* and *4* (not shown) they exceed it. Clearly, initialization will be required if we hope to make use of the FMP filters.

We will see later – when we discuss the composite EMP/FMP filters – how the EMP filters are used to initialize their FMP counterparts so that they start up without transients.

Note: To view live examples of the FMP start-up transients, see Project 12.2.

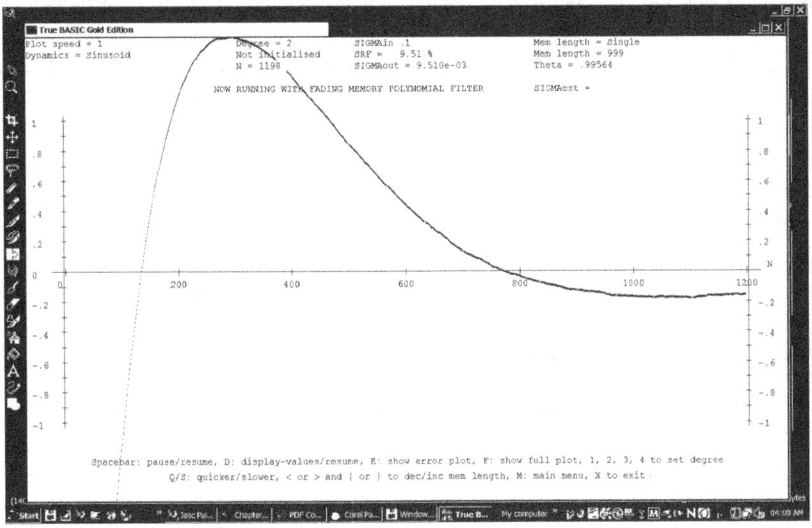

Figure 12.13 Start-up transient, 2nd-degree 1-step predictor FMP filter

12.2.14 The EMP filters have an expanding memory

In the next chapter we will see that the EMP algorithms are based on fitting a polynomial by least-squares at time t_n to the following observation vector:

$$\mathbf{Y}_n = (y_n, y_{n-1}, \ldots y_0)^T \tag{12.2.27}$$

On the next cycle at time t_{n+1} the algorithms fit a *completely new* polynomial by least-squares to the expanded total observation vector

$$\mathbf{Y}_{n+1} = (y_{n+1}, y_n, y_{n-1}, \ldots y_0)^T \tag{12.2.28}$$

and so on.

(It has been the author's experience that some people have difficulty believing that the algorithms really *do* fit a completely new polynomial to each $\mathbf{Y}$ vector on each cycle, and that each of the new polynomials is fitted all the way back to y_0. This may not be immediately obvious from the extremely compact recursive structure of the algorithms, but in Chapter 13 we will see that it definitely is true.)

Because the input sequence always starts with y_0 and ends with the most recent observation, the number of observations that goes into forming each of the EMP estimates increases steadily with n. Thus the EMP filters have an expanding memory.

The EMP least-squares weight-function at time t_n is depicted in Figure 12.14.

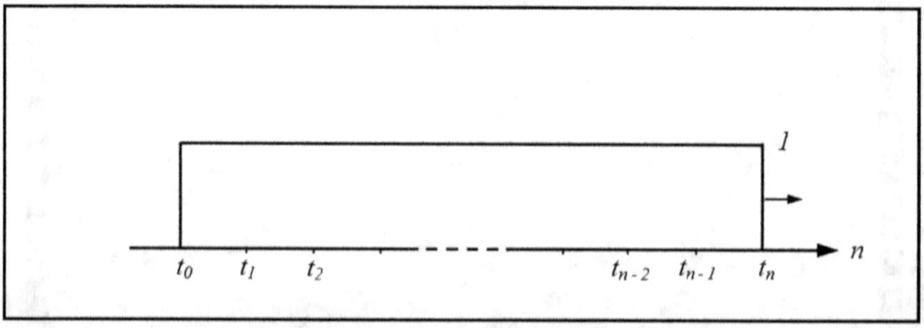

Figure 12.14 Expanding uniform weight-functions of the EMP filters

Observe that its trailing edge is always fixed at t_0 and its leading edge is at t_n. On the next cycle the leading edge will move to t_{n+1}. Its height is always *1*. All observations play an equal role in the determination of the estimate.

◊ If an *m*th-degree EMP filter is smoothing data that comes from an observed trajectory that is a polynomial of degree *m* or less, the filter model will be perfectly matched to the external model.

  The EMP filter's estimate will then be free of bias errors no matter how many times the filter is cycled. Under these circumstances the expanding memory serves to produce a sequence of estimates whose error variances decrease steadily to zero.

◊ However, if the observed trajectory is not a polynomial of degree *m* or less, then the filter model will not be perfectly matched to the external model.

  Because it is based on least squares, initially the EMP filter will track the observed trajectory with acceptable bias errors. However, because of the filter's expanding memory, sooner or later the estimate that it produces *must* develop bias errors.

To verify these statements, see Project 12.3.

12.2.15 The FMP filters have a fading memory

The FMP algorithms are based on fitting a polynomial by least-squares at time t_n to the following (conceptually) infinite-dimensional observation vector:

$$\boldsymbol{Y}_n = (y_n, y_{n-1}, y_{n-2}, \ldots)^T \tag{12.2.29}$$

On the next cycle at time t_{n+1} the algorithms fit a completely new polynomial by least-squares to the expanded infinite-dimensional total observation vector

$$\mathbf{Y}_{n+1} = (y_{n+1},\ y_n,\ y_{n-1},\ \ldots)^T \tag{12.2.30}$$

In Chapter 13 we will see that the FMP least-squares procedure uses the exponentially decaying weight-function depicted in Figure 12.15.

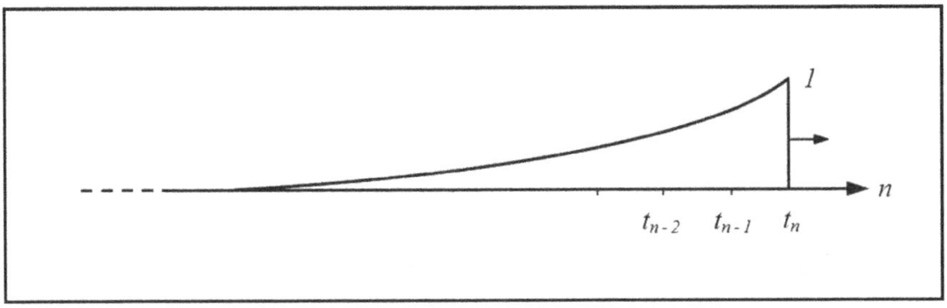

Figure 12.15 Exponentially decaying weight-function of the FMP filters

The leading edge of the function moves forward each time that the filter is cycled. The most recent observation is weighted by *1* and the preceding observations are weighted by successively smaller exponentially-decaying values, and so the FMP filters have a fading memory (see Project 12.4).

The variable n does not appear in the FMP filter weights $\alpha,\ \beta,\ \gamma,\ \ldots$ as it does in the EMP filters. Instead, the FMP filter weights are functions of the *fading parameter* θ. This is a real number selected by the user that lies in the range

$$0 < \theta < 1 \tag{12.2.31}$$

and controls the properties of the filter by controlling the exponential rate of decay of its memory.

This exponentially fading weight function gives the FMP filters an effective *fixed-length* memory (see Figure 12.16), and because of this they can be used to track data that come from *any* reasonably smooth process – polynomial or not – without significant bias errors and regardless of the number of times that they are cycled.

The end result is that, no matter what (reasonably smooth) process is being tracked, and no matter how many times an FMP filter is cycled, if its degree and fading parameter θ are properly selected, the bias errors can be held to negligible levels (see Project 12.5).

Later on we will see how the length of the FMP filter's effective fixed-length memory can be quantified, i.e. expressed as a number.

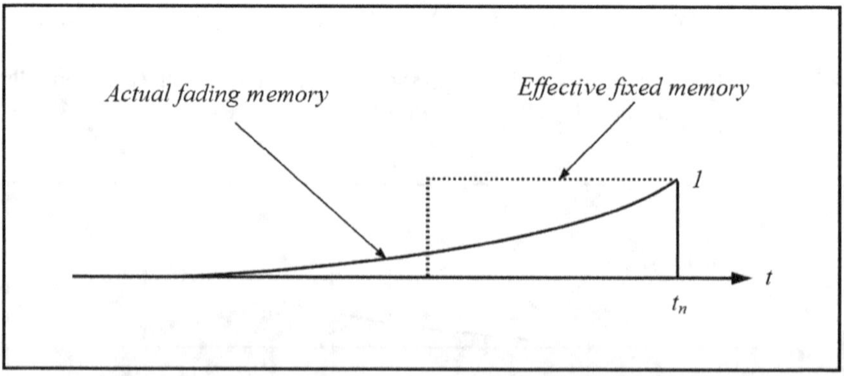

Figure 12.16 Effective fixed-length memory of the FMP filters

Example 12.4: In Figure 12.17 below we show the *z*-coordinate of a simulated military jet that is diving and spiraling simultaneously. The FMP filter can readily smooth observations obtained from such a flight-path while at the same maintaining minimal bias errors.

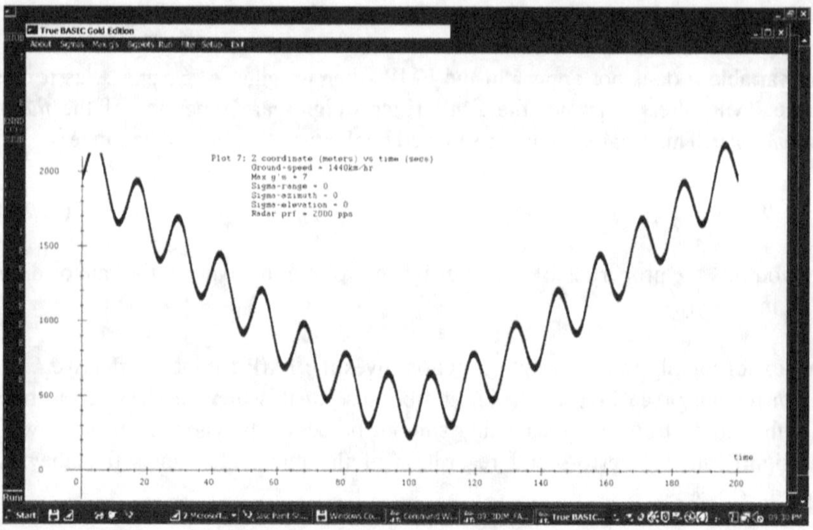

Figure 12.17 z-coordinate of a simulated military jet diving and flying in a spiral

As an example: For a target pulling *7g's* at the lowest point ($t = 100$) while also spiraling throughout the dive, and using a 3rd-degree FMP filter with an effective memory length of *2500* filter cycles, there will be no observable bias errors, and the *SRF* achieved by the filter will be *6%*.

Note: *SRF* means *sigma reduction factor*. An *SRF* of *6%* means that the sigma of the output errors is *6%* of the sigma of the input errors. *SRF* is discussed in the next subsection (see Project 12.6). ■■

12.2.16 Variance reduction

The smoothing capability for the EMP and FMP filters is quantified by what is called their *variance reduction factors* or *VRFs*. The meaning of *VRF* is illustrated by an example.

Example 12.5:
The covariance matrix of the vector $\boldsymbol{Z}^*_{n+1,n} = (z_0{}^*_{n+1,n}, z_1{}^*_{n+1,n})^T$ for the 1st-degree *1*-step predictor EMP filter is as follows (we will see in Chapter 13 how this matrix is obtained):

$$\boldsymbol{S}^*(\boldsymbol{Z}^*_{n+1,n}) = \sigma_\nu^2 \begin{pmatrix} 2(2n+3)/(n+1)^{(2)} & 6/(n+1)n \\ 6/(n+1)n & 12/(n+2)^{(3)} \end{pmatrix} \tag{12.2.32}$$

Observe how we show $\boldsymbol{S}^*$ with its functional argument $\boldsymbol{Z}^*_{n+1,n}$.

From the upper-left element of the matrix we see that the variance of $z_0{}^*_{n+1,n}$ is

$$\sigma(z_0{}^*_{n+1,n})^2 = \sigma_\nu^2 2(2n+3)/(n+1)^{(2)} \tag{12.2.33}$$

Forming the ratio $\sigma(z_0{}^*_{n+1,n})^2/\sigma_\nu^2$ from (12.2.33) we obtain what is called the *VRF* for $z_0{}^*_{n+1,n}$ (in this case for the 1st-degree EMP filter), namely

$$VRF(z_0{}^*_{n+1,n}) \equiv \sigma(z_0{}^*_{n+1,n})^2/\sigma_\nu^2 = 2(2n+3)/(n+1)^{(2)} \tag{12.2.34}$$

In the same way we can form the ratio $\sigma(z_1{}^*_{n+1,n})^2/\sigma_\nu{}^2$ and obtain

$$VRF(z_1{}^*_{n+1,n}) \equiv \sigma(z_1{}^*_{n+1,n})^2/\sigma_\nu^2 = 12/(n+2)^{(3)} \tag{12.2.35}$$

■■

Thus the *VRFs* of the elements of the vector $\boldsymbol{Z}^*$ are the diagonal elements of its covariance matrix $\boldsymbol{S}^*(\boldsymbol{Z}^*_{n+1,n})$ when $\sigma_\nu^2 = 1$. This gives us the following:

Definition 12.2:

a. For the *1*-step predictor of degree m, the variance of the errors in the output $z_j{}^*_{n+1,n}$ divided by the variance of the input errors σ_ν^2 is called the jth **normalized *1*-step-predictor *VRF*** and is written as $VRF(z_j{}^*_{n+1,n})$.

b. For the current-estimate filter of degree m, the variance of the errors in the output $z_j{}^*{}_{n,n}$ divided by the variance of the input errors σ_ν^2 is called the jth **normalized current-estimate *VRF*** and is written as $VRF(z_j{}^*{}_{n,n})$. ■■

Definition 12.3: The square root of a *VRF* is a **sigma reduction factor** or ***SRF***. ■■

In practice we require the *VRFs* and *SRFs* of the *de*normalized vectors, namely of

$$\boldsymbol{X}^*{}_{n+1,n} = (x_0{}^*, x_1{}^*, x_2{}^*, \ldots x_m{}^*)^T_{n+1,n} \tag{12.2.36}$$

and

$$\boldsymbol{X}^*{}_{n,n} = (x_0{}^*, x_1{}^*, x_2{}^*, \ldots x_m{}^*)^T_{n,n} \tag{12.2.37}$$

These *VRFs* and *SRFs* are called the *denormalized VRFs* and *denormalized SRFs*. For the *1*-step predictors we obtain them as follows:

From (12.2.10) we recall that

$$\boldsymbol{X}^*{}_{n+1,n} = \boldsymbol{D}(\tau)\boldsymbol{Z}^*{}_{n+1,n} \tag{12.2.38}$$

in which the diagonal matrix $\boldsymbol{D}(\tau)$ was defined in (12.2.11). From (12.2.38) it follows by Theorem 4.1 that the covariance matrices of $\boldsymbol{X}^*{}_{n+1,n}$ and $\boldsymbol{Z}^*{}_{n+1,n}$ will be related as follows:

$$\boldsymbol{S}^*(\boldsymbol{X}^*{}_{n+1,n}) = \boldsymbol{D}(\tau)\boldsymbol{S}^*(\boldsymbol{Z}^*{}_{n+1,n})\boldsymbol{D}(\tau) \tag{12.2.39}$$

Because the diagonal elements of these matrices contain the *VRFs* in which we are interested, we can use (12.2.39) to write

$$VRF(x_j{}^*{}_{n+1,n}) = (j!/\tau^j)^2 VRF(z_j{}^*{}_{n+1,n}) \qquad 0 \le j \le m \tag{12.2.40}$$

in which $VRF(x_j{}^*{}_{n+1,n})$ and $VRF(z_j{}^*{}_{n+1,n})$ are, respectively, the denormalized and normalized *VRFs* for the jth derivatives of the *1*-step predictions.

To obtain the denormalized *SRF* we then take the square root of the denormalized *VRF* in (12.2.40), obtaining

$$SRF(x_j{}^*{}_{n+1,n}) = \left(VRF(x_j{}^*{}_{n+1,n})\right)^{1/2} = (j!/\tau^j)SRF(z_j{}^*{}_{n+1,n}) \tag{12.2.41}$$

Similarly we obtain the *VRFs* and *SRFs* for the current-estimate filters from

$$VRF(x_j{}^*{}_{n,n}) = (j!/\tau^j)^2 VRF(z_j{}^*{}_{n,n}) \qquad 0 \le j \le m \tag{12.2.42}$$

and

$$SRF(x_j{}^*{}_{n,n}) = \left(VRF(x_j{}^*{}_{n,n})\right)^{1/2} = (j!/\tau^j)SRF(z_j{}^*{}_{n,n}) \tag{12.2.43}$$

■■

The *VRFs* and *SRFs* tell us how well the filter is smoothing the data, and we want them to be as small as possible. We now consider an example that demonstrates how they can be used.

Example 12.6:

Problem: A *1*-step predictor 1st-degree EMP filter is being operated in conjunction with a radar that has a *prf* of *2000 pps* ($\tau = 5e\text{-}4\,sec$). Assume that $\sigma_\nu = 2$ meters. Find the *sigma* values of the denormalized estimates x^* and $\dot{x}^*$ when $n = 100$.

Solution: From (12.2.34)

$$\begin{aligned} VRF(z_0{}^*{}_{101,100}) &= 2(2n+3)/(n+1)^{(2)}\big|_{n=100} \\ &= 2(203)/(101)^{(2)} = 4.0198e\text{-}2 \approx 4.0e\text{-}2 \end{aligned} \tag{12.2.44}$$

and from (12.2.35)

$$\begin{aligned} VRF(z_1{}^*{}_{101,100}) &= 12/(n+2)^{(3)}\big|_{n=100} \\ &= 12/(102)^{(3)} = 1.1648e\text{-}5 \approx 1.2e\text{-}5 \end{aligned} \tag{12.2.45}$$

When $n = 100$ the errors in $z_0{}^*$ and $z_1{}^*$ will then have variances equal to

$$\sigma(z_0{}^*)^2 = \sigma_\nu{}^2 \times VRF(z_0{}^*) = 4 \times 4.0e\text{-}2 = 16.0e\text{-}2\,m^2 \tag{12.2.46}$$

$$\sigma(z_1{}^*)^2 = \sigma_\nu{}^2 \times VRF(z_1{}^*) = 4 \times 1.2e\text{-}5 = 4.8e\text{-}5\,m^2 \tag{12.2.47}$$

We note that the *VRF* in both of these equations is dimensionless, and so the dimensions of both $\sigma(z_0{}^*)^2$ and $\sigma(z_1{}^*)^2$ are the same as those of $\sigma_\nu{}^2$, namely metres-squared.

These variances must now be denormalized. From (12.2.40) we obtain

$$\sigma(x_0{}^*)^2 = (0!/\tau^0)^2\sigma(z_0{}^*)^2 = 16.0e\text{-}2\,m^2 \tag{12.2.48}$$

and

$$\sigma(x_1{}^*)^2 = (1!/\tau^1)^2\sigma(z_1{}^*)^2 = 4.8e\text{-}5/\tau^2\,m^2/sec^2 \tag{12.2.49}$$

The required denormalized output *sigma* values using $\tau = 5e{-}4\ sec$ will then be

$$\sigma(x_0{}^*{}_{101,100}) \approx 0.4\,m \tag{12.2.50}$$

and

$$\sigma(x_1{}^*{}_{101,100}) \approx 13.9\,m/sec \tag{12.2.51}$$

■■

12.2.17 The VRF expressions for the 1-step predictor EMP filters

Figures 12.18(a) and 12.18(b) below show the expressions for the denormalized *VRFs* for the *1*-step predictor EMP filters up to 4th degree.

In the file ***1-step_predictors_up_to_8th_degree*** in the supplementary material we display the same expressions up to 8th degree.

In Figure 12.19 below we show plots of the denormalized *VRFs* of the *1*-step predictor EMP 0th-derivative elements for 0th to 4th degree.

◊ Observe from both Figures 12.18(a) and 12.18(b), and also from Figure 12.19 how the *VRFs* for the 0th-derivatives of the EMP filters all go to zero as *n* increases. (Note that the vertical scale in the Figure 12.19 is logarithmic.)

◊ From Figure 12.18 we see that the *VRF*s all tend to zero as *n* increases and that they do so at the following rates:

  - ◊ For x_0^* like $1/n$
  - ◊ For x_1^* like $1/n^3$
  - ◊ For x_2^* like $1/n^5$ and so on.

◊ Observe also: The *VRF* of the 0th-degree filter is computable for $n \geq 0$. However, the *VRF* of the 1st-degree filter is not computable when $n = 0$ because there is a zero in its denominator, and it is only when $n = 1$ that the zero disappears. This is because the 1st-degree filter is self-initializing when $n = 0$ and so its output is meaningless.

Degree	*Output*	*Denormalized VRFs for the 1-step-predictor EMP filters*
0	$x_0{}^*{}_{n+1,n}$	$1/(n + 1)^{(1)}$
1	$x_0{}^*{}_{n+1,n}$	$2(2n + 3)/(n + 1)^{(2)}$
	$x_1{}^*{}_{n+1,n}$	$12/\tau^2(n + 2)^{(3)}$
2	$x_0{}^*{}_{n+1,n}$	$3(3n^2 + 9n + 8)/(n + 1)^{(3)}$
	$x_1{}^*{}_{n+1,n}$	$12(16n^2 + 62n + 57)/\tau^2(n + 3)^{(5)}$
	$x_2{}^*{}_{n+1,n}$	$720/\tau^4(n + 3)^{(5)}$

Figure 12.18(a) Denormalized VRFs for the 1-step predictor EMP filters

Degree	*Output*	*Denormalized VRFs for the 1-step-predictor EMP filters*
3	$x_0{}^*_{n+1,n}$	$4(4n^3+18n^2+38n+30)/(n+1)^{(4)}$
	$x_1{}^*_{n+1,n}$	$200(6n^4+51n^3+159n^2+219n+116)/\tau^2(n+4)^{(7)}$
	$x_2{}^*_{n+1,n}$	$80(324n^2+1278n+1188)/\tau^4(n+4)^{(7)}$
	$x_3{}^*_{n+1,n}$	$100800/\tau^6(n+4)^{(7)}$
4	$x_0{}^*_{n+1,n}$	$5(5n^4+30n^3+115n^2+210n+144)/(n+1)^{(5)}$
	$x_1{}^*_{n+1,n}$	$100(48n^6+666n^5+3843n^4+11982n^3+21727n^2+21938n+9516)/\tau^2(n+5)^{(9)}$
	$x_2{}^*_{n+1,n}$	$720(441n^4+3724n^3+11711n^2+16464n+9065)/\tau^4(n+5)^{(9)}$
	$x_3{}^*_{n+1,n}$	$100800(64n^2+254n+237)/\tau^6(n+5)^{(9)}$
	$x_4{}^*_{n+1,n}$	$25401600/\tau^8(n+5)^{(9)}$

Figure 12.18(b) Denormalized VRFs for the 1-step predictor EMP filters

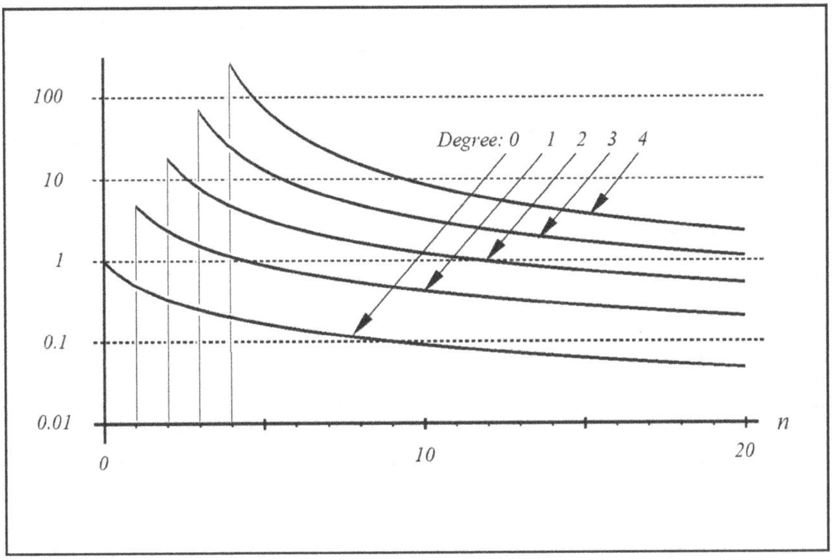

Figure 12.19 Denormalized VRFs of the 1-step predictor EMP 0th-derivative elements

In the same way the *VRF* of the 2nd-degree filter is not computable until $n = 2$, the *VRF* of the 3rd-degree filter is not computable until $n = 3$, and so on.

Thus in each case the mathematics prevents the *VRF* from being computed as long as the filter is self-initializing and the outputs have no meaning. These delays in the computability are clearly evident in Figure 12.19.

Example 12.7: A 2nd-degree *1*-step predictor EMP filter ($m = 2$) is being cycled *1000* times/second ($\tau = 1/1000$) where $\sigma_\nu = 1.5\,meters$. Then, from Figure 12.18(a) when $n = 1000$:

$$\sigma(x_0{}^*{}_{101,\,100}) = 1.5\big(3(3n^2 + 9n + 8)/(n + 1)^{(3)}\big)^{1/2} \approx 0.14\,m$$

$$\sigma(x_1{}^*{}_{101,\,100}) = 1.5\big(12(16n^2 + 62n + 57)/\tau^2(n + 3)^{(5)}\big)^{1/2} = 0.66\,m/sec$$

$$\sigma(x_2{}^*{}_{101,\,100}) = 1.5\big(720/\tau^4(n + 3)^{(5)}\big)^{1/2} = 1.3\,m/sec^2$$

■■

In Figure 12.20 we show the expressions for the denormalized *VRF*s of the current-estimate EMP 0th-derivative elements. These expressions were obtained from the recursion formula derived in Project 13.4.

Degree	Output	Denormalized VRFs for the current-estimate EMP filters
0	$x_0{}^*{}_{n,n}$	$1/(n+1)^{(1)}$
1	$x_0{}^*{}_{n,n}$	$2(2n+1)/(n+2)^{(2)}$
2	$x_0{}^*{}_{n,n}$	$3(3n^2+3n+2)/(n+3)^{(3)}$
3	$x_0{}^*{}_{n,n}$	$4(4n^3+6n^2+14n+6)/(n+4)^{(4)}$
4	$x_0{}^*{}_{n,n}$	$5(5n^4+10n^3+55n^2+50n+24)/(n+5)^{(5)}$

Figure 12.20 Denormalized VRFs of the current-estimate EMP 0th-derivative elements

From the expressions in Figure 12.20 we can see that the asymptotic behavior of the current-estimate elements when n is large is the same as it was for the *1*-step predictors in Figures 12.18(a) and 12.18(b).

This means that the plots appearing in Figure 12.19 (for the *1*-step predictor EMP's) are very similar to what the plots would be for the current-estimate EMP's. We will see later why this is important.

12.2.18 The VRF expressions for the 1-step predictor FMP filters

Figures 12.21(a) and 12.21(b) below show the expressions for the denormalized *VRFs* of the *1*-step predictor FMP filters up to 4th degree.

In ***1-step_predictors_up_to_8th_degree*** we show the expressions up to 8th degree.

In Figure 12.22 below we show plots of the denormalized *VRFs* of the *1*-step predictor FMP 0th-derivative elements, using $\theta = 0.9$.

Observe from both Figures 12.21(a) and 12.21(b), and also from Figure 12.22 that the *VRFs* of the FMP filters are constants whose values depend on θ. The fact that they are constant comes about because of their effective fixed-length memories.

◊ In practice θ is usually very close to *1*, and from Figure 12.21 we see the following:

- ◊ For x_0^* the *VRFs* are proportional to $(1 - \theta)$
- ◊ For x_1^* they are proportional to $(1 - \theta)^3$
- ◊ For x_2^* they are proportional to $(1 - \theta)^5$ and so on.

Degree	**Output**	***Denormalized VRFs for the 1-step-predictor FMP filters***
0	${x_0}^*_{n+1,n}$	$(1 - \theta)/(1 + \theta)$
1	${x_0}^*_{n+1,n}$	$(\theta^2 + 4\theta + 5)(1 - \theta)/(1 + \theta)^3$
	${x_1}^*_{n+1,n}$	$2(1 - \theta)^3/\tau^2(1 + \theta)^3$
2	${x_0}^*_{n+1,n}$	$(\theta^4 + 6\theta^3 + 16\theta^2 + 24\theta + 19)(1 - \theta)/(1 + \theta)^5$
	${x_1}^*_{n+1,n}$	$(13\theta^2 + 50\theta + 49)(1 - \theta)^3/2\tau^2(1 + \theta)^5$
	${x_2}^*_{n+1,n}$	$6(1 - \theta)^5/\tau^4(1 + \theta)^5$

Figure 12.21(a) Denormalized VRFs for the 1-step predictor FMP filters

Degree	*Output*	*Denormalized VRFs for the 1-step-predictor FMP filters*
3	$x_0{}^*{}_{n+1,n}$	$(\theta^6 + 8\theta^5 + 29\theta^4 + 64\theta^3 + 97\theta^2 + 104\theta + 69)(1 - \theta)/(1 + \theta)^7$
	$x_1{}^*{}_{n+1,n}$	$5(53\theta^4 + 298\theta^3 + 762\theta^2 + 970\theta + 581)(1 - \theta)^3/18\tau^2(1 + \theta)^7$
	$x_2{}^*{}_{n+1,n}$	$2(23\theta^2 + 76\theta + 63)(1 - \theta)^5/\tau^4(1 + \theta)^7$
	$x_3{}^*{}_{n+1,n}$	$20(1 - \theta)^7/\tau^2(1 + \theta)^7$
4	$x_0{}^*{}_{n+1,n}$	$(\theta^8+10\theta^7+46\theta^6+130\theta^5+256\theta^4+380\theta^3+446\theta^2+410\theta+251)(1 - \theta)/(1 + \theta)^9$
	$x_1{}^*{}_{n+1,n}$	$5(449\theta^6+2988\theta^5+10013\theta^4+21216\theta^3+28923\theta^2+25588\theta+12199)(1 - \theta)^3$ $72\tau^2(1 + \theta)^9$
	$x_2{}^*{}_{n+1,n}$	$7(2021\theta^4+10144\theta^3+22746\theta^2+25144\theta+12521)(1 - \theta)^5/72\tau^4(1 + \theta)^9$
	$x_3{}^*{}_{n+1,n}$	$5(113\theta^2+338\theta+253)(1 - \theta)^7/2\tau^6(1 + \theta)^9$
	$x_4{}^*{}_{n+1,n}$	$70(1 - \theta)^9/\tau^8(1 + \theta)^9$

Figure 12.21(b) Denormalized VRFs for the 1-step predictor FMP filters

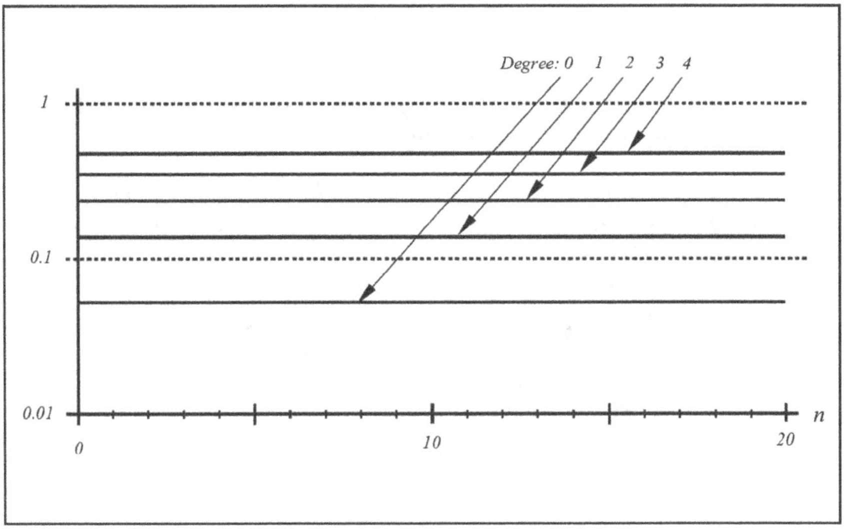

Figure 12.22 Denormalized VRFs of the 1-step predictor FMP 0th-derivative elements ($\theta = 0.9$)

◊ When an mth-degree FMP filter is smoothing data that comes from an observed trajectory that is not a polynomial of degree *m* or less, then the filter can still track with acceptably small bias errors, but now we are forced to trade off the smoothing capability of the filter against the filter-model/external-process mismatch.

If we make θ closer to unity, then we obtain lower (constant) values for the FMP *VRFs*, thereby obtaining more smoothing, but risk incurring bias errors.

On the other hand, if we make θ closer to zero, then we lower the risk of bias errors but we must accept higher (constant) values for the *VRF,* i.e. less smoothing.

It is this ability of the FMP filters to track arbitrary (non-polynomial) processes that makes them so useful. We discuss this further when we examine the composite EMP/FMP filters lower down in this chapter.

12.2.19 Approximate expressions for the denormalized covariance matrices

In Problem 13.2 we consider how one can obtain approximate expressions for the entire denormalized covariance matrices of the *1*-step predictor EMP filters. These approximations are quite accurate for $n = 100$ and are extremely accurate for $n \geq 1000$, and since these filters are usually being cycled at least *1000* times/sec, a value of $n = 1000$ is attained in one second or less.

Note: For large *n*, the approximations are valid for both the *1*-step predictor and the current estimate EMP filters.

The first three approximations appear below, and in the solution to Problem 13.2 you will find the approximate matrices up to 4th-degree.

(In the folder ***Miscellaneous\Code_for_Solutions_To_Chapter_13*** you will find the program ***Problem_13.2.exe*** which enables you to generate these matrices up to 8th degree.)

$$m = 0: \quad \boldsymbol{VRF}(\boldsymbol{X}^*{}_{n+1,n}) = 1/n \tag{12.2.52}$$

$$m = 1: \quad \boldsymbol{VRF}(\boldsymbol{X}^*{}_{n+1,n}) = \begin{pmatrix} 4/n & 6/\tau n^2 \\ 6/\tau n^2 & 12/\tau^2 n^3 \end{pmatrix} \tag{12.2.53}$$

$$m = 2: \quad \boldsymbol{VRF}(\boldsymbol{X}^*{}_{n+1,n}) = \begin{pmatrix} 9/n & 36/\tau n^2 & 60/\tau^2 n^3 \\ 36/\tau n^2 & 192/\tau^2 n^3 & 360/\tau^3 n^4 \\ 60/\tau^2 n^3 & 360/\tau^3 n^4 & 720/\tau^4 n^5 \end{pmatrix} \tag{12.2.54}$$

It is of course also possible to compute the exact values for the covariance matrices (as was done in a number of the programs in the supplementary material), and in Chapter 13 we discuss how that is carried out.

However, that approach yields matrices of plain numbers for each specified value of *n* which would be difficult to use in pencil-and-paper engineering studies. The approximate matrices shown in (12.2.52), (12.2.53) and (12.2.54), plus the others that we generate in Problem 13.2 are therefore much more useful because *n* is retained as a variable.

◀▶

In Problem 13.3 we consider how we can obtain approximate expressions for the entire covariance matrices of the FMP filters. These approximations are quite accurate for $\theta = 0.99$ and become extremely accurate when θ increases beyond *0.99* up to *1*. In practice θ almost always lies in this latter range.

Note: For θ close to *1*, the approximations are valid for both the *1*-step predictors and the current estimate FMP filters.

The first three approximations appear below, and in the solution to Problem 13.3 you will find the approximate matrices up to 4th-degree. In the folder ***Miscellaneous\Code_for_Solutions_To_Chapter_13*** you will find the program ***Problem_13.3.exe*** which enables you to derive the matrices up to 8th degree.

$$m = 0\text{: } \boldsymbol{VRF}(\boldsymbol{X}^*_{n+1,n}) = 0.5(1 - \theta) \tag{12.2.55}$$

$$m = 1\text{: } \boldsymbol{VRF}(\boldsymbol{X}^*_{n+1,n}) = \begin{pmatrix} 1.25(1-\theta) & 6(1-\theta)^2/\tau \\ 6(1-\theta)^2/\tau & 12(1-\theta)^3/\tau^2 \end{pmatrix} \tag{12.2.56}$$

$$m = 2\text{: } \boldsymbol{VRF}(\boldsymbol{X}^*_{n+1,n})$$

$$= \begin{pmatrix} 2.0625(1-\theta) & 1.6875(1-\theta)^2/\tau & 0.5(1-\theta)^3/\tau^2 \\ 1.6875(1-\theta)^2\tau & 1.75(1-\theta)^3/\tau^2 & 0.5625(1-\theta)^4/\tau^3 \\ 0.5(1-\theta)^3/\tau^2 & 0.5625(1-\theta)^4/\tau^3 & 1.1875(1-\theta)^5/\tau^4 \end{pmatrix} \tag{12.2.57}$$

It is of course also possible to compute the exact values for the FMP covariance matrices (as was done in a number of the programs), and in Chapter 13 we discuss how that is done.

However, that yields matrices of plain numbers for each specified value of θ which would be difficult to use in pencil-and-paper engineering studies. The approximate matrices that we generate in Problem 13.3 are therefore much more useful because θ is retained as a variable.

Example 12.8: Consider a 2nd-degree *1*-step predictor FMP filter ($m = 2$) that is being cycled *1000* times/second ($\tau = 0.001$). Let $\sigma_\nu = 1.5\ meters$ and let $\theta = 0.99564$. (We'll soon see where θ values like this come from.) Then from Figure 12.21:

$$\begin{aligned} \sigma(x_0{}^*_{n+1,n}) &= 1.5\left((\theta^4 + 6\theta^3 + 16\theta^2 + 24\theta + 19)(1-\theta)/(1+\theta)^5\right)^{1/2} \\ &\approx 0.14\,m \\ \sigma(x_1{}^*_{n+1,n}) &= 1.5\left((13\theta^2 + 50\theta + 49)(1-\theta)^3 2\tau^2(1+\theta)^5\right)^{1/2} \\ &\approx 0.57\,m/sec \\ \sigma(x_2{}^*_{n+1,n}) &= 1.5\left(6(1-\theta)^5/\tau^4(1+\theta)^5\right)^{1/2} \approx 0.82\,m/sec^2 \end{aligned} \tag{12.2.58}$$

We can obtain exactly the same answers with far less effort by using the approximate values contained in the covariance matrix in (12.2.57) above. ■■

12.2.20 Tracking ability, ECM consistency and CR consistency

From Figures 12.19 and 12.22 we see that the *VRFs* of both the EMP and the FMP filters for a given value of n increase as the filter degree m increases. This means that the higher-degree EMP and FMP filters do not smooth as well as do the lower-degree ones. (This is because they are more 'flexible' and so they tend to 'follow' the observation errors.)

However, the higher degree filters offer better protection against bias errors because they can adapt themselves better to dynamically changing inputs, and so we must also make a trade-off of the following two items:

◊ Smoothing requirements that call for lower degree.

◊ Bias-error protection that calls for higher degree.

This, in turn, calls for an analysis of the type of data that the filters will be required to smooth, as well as simulation studies and user judgment. The final outcome of such activities will be the selection of the degree of the EMP and FMP filters, as well as the value of θ for the FMP.

We will return again to this question of the filter degree and the parameter θ when we discuss the composite EMP/FMP filter further down.

The eternal tracking ability and the ECM consistency of the EMP and FMP filters are summarized in Figures 12.23 and 12.24.

Eternal tracking ability and ECM consistency of the EMP filters						
	Linear	*Quadratic*	*Cubic*	*Quartic*	*Quintic*	*Sinusoid*
1st degree	*Yes*	*No*	*No*	*No*	*No*	*No*
2nd degree	*Yes*	*Yes*	*No*	*No*	*No*	*No*
3rd degree	*Yes*	*Yes*	*Yes*	*No*	*No*	*No*
4th degree	*Yes*	*Yes*	*Yes*	*Yes*	*No*	*No*
5th degree	*Yes*	*Yes*	*Yes*	*Yes*	*Yes*	*No*

Figure 12.23 Eternal tracking ability and ECM consistency of the EMP filters

Eternal tracking ability of the FMP filters (with θ properly selected)						
	Linear	*Quadratic*	*Cubic*	*Quartic*	*Quintic*	*Sinusoid*
1st degree	*Yes*	*Yes*	*Yes*	*Yes*	*Yes*	*Yes*
2nd degree	*Yes*	*Yes*	*Yes*	*Yes*	*Yes*	*Yes*
3rd degree	*Yes*	*Yes*	*Yes*	*Yes*	*Yes*	*Yes*
4th degree	*Yes*	*Yes*	*Yes*	*Yes*	*Yes*	*Yes*
5th degree	*Yes*	*Yes*	*Yes*	*Yes*	*Yes*	*Yes*

Figure 12.24 Eternal tracking ability of the FMP filters (with θ properly selected)

To understand fully what these figures mean, follow the instructions given in Projects 12.7a and b.

◊ The EMP filters are CR consistent when they are properly matched to the external process for the following reason:

At the start of this chapter it was stated that these filters are based on the assumption that the errors in the observations are of constant variance σ_ν^2 and stage-wise uncorrelated.

We will see in Chapter 13 that the EMP filters are based on fitting a polynomial to the observations by equally-weighted least squares. That is the same as using the CR-consistent Gauss–Aitken algorithm

$$X^*_{n,n} = (\boldsymbol{T}_n^T \boldsymbol{R}_n^{-1} \boldsymbol{T}_n)^{-1} \boldsymbol{T}_n^T \boldsymbol{R}_n^{-1} \boldsymbol{Y}_n \tag{12.2.59}$$

in which $\boldsymbol{R}_n$ has the form $\sigma_\nu^2 \boldsymbol{I}$ and $\boldsymbol{T}_n$ here has the same structure as $\boldsymbol{T}_n$ in (3.2.12).

The EMP filters are consistent with (12.2.59), and so they are CR consistent.

◊ The FMP filters are not CR consistent because – as we pointed out earlier – they are based on fitting a polynomial to the data using a decaying-exponential weight function. Thus they are not based on (12.2.59) and so they cannot be CR consistent.

However, because of their other extremely useful properties, this is a price that we are happy to pay.

In Project 12.16 we discuss this issue in greater depth.

12.2.21 Choice of coordinate system

The EMP and FMP filters are well suited to smoothing data that come from tracking radars, and because of their compactness, difficulties are seldom encountered when cycling them at the high data-rates of such radars.

The radar observations that require smoothing are ρ, θ, ψ, where ρ is the slant range, θ the elevation angle and ψ the azimuth angle.

As long as the target giving rise to these observations is at a considerable distance from the radar, the data streams of these three observables will be well-behaved and so properly configured polynomial filters will be able to track them easily.

However, if the target approaches close to the radar, then all three of the data streams will exhibit evidence of the singularities that are inherent in radar coordinates, the severity of which depends on the proximity of the point of closest approach to the radar.

We are referring here to the fact that, in radar coordinates:

◊ When an object passes directly over the origin (i.e. the assumed location of the radar) its azimuth angle ψ will make an instantaneous jump of π radians.

◊ If a distant object is relatively close to the x, y plane and moves on a course that takes it over the origin, then its elevation angle θ will start at approximately zero. As it passes over the origin, θ exhibits an abrupt rise to $\pi/2$ radians followed by an abrupt fall. Finally, as it recedes into the distance θ once again tends to zero.

◊ An object such as the one just described will have a slant-range ρ that first decreases approximately linearly until it reaches a point over the origin, after which it abruptly begins to increase linearly.

In Figure 12.25 below we show plots created by the program ***08_Radar_Cartesian.exe.*** In this case a constant-speed target came close to the radar, but passed it on one side rather than directly overhead. (The horizontal axis shows the time in seconds.)

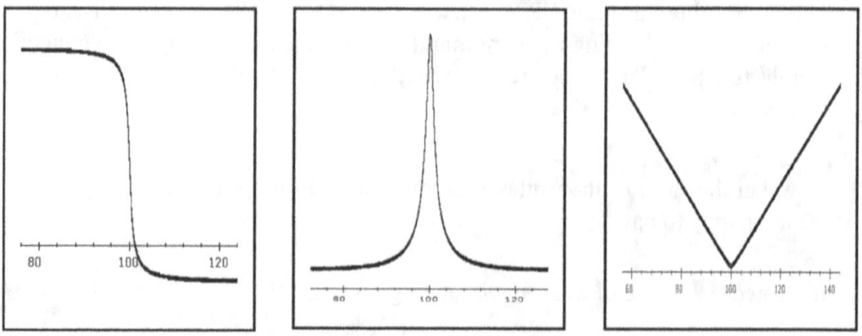

Figure 12.25 Singularities in azimuth(ψ), elevation(θ) and range(ρ) for close approach

◊ On the left the value of ψ changes abruptly.

◊ In the centre θ goes through a sharp peak.

◊ On the right ρ first diminishes linearly and then abruptly begins to increase linearly.

If data like those in the figure were fed to polynomial filters (or any other filter for that matter), then the abrupt changes will create serious filtering problems that could easily result in loss of track.

These singularities only arise because of the idiosyncrasies of the radar coordinate system, and do not occur if the flight-path is observed in Cartesian coordinates. Thus, if at all possible, smoothing of such data should be done in Cartesian coordinates rather than in radar coordinates.

Referring to Figure 12.26 below, in order to smooth in Cartesian coordinates we first transform the raw data from radar to Cartesian coordinates. The outputs of the Cartesian transformation are then fed to three polynomial filters.

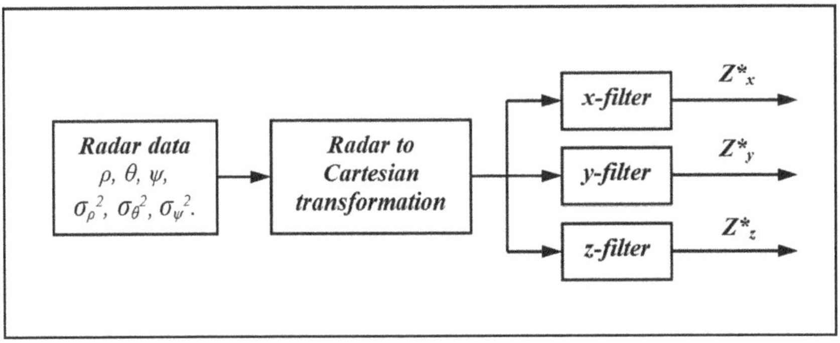

Figure 12.26 Smoothing radar data in Cartesian coordinates

Smoothing in Cartesian coordinate solves another problem as well. Radar systems specify the azimuth angle ψ by its principal value, which is either in the range $-\pi < \psi \leq \pi$ or else $0 \leq \psi < 2\pi$.

As an example, in the first of these two cases if a target flies from north to south across the line running due west from the radar, then its azimuth angle will appear to jump instantaneously from $+\pi$ to $-\pi$. (See Figures 12.27 and 12.28.)

This would cause severe problems for any filter that is attempting to smooth such data. However, this situation does not arise when smoothing is done in Cartesian coordinates, because neither the x nor the y coordinates experience any discontinuities.

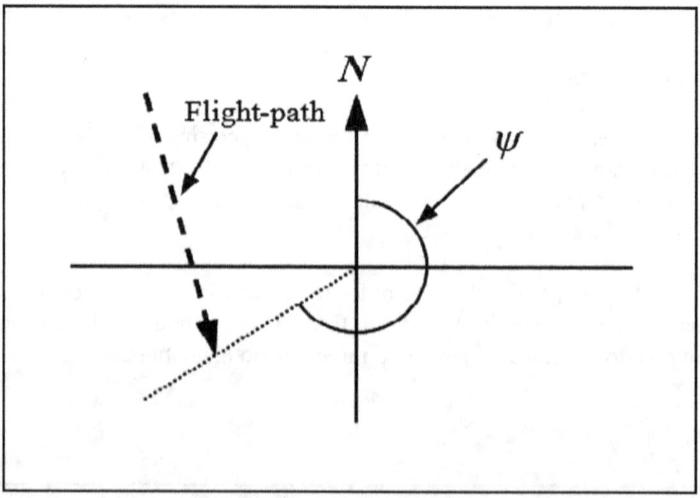

Figure 12.27 Flight-path

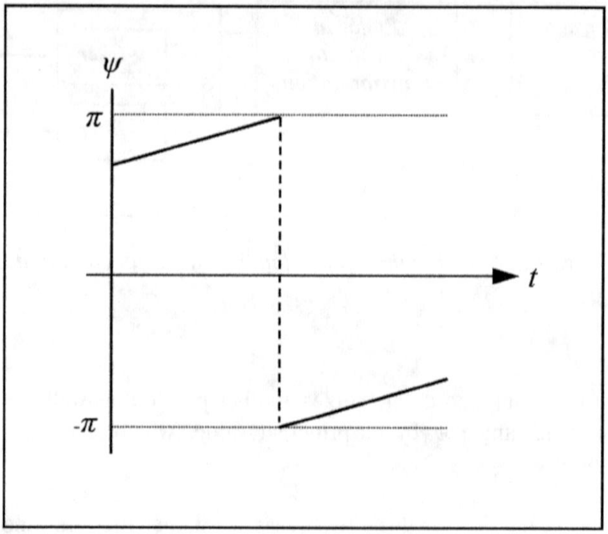

Figure 12.28 Azimuth angle

Each time that radar data are sent to the filters they will consist of at least the following six numbers:

$$\rho,\ \theta,\ \psi,\ \sigma_\rho^2,\ \sigma_\theta^2,\ \sigma_\psi^2$$

Following the procedure discussed in Section 5.1, the vector $(\rho, \theta, \psi)_n^T$ is transformed, immediately upon receipt, from radar coordinates to the Cartesian vector $(x, y, z)_n^T$. The radar covariance matrix $\boldsymbol{R}_r = diag(\sigma_\rho^2, \sigma_\theta^2, \sigma_\psi^2)_n$ is likewise transformed to a Cartesian-coordinate covariance matrix $\boldsymbol{R}_c$ (see equations (5.1.3) and (5.1.20)).

The three sequences of unfiltered values x_n, y_n, z_n so obtained are then each sent to its own polynomial filter for smoothing.

Three issues now arise:

1. We no longer have constant error variances in the data being fed to the three polynomial filters, which is one of the assumptions on which the derivation of the EMP and FMP filters was based. Instead those variances are now functions of the instantaneous values of the radar coordinates.

2. We also no longer have a diagonal covariance matrix, as was the case in radar coordinates, and so in Cartesian coordinates the observation errors are correlated across x, y and z.

3. The polynomial filter outputs are stage-wise correlated (because the EMP and FMP filters are recursive).

◊ Regarding Item 1: The filters will usually be operating in FMP mode with its fading memory, and therefore with an effective fixed-length memory. The change in the data variances as the target changes position will thus be small over the span of the FMP filter's memory.

  Accordingly, we will not be seriously violating the assumptions on which these filters are based, and so smoothing will still take place without problems (see Project 12.8).

◊ Regarding Items 2 and 3: We will see in Section (12.2.28) near the end of this chapter that their effects are also negligible.

12.2.22 Outlier exclusion

An outlier is a datum or observation that lies clearly outside the line of fit to the data that have preceded it.[4] Outliers will unnecessarily contaminate the curve that is being fitted to the data and are best excluded.

The EMP and FMP filters are perfectly structured for outlier detection and exclusion. This is discussed further in Projects 12.14 and 12.15.

[4] For a detailed discussion on outliers, see Reference 172.

12.2.23 Missed detections

On occasions a radar return is not received, and we have what is called a *missed detection*. When this occurs, it is necessary to let the filter extrapolate through the observation instant. Hopefully missed detections will not occur too frequently.

In Project 12.14 we show that the EMP and FMP filters are perfectly suited to extrapolation, and all that one has to do is

◊ Set $e_n = 0$ in the filter equations.

◊ Cycle the filter.

The filter will then extrapolate. In the case of the EMP filter, the value of n must be incremented by 1 each time that such an extrapolation is done.

12.2.24 Quantifying the memory lengths of the FMP filters

Assume that we are required to filter radar data coming from a manoeuvring target. Based on what was said above regarding choice of coordinates, we convert the raw radar data (ρ, θ, ψ) to Cartesian coordinates (x, y, z).

As was shown in Figure 12.26, the Cartesian values are then smoothed by three polynomial filters. However, first we must decide whether those filters will be EMP or FMP.

◊ The EMP filters possess the extremely desirable attribute of being able to self-initialize. However, they possess the drawback that their memories expand steadily, and because non-cooperative potentially manoeuvring targets cannot be depended on to follow polynomial flight paths of a specified degree, it follows that the EMP filters cannot be used for sustained tracking. Thus we can only use them for a small initial time interval and must rule them out for steady-state filtering.

◊ By contrast, the FMP filters are not self-initializing. However, as was shown in Figure 12.16, their fading memories endow them with memories that are effectively of fixed length. This means that, if properly initialized, they *can* be used for sustained tracking of such targets.

In what now follows we focus only on the effective fixed-length memories of the FMP filters. Their initialization requirements will be addressed later.

◆▶

In order to make use of the effective fixed-length memories of the FMP filters it is essential that we be able to quantify them, i.e. be able to specify their effective memory lengths in *seconds*. We now show how that is done.

The *VRF* formulas for the *1*-step predictor EMP and FMP filters are shown in Figures 12.18(a), 12.18(b), 12.21(a) and 12.21(b).

If we equate the *VRF* of an EMP filter to the *VRF* of the FMP filter of the same degree, then we are saying that those two filters have the same smoothing power, and if they have the same smoothing power then they must have the same effective memory lengths.

In the following example we show how this argument can be applied, but before doing so we make two simplifying assumptions:

1. In most practical situations the value of the EMP cycle count n will be large enough so that n plus a small integer is approximately equal to n.
2. Lower down we will find that the value of θ in the FMP filters almost always exceeds 0.99, and because θ must be less than 1 we will make the assumption that $\theta \approx 1$. However, we must always retain the term $1 - \theta$ rather than setting it equal to *zero*.

Example 12.9*:* From Figure 12.18, the *VRF* for $z_0{}^*_{n+1,n}$ for the 1st-degree *1*-step predictor EMP filter is

$$VRF_{EMP} = 2(2n+3)/(n+1)^{(2)} \tag{12.2.60}$$

and from Figure 21 the corresponding *VRF* for the FMP filter is

$$VRF_{FMP} = (\theta^2 + 4\theta + 5)(1 - \theta)/(1+\theta)^3 \tag{12.2.61}$$

Equating them gives the following:

$$2(2n+3)/(n+1)n = (\theta^2 + 4\theta + 5)(1 - \theta)/(1+\theta)^3 \tag{12.2.62}$$

By Assumption (1) above, the left-hand side of (12.2.62) now reduces as follows:

$$2(2n+3)/(n+1)n \approx 4n/n^2 = 4/n \tag{12.2.63}$$

Similarly, by Assumption (2):

$$\begin{aligned}(\theta^2 + 4\theta + 5)(1 - \theta)/(1+\theta)^3 &\approx (1+4+5)(1-\theta)/(1+1)^3 \\ &= 1.25(1-\theta)\end{aligned} \tag{12.2.64}$$

Equating these two results we obtain

$$n = 3.2/(1 - \theta) \tag{12.2.65}$$

To convert the cycle count n into memory length in *seconds* we now multiply both sides by the sampling interval τ. Then (12.2.65) becomes

$$L_{eff} = n\tau = 3.2\tau/(1-\theta) \tag{12.2.66}$$

Note: We used the exact expressions for the EMP and FMP *VRFs* from Figures 18 and 21. We could equally well have used the approximate expression for the *VRFs* from the matrices in (12.2.53) and (12.2.56) and obtained the same answer (with much less effort).

Example 12.10: Let $\theta = 0.999$ and $\tau = 0.5\ msec$ (corresponding to a *prf* of *2000 pps*). Then from (12.2.66)

$$L_{eff} = 1.6\ sec \tag{12.2.67}$$

Equation (12.2.66) is the required expression for quantifying the effective memory length of the 1st-degree FMP filter. Repeating the above calculation for the other four degrees ($m = 0, 2, 3, 4$) gives the expressions shown in Figure 12.29.

Degree	*0*	*1*	*2*	*3*	*4*
L_{eff} *(sec)*	$2\tau/(1-\theta)$	$3.2\tau/(1-\theta)$	$4.36\tau/(1-\theta)$	$5.51\tau/(1-\theta)$	$6.63\tau/(1-\theta)$

Figure 12.29 FMP effective memory length

Quantifying the effective length of the FMP filters' fixed-length memories has now been accomplished and Figure 12.29. provides us with what we need.

Note: In document ***Miscellaneous\Polynomial_filters\1-step_predictors_up_to_8th_degree*** we display values for L_{eff} up to 8th-degree. ■■

Example 12.11:

a) We wish to perform 3rd-degree filtering in conjunction with a radar that has a PRF of *1000 pps*. What value should we use for θ in order that the FMP memory has an effective length of *10 sec?*

   From Figure 12.29 we see that θ must satisfy $5.51\tau/(1-\theta) = 10$. Using $\tau = 1/1000$ gives $\theta = 0.999449$.

b) We wish to perform 4th-degree filtering in conjunction with a radar that has a PRF of *2000 pps*. What value should we use for θ in order that the FMP memory has an effective length of *15 sec?*

From Figure 12.29 the value of θ must satisfy $6.63\tau/(1-\theta)=15$. Using $\tau=1/2000$ gives $\theta=0.999779$. ■■

◀▶

The fact that the single parameter θ controls the effective memory length of the FMP filters opens up the possibility of adaptively changing θ according to the observed performance of the filter.

Scrutiny of the quantity $e_n = y_n - z_0{}^*_{n,n-1}$ that is present in the FMP filters could provide an indication that biases are developing when using the current value of θ, (see Project 12.14) signaling that the memory length is too large. This would suggest that θ should be reduced.

While there is no mathematical theory to support it, if θ is changed slowly then the FMP filters work perfectly well, and continue to smooth the data just as they would if θ were held constant (see Project 12.9).

12.2.25 *Quick settling*

When we first derived the polynomial algorithms we were confronted by the fact that – even though self-initializing – the higher-degree EMP filters exhibit *spattering* in their estimates for some time after start-up, making them impossible to use.

Figures 12.30 and 12.31 below show typical error plots for the 2nd and 4th-degree *1*-step predictor EMP filters.

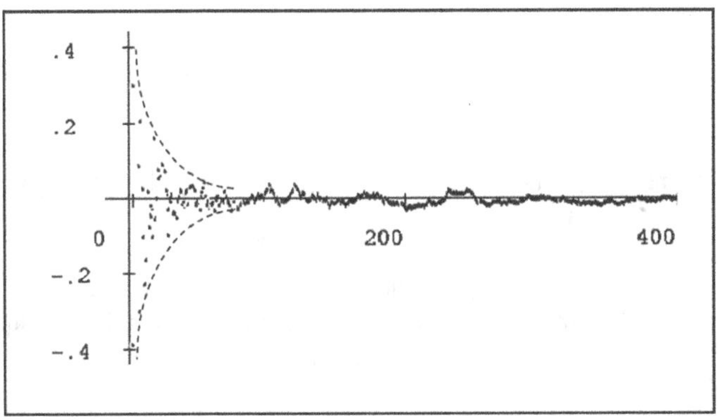

Figure 12.30 Error plot for 2nd-degree 1-step predictor EMP filter showing spattering at start-up

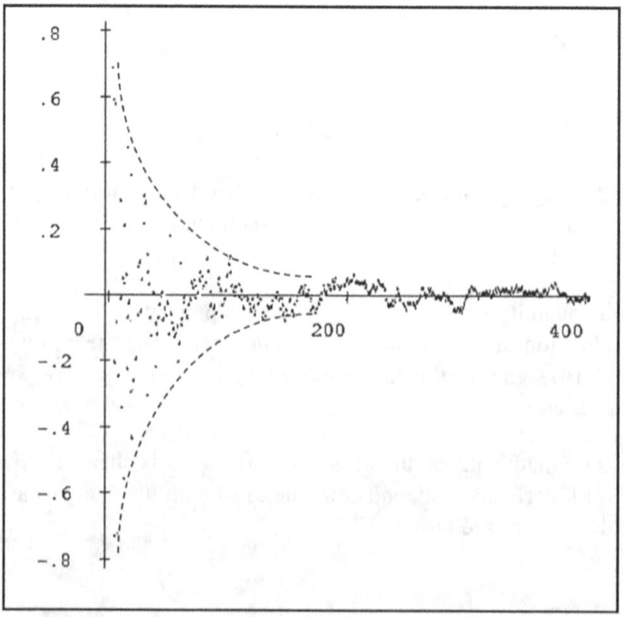

Figure 12.31 Error plot for 4th-degree 1-step predictor EMP filter showing spattering at start-up

◊ The horizontal axis is *n*, the cycle number of the filter.

◊ The ragged black line running out towards the right of the figures shows the filter's 0th derivative output minus the true value of the input, and so it shows the estimation errors.

  ◊ If the filter were working perfectly then the ragged line would be dead straight, and would lie directly on the horizontal axis.

  ◊ If there were bias errors present, the line would be above or below the horizontal axis.

The dashed lines in the figures show the envelopes of the spattering, and are similar to plots of three times the *SRF* values of the filters, i.e. the square roots of the *VRF* formulas in Figure 12.18(a) and 12.18(b).

◊ For Figure 12.30:

$$3 \times SRF = 3\left(3(3n^2 + 9n + 8)/(n+1)^{(3)}\right)^{1/2} \approx 3(9/n)^{1/2} \qquad (12.2.68)$$

◊ For Figure 12.31:

$$3 \times SRF = 3\big(5(5n^4 + 30n^3 + 115n^2 + 210n + 144)/(n+1)^{(5)}\big)^{1/2}$$
$$\approx 3(25/n)^{1/2} \tag{12.2.69}$$

Thus the filters are performing exactly as they should be – initially their outputs are extremely noisy when their *SRFs* are significantly larger than *1*, and then as the *SRFs* fall the outputs settle down and become smoother.

It is the initially-large *SRFs* that cause the spattering.

◄►

Now let's examine what we call *quick settling*. In Figures 12.32 and 12.33 below we show the outputs of the same 2nd and 4th-degree EMP filters, but now with quick settling.

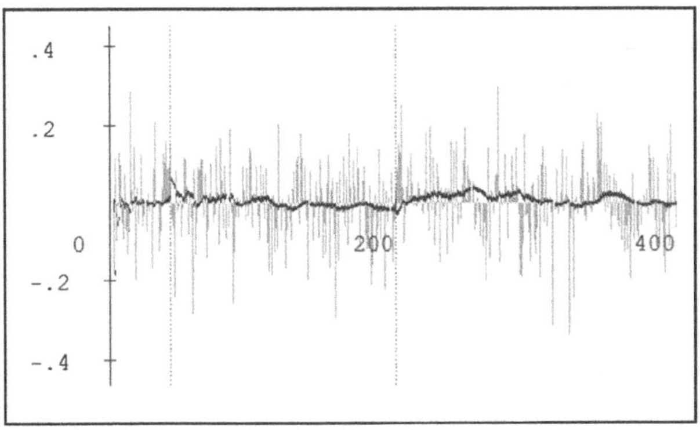

Figure 12.32 Error plot for 2nd-degree EMP filter with quick settling

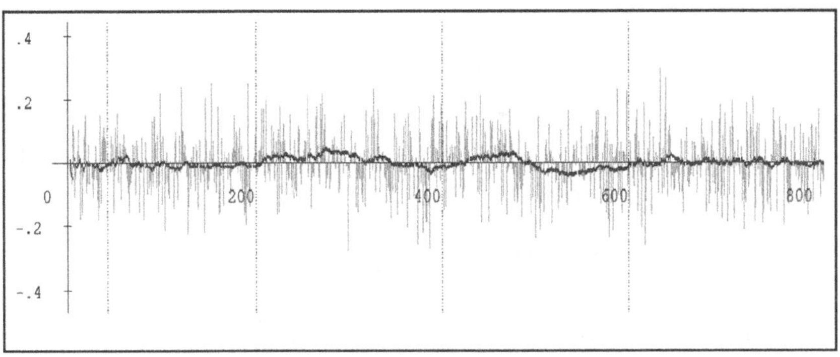

Figure 12.33 Error plot for 4th-degree EMP filter with quick settling

◊ The hash in the background is the input minus the true value, and so it shows the input errors.

◊ The heavy black lines in the foreground are plots of the 0th-derivative filter outputs minus the true value, and so they show the output errors.

◊ The spattering is gone.

◊ The outputs are well within the extremes of the hash, and so effective smoothing is taking place from the very first moment that the filter starts operating.

Quick settling makes it possible to implement EMP filters of any degree, and should always be used. We now explain how quick settling works.

As an example, suppose that we wish to implement a 4th-degree EMP filter. Figure 12.34 is a block diagram of the quick-settling procedure, and shows that for 4th-degree filtering five filters from 0th to 4th degree are involved

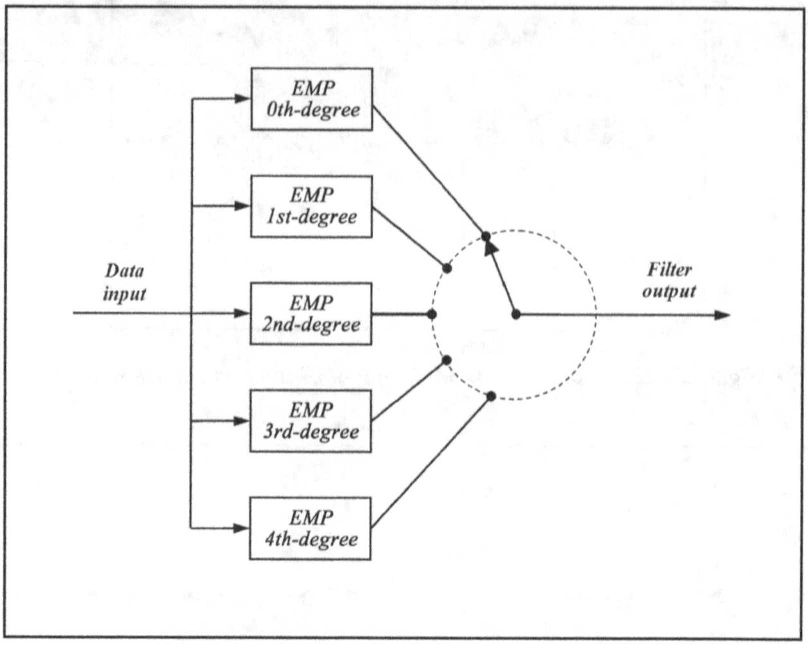

Figure 12.34 Quick-settling block diagram

◊ At start-up all five filters are activated, and commence smoothing the *same* input stream. Each is kept running until the rotary switch moves past its output, whereupon it is deactivated.

◊ The 0th-degree filter settles almost immediately (without spattering), and so at start-up the rotary switch is connected to its output. The output on the right of the figure will be the *1*-element vector $\boldsymbol{Z}^* = z^*_0$.

◊ After *20* cycles, the 0th-degree filter could begin to produce biased estimates that make its output increasingly unusable, and so at that point we switch to the 1st-degree output. This filter has been running since start-up and has had *20* cycles in which to settle, and so its output is now free of spattering and totally usable. The output on the right of the figure will now be the *2*-element vector $\boldsymbol{Z}^* = (z^*_0, z^*_1)^T$.

◊ After a further *180* cycles the 1st-degree filter could start to develop bias errors, and so at that time we switch to the 2nd-degree filter that has also been running since start-up. Since it has had *200* cycles in which to settle, its output is now spatter-free and completely usable. The output on the right of the figure will now be the *3*-element vector $\boldsymbol{Z}^* = (z^*_0, z^*_1, z^*_2)^T$.

◊ After a further *200* cycles the 2nd-degree filter could start to develop bias errors, and so at that time we switch to the 3rd-degree filter that has also been running since start-up and has, by now, settled completely. The output on the right of the figure will now be the *4*-element vector $\boldsymbol{Z}^* = (z^*_0, z^*_1, z^*_2, z^*_3)^T$.

◊ After a further *200* cycles the 3rd-degree filter could start to develop bias errors, and so at that time we switch to the 4th-degree filter that has also been running since start-up and has, by now, settled completely. The output on the right of the figure will now be the *5*-element vector $\boldsymbol{Z}^* = (z^*_0, z^*_1, z^*_2, z^*_3, z^*_4)^T$ as required.

In this way we have arrived at the 4th-degree EMP filter that has had *600* cycles in which to settle, and so it is now completely spatter-free and ready for use.

The result is shown in Figure 12.35 where we see that the spattering that was present in Figure 12.31 is entirely gone.

The reader may wonder how we arrived at the switching counts of *20, 200, 200* and *200*. The answer is as follows:

◊ Starting out by using a 0th-degree filter is undesirable because of its poor ability to track changing data, but we have no choice – the 1st-degree filter must first be allowed to settle before its output can be used.

◊ The 1st-degree filter's *SRF* is obtained from Figure 12.18(a) as

$$SRF = (2(2n+3)/(n+1)^{(2)})^{1/2} \tag{12.2.70}$$

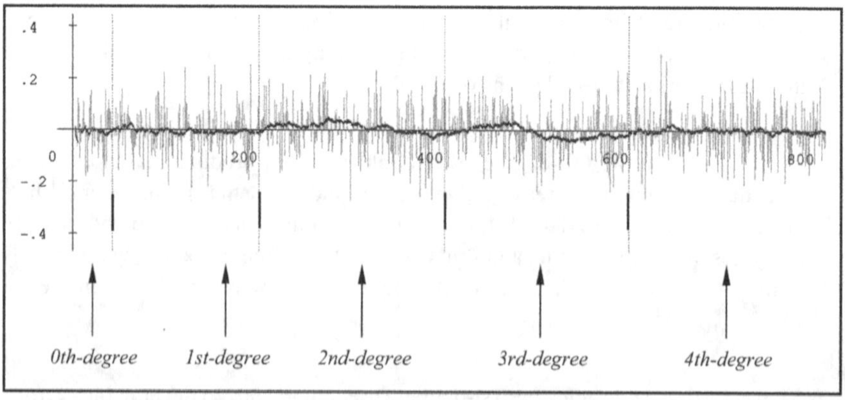

Figure 12.35 Quick settling in operation

and when $n = 20$ this evaluates to 0.452 which means that it has settled fairly well. We made the decision that we would accept this and that the 1st-degree filter's output would then be used. (It might not be what we would really like, but it gets the 0th-degree filter out of the way.)

◊ Once the 1st-degree filter is providing the output we have more breathing space, because it is a much better tracker than 0th-degree. We now examine the 2nd-degree *SRF* and find from Figure 12.18(a) that it is

$$SRF = (3(3n^2 + 9n + 8)/(n + 1)^{(3)})^{1/2} \tag{12.2.71}$$

When $n = 200$ this evaluates to 0.214 which is a number that is more to our liking, and so we adopted 200 as the value for switching to the 2nd-degree output.

◊ Waiting another 200 counts takes us to $n = 400$ by which time the *SRF* formula for 3rd-degree has dropped to 0.202 – again a number that we like. Hence we switch to 3rd-degree at $n = 400$.

◊ Waiting another 200 counts takes us to $n = 600$ by which time the *SRF* formula for 4th-degree has dropped to 0.206 – again a number that we like. Hence the final switch to 4th-degree is at $n = 600$.

For a live demonstration of EMP filters operating without and with quick settling, see Project 12.10.

12.2.26 Fixed-length and variable-length EMP filtering

We have seen that the EMP filters operate naturally in expanding-memory mode. In this section we describe how the EMP filters can also be made to operate in either fixed or variable-length mode.

The required configuration is depicted in Figure 12.36 below where we show a circular stack that serves as a data reservoir for an EMP filter of selected degree.

Operation is as follows:

◊ The switch is opened.

◊ When the datum y_n arrives the pointer P_1 shows the location where it is to be stored.

◊ The data are then processed sequentially by the (self-initializing) EMP filter commencing with the oldest datum at P_2 (y_{n-5} in this example) and ending with the most recent at P_1 namely y_n.

◊ The switch is then closed and the output of the filter – in this case an estimate based on the six input values $y_{n-5}, y_{n-4}, \ldots y_n$ – is made available to the next stage of the system. The filter now has a memory length $T_{mem} = 5\tau$.

◊ The switch is opened, pointer P_1 is incremented by *1*, and the filter awaits the arrival of the next datum.

◊ Pointer P_2 can either be incremented by *1*, left where it is, or relocated arbitrarily.

  ◊ If P_2 is always incremented by *1* then the filter will operate in fixed-memory mode with $T_{mem} = 5\tau$.

  ◊ If P_2 is left where it is then the filter will operate in expanding-memory mode.

  ◊ If P_2 is incremented or decremented by any other value then the filter will be operating in variable-memory mode.

When configured in this way, the EMP filter will run more slowly than it will in its native expanding-memory mode (see Figures 12.7 and 12.8). As an example, if it is running in fixed-memory mode with a memory length of 100τ then it will have to be cycled *101* times to produce a single output vector.

The fixed-memory throughput rate will then be about *1/101* of what it would be in its natural expanding-memory mode, and the cycle times in Figure 12.7 and 12.8 would have to be multiplied by *101*.

This slow-down can be somewhat mitigated by pre-computing the EMP filter weights for all values from $n = 0$ up to the maximum value for n that has been selected for the fixed memory length, and then retrieving them by look-up as the filter is being cycled. (This applies to both fixed-memory and variable-memory mode.)

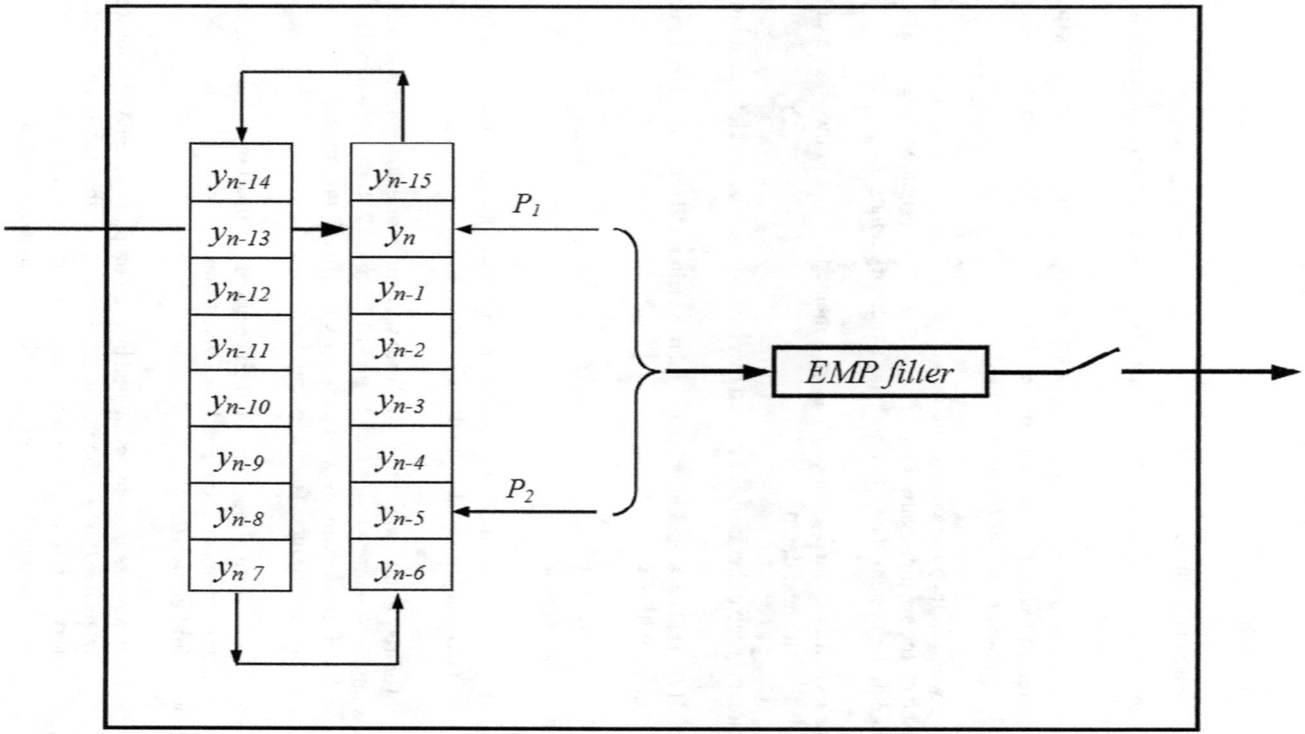

Figure 12.36 Fixed or variable memory EMP filter

In Chapter 7 of Reference 6 we derived the algorithms for the (nonrecursive) *fixed-length* memory polynomial filters. (They formed the starting point for the recursive EMP filters that we are now considering.)

One might be tempted to conclude that those filters will run faster than the fixed-length memory scheme based on the recursive EMP filters that are shown in Figure 12.36.

However, that is not the case. A careful count of the number of multiplications, additions and subtractions that are required for both schemes shows that the differences are only slight. Accordingly, we may as well stay with the recursive EMP filters with which we are now gaining familiarity, and use them as shown in Figure 12.36 to operate in fixed or variable memory mode.

◀▶

What we have been discussing gives us a way to operate *literally* in fixed-memory mode, at the cost of reduced throughput. But that is only one option.

In what now follows we discuss how polynomial filtering can be performed in *effective* fixed-memory mode, in a way that fully achieves the cycle times listed in Figures 12.7 and 12.8.

12.2.27 The composite EMP/FMP filter

The properties of the two filters – EMP and FMP – make them an ideal combination, and used together they form what we call the *composite EMP/FMP filter* shown in Figure 12.37.

In the following discussion we have the *1*-step predictors in mind, but everything applies equally well to the current-estimate filters.

The composite EMP/FMP filter combines the best of both worlds:

◊ The EMP filters are self-initializing but cannot track an arbitrary process indefinitely because of their expanding memories.

◊ The FMP filters are not self-initializing but *can* in track an arbitrary process indefinitely because of their effective fixed-length memories.

When operated as we now describe, the cycle times of Figures 12.7 and 12.8 are fully achieved, and so we can accomplish fixed-memory filtering without any slowdown.

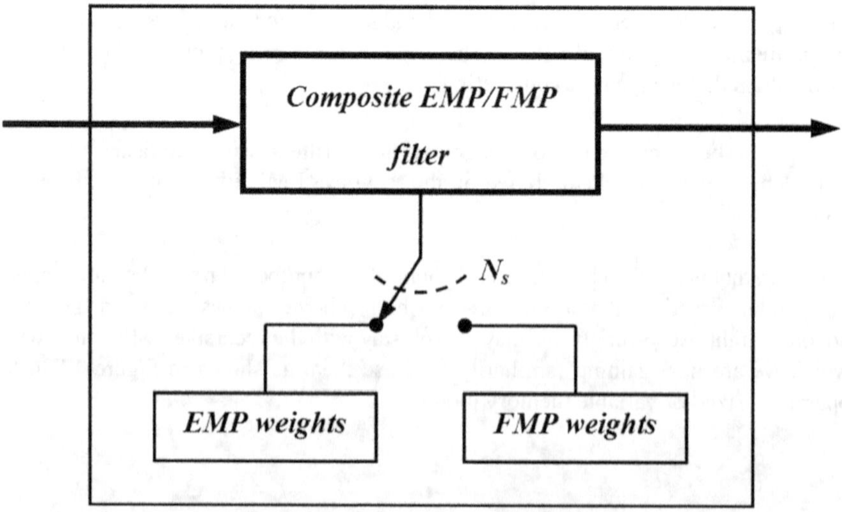

Figure 12.37 Composite EMP/FMP filter

Operation of the scheme is as follows:

◊ At start-up the switch in Figure 12.37 is set to the EMP weights $(\alpha(n), \beta(n), \gamma(n), \ldots)$, and so the composite filter's properties are those of the self-initializing EMP filter with quick settling. (The latter is *always* used.) However, the filter cannot track an arbitrary process indefinitely.

◊ At some point (well after quick settling has ended) the switch is moved to the FMP weights $(\alpha(\theta), \beta(\theta), \gamma(\theta), \ldots)$, and the composite filter's properties then become those of the FMP filter with its effective fixed memory and its ability to track arbitrary, non-polynomial, processes indefinitely.

The cycle number when the switch is moved to the right is called N_S (read as *N-switch*), and if the latter is properly selected then the move from EMP to FMP takes place seamlessly and without transients. We now discuss how one arrives at the required value of N_S.

Assume that an FMP filter has been in operation for some time, and is now running in a steady-state, transient-free mode. Then one way to view this is as follows:

◊ The previous vector $\boldsymbol{Z}^*_{n-1,n-2}$ was able to 'initialize' the FMP filter so that it produces the next vector $\boldsymbol{Z}^*_{n,n-1}$ without a transient being excited.

There are three attributes that distinguish that previous $\boldsymbol{Z}^*_{n-1,n-2}$ from any other vector of the same dimension:

1. The values of its elements are close to the error-free values of the normalized observed state-vector.

2. Its elements have in them random errors that are drawn from a population whose covariance matrix is $\boldsymbol{S}^*$ for the FMP filter with the selected degree and value of θ.

3. They also have in them a particular set of bias errors. (Hopefully these are small.)

If we could obtain a vector that is similar to $\boldsymbol{Z}^*_{n-1,n-2}$ in the sense of (1) and (2), then it could be used to initialize the FMP filter when the switch in the composite filter is moved to the right. Transients should then not occur.

Such a vector *can* indeed be found, and we now described how that is done.

In Figure 12.38 we show two *VRF* curves – one for the EMP filter of the selected degree, and the other for the FMP filter of the same degree and the selected value of θ.

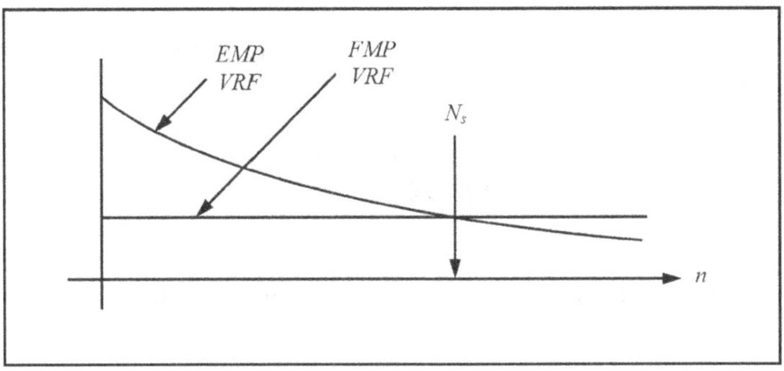

Figure 12.38 EMP and FMP VRF curves

The *VRF* curves for the EMP filters all go to zero asymptotically as n increases (see Figure 12.19), whereas the *VRF* curves for the FMP filters are all constants whose values depend on the value of θ (see Figure 12.22). Thus the EMP curve will *always* cross the corresponding FMP curve at a certain cycle number.

We call that cycle number N_S and we now show that the EMP estimate vector at that instant is very similar to the FMP vector $\boldsymbol{Z}^*_{n-1,n-2}$ that we discussed above. Thus if the EMP estimate vector is used to initialize the FMP filter when $n = N_S$, then transients will not arise.

◊ Assume that the EMP filter in Figure 12.37 has been running from $n = 0$ until $n = N_S$.

◊ Assume also that a hypothetical transient-free FMP filter has been running in the background. (When the switch is thrown the EMP filter will be halted, and the FMP filter will move from background to foreground.)

On completion of the cycle for which $n = N_S$ we compare the foreground EMP and background FMP filters, and here's what we find:

◊ The values of the elements of their estimate vector $\boldsymbol{X}^*$ are close to the error-free values of the normalized observed state-vector.

◊ The *1, 1* elements of their covariance matrices will be equal because that was how N_S was selected. That in turn means that their entire covariance matrices will be similar because both are now based on fitting a polynomial to the same data set.

  This is better understood if we keep in mind that the value of θ is always very close to *1*, which in turn means that the decay of the FMP memory is very slow and so its smoothing properties are quite similar to those of the EMP filter.

◊ For the same reasons their output vectors will contain similar (hopefully small) bias errors.

Thus throwing the switch in Figure 12.37 when n is equal to N_S selected in this way, should result in a transient-free start-up of the FMP filter.

In Figures 12.39 to 12.42 below we show screen captures of the composite filter in operation. The input to all of the filters was a sequence of samples from a sinusoid whose peak values were ± 1.

As we see from the figures, the transition from EMP to FMP is completely seamless and transients are entirely avoided.

◊ The grey hash in the large figures shows the errors in the input, and the heavy black lines show the errors in the 0th-derivative outputs.

◊ The four large figures show the start-up of the EMP filters. They are spatter-free because quick settling has been used.

◊ The insets below them are enlargements of what takes place at the instant of switch-over from EMP to FMP.

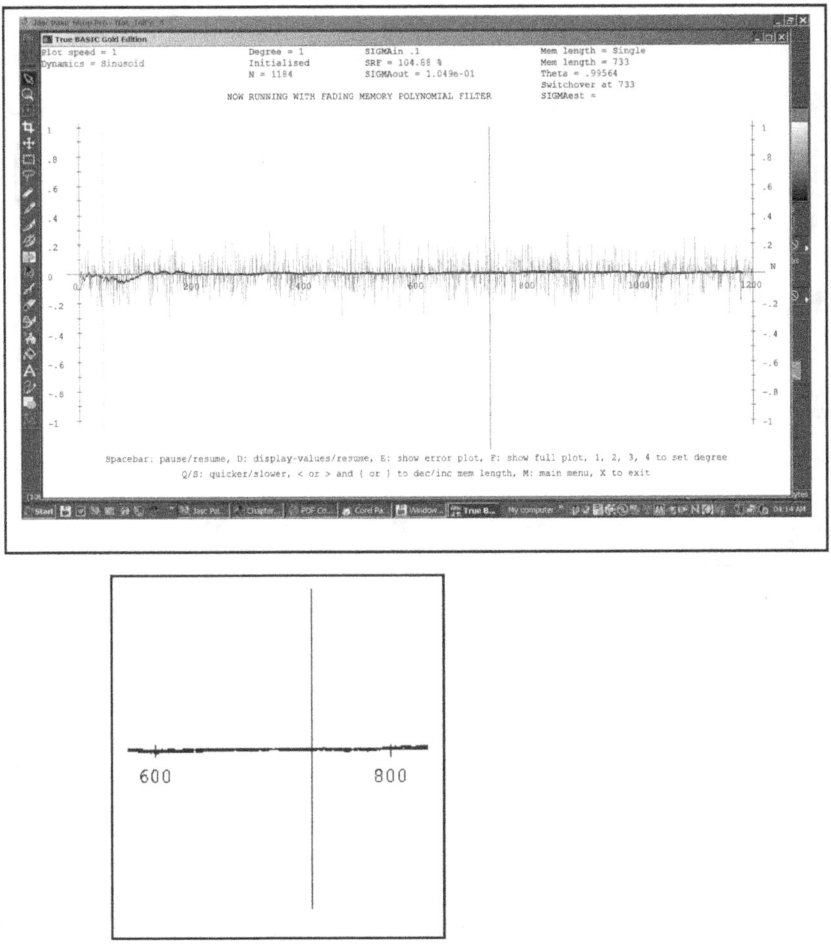

Figure 12.39 1st-degree

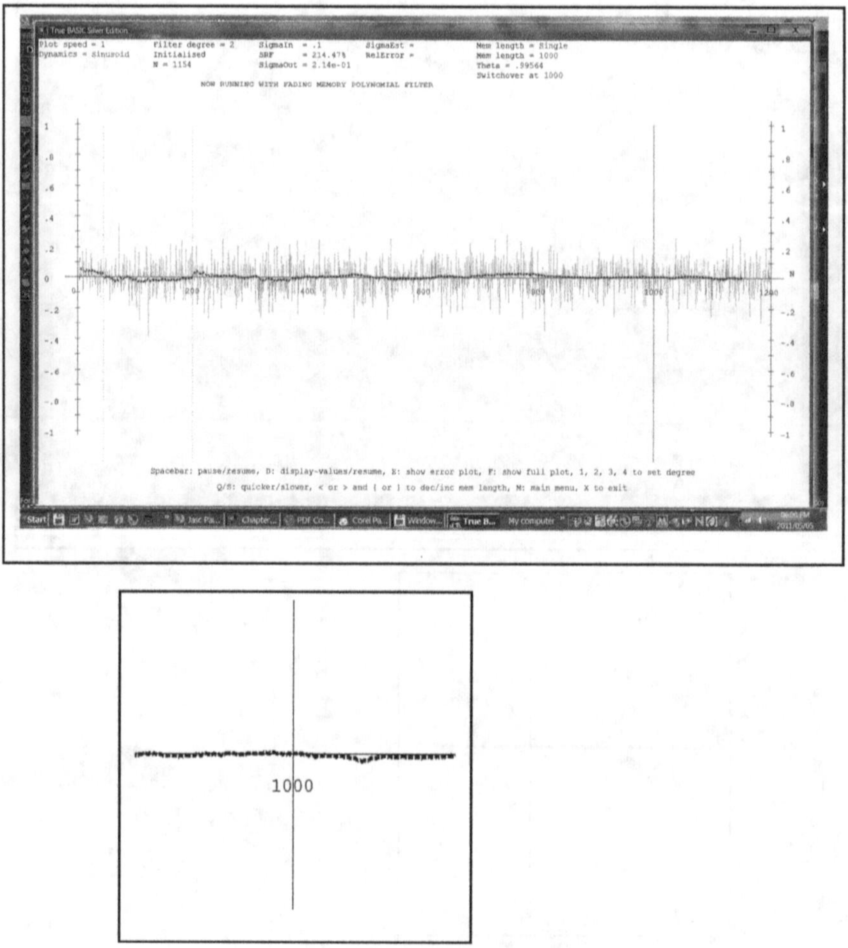

Figure 12.40 2nd-degree

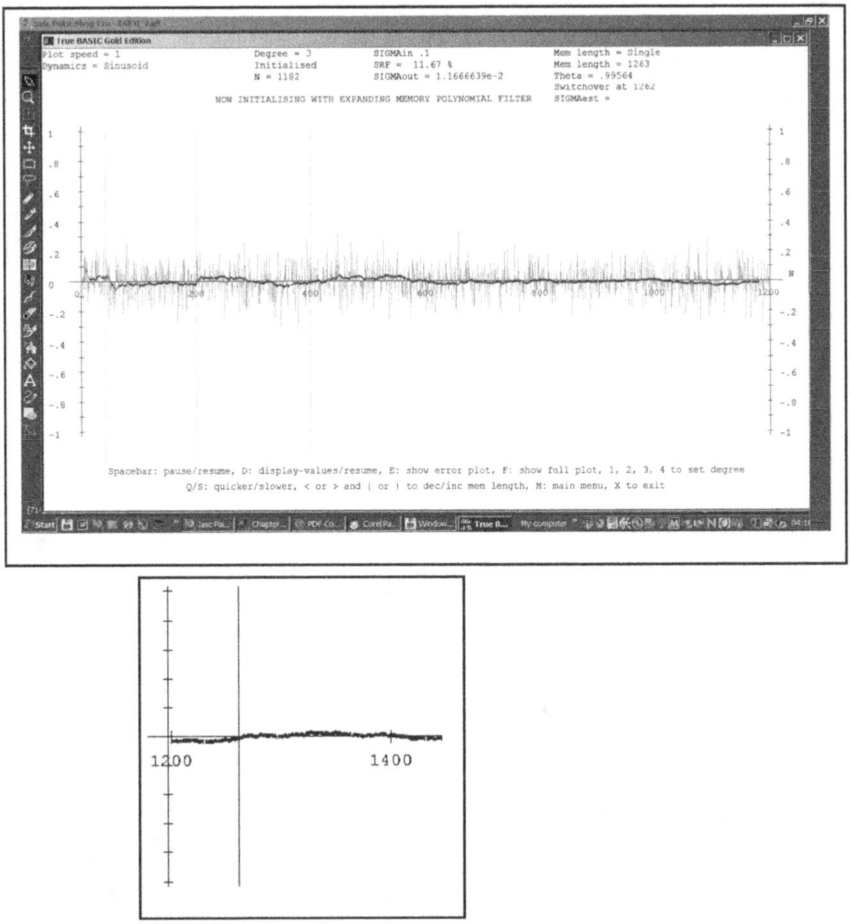

Figure 12.41 3rd-degree

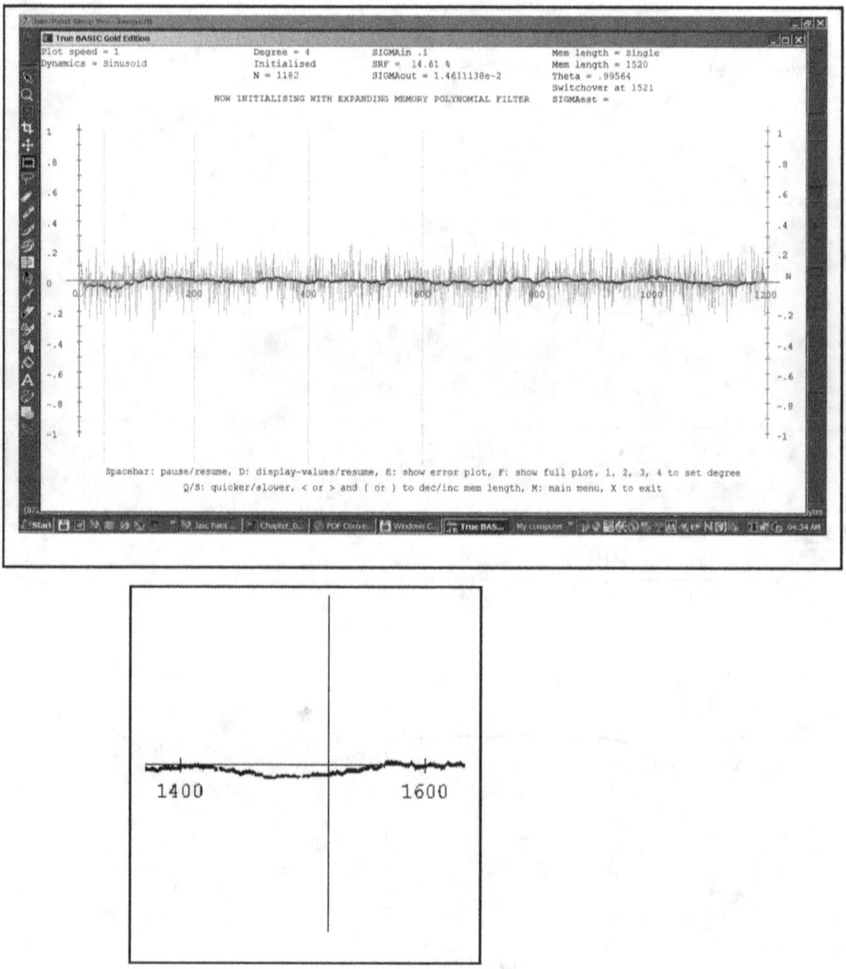

Figure 12.42 4th-degree

◊ The vertical lines in the enlargements are the switching instants when $n = N_S$.

◊ On the left of the vertical lines we see the EMP error plots.

◊ On the right of the vertical lines are the FMP error plots.

We observe that the transitions are all seamless, and that transients are entirely avoided.

◀▶

The following is an example of how the value of N_S is obtained:

Example 12.12: Let the EMP and FMP filter degrees be *2*. From Figures 12.18 and 12.21 we have the expressions for the *VRFs* of the 0th derivatives of the two filters. Equating them gives:

$$3(3n^2 + 9n + 8)/(n + 1)^{(3)} = (\theta^4 + 6\theta^3 + 16\theta^2 + 24\theta + 19)(1 - \theta)/(1 + \theta)^5 \qquad (12.2.72)$$

We now make the same two simplifying assumptions that we made earlier, namely that

◊ *n is large* and so $3(3n^2 + 9n + 8)/(n + 1)^{(3)} \approx 9/n$

◊ *θ is close to 1* and so $(\theta^4 + 6\theta^3 + 16\theta^2 + 24\theta + 19)(1 - \theta)/(1 + \theta)^5 \approx 66(1 - \theta)/32$

Then (12.2.72) reduces to

$$9/n \approx 66(1 - \theta)/32 \qquad (12.2.73)$$

Solving for *n* gives us the following value at which the two *VRFs* are equal:

$$N_S \approx 4.36/(1 - \theta) \qquad (12.2.74)$$

Repeating the calculation for the remaining degrees $(0, 1, 3, 4)$, we obtain Figure 12.10.

Degree	*0*	*1*	*2*	*3*	*4*
N_s	$2/(1-\theta)$	$3.2/(1-\theta)$	$4.36/(1-\theta)$	$5.51/(1-\theta)$	$6.63/(1-\theta)$

Figure 12.43 N_S vs. degree

Once θ has been specified we can solve for N_S and so Figure 12.43 provides us with all that we need. ■■

Note 1: Figure 12.43 is similar to Figure 12.29 because N_S and L_{eff} are related as follows:

$$L_{eff} = N_S \tau \tag{12.2.75}$$

Note 2: In the document

Miscellaneous\Polynomial_filters\1-step_predictors_up_to_8th_degree

we display values for N_s up to 8th-degree.

Example 12.13:

a) Assume that we have selected 3rd-degree for filtering, and from the computation in Example 12.11 we have arrived at a value of $\theta = 0.999449$.

   From Figure 12.43 we see that $N_S = 5.51/(1 - \theta)$. For $\theta = 0.999449$ this evaluates to $N_S = 10{,}000$. We would thus cycle the composite filter *10,000* times using the EMP weights, at which point we would switch to the FMP weights.

b) We have selected 4th degree, and from the computation in Example 12.11 we have arrived at a value of $\theta = 0.999779$.

   From Figure 12.43 we see that $N_S = 6.63/(1 - \theta)$. For $\theta = 0.999779$ this evaluates to $N_S = 30{,}000$. We would thus cycle the composite filter *30,000* times using the EMP weights, at which point we would switch to the FMP weights. ■■

Example 12.14:

a) We have selected 3rd degree and an effective memory length for the FMP filter of $L_{eff} = 7\,sec$. The radar PRF is *1000 pps*. What is the required value of θ, and what value should we use for N_S when we switch the composite filter from EMP to FMP?

From Figure 12.29 we see that θ must satisfy

$$7 = 5.51\tau/(1 - \theta) \tag{12.2.76}$$

where $\tau = 1/1000$. This evaluates to $\theta = 0.99921$. From Figure 12.43, we then see that N_S and θ must satisfy

$$N_S = 5.51/(1 - \theta) \tag{12.2.77}$$

For $\theta = 0.99921$ this evaluates to $N_S = 7000$ ■■

Note 1: The switching count N_S derived by the above method is a *safe* number. However, it is possible to use a smaller value, such as for example $N_S/2$. The user should investigate this possibility further during simulation studies.

Note 2: In the program ***14_Orbit_Track*** we used a composite *2*nd-degree EMP/FMP filter as a prefilter and we wanted the EMP filter to run for $T_S = 50$ *seconds* before switching to the FMP. The radar has a prf of 2000/sec. Using the above method, we arrived at a value of $\theta = 0.9999564$. Prefiltering is discussed in the next section.

12.2.28 Prefiltering

The central idea of prefiltering is that one or more composite EMP/FMP polynomial filters act as front ends for one of the more powerful main filters that have been discussed earlier, namely Gauss–Newton, Swerling or Kalman.

The prefilters are cycled at their usual high data rates, and from time to time a set of their outputs is extracted and sent to the main filter, where more sophisticated computations are carried out.

Prefiltering is a win-win situation:

◊ The highly compact polynomial prefilters make only small computational demands on the host machine.

◊ If a tracking radar is involved, the prefilters could be those that are also being used to control the radar pedestal.

◊ The more powerful main filter operates at a lower cycle rate, thereby also reducing the computational load on the host machine.

◊ The main filter receives data of improved quality.

The following discussion examines prefiltering, and brings together most of the concepts discussed in the book.

Figure 12.44 shows a simplified block diagram of a radar system that is tracking a satellite. (The system is discussed in much greater detail in the documentation for the program ***14_Orbit_Track***.) The objective is to extract estimates of the satellite's Keplerians from the radar data.

The system uses three prefilters.

◊ On the left we see the radar data (ρ, θ, ψ) with variances $(\sigma_\rho^2, \sigma_\theta^2, \sigma_\psi^2)$ arriving at the data-rate of the radar's *prf*. The time between data arrivals, namely τ, is equal to the reciprocal of the prf. In this example we assume that the prf is equal to *2000 pps* and so $\tau = 0.5$ *msec.*

◊ The radar data are transformed from radar coordinates to Cartesian coordinates using (5.1.3), followed by a number of other transformations (see documentation for ***14_Orbit_Track***) to produce the vector $(x_F, y_F, z_F)^T$ which is in fixed inertial coordinates.

◊ The radar's diagonal covariance matrix is transformed in a similar way into the covariance matrix whose diagonal elements are $\sigma(x_F)^2$, $\sigma(y_F)^2$, $\sigma(z_F)^2$.

◊ The values $(x_F, y_F, z_F)^T$ become the inputs to three 2nd-degree EMP/FMP composite filters. These are the prefilters.

◊ The outputs of the prefilters are

$$\boldsymbol{X}^* = (x_F^*, \dot{x}_F^*, \ddot{x}_F^*)^T \tag{12.2.78a}$$

with covariance matrix $\boldsymbol{S}^*(\boldsymbol{X}^*)$ whose *1, 1* element is $\sigma(x_F^*)^2$

$$\boldsymbol{Y}^* = (y_F^*, \dot{y}_F^*, \ddot{y}_F^*)^T \tag{12.2.78b}$$

with covariance matrix $\boldsymbol{S}^*(\boldsymbol{Y}^*)$ whose *1, 1* element is $\sigma(y_F^*)^2$

$$\boldsymbol{Z}^* = (z_F^*, \dot{z}_F^*, \ddot{z}_F^*)^T \tag{12.2.78c}$$

with covariance matrix $\boldsymbol{S}^*(\boldsymbol{Z}^*)$ whose *1, 1* element is $\sigma(z_F^*)^2$

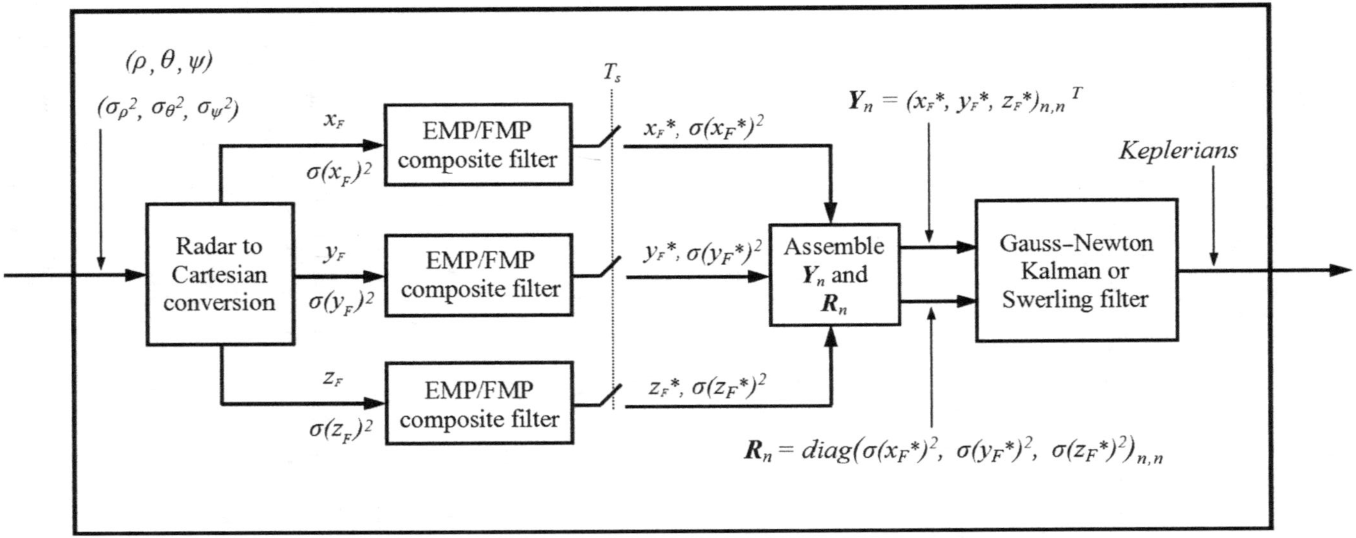

Figure 12.44 Prefiltering block diagram

◊ Three main filters are implemented in ***14_Orbit_Track*** – a Kalman filter, a Swerling filter and a Gauss–Newton filter, all of them operating in the x_F, y_F, z_F system. These filters estimate the Cartesian coordinates of the satellite which are then converted to Keplerians. The main filters are initialized using the estimates and covariance matrices of the three EMP prefilters. Initialization coincides with the instant of switch from EMP to FMP.

◊ After the main filters have been initialized, the FMP prefilters are constantly sampled every T_s seconds, and the following six items are extracted from (12.2.78):

x_F^* and its variance $\sigma\,(x_F^*)^2$

y_F^* and its variance $\sigma\,(y_F^*)^2$

z_F^* and its variance $\sigma\,(z_F^*)^2$

◊ These items are assembled into the vector

$$\boldsymbol{Y}_n = (x_F^*,\, y_F^*,\, z_F^*)_{n,n}^T \tag{12.2.79}$$

and its covariance matrix

$$\boldsymbol{R}_n = diag\left(\sigma\,(x_F^*)^2,\ \sigma\,(y_F^*)^2,\ \sigma\,(z_F^{*})^2\right)_{n,n} \tag{12.2.80}$$

◊ $\boldsymbol{Y}_n$ and $\boldsymbol{R}_n$ then become the inputs to the main filters.

- ◊ If the main filter is Gauss–Newton, then $\boldsymbol{Y}_n$ and $\boldsymbol{R}_n$ are appended to the top of the filter's memory stack and become part of $\mathsf{\boldsymbol{Y}}_n$ and $\mathsf{\boldsymbol{R}}_n$ that are then processed to give estimates for the Keplerians.

- ◊ If the main filter is Kalman or Swerling, then $\boldsymbol{Y}_n$ and $\boldsymbol{R}_n$ are processed directly to give estimates for the Keplerians.

Comments on correlation

◊ The radar data $(\rho,\, \theta,\, \psi)$ are uncorrelated, and so their covariance matrix is diagonal.

◊ After transformation from radar to Cartesian coordinates, the resulting covariance matrix is no longer diagonal, meaning that the inputs to the three prefilters are correlated between x, y and z.

◊ The outputs of each of the prefilters are stage-wise correlated, because those filters are recursive.

We can ignore both of these correlation factors for the reasons discussed in Project 13.2.

◂▸

We now do the filter engineering for the prefiltering system:

◊ Assume that the EMP/FMP prefilters are of 2nd-degree.

◊ Select an effective memory length of $L_{eff} = 50\,sec$. This means that we are approximating the orbit by 2nd-degree polynomials that are fitted over *50 sec*, and so the bias errors will be negligible. (See Problem 12.11.) It also means that we are initializing by using a *50-second* 2nd-degree polynomial fit to the orbit.

◊ From Figure 12.29 we have $L_{eff} = 4.36\tau/(1 - \theta)$. For $L_{eff} = 50\,sec$ and $\tau = 0.5\,msec$ this evaluates to $\theta = 0.9999564$ which will then be used in the FMP prefilters.

◊ From Figure 12.43 we derive the value of *n* when the switching of the prefilters from EMP to FMP takes place. The formula from the figure is $N_S = 4.36/(1 - \theta)$. For $\theta = 0.9999564$ this evaluates to $N_S = 100{,}000$.

◊ The outputs of the prefilters are sampled once every T_S seconds. We select $T_S = {}^1\!/_2\ sec$.

◊ The prefilters are implemented as *1*-step predictors because of their superior computational efficiency. Whenever they are sampled, we use the transition matrix appearing in (2.3.7) to back-shift the validity instant in each of their output vectors, thereby obtaining the current-estimate vectors. We also back-shift their covariance matrices to obtain the current-estimate covariance matrices.

◊ From each of the samples from the prefilter outputs we extract the 0th derivative values to form

$$\boldsymbol{Y}_n = (x_F^*,\, y_F^*,\, z_F^*)_{n,n}^T \tag{12.2.81}$$

◊ The covariance matrices of the composite prefilters are 3×3. For every sample, we extract the *1, 1* item from each of those matrices. Then these are the variances of the quantities appearing in (12.2.81).

Let those three variances be called $\sigma(x^*_F)^2$, $\sigma(y^*_F)^2$, $\sigma(z^*_F)^2$. These are assembled into the following 3×3 diagonal covariance matrix:

$$\boldsymbol{R}_n = diag\big(\sigma(x^*_F)^2,\ \sigma(y^*_F)^2,\ \sigma(z^*_F)^2\big)_{n,n} \tag{12.2.82}$$

Note: Because we are neglecting the correlation between x_F, y_F, z_F going into the prefilters, it follows that their outputs are being regarded as uncorrelated from one prefilter to the other, and for that reason the covariance matrix $\boldsymbol{R}_n$ in (12.2.82) is diagonal.

◊ The vector $\boldsymbol{Y}_n$ in (12.2.81) and its covariance matrix $\boldsymbol{R}_n$ in (12.2.82) are sent to the main filters.

 ◊ If the main filter is Gauss–Newton, then successive short-vectors $\boldsymbol{Y}_{n-k}$ $(0 \le k \le L)$ are concatenated to form the total observation vector $\boldsymbol{Y}_n$ which becomes the input to that filter. Further down we state that the short-vectors $\boldsymbol{Y}_{n-k}$ can be regarded as being stage-wise uncorrelated, and so the covariance matrix of $\boldsymbol{Y}_n$ will be diagonal and made up of the diagonal matrices $\boldsymbol{R}_{n-k}$ $(0 \le k \le L)$.

 ◊ If the main filter is either Kalman or Swerling then concatenation is not required, and $\boldsymbol{Y}_n$ and $\boldsymbol{R}_n$ are the inputs to those filters.

In conclusion, we note the following:

1. Because $T_s >> \tau$ it follows that the cycling rate of the main filters is a small fraction of the radar's PRF, and so reduced machine resources are required. (This is the first objective of prefiltering.)

2. The inputs to the main filters have been prefiltered by the composite prefilters, and so the data that the main filters receive is of improved quality compared to the raw radar data. (This is the second objective of prefiltering.)

3. Each of the composite prefilters produces successive outputs from their FMP filters that are stage-wise correlated (because those FMP filters are recursive) – stage-wise correlation which diminishes as the sampling interval T_s is increased. However, the effective memory length of the composite filters is *50 sec*, which is well in excess of the *1-second* sampling interval T_s and so we must acknowledge that stage-wise correlation will definitely be present between the successive prefilter samples that are being fed to the main filters.

 This violates the assumption on which the Kalman and Swerling filters are based, namely that their successive inputs be stage-wise uncorrelated.

The Gauss–Newton filter can accept input data that are stage-wise correlated, and if we wish to take that stage-wise correlation into account then we can do so by deriving the full expression for the total covariance matrix $\boldsymbol{R}_n$ that includes all of the stage-wise correlation terms.

We mentioned earlier (see comments on correlation) that the correlation between the prefilters is being ignored in the system depicted in Figure 12.44. We now also recognize that the stage-wise correlation in each of their outputs is being ignored.

In Project 13.2 we examine all of this in considerable detail, and we show that, while there is a price for ignoring the correlation, the price is manageable.

4. The program ***14_Orbit_Track*** uses prefiltering in a manner which is very similar to what we have just described. In the documentation for the program we discuss prefiltering and also discuss how the Keplerians are extracted from the main-filter Cartesian outputs.

12.2.29 The sigma monitoring test

In Chapter 1 we spoke about eight performance monitoring tests.

◊ The first three were tests for ECM consistency – the matrix-to-matrix, the *3*-sigma and the Chi-squared ECM-consistency tests. These were discussed in detail in Chapter 7, and as pointed out there, they can only be used in simulations.

◊ The next three were tests for CR-consistency – the matrix-to-matrix, *3*-sigma and Chi-squared CR-consistency tests. These were discussed in Chapter 9 and as noted there, they can also only be used in simulations.

◊ The seventh test – called the goodness-of-fit (GOF) test – played a central role in the master control algorithm (MCA), two versions of which were discussed in Chapter 10. The GOF test can also be implemented stand-alone, i.e. unrelated to the MCA. Because the GOF test is based on the residuals and does not require values of the true state vector, it can be used both in simulations as well as in the field.

We come now to the eighth test which we call the *sigma monitoring (SM) test*. Like the GOF test, the SM test is also based on the residuals and does not require values of the true state vector, and so it too can be used both in simulations as well as in the field.

◂▸

Referring to Figure 12.45 below, we assume that we are running one or more FMP filters that are receiving data from an observation instrument.

In the figure we show only a single observation channel, whereas in a radar system, for example, there would typically be at least three, each with its own SM test.

For the single channel shown in the figure, the data stream from the observation instrument, whose generic name is y_n, is being smoothed by an FMP filter whose output is the vector with the generic name $\boldsymbol{X}^*$. The covariance matrix of $\boldsymbol{X}^*$ is shown as $\boldsymbol{S}^*(\boldsymbol{X}^*)$.

We are assuming that the data stream y_n has an actual standard deviation called $\sigma(y_n)_{actual}$. This is (hopefully) the same as what the engineers who built the instrument have told us, but it may not be.

The SM test uses y_n and the first element of $\boldsymbol{X}^*$ namely $x_0{}^*_{n,n}$ (assuming that we are using a current-estimate FMP filter) to compute the quantity

$$r_n^2 = (y_n - x_0{}^*_{n,n})^2 \tag{12.2.83}$$

Then r_n^2 is the squared residual at time t_n. Successive values of r_n^2 are then passed through a 0th-degree EMP filter, and after square-rooting, what emerges are *estimates* of the quantity $\sigma(y_n)_{actual}$ which we call $\sigma(y_n)_{estimated}$.

Note: If the FMP filter is being operated as a *1*-step predictor, then the output $\boldsymbol{X}^*_{n+1,n}$ must first be shifted back (inside the SM test), using the polynomial transition equation

$$\boldsymbol{X}^*_{n,n} = \boldsymbol{\Phi}(-\tau)\boldsymbol{X}^*_{n+1,n} \tag{12.2.84}$$

In practice the value $\sigma(y_n)_{actual}$ is unknown to us, and in its place we use the value that the engineers who built the observation instrument provide, which we call $\sigma(y_n)_{engineers}$.

The SM test then operates as follows:

◊ If the $\sigma(y_n)_{engineers}$ is acceptably close to the actual value $\sigma(y_n)_{actual}$ then the estimate $\sigma(y_n)_{estimated}$ from the SM test should be close to $\sigma(y_n)_{engineers}$ and all is then in order.

◊ If the estimate $\sigma(y_n)_{estimated}$ from the SM test is not close to $\sigma(y_n)_{engineers}$ then we can infer that something is wrong, and that there is a problem needs to be resolved.

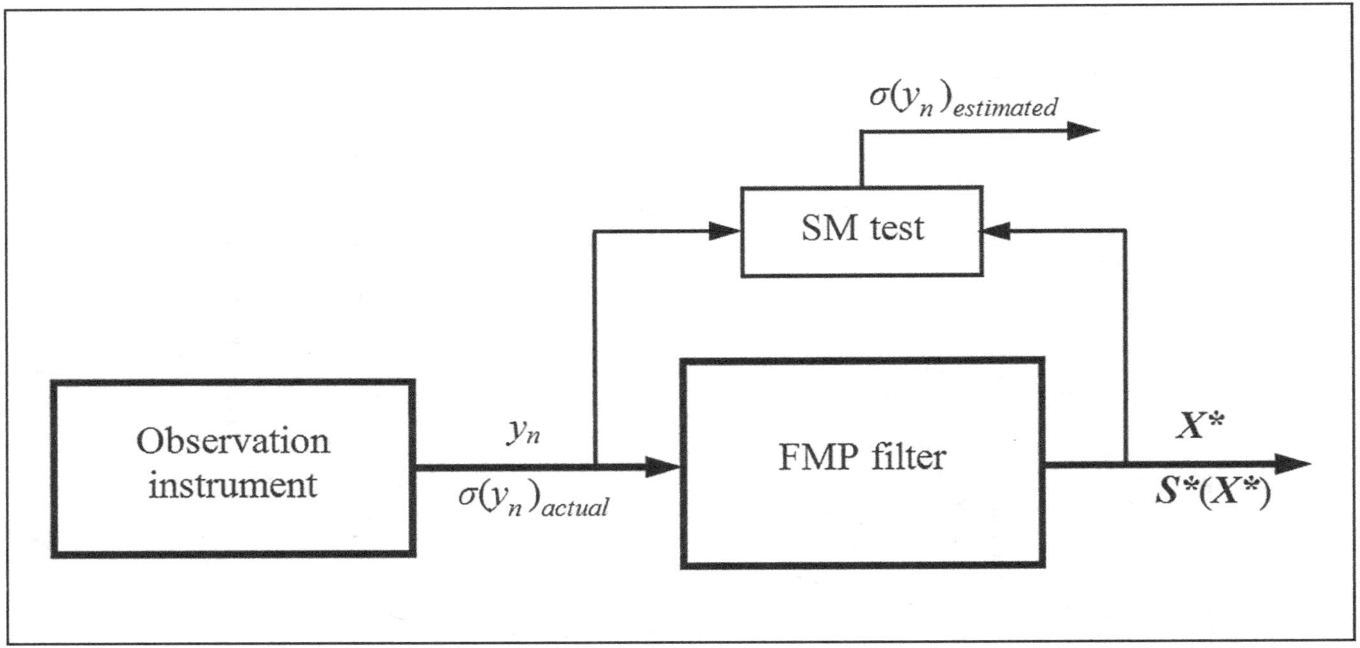

Figure 12.45 The SM test

Such problems might be present during the very first time that the system is used, or they could arise later when the system is already in operation.

◆

The complete theory underlying the SM test is given in Project 12.17.

The actual code for the SM test is surprisingly short, and in Project 12.17 we list *all four lines* of that code as it appears in the subroutine ***Sub_EstimateVar*** in the program ***06_Pol_Filter_FMP***.

The SM test can be seen in action in three of the programs. To do that, follow the instructions given in Project 12.11.

◆

If radar data are converted to Cartesians before entering the FMP filters and we are using an architecture like the one shown in Figure 12.44, then the SM test can still play a valuable role by estimating the three values $\sigma(x)^2$, $\sigma(y)^2$, $\sigma(z)^2$ shown in that figure.

In the text listing ***14_Orbit_Track\Seq_Gauss_Newton.txt*** you will find the subroutine ***EstimateVar*** that derives the estimates of $\sigma(x_F)$, $\sigma(y_F)$, $\sigma(z_F)$.

12.2.30 Conclusion

This completes our first look the properties of the EMP and FMP filters. In the next chapter we consider how their equations are derived, and we also derive the expressions for their covariance matrices.

The following items are in the supplementary material

Problems for Chapter 12

Solutions to Problems for Chapter 12

Projects for Chapter 12

Appendix 12.1
One-step predictor EMP algorithms, degrees 0 to 4

In this appendix we display the *1*-step predictor EMP algorithms for degrees *0* to *4*.

In the folder ***Miscellaneous*** there is a file called ***1-step_predictors_up_to_8th_degree*** in which we display the EMP algorithms for degrees *0* to *8*, and also give a brief summary of how they were derived.

◀▶

$$e_n = y_n - z_0{}^*_{n,n-1} \qquad \textit{In all cases n starts at 0}$$

$$(n+2)^{(2)} \textit{ means } (n+2)(n+1) \qquad (n+3)^{(3)} \textit{ means } (n+3)(n+2)(n+1) \textit{ etc.}$$

0th-degree

$$z_0{}^*_{n+1,n} = z_0{}^*_{n,n-1} + \alpha e_n \qquad \alpha = 1/(n+1)$$

1st-degree

$$z_1{}^*_{n+1,n} = z_1{}^*_{n,n-1} + \beta e_n \qquad \beta = 6/(n+2)^{(2)}$$

$$z_0{}^*_{n+1,n} = z_0{}^*_{n,n-1} + z_1{}^*_{n+1,n} + \alpha e_n \qquad \alpha = 2(2n+1)/(n+2)^{(2)}$$

2nd-degree

$$z_2{}^*_{n+1,n} = z_2{}^*_{n,n-1} + \gamma e_n \qquad \gamma = 30/(n+3)^{(3)}$$

$$z_1{}^*_{n+1,n} = z_1{}^*_{n,n-1} + 2z_2{}^*_{n+1,n} + \beta e_n \qquad \beta = 18(2n+1)/(n+3)^{(3)}$$

$$z_0{}^*_{n+1,n} = z_0{}^*_{n,n-1} + z_1{}^*_{n+1,n} - z_2{}^*_{n+1,n} + \alpha e_n \qquad \alpha = 3(3n^2+3n+2)/(n+3)^{(3)}$$

3rd-degree

$$z_3{}^*_{n+1,n} = z_3{}^*_{n,n-1} + \delta e_n$$

$$z_2{}^*_{n+1,n} = z_2{}^*_{n,n-1} + 3z_3{}^*_{n+1,n} + \gamma e_n$$

$$z_1{}^*_{n+1,n} = z_1{}^*_{n,n-1} + 2z_2{}^*_{n+1,n} - 3z_3{}^*_{n+1,n} + \beta e_n$$

$$z_0{}^*_{n+1,n} = z_0{}^*_{n,n-1} + z_1{}^*_{n+1,n} - z_2{}^*_{n+1,n} + z_3{}^*_{n+1,n} + \alpha e_n$$

$$\delta = 140/(n+4)^{(4)}$$

$$\gamma = 120(2n+1)/(n+4)^{(4)}$$

$$\beta = 20(6n^2+6n+5)/(n+4)^{(4)}$$

$$\alpha = 8(2n^3+3n^2+7n+3)/(n+4)^{(4)}$$

4th-degree

$$z_4{}^*_{n+1,n} = z_4{}^*_{n,n-1} + \varepsilon e_n$$

$$z_3{}^*_{n+1,n} = z_3{}^*_{n,n-1} + 4z_4{}^*_{n+1,n} + \delta e_n$$

$$z_2{}^*_{n+1,n} = z_2{}^*_{n,n-1} + 3z_3{}^*_{n+1,n} - 6z_4{}^*_{n+1,n} + \gamma e_n$$

$$z_1{}^*_{n+1,n} = z_1{}^*_{n,n-1} + 2z_2{}^*_{n+1,n} - 3z_3{}^*_{n+1,n} + 4z_4{}^*_{n+1,n} + \beta e_n$$

$$z_0{}^*_{n+1,n} = z_0{}^*_{n,n-1} + z_1{}^*_{n+1,n} - z_2{}^*_{n+1,n} + z_3{}^*_{n+1,n} - z_4{}^*_{n+1,n} + \alpha e_n$$

$$\varepsilon = 630/(n+5)^{(5)}$$

$$\delta = 700(2n+1)/(n+5)^{(5)}$$

$$\gamma = 1050(n^2+n+1)/(n+5)^{(5)}$$

$$\beta = 50(6n^3+9n^2+23n+10)/(n+5)^{(5)}$$

$$\alpha = 5(5n^4+10n^3+55n^2+50n+24)/(n+5)^{(5)}$$

Appendix 12.2

One-step predictor FMP algorithms, degrees 0 to 4

In this appendix we display the *1*-step predictor FMP algorithms for degrees *0* to *4*.

In the folder ***Miscellaneous*** there is a file called ***1-step_predictors_up_to_8th_degree*** in which we display the FMP algorithms for degrees *0* to *8*, and also give a brief summary of how they were derived.

◀▶

$$e_n = y_n - z_0{}^*_{n,n-1}$$

0th-degree

$$z_0{}^*_{n+1,n} = z_0{}^*_{n,n-1} + \alpha e_n \qquad \alpha = 1 - \theta$$

1st-degree

$$z_1{}^*_{n+1,n} = z_1{}^*_{n,n-1} + \beta e_n \qquad \beta = (1 - \theta)^2$$

$$z_0{}^*_{n+1,n} = z_0{}^*_{n,n-1} + z_1{}^*_{n+1,n} + \alpha e_n \qquad \alpha = 1 - \theta^2$$

2nd-degree

$$z_2{}^*_{n+1,n} = z_2{}^*_{n,n-1} + \gamma e_n \qquad \gamma = {}^1/_2(1 - \theta)^3$$

$$z_1{}^*_{n+1,n} = z_1{}^*_{n,n-1} + 2z_2{}^*_{n+1,n} + \beta e_n \qquad \beta = {}^3/_2(1 - \theta)^2(1 + \theta)$$

$$z_0{}^*_{n+1,n} = z_0{}^*_{n,n-1} + z_1{}^*_{n+1,n} - z_2{}^*_{n+1,n} + \alpha e_n \qquad \alpha = 1 - \theta^3$$

3rd-degree

$$z_3{}^*_{n+1,n} = z_3{}^*_{n,n-1} + \delta e_n$$

$$z_2{}^*_{n+1,n} = z_2{}^*_{n,n-1} + 3z_3{}^*_{n+1,n} + \gamma e_n$$

$$z_1{}^*_{n+1,n} = z_1{}^*_{n,n-1} + 2z_2{}^*_{n+1,n} - 3z_3{}^*_{n+1,n} + \beta e_n$$

$$z_0{}^*_{n+1,n} = z_0{}^*_{n,n-1} + z_1{}^*_{n+1,n} - z_2{}^*_{n+1,n} + z_3{}^*_{n+1,n} + \alpha e_n$$

$$\delta = {}^1/_6(1-\theta)^4$$

$$\gamma = (1-\theta)^3(1+\theta)$$

$$\beta = {}^1/_6(1-\theta)^2(11 + 14\theta + 11\theta^2)$$

$$\alpha = 1 - \theta^4$$

4th-degree

$$z_4{}^*_{n+1,n} = z_4{}^*_{n,n-1} + \varepsilon e_n$$

$$z_3{}^*_{n+1,n} = z_3{}^*_{n,n-1} + 4z_4{}^*_{n+1,n} + \delta e_n$$

$$z_2{}^*_{n+1,n} = z_2{}^*_{n,n-1} + 3z_3{}^*_{n+1,n} - 6z_4{}^*_{n+1,n} + \gamma e_n$$

$$z_1{}^*_{n+1,n} = z_1{}^*_{n,n-1} + 2z_2{}^*_{n+1,n} - 3z_3{}^*_{n+1,n} + 4z_4{}^*_{n+1,n} + \beta e_n$$

$$z_0{}^*_{n+1,n} = z_0{}^*_{n,n-1} + z_1{}^*_{n+1,n} - z_2{}^*_{n+1,n} + z_3{}^*_{n+1,n} - z_4{}^*_{n+1,n} + \alpha e_n$$

$$\varepsilon = {}^1/_{24}(1-\theta)^5$$

$$\delta = {}^5/_{12}(1-\theta)^4(1+\theta)$$

$$\gamma = {}^5/_{24}(1-\theta)^3(7 + 10\theta + 7\theta^2)$$

$$\beta = {}^5/_{12}(1-\theta)^2(5 + 7\theta + 7\theta^2 + 5\theta^3)$$

$$\alpha = 1 - \theta^5$$

Appendix 12.3

Current estimate EMP algorithms, degrees 0 to 4

In this appendix we display the current estimate EMP algorithms for degrees *0* to *4*.

In the folder ***Miscellaneous*** there is a file called ***Current_estimate_up_to_8th_degree*** in which we display the EMP algorithms for degrees *0* to *8*, and also give a brief summary of how they were derived.

In all cases n starts at 0

$(n+2)^{(2)}$ *means* $(n+2)(n+1)$ $\quad$ $(n+3)^{(3)}$ *means* $(n+3)(n+2)(n+1)$, *etc.*

0th-degree

$$e_n = y_n - z_{0\,n-1,n-1}^{*}$$

$$z_{0\,n,n}^{*} = z_{0\,n-1,n-1}^{*} + \alpha e_n \qquad \alpha = 1/(n+1)$$

1st-degree

$$e_n = y_n - (z_{0\,n-1,n-1}^{*} + z_{1\,n-1,n-1}^{*})$$

$$z_{1\,n,n}^{*} = z_{1\,n-1,n-1}^{*} + \beta e_n \qquad \beta = 6/(n+2)^{(2)}$$

$$z_{0\,n,n}^{*} = z_{0\,n-1,n-1}^{*} + z_{1\,n-1,n-1}^{*} + \alpha e_n \qquad \alpha = 2(2n+1)/(n+2)^{(2)}$$

2nd-degree

$$e_n = y_n - (z_{0\,n-1,n-1}^{*} + z_{1\,n-1,n-1}^{*} + z_{2\,n-1,n-1}^{*})$$

$$z_{2\,n,n}^{*} = z_{2\,n-1,n-1}^{*} + \gamma e_n \qquad \gamma = 30/(n+3)^{(3)}$$

$$z_{1\,n,n}^{*} = z_{1\,n-1,n-1}^{*} + 2z_{2\,n-1,n-1}^{*} + \beta e_n \qquad \beta = 18(2n+1)/(n+3)^{(3)}$$

$$z_{0\,n,n}^{*} = z_{0\,n-1,n-1}^{*} + z_{1\,n-1,n-1}^{*} + z_{2\,n-1,n-1}^{*} + \alpha e_n \qquad \alpha = 3(3n^2+3n+2)/(n+3)^{(3)}$$

3rd-degree

$$e_n = y_n - (z_0{}^*_{n-1,n-1} + z_1{}^*_{n-1,n-1} + z_2{}^*_{n-1,n-1} + z_3{}^*_{n-1,n-1})$$

$$z_3{}^*_{n,n} = z_3{}^*_{n-1,n-1} + \delta e_n$$

$$z_2{}^*_{n,n} = z_2{}^*_{n-1,n-1} + 3z_3{}^*_{n-1,n-1} + \gamma e_n$$

$$z_1{}^*_{n,n} = z_1{}^*_{n-1,n-1} + 2\,z_2{}^*_{n-1,n-1} + 3z_3{}^*_{n-1,n-1} + \beta e_n$$

$$z_0{}^*_{n,n} = z_0{}^*_{n-1,n-1} + z_1{}^*_{n-1,n-1} + z_2{}^*_{n-1,n-1} + z_3{}^*_{n-1,n-1} + \alpha e_n$$

$$\delta = 140/(n+4)^{(4)}$$

$$\gamma = 120(2n+1)/(n+4)^{(4)}$$

$$\beta = 20(6n^2+6n+5)/(n+4)^{(4)}$$

$$\alpha = 8(2n^3+3n^2+7n+3)/(n+4)^{(4)}$$

4th-degree

$$e_n = y_n - (z_0{}^*_{n-1,n-1} + z_1{}^*_{n-1,n-1} + z_2{}^*_{n-1,n-1} + z_3{}^*_{n-1,n-1} + z_4{}^*_{n-1,n-1})$$

$$z_4{}^*_{n,n} = z_4{}^*_{n-1,n-1} + \varepsilon e_n$$

$$z_3{}^*_{n,n} = z_3{}^*_{n-1,n-1} + 4z_4{}^*_{n-1,n-1} + \delta e_n$$

$$z_2{}^*_{n,n} = z_2{}^*_{n-1,n-1} + 3z_3{}^*_{n-1,n-1} + 6z_4{}^*_{n-1,n-1} + \gamma e_n$$

$$z_1{}^*_{n,n} = z_1{}^*_{n-1,n-1} + 2\,z_2{}^*_{n-1,n-1} + 3z_3{}^*_{n-1,n-1} + 4z_4{}^*_{n-1,n-1} + \beta e_n$$

$$z_0{}^*_{n,n} = z_0{}^*_{n-1,n-1} + z_1{}^*_{n-1,n-1} + z_2{}^*_{n-1,n-1} + z_3{}^*_{n-1,n-1} + z_4{}^*_{n-1,n-1} + \alpha e_n$$

$$\varepsilon = 630/(n+5)^{(5)}$$

$$\delta = 700(2n+1)/(n+5)^{(5)}$$

$$\gamma = 1050(n^2+n+1)/(n+5)^{(5)}$$

$$\beta = 25(12n^3+18n^2+46n+20)/(n+5)^{(5)}$$

$$\alpha = 5(5n^4+10n^3+55n^2+50n+24)/(n+5)^{(5)}$$

Appendix 12.4

Current estimate FMP algorithms, degrees 0 to 4

In this appendix we display the current estimate FMP algorithms for degrees *0* to *4*.

In the folder ***Miscellaneous*** there is a file called ***Current_estimate_up_to_8th_degree*** in which we display the FMP algorithms for degrees *0* to *8*, and also give a brief summary of how they were derived.

0th-degree

$$e_n = y_n - z_0{}^*_{n-1,n-1}$$

$$z_0{}^*_{n,n} = z_0{}^*_{n-1,n-1} + \alpha e_n \qquad \alpha = 1 - \theta$$

1st-degree

$$e_n = y_n - (z_0{}^*_{n-1,n-1} + z_1{}^*_{n-1,n-1})$$

$$z_1{}^*_{n,n} = z_1{}^*_{n-1,n-1} + \beta e_n \qquad \beta = (1-\theta)^2$$

$$z_0{}^*_{n,n} = z_0{}^*_{n-1,n-1} + z_1{}^*_{n-1,n-1} + \alpha e_n \qquad \alpha = 1 - \theta^2$$

2nd-degree

$$e_n = y_n - (z_0{}^*_{n-1,n-1} + z_1{}^*_{n-1,n-1} + z_2{}^*_{n-1,n-1})$$

$$z_2{}^*_{n,n} = z_2{}^*_{n-1,n-1} + \gamma e_n \qquad \gamma = {}^1/_2(1-\theta)^3$$

$$z_1{}^*_{n,n} = z_1{}^*_{n-1,n-1} + 2z_2{}^*_{n-1,n-1} + \beta e_n \qquad \beta = {}^3/_2(1-\theta)^2(1+\theta)$$

$$z_0{}^*_{n,n} = z_0{}^*_{n-1,n-1} + z_1{}^*_{n-1,n-1} + z_2{}^*_{n-1,n-1} + \alpha e_n \qquad \alpha = 1 - \theta^3$$

3rd-degree

$$e_n = y_n - (z_0{}^*_{n-1,n-1} + z_1{}^*_{n-1,n-1} + z_2{}^*_{n-1,n-1} + z_3{}^*_{n-1,n-1})$$

$$z_3{}^*_{n,n} = z_3{}^*_{n-1,n-1} + \delta e_n$$

$$z_2{}^*_{n,n} = z_2{}^*_{n-1,n-1} + 3z_3{}^*_{n-1,n-1} + \gamma e_n$$

$$z_1{}^*_{n,n} = z_1{}^*_{n-1,n-1} + 2z_2{}^*_{n-1,n-1} + 3z_3{}^*_{n-1,n-1} + \beta e_n$$

$$z_0{}^*_{n,n} = z_0{}^*_{n-1,n-1} + z_1{}^*_{n-1,n-1} + z_2{}^*_{n-1,n-1} + z_3{}^*_{n-1,n-1} + \alpha e_n$$

$$\delta = {}^1/_6(1-\theta)^4$$

$$\gamma = (1-\theta)^3(1+\theta)$$

$$\beta = {}^1/_6(1-\theta)^2(11+14\theta+11\theta^2)$$

$$\alpha = 1-\theta^4$$

4th-degree

$$e_n = y_n - (z_0{}^*_{n-1,n-1} + z_1{}^*_{n-1,n-1} + z_2{}^*_{n-1,n-1} + z_3{}^*_{n-1,n-1} + z_4{}^*_{n-1,n-1})$$

$$z_4{}^*_{n,n} = z_4{}^*_{n-1,n-1} + \varepsilon e_n$$

$$z_3{}^*_{n,n} = z_3{}^*_{n-1,n-1} + 4z_4{}^*_{n-1,n-1} + \delta e_n$$

$$z_2{}^*_{n,n} = z_2{}^*_{n-1,n-1} + 3z_3{}^*_{n-1,n-1} + 6z_4{}^*_{n-1,n-1} + \gamma e_n$$

$$z_1{}^*_{n,n} = z_1{}^*_{n-1,n-1} + 2z_2{}^*_{n-1,n-1} + 3z_3{}^*_{n-1,n-1} + 4z_4{}^*_{n-1,n-1} + \beta e_n$$

$$z_0{}^*_{n,n} = z_0{}^*_{n-1,n-1} + z_1{}^*_{n-1,n-1} + z_2{}^*_{n-1,n-1} + z_3{}^*_{n-1,n-1} + z_4{}^*_{n-1,n-1} + \alpha e_n$$

$$\varepsilon = {}^1/_{24}(1-\theta)^5$$

$$\delta = {}^5/_{12}(1-\theta)^4(1+\theta)$$

$$\gamma = {}^5/_{24}(1-\theta)^3(7+10\theta+7\theta^2)$$

$$\beta = {}^5/_{12}(1-\theta)^2(5+7\theta+7\theta^2+5\theta^3)$$

$$\alpha = 1-\theta^5$$

Chapter 13
Polynomial filtering – 2

Chapter 12 was an introduction to the EMP and FMP filters and to some of the ways in which they can be used.

In this chapter we derive the equations for the filters and the expressions for their covariance matrices:

13.1 Deriving the EMP equations

13.1.1 The approximating polynomial

Referring to Figure 13.1 on the following page:

◊ The time axis starts at $t_0 = 0$ on the left and runs towards the right.

◊ The s axis starts at $s = 0$ on the left and runs towards the right.

◊ When t increases by τ seconds then s increases by 1. This means that

$$t = s\tau \tag{13.1.1}$$

◊ The observations are shown as vertical lines with small black circles on their tops. From right to left they form the total observation vector

$$\mathbf{Y}_n = (y_n, y_{n-1}, \ldots y_0)^T. \tag{13.1.2}$$

Observe that the most recent observation y_n is shown as the first element of $\mathbf{Y}_n$, namely at the 'top' of the column vector. That is important for the following reason:

In a non-recursive Gauss filter, the filter estimation instant will always be the same as the time-stamp of the first element in $\mathbf{Y}_n$.

- ◊ If the first time stamp in $\mathbf{Y}_n$ is t_n then the filter's estimation instant will be at the leading edge of the observation window.

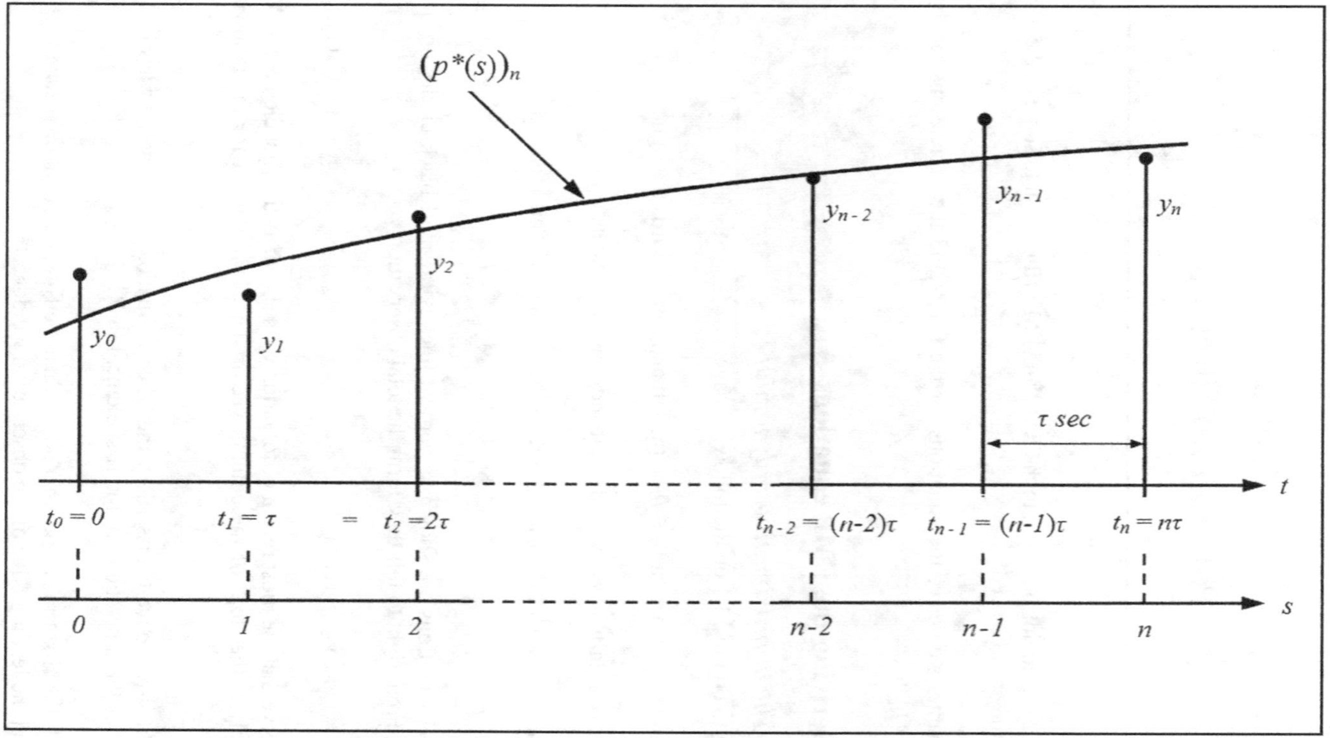

Figure 13.1 The EMP approximating polynomial

- If the first time stamp in $\mathbf{Y}_n$ is t_0 and so $\mathbf{Y}_n$ is as follows:

$$\mathbf{Y}_n = (y_0, y_1, \ldots y_n)^T. \tag{13.1.3}$$

then the filter's estimation instant will be at t_0, namely at the trailing edge of the observation window.

However, if the filter is recursive, like the ones that we are considering in this chapter, then the observations are submitted to the filter in the following order:

y_0 then y_1 then y_2 and so on.

- In Figure 13.1 the observed trajectory is not shown.

- The solid black curve shows the polynomial $(p^*(s))_n$ that has been fitted to $\mathbf{Y}_n$ at time t_n by the method of least squares. Observe the following:

  - The fitted polynomial is called p^*.

  - The polynomial's independent variable is called s.

  - The subscript n signifies that this was the polynomial that was fitted to the observations just after the nth observation was received.

- On the next cycle at time t_{n+1} the total observation vector will expand and will become

$$\mathbf{Y}_{n+1} = (y_{n+1},\ y_n,\ y_{n-1},\ \ldots\ y_0)^T \tag{13.1.4}$$

A completely new polynomial will then be fitted to $\mathbf{Y}_{n+1}$ and will be called $(p^*(s))_{n+1}$ which is different from $(p^*(s))_n$.

- The degrees of the successive polynomials $(p^*(s))_n$, $(p^*(s))_{n+1}, \ldots$ are all equal to m. As time passes, we are therefore fitting a sequence of differing polynomials of degree m to an expanding set of observations.

13.1.2 Classical least squares

Classical least squares starts by defining the polynomial $(p^*(s))_n$ as follows:

$$(p^*(s))_n = \beta_{0,n}s^0 + \beta_{1,n}s^1 + \beta_{2,n}s^2 + \cdots + \beta_{m,n}s^m \tag{13.1.5}$$

In this expression the polynomial is a sum of successively higher powers of the independent variable s, and $\beta_{0,n}, \beta_{1,n}, \ldots \beta_{m,n}$ are the polynomial's coefficients. On the left we are subscripting $(p^*(s))_n$ with an n because it is the polynomial that was fitted at time t_n and so we also subscript each of the β's with an n.

In order to obtain expressions for the β's in (13.1.5) we start from the sum of the squared residuals, namely

$$e_n = \sum_{k=0}^{n} \left(y_k - (p^*(k))_n \right)^2 \tag{13.1.6}$$

Summation in this equation is over the discrete set $k = 0$ to n, and so the continuous variable s in (13.1.5) has been replaced by the integer k as the polynomial argument.

The polynomial $(p^*(s))_n$ in (13.1.5) can be thought of as a vector that is made up of a linear combination of the set of basis vectors $\{s^0, s^1, s^2, \ldots s^m\}$.

This is perhaps the simplest basis that we might choose, but it has the following serious drawback:

◊ It will *always* lead to the need for a matrix inversion in order to find the required expressions for the β's.

To see that this is so, we refer you to Project 6.1 where the method of classical least squares was used.

Inverting a matrix is generally not a problem if its elements are numbers, but here we will have a matrix whose elements are functions of n and so its inversion can only be carried out (beyond a 2×2) by assigning numerical values to n.

However, assigning numerical values to n is something that we expressly do not wish to do, because it immediately puts an end to further theoretical analysis.

Accordingly, we start again, but this time we make use of a property called *orthogonality*, and we will see that it enables us to avoid matrix inversion entirely.

13.1.3 The discrete Legendre orthogonal polynomials

If you are already familiar with the concept of orthogonality in relation to polynomials then continue reading.

However, if the concept is new to you then we suggest that you first take a look at Project 13.1 in which the orthogonality of the discrete Legendre polynomials is explained and demonstrated by use of the program ***01_Orthogonality***. You will then have no difficulty understanding what lies ahead.

The discrete Legendre orthogonal polynomials satisfy the following orthogonality condition:

$$\sum_{s=0}^{n} p_i(s, n) p_j(s, n) = 0 \quad i \neq j \tag{13.1.7}$$

The quantities $p_i(s, n)$ and $p_j(s, n)$ in this equation are polynomials in s of degree i and j. Because summation in (13.1.7) is from 0 to n in which n can vary, it follows that the polynomials must also be dependent on n and so we show n as a parameter in the polynomial arguments.

What (13.1.7) states is the following:

◊ If we multiply together any two of the Legendre polynomials of differing degrees $(i \neq j)$ and sum the result from 0 to n, we obtain zero.

If the two polynomials have the same degree $(i = j)$ then clearly the sum in (13.1.7) cannot be zero.

Observe that the orthogonality condition in (13.1.7) involves a sum over a *discreate* set rather than an integral over a continuous interval – hence the use of the word *discrete* in the name of the polynomials. From now on when speaking about these polynomials we will drop the word discrete and its presence will always be assumed.[1]

It is shown in Chapter 3 of Reference 6 that the set of polynomials satisfying condition (13.1.7) has the following general expression:

$$p_j(s, n) = \sum_{\nu=0}^{j} (-1)^{\nu} \binom{j}{\nu} \binom{j+\nu}{\nu} \frac{s^{(\nu)}}{n^{(\nu)}} \qquad j = 0,\ 1,\ 2, \ldots n \tag{13.1.8}$$

For the meanings of $s^{(\nu)}$ and $n^{(\nu)}$ see Definition 12.1.

Note 1: In equation (13.1.8), ν is an integer, and because $n^{(\nu)}$ in the denominator is equal to 0 when $\nu > n$ it follows that ν cannot be allowed to exceed n. This in turn means that the upper limit of summation j (the polynomial degree) also cannot exceed n, and so this is a finite set of polynomials with $n + 1$ members. Their degrees are $0,\ 1,\ 2, \ldots n$.

Note 2: The definition for $p_j(s, n)$ in (13.1.8) looks formidable. To see what the polynomials actually look like, go to Appendix 13.1 where we display them as polynomials in their more usual form up to degree $j = 4$.

[1] For a discussion of the Legendre polynomials over a continuous interval, see References 74, 136 or 137. See also References 119 and 121 for further information on the discrete Legendre polynomials and References 138, 139, 144, 145, 146 and 150 on issues relating to approximation by polynomials.

Note 3: Although the orthogonality condition in (13.1.7) involves a discrete set of values, the independent variable s in (13.1.8) is a continuous one, and so these are polynomials in the ordinary sense.

Note 4: Later on we will need to differentiate $p_i(s, n)$ repeatedly with respect to s. However, the terms $s^{(\nu)} \equiv s(s - 1)(s - 2)\ldots(s - \nu + 1)$ are not in a form in which differentiation can be readily carried out. In Project 13.6 we provide you with the expressions that you need to convert $s^{(\nu)}$ into a simple polynomial which can then easily be differentiated repeatedly.

◄►

If we multiply together two of the Legendre polynomials with the same degrees $(i = j)$ and sum the result as we did in (13.1.7), we obtain the quantity $c_j(n)$ whose square is defined by

$$(c_j(n))^2 \equiv \sum_{s=0}^{n} (p_i(s, n))^2 \tag{13.1.9}$$

The general expression for $(c_j(n))^2$ for the Legendre polynomials is as follows (see Chapter 3 in Reference 6):

$$(c_j(n))^2 \equiv \frac{(n + j + 1)^{(j+1)}}{(2j + 1)n^{(j)}} \tag{13.1.10}$$

In Appendix 13.1 we also display simpler expressions for $(c_j(n))^2$ up to degree $j = 4$.

We now form the polynomials $\varphi_j(s, n)$ defined by

$$\varphi_j(s, n) \equiv \frac{1}{c_j(n)} p_j(s, n) \qquad j = 0, 1, 2, \ldots n \tag{13.1.11}$$

Then these polynomials satisfy the following two conditions:

◊ By (13.1.7) when $i \neq j$:

$$\sum_{s=0}^{n} \varphi_i(s, n)\varphi_j(s, n) = \sum_{s=0}^{n} \frac{1}{c_i(n)} p_i(s, n) \frac{1}{c_j(n)} p_j(s, n) = 0 \tag{13.1.12}$$

◊ By (13.1.9) when $i = j$:

$$\sum_{s=0}^{n} \left(\varphi_j(s,n)\right)^2 = \sum_{s=0}^{n} \frac{1}{\left(c_j(n)\right)^2} \left(p_j(s,n)\right)^2 = 1 \tag{13.1.13}$$

Taking (13.1.12) and (13.1.13) together enables us to write the following:

$$\sum_{s=0}^{n} \varphi_i(s,n)\varphi_j(s,n) = \delta_{i,j} \tag{13.1.14}$$

in which $\delta_{i,j}$ is the Kronecker delta which is 1 when $i = j$ and zero otherwise.

Equation (13.1.14) tells us that the polynomials $\varphi_j(s,n)$ form what is called an *orthonormal set* (see Project 13.1).

13.1.4 Using the Legendre orthogonal polynomials as the basis

We now return to the problem that we were considering earlier, namely that of finding the expression for the approximating polynomial in Figure 13.1 by the method of least squares. However, rather than using the definition for $\left(p^*(s)\right)_n$ that appeared in (13.1.5) we will use instead the following:

$$\left(p^*(s)\right)_n = \beta_{0,n}\varphi_0(s,n) + \beta_{1,n}\varphi_1(s,n) + \cdots + \beta_{m,n}\varphi_m(s,n) \tag{13.1.15}$$

in which φ was defined in (13.1.11). Equation (13.1.15) is the same as

$$\left(p^*(s)\right)_n = \sum_{j=0}^{m} \beta_{j,n}\varphi_j(s,n) \tag{13.1.16}$$

The polynomial $\left(p^*(s)\right)_n$ can again be viewed as a linear combination of a set of basis vectors $\{\varphi_0(s,n),\ \varphi_1(s,n), \ldots \varphi_m(s,n)\}$, and because these are polynomials whose degrees go from 0 to m, it follows that $\left(p^*(s)\right)_n$ is also a polynomial of degree m.

The basis that we are now using is a set of orthonormal polynomials, and as we will soon see, the matrix inversion that we spoke about earlier will not be required.

13.1.5 Least squares using the Legendre polynomials

In order to obtain the β's in (13.1.15) we again use the method of least squares and start from the sum of the squared residuals:

$$e_n = \sum_{k=0}^{n} \left(y_k - \left(p^*(k)\right)_n\right)^2 \tag{13.1.17}$$

As before, we are summing over the discrete set $k = 0$ to n, and so the integer k has replaced the continuous variable s as the argument of p^*.

We now use (13.1.16) to write e_n as follows:

$$e_n = \sum_{k=0}^{n} \left(y_k - \sum_{j=0}^{m} \beta_{j,n} \varphi_j(k, n) \right)^2 \tag{13.1.18}$$

As an example, let $m = 1$. Then (13.1.18) becomes

$$e_n = \sum_{k=0}^{n} (y_k - \beta_{0,n}\varphi_0(k, n) - \beta_{1,n}\varphi_1(k, n))^2 \tag{13.1.19}$$

Following the method of least squares, we now differentiate (13.1.18) with respect to each of the β's and set the results equal to zero. For this example where $m = 1$, we obtain the following two equations:

$$\sum_{k=0}^{n} (y_k - \beta_{0,n}\varphi_0(k, n) - \beta_{1,n}\varphi_1(k, n))\, \varphi_0(k, n) = 0 \tag{13.1.20}$$

and

$$\sum_{k=0}^{n} (y_k - \beta_{0,n}\varphi_0(k, n) - \beta_{1,n}\varphi_1(k, n))\, \varphi_1(k, n) = 0 \tag{13.1.21}$$

and in general, for any m, we would obtain the following $m + 1$ equations

$$\sum_{k=0}^{n} \left(y_k - \sum_{j=0}^{m} \beta_{j,n}\varphi_j(k, n) \right) \varphi_i(k, n) = 0 \qquad 0 \le i \le m \tag{13.1.22}$$

Reversing the order of summation in (13.1.22) and moving the terms involving y to the right, gives us

$$\sum_{j=0}^{m} \beta_{j,n} \sum_{k=0}^{n} \varphi_i(k, n)\varphi_j(k, n) = \sum_{k=0}^{n} \varphi_i(k, n) y_k \qquad 0 \le i \le m \tag{13.1.23}$$

Then, by virtue of (13.1.14) this becomes

$$\sum_{j=0}^{m} \beta_{j,n}\delta_{i,j} = \sum_{k=0}^{n} \varphi_i(k,n)y_k \qquad 0 \le i \le m \tag{13.1.24}$$

As an example, assume again that $m = 1$. Writing out (13.1.24) in full we obtain the following two equations:

$$i = 0: \quad \beta_{0,n}\delta_{0,0} + \beta_{1,n}\delta_{0,1} = \sum_{k=0}^{n} \varphi_0(k,\, n)y_k \tag{13.1.25}$$

$$i = 1: \quad \beta_{0,n}\delta_{1,0} + \beta_{1,n}\delta_{1,1} = \sum_{k=0}^{n} \varphi_1(k,\, n)y_k \tag{13.1.26}$$

These can now be written in matrix/vector format as

$$\begin{pmatrix} 1 & 0 \\ 0 & 1 \end{pmatrix} \begin{pmatrix} \beta_{0,n} \\ \beta_{1,n} \end{pmatrix} = \begin{pmatrix} \sum\limits_{k=0}^{n} \varphi_0(k,\, n)y_k \\ \sum\limits_{k=0}^{n} \varphi_1(k,\, n)y_k \end{pmatrix} \tag{13.1.27}$$

from which it is clear that matrix inversion will not be required to solve for the β's.

The above was for $m = 1$. For m any positive integer, equation (13.1.24) reduces to the following:

$$\beta_{i,n} = \sum_{k=0}^{n} \varphi_i(k,\, n)y_k \qquad 0 \le i \le m \tag{13.1.28}$$

By the use of orthogonality we have obtained the general expression for the β's without the need for matrix inversion, and so the analysis can continue.

Note: On the right of (13.1.28) the variable k sums out, and so the result is a function of n. This is consistent with the notation on the left.

13.1.6 Using the β's to write the approximating polynomial

The expression for the coefficients $\beta_{i,n}$ in (13.1.28) now enables us to write the least-squares approximating polynomial in (13.1.16) as follows:

$$\left(p^*(s)\right)_n = \sum_{j=0}^{m} \left(\sum_{k=0}^{n} \varphi_j(k,\, n)y_k \right) \varphi_j(s,n) \tag{13.1.29}$$

Equation (13.1.29) is a sum of the polynomials $\varphi_j(s, n)$ whose degrees run from $j = 0$ to m, and so it is a polynomial of degree m.

13.1.7 Estimating the true state vector

Returning now to Figure 13.1 we see that the polynomial $(p^*(s))_n$ in (13.1.29) can be used to provide estimates of $x(t)$, the true state vector's 0th derivative, by assigning various values to s.

For example, if we set $s = n$ then we obtain the current-estimate, i.e. an estimate of what $x(t)$ is at time t_n based on observations up to time t_n. Calling that estimate $x_0{}^*{}_{n,n}$ we are then able to write the following:

$$\left(x(t_n) \text{ based on observations up to } t_n\right) \approx \left(x_0{}^*{}_{n,n} = (p^*(n))_n\right) \tag{13.1.30}$$

The subscript 0 in $x_0{}^*{}_{n,n}$ signifies that this is for the 0th derivative.

Similarly, for a 1-step prediction we would set $s = n + 1$ obtaining

$$\left(x(t_{n+1}) \text{ based on observations up to } t_n\right) \approx \left(x_0{}^*{}_{n+1,n} = (p^*(n+1))_n\right) \tag{13.1.31}$$

and for *any* validity instant $t = s\tau$ we would have

$$\left(x(s\tau) \text{ based on observations up to } t_n\right) \approx \left(x_0{}^*{}_{s,n} = (p^*(s))_n\right) \tag{13.1.32}$$

Note: The subscript s, n in $x_0{}^*{}_{s,n}$ means that the validity instant is whatever numerical value is assigned to s, and that $x_0{}^*$ was derived from data whose most recent time stamp was t_n.

We also require estimates of the time-derivatives of $x(t)$. To obtain these we recall from (13.1.1) that t and s are related by

$$t = s\tau \tag{13.1.33}$$

which means that

$$dt = \tau\, ds \tag{13.1.34}$$

from which

$$d/dt = (1/\tau)\, d/ds \tag{13.1.35}$$

This, in turn, means that

$$d/dt\left(p^*(s)\right)_n = (1/\tau)d/ds\left(p^*(s)\right)_n \tag{13.1.36}$$

and in general that

$$d^i/dt^i\left(p^*(s)\right)_n = (1/\tau^i)\, d^i/ds^i\left(p^*(s)\right)_n \tag{13.1.37}$$

If we now let D mean d/dt and assume for simplicity that $m = 3$, we obtain the following vector of estimates based on observations up to t_n, in terms of the derivatives of $\left(p^*(s)\right)_n$:

$$\begin{pmatrix} x(s\tau) \\ Dx(s\tau) \\ D^2x(s\tau) \\ D^3x(s\tau) \end{pmatrix} \approx \begin{pmatrix} x_0* \\ x_1* \\ x_2* \\ x_3* \end{pmatrix}_{s,\,n} = \begin{pmatrix} \left(p^*(s)\right)_n \\ (1/\tau)\, d/ds\left(p^*(s)\right)_n \\ (1/\tau^2) d^2/ds^2\left(p^*(s)\right)_n \\ (1/\tau^3) d^3/ds^3\left(p^*(s)\right)_n \end{pmatrix} \tag{13.1.38}$$

Using (13.1.29) for $\left(p^*(s)\right)_n$ in (13.1.38) then gives us the following general expression for any degree m and any derivative i from 0 to m:

$$x_i*_{s,\,n} = 1/\tau^i \sum_{j=0}^{m}\left(\sum_{k=0}^{n}\varphi_j(k,\,n)\,y_k\right) d^i/ds^i\varphi_j(s,n) \quad 0 \le i \le m \tag{13.1.39}$$

We now rearrange the terms in (13.1.39) and multiply both sides by $\tau^i/i!$ obtaining

$$(\tau^i/i!)\, x_i*_{s,\,n} = \sum_{j=0}^{m} 1/i!\; d^i/ds^i\; \varphi_j(s,n)\left(\sum_{k=0}^{n}\varphi_j(k,\,n)\,y_k\right) \quad 0 \le i \le m \tag{13.1.40}$$

The EMP filter does its internal computing using the normalized vector $\boldsymbol{Z}^*$, and we recall from (12.2.5) or (12.2.6) that the elements of $\boldsymbol{Z}^*$ and $\boldsymbol{X}^*$ are related by

$$z_i* = (\tau^i/i!)\, x_i* \qquad 0 \le i \le m \tag{13.1.41}$$

This means that (13.1.40) can be written as

$$z_i{*}_{s,n} = \sum_{j=0}^{m} 1/i!\; d^i/ds^i\; \varphi_j(s,n)\left(\sum_{k=0}^{n} \varphi_j(k,n) y_k\right) \quad 0 \le i \le m \qquad (13.1.42)$$

From now on we will focus on the normalized estimate vector $\boldsymbol{Z}^*$, and only when we are completely done will we denormalize.

Setting $s = n$ in (13.1.42) gives the elements of the current-estimate vector $\boldsymbol{Z}^*_{n,n}$ as

$$z_i{*}_{n,n} = \sum_{j=0}^{m} 1/i!\; d^i/ds^i\; \varphi_j(s,n)\Big|_{s=n}\left(\sum_{k=0}^{n} \varphi_j(k,n) y_k\right) \quad 0 \le i \le m \qquad (13.1.43)$$

and setting $s = n + 1$ we obtain elements of the 1-step prediction vector $\boldsymbol{Z}^*_{n+1,n}$ as

$$z_i{*}_{n+1,n} = \sum_{j=0}^{m} 1/i!\; d^i/ds^i \varphi_j(s,n)\Big|_{s=n+1}\left(\sum_{k=0}^{n} \varphi_j(k,n) y_k\right) \quad 0 \le i \le m$$

(13.1.44)

Equations (13.1.43) and (13.1.44) are the non-recursive expressions for the current-estimate and the *1*-step-predictor EMP filters. We will return to them further down when we derive the expressions for their covariance matrices.

13.1.8 The EMP recursive formulation

The true power of the EMP filters lies in their recursive structure. However, the road from equations (13.1.43) and (13.1.44) to the recursive formulations in Appendices 12.1 and 12.3 is long and arduous, and so we will not include it here. The reader who is interested in seeing how it was done should consult Reference 6, starting on p. 348.

In addition to deriving the recursive algorithms themselves, the following three results of importance are also derived in Reference 6. We touch on these only briefly, and the interested reader can find the complete details there.

13.1.8.1 The general expression for the EMP weights for any degree *m*

Define the $(m + 1) \times (m + 1)$ matrix $\boldsymbol{P}(n,n)$ whose i, jth element is (see Appendix 13.2)

$$[\boldsymbol{P}(n,n)]_{i,j} = 1/i!\; d^i/ds^i p_j(s,n)\Big|_{s=n} \qquad 0 \le i, j \le m \qquad (13.1.45)$$

Also define the vector $\boldsymbol{K}(n)$ whose jth term is

$$[\boldsymbol{K}(n)]_j = (-1)^j / \left(c_j(n)\right)^2 \quad 0 \le j \le m \qquad (13.1.46)$$

Then the general expressions for the weights $\alpha,\ \beta,\ \gamma, \ldots$ in the EMP algorithms (both *1*-step-predictor and current-estimate) for any degree m is

$$\boldsymbol{H}(n) = \boldsymbol{P}(n\,, n)\boldsymbol{K}(n) \tag{13.1.47}$$

in which $\boldsymbol{H}(n) = (\alpha,\ \beta,\ \gamma, \ldots)^T$ is called the *weight vector.*

For a proof of this statement see Problem 13.7.

13.1.8.2 The EMP filters of any degree are self-initializing

We discussed the self-initializing property of the EMP filters in Chapter 12 and the fact that this is highly unusual for a recursive algorithm. A general proof of this property is given Reference 6, starting on p. 367.

13.1.8.3 Stability properties of the EMP filters

The EMP filters are recursive, and so we need to show that they are stable, i.e. that any unwanted disturbances will die out. This is done in Section 9.5 of Reference 6.

13.2 The EMP covariance matrices

The filter models in powerful algorithms like Gauss–Newton, Kalman and Swerling can be many and varied, and so the best that we can expect for their covariance matrices are algorithmic expressions that give us numerical values at run-time.

However, the filter model for the EMP algorithms is always a polynomial and, as we now show, closed-form algebraic expressions for their covariance matrices can be derived.

◀▶

We start from (13.1.42) and assume for simplicity that $m = 1$. In matrix/vector format we then have

$$\begin{pmatrix} z_0{}^* \\ z_1{}^* \end{pmatrix}_{s\,,\,n} = \begin{pmatrix} \varphi_0(s\,, n) & \varphi_1(s\,, n) \\ 0 & d/ds\ \varphi_1(s\,, n) \end{pmatrix} \begin{pmatrix} \sum\limits_{k=0}^{n} \varphi_0(k\,,\ n) y_k \\ \sum\limits_{k=0}^{n} \varphi_1(k\,,\ n) y_k \end{pmatrix} \tag{13.2.1}$$

Note: In (13.1.42) the terms $1/i!$ are equal to 1 when $i = 0$ and $i = 1$, and so $1/i!$ does not appear in (13.2.1). However, at a later stage when $i > 1$ the appropriate terms must be included.

The final vector on the right in (13.2.1) can now be expanded as follows:

$$\begin{pmatrix} \sum_{k=0}^{n} \varphi_0(k, n)\, y_k \\ \sum_{k=0}^{n} \varphi_1(k, n)\, y_k \end{pmatrix} = \begin{pmatrix} \varphi_0(n, n) & \varphi_0(n-1, n) & \cdots & \varphi_0(0, n) \\ \varphi_1(n, n) & \varphi_1(n-1, n) & \cdots & \varphi_1(0, n) \end{pmatrix} \begin{pmatrix} y_n \\ y_{n-1} \\ \vdots \\ y_0 \end{pmatrix} \tag{13.2.2}$$

For the general case we will use the symbol $\boldsymbol{Z}^*_{s,n}$ for the vector on the left of (13.2.1) and the symbol $\boldsymbol{Y}_n$ for the vector on the far right of (13.2.2). Then we see from those two equations that $\boldsymbol{Z}^*_{s,n}$ and $\boldsymbol{Y}_n$ are related by a linear transformation of the form

$$\boldsymbol{Z}^*_{s,n} = \boldsymbol{W}(s)\boldsymbol{Y}_n \tag{13.2.3}$$

For the case where $m = 1$, from (13.2.1) and (13.2.2) the matrix $\boldsymbol{W}(s)$ will be

$$\boldsymbol{W}(s) = \begin{pmatrix} \varphi_0(s,n) & \varphi_1(s,n) \\ 0 & d/ds\, \varphi_1(s,n) \end{pmatrix} \begin{pmatrix} \varphi_0(n, n) & \varphi_0(n-1, n) & \cdots & \varphi_0(0, n) \\ \varphi_1(n, n) & \varphi_1(n-1, n) & \cdots & \varphi_1(0, n) \end{pmatrix} \tag{13.2.4}$$

and from (13.2.3) and Theorem 4.1 the expression for the covariance matrix of $\boldsymbol{Z}^*_{s,n}$ will then be

$$\boldsymbol{S}^*(\boldsymbol{Z}^*_{s,n}) = \boldsymbol{W}(s)\boldsymbol{R}_{\boldsymbol{Y}_n}\boldsymbol{W}(s)^T \tag{13.2.5}$$

in which $\boldsymbol{R}_{\boldsymbol{Y}_n}$ is the covariance matrix of $\boldsymbol{Y}_n$.

At the start of Chapter 12 we laid out the assumptions on which the polynomial filters are based, and we stated that the errors in $\boldsymbol{Y}_n$ are assumed to be stage-wise uncorrelated and of constant variance σ_ν^2. It thus follows that

$$\boldsymbol{R}_{\boldsymbol{Y}_n} = \sigma_\nu^2 \boldsymbol{I} \tag{13.2.6}$$

and so (13.2.5) reduces to

$$\boldsymbol{S}^*(\boldsymbol{Z}^*_{s,n}) = \sigma_\nu^2\, \boldsymbol{W}(s)\boldsymbol{W}(s)^T \tag{13.2.7}$$

Note: If equation (13.2.6) does not hold and the covariance matrix $\boldsymbol{R}_{\boldsymbol{Y}_n}$ differs from $\sigma_\nu^2 \boldsymbol{I}$, then (13.2.5) becomes the expression for $\boldsymbol{S}^*(\boldsymbol{Z}^*_{s,n})$ without further simplification.

Assuming that (13.2.6) holds, we now write out the full expression (13.2.7) for degree $m = 1$ in (13.2.8) below:

Further simplification now takes place because the polynomials $\varphi_j(s, n)$ form an orthonormal set. The product of the two matrices in the centre of (13.2.8) becomes the identity matrix and so (13.2.8) reduces to

$$\boldsymbol{S}^*(\boldsymbol{Z}^*_{s,n}) = \sigma_\nu^2 \begin{pmatrix} \varphi_0(s,n) & \varphi_1(s,n) \\ 0 & d/ds\, \varphi_1(s,n) \end{pmatrix} \begin{pmatrix} \varphi_0(s,n) & 0 \\ \varphi_1(s,n) & d/ds\, \varphi_1(s,n) \end{pmatrix}$$

$$= \sigma_\nu^2 \begin{pmatrix} \varphi_0(s,n)^2 + \varphi_1(s,n)^2 & \varphi_1(s,n) d/ds\, \varphi_1(s,n) \\ \varphi_1(s,n) d/ds\, \varphi_1(s,n) & (d/ds\, \varphi_1(s,n))^2 \end{pmatrix} \qquad (13.2.9)$$

The above was for $m = 1$. For $m = 2$, Equation (13.2.7) would be as shown in equation (13.2.10) further down.

From this it is clear what the general expression for the covariance matrix of $\boldsymbol{Z}^*$ will be for any degree m and any validity instant s. We return to that presently.

13.2.1 Algebraic expressions for the diagonal elements

Figures 12.18(a) and 12.18(b) in Chapter 12 showed the algebraic expressions for the diagonal elements of the denormalized *1*-step-predictor EMP covariance matrices for degrees *0* to *4*.

Those expressions and some of their plots in Figure 12.19 contain information that is essential for a full understanding of the EMP filters and their properties, and it is important that we understand how they were obtained.

From (13.2.10) we see that the diagonal elements of the normalized covariance matrix $[\boldsymbol{S}^*(\boldsymbol{Z}^*_{s,n})]_{i,i}$ of degree m have the following general expressions:

$$[\boldsymbol{S}^*(\boldsymbol{Z}^*_{s,n})]_{i,i} = \sigma_\nu^2 \sum_{j=0}^{m} \left(1/i!\; d^i/ds^i\, \varphi_j(s,n) \right)^2 \quad 0 \le i \le m \qquad (13.2.11)$$

We now recall from (13.1.11) that

$$\varphi_j(s,n) \equiv \frac{1}{c_j(n)} p_j(s,n) \qquad (13.2.12)$$

$$\boldsymbol{S}^*(\boldsymbol{Z}^*{}_{s,n}) =$$

$$\sigma_v^2 \begin{pmatrix} \varphi_0(s,n) & \varphi_1(s,n) \\ 0 & d/ds\varphi_1(s,n) \end{pmatrix} \begin{pmatrix} \varphi_0(n,n) & \varphi_0(n-1,n) & \cdots & \varphi_0(0,n) \\ \varphi_1(n,n) & \varphi_1(n-1,n) & \cdots & \varphi_1(0,n) \end{pmatrix} \begin{pmatrix} \varphi_0(n,n) & \varphi_1(n,n) \\ \varphi_0(n-1,n) & \varphi_1(n-1,n) \\ \vdots & \vdots \\ \varphi_0(0,n) & \varphi_1(0,n) \end{pmatrix} \begin{pmatrix} \varphi_0(s,n) & 0 \\ \varphi_1(s,n) & d/ds\varphi_1(s,n) \end{pmatrix} \tag{13.2.8}$$

$$\boldsymbol{S}^*(\boldsymbol{Z}^*_{s,n}) = \sigma_v^2 \begin{pmatrix} \varphi_0(s,n) & \varphi_1(s,n) & \varphi_2(s,n) \\ 0 & d/ds\,\varphi_1(s,n) & d/ds\,\varphi_2(s,n) \\ 0 & 0 & 1/2!d^2/ds^2\,\varphi_2(s,n) \end{pmatrix} \begin{pmatrix} \varphi_0(s,n) & 0 & 0 \\ \varphi_1(s,n) & d/ds\,\varphi_1(s,n) & 0 \\ \varphi_2(s,n) & d/ds\,\varphi_2(s,n) & 1/2!d^2/ds^2\,\varphi_{2s}(s,n) \end{pmatrix} \tag{13.2.10}$$

and if we define the matrix $\boldsymbol{P}(s, n)$ whose i, jth element is

$$[\boldsymbol{P}(s, n)]_{i,j} \equiv (1/i!)\, d^i/ds^i p_j(s, n) \tag{13.2.13}$$

then (13.2.11) becomes

$$[\boldsymbol{S}^*(\boldsymbol{Z}^*_{s,n})]_{i,i} = \sigma_\nu^2 \sum_{j=0}^{m} \left([\boldsymbol{P}(s, n)]_{i,j}\right)^2 / (c_j(n))^2 \quad 0 \le i \le m \tag{13.2.14}$$

From this we have the following:

◊ To obtain the *1*-step-predictor diagonal elements $[\boldsymbol{S}^*(\boldsymbol{Z}^*_{n+1,n})]_{i,i}$ we must evaluate (13.2.14) using $s = n + 1$ in the matrix $\boldsymbol{P}(s, n)$, which then becomes $\boldsymbol{P}(n + 1, n)$.

◊ To obtain the current-estimate diagonal elements $[\boldsymbol{S}^*(\boldsymbol{Z}^*_{n,n})]_{i,i}$ we must evaluate (13.2.14) using $s = n$ in the matrix $\boldsymbol{P}(s, n)$ which then becomes $\boldsymbol{P}(n, n)$

The three matrices that are needed for this undertaking, namely

$$\boldsymbol{P}(s, n) \quad \boldsymbol{P}(n + 1, n) \equiv \boldsymbol{P}(s, n)\big|_{s=n+1} \quad \text{and} \quad \boldsymbol{P}(n, n) \equiv \boldsymbol{P}(s, n)\big|_{s=n}$$

are all displayed up to 4th-degree in Appendix 13.2.

Once the normalized expressions $[\boldsymbol{S}^*(\boldsymbol{Z}^*_{s,n})]_{i,i}$ have been obtained – whether current-estimate or *1*-step-predictor – they must then be denormalized to give $[\boldsymbol{S}^*(\boldsymbol{X}^*_{s,n})]_{i,i}$ which is what we really want. To do that, we recall from (12.2.10) that

$$\boldsymbol{X}^*_{s,n} = \boldsymbol{D}(\tau)\boldsymbol{Z}^*_{s,n} \tag{13.2.15}$$

where $\boldsymbol{D}(\tau)$ was defined in (12.2.11). It therefore follows that

$$\boldsymbol{S}^*(\boldsymbol{X}^*_{s,n}) = \boldsymbol{D}(\tau)\boldsymbol{S}^*(\boldsymbol{Z}^*_{s,n})\boldsymbol{D}(\tau) \tag{13.2.16}$$

which means that for the diagonal elements of the *1*-step-predictor

$$[\boldsymbol{S}^*(\boldsymbol{X}^*_{n+1,n})]_{i,i} = (i!/\tau^i)^2 [\boldsymbol{S}^*(\boldsymbol{Z}^*_{n+1,n})]_{i,i} \qquad 0 \le i \le m \tag{13.2.17}$$

and for the diagonal elements of the current-estimate filter

$$[\boldsymbol{S}^*(\boldsymbol{X}^*_{n,n})]_{i,i} = (i!/\tau^i)^2 [\boldsymbol{S}^*(\boldsymbol{Z}^*_{n,n})]_{i,i} \qquad 0 \le i \le m \tag{13.2.18}$$

Example 13.1: We use (13.2.17) to derive the expression for the diagonal element $[\boldsymbol{S}^*(\boldsymbol{X}^*_{n+1,n})]_{0,0}$ for $m = 2$.

First we require $[\boldsymbol{S}^*(\boldsymbol{Z}^*_{n+1,n})]_{0,0}$ for $m = 2$. From Appendix 13.2 we have

$$[\boldsymbol{P}(n+1,n)]_{0,0} = 1, \quad [\boldsymbol{P}(n+1,n)]_{0,1} = -(n+2)/n,$$
$$[\boldsymbol{P}(n+1,n)]_{0,2} = (n+3)^{(2)}/n^{(2)}$$

and from Appendix 13.1

$$(c_0(n))^2 = n+1, \ (c_1(n))^2 = (n+2)^{(2)}/3n, \ (c_2(n))^2 = (n+3)^{(3)}/5n^{(2)}$$

We now let $s = n+1$ in (13.2.14) and obtain the following:

$$\begin{aligned}
&[\boldsymbol{S}^*(\boldsymbol{Z}^*_{n+1,n})]_{0,0} \\
&= \sigma_v^2\left(\frac{([\boldsymbol{P}(n+1,n)]_{0,0})^2}{(c_0(n))^2} + \frac{([\boldsymbol{P}(n+1,n)]_{0,1})^2}{(c_1(n))^2} + \frac{([\boldsymbol{P}(n+1,n)]_{0,2})^2}{(c_2(n))^2}\right) \\
&= \sigma_v^2\left(\frac{1}{n+1} + \frac{(-(n+2/n)^2}{(n+2)^{(2)}/3n} + \frac{((n+3)^{(2)}/n^{(2)})^2}{(n+3)^{(3)}/5n^{(2)}}\right) \\
&= \sigma_v^2 3(3n^2+9n+8)/(n+1)^{(3)} \qquad (13.2.19)
\end{aligned}$$

From (13.2.17) denormalization now gives

$$\begin{aligned}
[\boldsymbol{S}^*(\boldsymbol{X}^*_{n+1,n})]_{0,0} &= (0!/\tau^0)^2[\boldsymbol{S}^*(\boldsymbol{Z}^*_{n+1,n})]_{0,0} \\
&= \sigma_v^2 3(3n^2+9n+8)/(n+1)^{(3)} \qquad (13.2.20)
\end{aligned}$$

This result appeared in Figure 12.18. The same result can also be obtained with slightly less effort by making use of the recursion formula in (13.2.41) further down. ■■

Example 13.2: We use (13.2.17) to derive the $1, 1$ element $[\boldsymbol{S}^*(\boldsymbol{Z}^*_{n+1,n})]_{1,1}$ for $m = 2$.

First we require $[\boldsymbol{S}^*(\boldsymbol{Z}^*_{n+1,n})]_{1,1}$. From Appendix 13.2 we have

$$[\boldsymbol{P}(n+1,n)]_{1,0} = 0, \quad [\boldsymbol{P}(n+1,n)]_{1,1} = -2/n,$$
$$[\boldsymbol{P}(n+1,n)]_{1,2} = 6(n+2)/n^{(2)}$$

and from Appendix 13.1

$$(c_0(n))^2 = n+1,\ (c_1(n))^2 = (n+2)^{(2)}/3n,\ (c_2(n))^2 = (n+3)^{(3)}/5n^{(2)}$$

We now let $s = n+1$ in (13.2.14) and obtain the following:

$$[\boldsymbol{S}^*(\boldsymbol{Z}^*_{n+1,\,n})]_{1,\,1}$$

$$= \sigma_v^2\left(\frac{([\boldsymbol{P}(n+1,\,n)]_{1,\,0})^2}{(c_0(n))^2} + \frac{([\boldsymbol{P}(n+1,\,n)]_{1,\,1})^2}{(c_1(n))^2} + \frac{([\boldsymbol{P}(n+1,\,n)]_{1,\,2})^2}{(c_2(n))^2}\right)$$

$$= \sigma_v^2\left(0 + \frac{((-2)/n)^2}{(n+2)^{(2)}/3n} + \frac{(6(n+2)/n^{(2)})^2}{(n+3)^{(3)}/5n^{(2)}}\right)$$

$$= \sigma_v^2\,12(16n^2+62n+57)/(n+3)^{(5)} \tag{13.2.21}$$

From (13.2.17) denormalization now gives

$$\begin{aligned}[\boldsymbol{S}^*(\boldsymbol{X}^*_{n+1,\,n})]_{1,\,1} &= (1!/\tau^1)^2[\boldsymbol{S}^*(\boldsymbol{Z}^*_{n+1,\,n})]_{1,\,1}\\ &= \sigma_v^2\,12(16n^2+62n+57)/\tau^2(n+3)^{(5)}\end{aligned} \tag{13.2.22}$$

This result also appeared in Figure 12.18.[2] ■■

In the file in the supplementary material

Miscellaneous\Polynomial_filters\1-step_predictors_up_to_8th_degree

we display the algebraic expressions for the diagonal elements of the denormalized EMP covariance matrices up to 8th-degree, obtained by the use of Symbolic Matlab.

In Problem 13.2 we consider how you can obtain approximate expressions for the entire covariance matrices of the EMP filters in a very useful form. These approximations are quite accurate for $n = 100$ and become extremely accurate when n increases beyond 100, and since these filters are usually being cycled at least 1000 times/sec, values of $n > 100$ are attained very quickly.

[2] Perhaps you can now get a feeling for the amount of manual labor that went into the preparation of Figure 12.18 and also how much effort would be needed to derive the expressions for the off-diagonal elements. These days we can make use of languages like Symbolic Matlab but when Figures 12.18 and 12.21 were first derived such languages did not yet exist, and so it all had to be done by hand.

For $n > 100$, the approximations are valid for the 1-step predictors as well as the current-estimate filters.

The program

Miscellaneous\Code_for_solutions_to_Chapter_13\Problem_13dot2.exe

enables you to display the matrices up to 8th-degree. See also Project 13.5 where we discuss this matter further.

13.2.2 Computing numerical values for the covariance matrices

Thus far we have considered only the algebraic expressions for the diagonal elements of the covariance matrices. We now derive a way in which to compute numerical values for the *entire* covariance matrices. These are needed if, for example, one wishes to implement the Chi-squared ECM-consistency tests when running the EMP filters.

Starting from (13.2.10) which was for degree $m = 2$, we call the first matrix on the right $\boldsymbol{G}(s)$. Then for any degree the i, jth element of $\boldsymbol{G}(s)$ is

$$[\boldsymbol{G}(s)]_{i,j} = 1/i!\, d^i/ds^i \varphi_j(s, n) = 1/i!\, d^i/ds^i p_j(s, n)/c_j(n) \quad 0 \le i, j \le m \tag{13.2.23}$$

This in turn means that

$$\boldsymbol{G}(s) = \boldsymbol{P}(s, n)\boldsymbol{C}(n) \tag{13.2.24}$$

in which $\boldsymbol{P}(s, n)$ was defined in (13.2.13) and $\boldsymbol{C}(n)$ is the diagonal matrix whose ith diagonal element is

$$[\boldsymbol{C}(n)]_{j,j} = 1/c_j(n) \tag{13.2.25}$$

From (13.2.10) it then follows that

$$\boldsymbol{S}^*(\boldsymbol{Z}^*_{s,n}) = \sigma_v^2 \boldsymbol{G}(s)\boldsymbol{G}(s)^T = \sigma_v^2 \boldsymbol{P}(s, n)\big(\boldsymbol{C}(n)\big)^2 \boldsymbol{P}(s, n)^T \tag{13.2.26}$$

For the 1-step-predictors we use $s = n + 1$ and so (13.2.26) becomes

$$\boldsymbol{S}^*(\boldsymbol{Z}^*_{n+1,\, n}) = \sigma_v^2 \boldsymbol{P}(n + 1,\, n)\big(\boldsymbol{C}(n)\big)^2 \boldsymbol{P}(n + 1,\, n)^T \tag{13.2.27}$$

and for the current-estimate filters we use $s = n$ and in which case (13.2.26) becomes

$$\boldsymbol{S}^*(\boldsymbol{Z}^*_{n,\, n}) = \sigma_v^2 \boldsymbol{P}(n,\, n)\big(\boldsymbol{C}(n)\big)^2 \boldsymbol{P}(n,\, n)^T \tag{13.2.28}$$

Note: In (13.2.26) to (13.2.28) we see the diagonal matrix $(\boldsymbol{C}(n))^2$ whose jth element is $1/c_j(n)^2$. The expressions $c_j(n)^2$ are listed in Appendix 13.1, and when setting up the matrix $(\boldsymbol{C}(n))^2$ remember to use the *reciprocals* of the expressions in that appendix.

Equations (13.2.26), (13.2.27) or (13.2.28) are what we implement when numerical values for the entire covariance matrices are required. (See Project 13.3 for further discussion.)

As the following example shows, when $m = 1$ it is possible to expand (13.2.27) or (13.2.28) manually and obtain the algebraic version of the entire covariance matrix. However, for degrees beyond that we either use a language like Symbolic Matlab to obtain the algebraic expressions or else we must be satisfied with numerical values.

Example 13.3: We use (13.2.28) to derive the algebraic expression for the complete covariance matrix $\boldsymbol{S}^*(\boldsymbol{Z}^*_{n,n})$ of the current-estimate 1st-degree EMP filter. From Appendix 13.1 we have

$$(\boldsymbol{C}(n))^2 = \begin{pmatrix} (1/c_0(n))^2 & 0 \\ 0 & (1/c_0(n))^2 \end{pmatrix} = \begin{pmatrix} 1/(n+1) & 0 \\ 0 & 3n/(n+2)(n+1) \end{pmatrix} \tag{13.2.29}$$

and from Appendix 13.2 (for $m = 1$)

$$\boldsymbol{P}(n, n) = \begin{pmatrix} 1 & -1 \\ 0 & -2/n \end{pmatrix} \tag{13.2.30}$$

Then the required covariance matrix will be

$$\begin{aligned} \boldsymbol{S}^*(\boldsymbol{Z}^*_{n,n}) &= \sigma_\nu^2 \boldsymbol{P}(n, n)(\boldsymbol{C}(n))^2 \boldsymbol{P}(n, n)^T \\ &= \sigma_\nu^2 \begin{pmatrix} 2(2n+1)/(n+2)^{(2)} & 6/(n+2)^{(2)} \\ 6/(n+2)^{(2)} & 12/(n+2)^{(3)} \end{pmatrix} \end{aligned} \tag{13.2.31}$$

■■

Once the matrix $\boldsymbol{S}^*(\boldsymbol{Z}^*_{s,n})$ has been obtained – whether current-estimate or 1-step-predictor – it must be denormalized in order to obtain $\boldsymbol{S}^*(\boldsymbol{X}^*_{s,n})$. Starting from (13.2.26), (13.2.27) or (13.2.28), the resulting equations are as follows:

◊ For any validity instant s:

$$S^*(X^*_{s,n}) = \sigma^2_\nu D(\tau)P(s, n)\big(C(n)\big)^2 P(s, n)^T D(\tau) \tag{13.2.32}$$

◊ For the *1*-step-predictor:

$$S^*(X^*_{n+1,n}) = \sigma^2_\nu D(\tau)P(n+1, n)\big(C(n)\big)^2 P(n+1, n)^T D(\tau) \tag{13.2.33}$$

◊ For the current-estimate filter:

$$S^*(X^*_{n,n}) = \sigma^2_\nu D(\tau)P(n, n)\big(C(n)\big)^2 P(n, n)^T D(\tau) \tag{13.2.34}$$

The matrix $D(\tau)$ was defined in (11.2.1).

Note: Extreme care must be exercised to ensure that denormalization is properly carried out, both for the estimate vectors as well as for their covariance matrices. See Project 12.2 in the supplementary material for comments regarding denormalization in the computer programs.

Example 13.4: We use (13.2.27) and (13.2.17) to derive the algebraic expression for the denormalized covariance matrix $S^*(X^*_{n+1,n})$ of the 1st-degree *1*-step-predictor EMP filter.

From Appendix 13.2 we have

$$P(n+1, n) = \begin{pmatrix} 1 & -(n+2)/n \\ 0 & -2/n \end{pmatrix} \tag{13.2.35}$$

Using $\big(C(n)\big)^2$ from (13.2.29) the required covariance matrix becomes

$$\begin{aligned} S^*(Z^*_{n+1,n}) &= \sigma^2_\nu P(n+1, n)\big(C(n)\big)^2 P(n+1, n)^T \\ &= \sigma^2_\nu \begin{pmatrix} 2(2n+3)/(n+1)^{(2)} & 6/(n+1)^{(2)} \\ 6/(n+1)^{(2)} & 12/(n+2)^{(3)} \end{pmatrix} \end{aligned} \tag{13.2.36}$$

Denormalization then gives

$$\begin{aligned} S^*(Z^*_{n+1,n}) &= D(\tau)S^*(Z^*_{n+1,n})D(\tau) \\ &= \sigma^2_\nu \begin{pmatrix} 2(2n+3)/(n+1)^{(2)} & 6/\tau(n+1)^{(2)} \\ 6/\tau(n+1)^{(2)} & 12/\tau^2(n+2)^{(3)} \end{pmatrix} \end{aligned} \tag{13.2.37}$$

■■

When $m = 0$ or 1 it is possible to form theses expansions by hand and arrive at explicit algebraic expressions for the covariance matrices. For degrees beyond that we either have to content ourselves with computer implementations that give us numerical values at run-time (see Project 13.3) or else we must use a language like Symbolic Matlab.

In closing, we point out that the 1-step-predictor matrix $\boldsymbol{S}^*(\boldsymbol{Z}^*_{n+1,n})$ can be obtained from the current-estimate matrix $\boldsymbol{S}^*(\boldsymbol{Z}^*_{n,n})$ as follows:

$$\boldsymbol{S}^*(\boldsymbol{Z}^*_{n+1,n}) = \boldsymbol{\Phi}(1)\boldsymbol{S}^*(\boldsymbol{Z}^*_{n,n})\boldsymbol{\Phi}(1)^T \tag{13.2.38}$$

where $\boldsymbol{\Phi}(1)$ is the transition matrix defined in (2.3.17). Similarly, $\boldsymbol{S}^*(\boldsymbol{X}^*_{n+1,n})$ can be obtained from $\boldsymbol{S}^*(\boldsymbol{X}^*_{n,n})$ as follows:

$$\boldsymbol{S}^*(\boldsymbol{X}^*_{n+1,n}) = \boldsymbol{\Phi}(\tau)\boldsymbol{S}^*(\boldsymbol{X}^*_{n+1,n})\boldsymbol{\Phi}(\tau)^T \tag{13.2.39}$$

where $\boldsymbol{\Phi}(\tau)$ is now the transition matrix defined in (2.3.6).

We also have the reverse equations that give us $\boldsymbol{S}^*(\boldsymbol{Z}^*_{n,n})$ and $\boldsymbol{S}^*(\boldsymbol{X}^*_{n,n})$ starting from $\boldsymbol{S}^*(\boldsymbol{Z}^*_{n+1,n})$ and $\boldsymbol{S}^*(\boldsymbol{X}^*_{n+1,n})$.

Example 13.4: We derive $\boldsymbol{S}^*(\boldsymbol{Z}^*_{n+1,n})$ in (13.2.38) starting from $\boldsymbol{S}^*(\boldsymbol{Z}^*_{n,n})$ in (13.2.31).

$$\begin{aligned}
&\boldsymbol{\Phi}(1)\boldsymbol{S}^*(\boldsymbol{Z}^*_{n,n})\boldsymbol{\Phi}(1)^T \\
&= \begin{pmatrix} 1 & 1 \\ 0 & 1 \end{pmatrix} \sigma_\nu^2 \begin{pmatrix} 2(2n+1)/(n+2)^{(2)} & 6/(n+2)^{(2)} \\ 6/(n+2)^{(2)} & 12/(n+2)^{(3)} \end{pmatrix} \begin{pmatrix} 1 & 0 \\ 1 & 1 \end{pmatrix} \\
&= \sigma_\nu^2 \begin{pmatrix} 2(2n+3)/(n+1)^{(2)} & 6/(n+1)^{(2)} \\ 6/(n+1)^{(2)} & 12/(n+2)^{(3)} \end{pmatrix} = \boldsymbol{S}^*(\boldsymbol{Z}^*_{n+1,n})
\end{aligned} \tag{13.2.40}$$

■■

The covariance matrices in this equation should be compared to those in (13.2.31) and (13.2.36).

13.2.3 *Recursion formulas for the variances of the 0th-derivative estimates*

In Project 13.4 we derive the following recursion formulas for the denormalized variances of the 0th-derivative EMP estimates, namely the $1, 1$ elements of the covariance matrices $\boldsymbol{S}^*(\boldsymbol{X}^*_{n+1,n})$ and $\boldsymbol{S}^*(\boldsymbol{X}^*_{n,n})$.

Note: $Var(x_0{*}_{n+1,n})_m$ means the variance of $x_0{*}_{n+1,n}$ for the 1-step-predictor filter of degree m.

1-step-predictor EMP:

$$Var(x_0{*}_{n+1,n})_0 = \sigma_\nu^2(1/(n+1))$$
$$Var(x_0{*}_{n+1,n})_m = Var(x_0{*}_{n+1,n})_{m-1} + \sigma_\nu^2\left((2m+1)(n+m+1)^{(m)}/(n+1)^{(m+1)}\right) \tag{13.2.41}$$

Current-estimate EMP:

$$Var(x_0{*}_{n,n})_0 = \sigma_\nu^2(1/(n+1))$$
$$Var(x_0{*}_{n,n})_m = Var(x_0{*}_{n,n})_{m-1} + \sigma_\nu^2\left((2m+1)n^{(m)}/(n+m+1)^{(m+1)}\right) \tag{13.2.42}$$

We have found these formulas to be extremely useful. We used (13.2.41) in the programs ***05_Pol_Filter_EMP*** and ***06_Pol_Filter_FMP***, and (13.2.42) was used to generate the expressions appearing in Figure 12.20.

13.2.4 Validating the expressions for the EMP covariance matrices

Verification 12.7a in ***Projects for Chapter 12*** runs a program that demonstrates that the EMP filters are ECM consistent. Running that verification, you will observe how – time after time and under almost all possible circumstances – the error/sigma ratios and the Chi-squared values lie perfectly within their prescribed bounds.

The Chi-squared test is an extremely sensitive one, and the slightest aberration can easily throw its value far beyond the allowable threshold.

Such performance can only take place if we have used the correct expressions for the covariance matrices when creating that program. Thus, verification that the EMP filters are ECM consistent also constitutes proof that the expressions for the covariance matrices that were derived above, are correct.

13.3 Deriving the FMP equations

In this section we derive the equations for the FMP filters. The derivations here differ slightly from those of the EMP filters above because here we have a weight function θ^k and summation from 0 to ∞ whereas previously the weight function was 1 and summation was from 0 to n.

After making a number of attempts at writing this section, we concluded that it would be better to carry out the *complete* derivation of the FMP equations, rather than trying to draw on parts of the earlier EMP derivation. The result may seem to be slightly repetitive in places, but the alternative would have been far worse.

In Section 13.4 we derive the expressions for the FMP covariance matrices. There are significant problems in doing so, problems that did not arise in the EMP case. Rather than carrying out the complete derivation we will provide only a brief summary and refer you instead to Reference 6 for full details.

13.3.1 The approximating polynomials

Referring to Figure 13.2 below:

◊ The time axis t runs from left to right. The point t_n represents the most recent time instant, and the points $t_{n-1}, t_{n-2}, \ldots$ are the preceding ones. The separation between the time instants is the constant time interval τ.

◊ The n axis runs from left to right and the point $n\tau$ lines up with t_n.

◊ The origin of the s axis *always* lines up with t_n, but the s axis runs *from right to left*. Accordingly, from the figure we see that

$$t = t_n - s\tau \tag{13.3.1}$$

◊ The observations are shown as vertical lines with small black circles on their tops. As a set they form the total observation vector

$$\mathbf{Y}_n = (y_n, y_{n-1}, \ldots)^T \tag{13.3.2}$$

We note that $\mathbf{Y}_n$ is conceptually an infinite-dimensional vector.

◊ The observed trajectory is not shown.

◊ The solid black curve shows the polynomial $(p^*(s))_n$ that has been fitted to $\mathbf{Y}_n$ at time t_n by the method of least squares. We observe the following:

◊ The polynomial is called p^*.

◊ The polynomial's independent variable is called s.

◊ The subscript n signifies that this was the polynomial that was fitted to the observations just after the nth observation was received.

◊ On the next cycle at time t_{n+1} the observation vector will expand and will become

$$\mathbf{Y}_{n+1} = (y_{n+1}, y_n, y_{n-1}, \ldots)^T \tag{13.3.3}$$

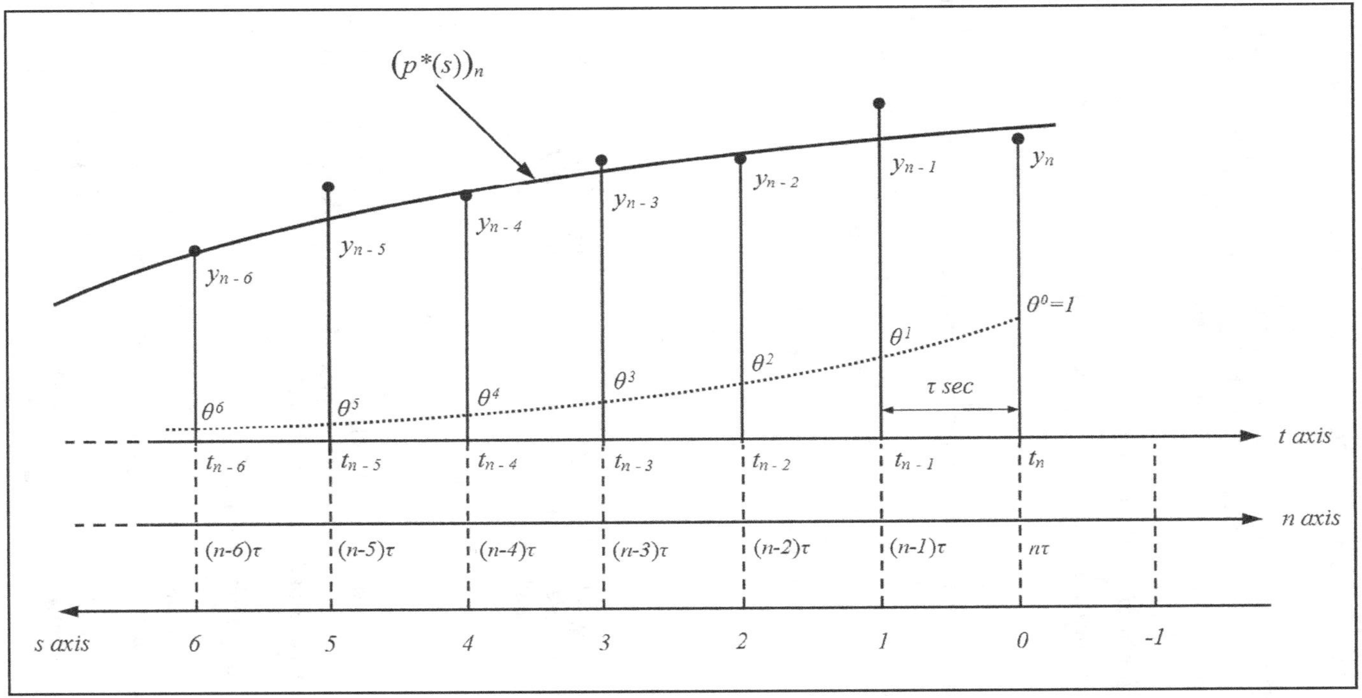

Figure 13.2 The FMP approximating polynomial

A completely new polynomial will then be fitted to $\mathbf{Y}_{n+1}$ and will be called $(p^*(s))_{n+1}$. The polynomial $(p^*(s))_{n+1}$ will be different from $(p^*(s))_n$.

◊ The degrees of the successive polynomials $(p^*(s))_n$, $(p^*(s))_{n+1}$, ... are all equal to m. As time passes, we are therefore fitting a sequence of differing polynomials of degree m to an expanding set of observations.

13.3.2 Classical least squares

Classical least squares starts by defining the polynomial $(p^*(s))_n$ as follows:

$$(p^*(s))_n = \beta_{0,n} s^0 + \beta_{1,n} s^1 + \beta_{2,n} s^2 + \cdots + \beta_{m,n} s^m \tag{13.3.4}$$

In this expression the polynomial is a sum of successively higher powers of the independent variable s, and $\beta_{0,n}, \beta_{1,n}, \ldots \beta_{m,n}$ are the polynomial's coefficients. On the left we are subscripting $(p^*(s))_n$ with an n because it is the polynomial that was fitted at time t_n and so we must also subscript each of the β's on the right with an n.

In order to obtain expressions for the β's in (13.3.4) we start from the following *weighted* sum of squared residuals:

$$e_n = \sum_{k=0}^{\infty} \left(y_{n-k} - (p^*(k))_n \right)^2 \theta^k \tag{13.3.5}$$

Observe in (13.3.5) that a weight function is now present in the form of θ^k – something that was not the case for the EMP in (13.1.6). The parameter θ lies in the range 0 to 1, exclusive at both ends.

◊ In Figure 13.2 the dotted line shows the function θ^s. Because s is increasing to the left and $0 < \theta < 1$, we see that θ^s is decaying to the left.

◊ Using θ^k as shown in (13.3.5) will put an exponentially diminishing emphasis on the residuals as they recede into the past. The filter that we obtain will therefore have an exponentially fading memory.

Observe also the following in (13.3.5):

◊ Summation is over the discrete set $k = 0$ to ∞, and so the continuous variable s has been replaced by the integer k as the polynomial argument.

◊ As k goes from 0 to ∞, incorporation of the observations y_{n-k} will start from the most recent and recede into the past.

The polynomial $(p^*(s))_n$ in (13.3.4) can be thought of as a vector that is made up of a linear combination of the set of basis vectors $\{s^0, s^1, s^2, \ldots s^m\}$.

This is perhaps the simplest basis that we might choose, but it has the following serious drawback:

◊ It will *always* lead to the need for a matrix inversion in order to find the required expressions for the β's.

(To see that this is so, we refer you to Project 6.1 where the method of classical least squares was also used.)

Inverting a matrix is generally not a problem if its elements are numbers, but here we will have a matrix whose elements are functions of θ and so its inversion can only be carried out (beyond a 2×2) by assigning numerical values to θ. However, assigning such numerical values is something that we expressly do not wish to do, because it would immediately put an end to further theoretical analysis.

For the EMP filters we were able to circumvent the matrix inversion problem by making use of orthogonality. We now do the same for the FMP filters, and we will see that we are again able to avoid the matrix inversion.

13.3.3 The discrete Laguerre orthogonal polynomials

In place of the discrete Legendre orthogonal polynomials that we used earlier, we now use the discrete Laguerre orthogonal polynomials that satisfy the following orthogonality condition:

$$\sum_{s=0}^{\infty} p_i(s,\theta)p_j(s,\theta)\theta^s = 0 \qquad i \neq j \qquad 0 < \theta < 1 \tag{13.3.6}$$

The quantities $p_i(s,\theta)$ and $p_j(s,\theta)$ are polynomials in s of degree i and j.

Comparing (13.3.6) to the orthogonality condition in (13.1.7) we see the following three differences:

◊ In place of the steadily expanding n as a parameter, we now have the *fixed* parameter θ.

◊ Instead of summation from 0 to n, we now sum from 0 to ∞.

◊ The orthogonality condition (13.3.6) now includes the weight function θ^s.

Equation (13.3.6) states the following:

◊ If we form the product of two of the discrete Laguerre polynomials of differing degrees ($i \neq j$), multiply by the weight function θ^s and then sum from 0 to ∞, we obtain zero.

If the two polynomials have the same degree ($i = j$) then clearly the sum in (13.3.6) cannot be zero.

Observe that the orthogonality condition in (13.3.6) involves a sum over a discrete set rather than an integral over a continuous interval – hence the use of the word *discrete* in the name of the polynomials. From now on when speaking about these polynomials we will drop the word discrete and its presence will always be assumed.[3]

It is shown in Chapter 3 of Reference 6 that the set of polynomials satisfying condition (13.3.6) has the following general expression:

$$p_j(s,\theta) = \theta^j \sum_{\nu=0}^{j} (-1)^\nu b^\nu \binom{j}{\nu}\binom{s}{\nu} \qquad j = 0, 1, 2, \ldots \qquad b = \frac{1-\theta}{\theta} \tag{13.3.7}$$

Note 1: We saw earlier that the discrete Legendre polynomials form a finite set with $n+1$ members. By contrast, the discrete Laguerre polynomials defined in (13.3.7) form a set with infinitely many members.

Note 2: The definition for $p_j(s,\theta)$ in (13.3.7) looks formidable. To see what the polynomials actually look like, go to Appendix 13.3 where we display them for degrees $j = 0$ to 4 as polynomials in their more usual form.

Note 3: Although the orthogonality condition in (13.3.6) involves a discrete set of values, the independent variable s in (13.3.7) is a continuous one, and so these are polynomials in the ordinary sense.

Note 4: Later on we will see that we need to differentiate $p_j(s,\theta)$ in (13.3.7) repeatedly with respect to s. However, the item

$$\binom{s}{\nu} = s^{(\nu)}/\nu! \tag{13.3.8}$$

[3] For a discussion of the Laguerre polynomials that are orthogonal over a continuous interval see Reference 74 or 137.

is not in a form in which that can be carried out. In Project 13.6 we provide you with the expressions that you need in order to convert $s^{(\nu)}$ into a simple polynomial which can then be easily differentiated.

◆▶

If we multiply together two of the polynomials with the same degrees ($i = j$), multiply by θ^s and then sum the result, we obtain the quantity $c_j(\theta)$ whose square is defined by

$$\left(c_j(\theta)\right)^2 \equiv \sum_{s=0}^{\infty} \left(p_j(s, \theta)\right)^2 \theta^s \tag{13.3.9}$$

The general expression for $\left(c_j(\theta)\right)^2$ for the Laguerre polynomials is as follows (see Chapter 3 Reference 6):

$$\left(c_j(\theta)\right)^2 = \frac{\theta^j}{1 - \theta} \tag{13.3.10}$$

In Appendix 13.3 we display the expressions for $\left(c_j(\theta)\right)^2$ for degrees from $j = 0$ to 4.

We now form the polynomials $\varphi_j(s, \theta)$ defined by

$$\varphi_j(s, \theta) \equiv \frac{1}{c_j(\theta)} p_j(s, \theta) \qquad j = 0, 1, 2, \ldots \tag{13.3.11}$$

Then these polynomials satisfy the following two conditions:

◊ By (13.3.6), when $i \neq j$:

$$\sum_{s=0}^{\infty} \varphi_i(s, \theta)\varphi_j(s, \theta)\theta^s = \sum_{s=0}^{\infty} \frac{1}{c_i(\theta)} p_i(s, \theta) \frac{1}{c_j(\theta)} p_j(s, \theta)\theta^s = 0 \tag{13.3.12}$$

◊ By (13.3.9) when $i = j$:

$$\sum_{s=0}^{\infty} \left(\varphi_j(s, \theta)\right)^2 \theta^s = \sum_{s=0}^{\infty} \frac{1}{\left(c_j(\theta)\right)^2} \left(p_j(s, \theta)\right)^2 \theta^s = 1 \tag{13.3.13}$$

Taking (13.3.12) and (13.3.13) together enables us to write the following:

$$\sum_{s=0}^{\infty} \varphi_i(s, \theta)\varphi_j(s, \theta)\theta^s = \delta_{i,j} \tag{13.3.14}$$

in which $\delta_{i,j}$ is the Kronecker delta that is *1* when $i = j$ and zero otherwise.

Equation (13.1.14) tells us that the polynomials $\varphi_j(s, \theta)$ form an *orthonormal set.*

13.3.4 Using the Laguerre orthogonal polynomials as the basis

We now return to the problem that we were considering earlier, namely that of finding the expression for the approximating polynomial in Figure 13.2 by the method of least squares.

However, rather that using the definition for $(p^*(s))_n$ that appeared in (13.3.4) we will use instead the following

$$(p^*(s))_n = \beta_{0,n}\, \varphi_0(s, \theta) + \beta_{1,n}\, \varphi_1(s, \theta) + \cdots + \beta_{m,n}\, \varphi_m(s, \theta) \tag{13.3.15}$$

in which φ was defined in (13.3.11). Equation (13.3.15) is the same as

$$(p^*(s))_n = \sum_{j=0}^{m} \beta_{j,n}\varphi_j(s, \theta) \tag{13.3.16}$$

The polynomial $(p^*(s))_n$ in this equation can be viewed as a linear combination of a set of basis vectors $\{\varphi_0(s, \theta), \varphi_1(s, \theta), \ldots \varphi_m(s, \theta)\}$, and because these are polynomials whose degrees go from *0* to *m*, it follows that $(p^*(s))_n$ is also a polynomial of degree *m*.

The basis that we are now using is a set of orthonormal polynomials, and as we will soon see, the matrix inversion that we spoke about earlier will not be required.

13.3.5 Least squares using the Laguerre polynomials

In order to obtain the β's in (13.3.15) we use the method of least squares and start from the weighted sum of the squared residuals:

$$e_n = \sum_{k=0}^{\infty} \left(y_{n-k} - (p^*(k))_n\right)^2 \theta^k \tag{13.3.17}$$

As before, we are summing over the discrete set $k = 0$ to ∞, and so the integer k has replaced the continuous variable s as the argument of p^*.

We now use (13.3.16) to write e_n as follows:

$$e_n = \sum_{k=0}^{\infty} \left(y_{n-k} - \sum_{j=0}^{m} (\beta_j)_n \varphi_j(k, \theta) \right)^2 \theta^k \tag{13.3.18}$$

Following the method of least squares we then differentiate (13.3.18) with respect to each of the β's and set the result equal to zero. In the same way that we obtained (13.1.22) from (13.1.18) we now obtain

$$\sum_{k=0}^{\infty} \left(y_{n-k} - \sum_{j=0}^{m} (\beta_j)_n \varphi_j(k, \theta) \right) \theta^k \varphi_i(k, \theta) = 0 \qquad 0 \le i \le m \tag{13.3.19}$$

Reversing the order of summation in (13.3.19) and moving the terms involving y to the right, gives us

$$\sum_{j=0}^{m} (\beta_j)_n \sum_{k=0}^{\infty} \varphi_i(k, \theta) \varphi_j(k, \theta) \theta^k = \sum_{k=0}^{\infty} \varphi_i(k, \theta) \theta^k y_{n-k} \qquad 0 \le i \le m \tag{13.3.20}$$

Then by virtue of (13.3.14) this becomes

$$\sum_{j=0}^{m} (\beta_j)_n \delta_{i,j} = \sum_{k=0}^{\infty} \varphi_i(k, \theta) \theta^k y_{n-k} \qquad 0 \le i \le m \tag{13.3.21}$$

which reduces to

$$\beta_{i,n} = \sum_{k=0}^{\infty} \varphi_i(k, \theta) \theta^k y_{-k} \qquad 0 \le i \le m \tag{13.3.22}$$

By the use of orthogonality we have thus obtained the general expressions for the β's without the need for matrix inversion, and so the analysis can continue.

Note: On the right of (13.3.22) the variable k sums out, and so the result is a function of θ and n. This is consistent with the notation on the left.

13.3.6 Using the β's to write the approximating polynomial

The expression for the coefficients $\beta_{i,n}$ in (13.3.22) now enables us to write the least-squares approximating polynomial in (13.3.16) as follows:

$$\left(p^*(s)\right)_n = \sum_{j=0}^{m}\left(\sum_{k=0}^{\infty}\varphi_j(k,\,\theta)\theta^k y_{n-k}\right)\varphi_j(s,\,\theta) \tag{13.3.23}$$

Equation (13.3.23) is a sum of the polynomials $\varphi_j(s,\,\theta)$ whose degrees run from $j = 0$ to m, and so it is a polynomial of degree m.

13.3.7 Estimating the true state vector

Returning to Figure 13.2, we see that the polynomial $\left(p^*(s)\right)_n$ in (13.3.23) can be used to provide estimates of $x(t)$, the true state vector's 0th-derivative, by assigning various values to s.

If we set $s = 0$, then we obtain the current-estimate, i.e. an estimate of what $x(t)$ is at time t_n based on observations up to time t_n. Calling that estimate $x_0{}^*_{n,\,n}$ we have the following:

$$\left(x(t_n)\text{ based on observations up to }t_n\right) \approx \left(x_0{}^*_{n,\,n} = \left(p^*(0)\right)_n\right) \tag{13.3.24}$$

The subscripted 0 in $x_0{}^*_{n,\,n}$ signifies that this is for the 0th derivative.

Similarly, for a 1-step-prediction we would set $s = -1$ obtaining

$$\left(x(t_{n+1})\text{ based on observations up to }t_n\right) \approx \left(x_0{}^*_{n+1,\,n} = \left(p^*(-1)\right)_n\right) \tag{13.3.25}$$

and for any validity instant $t = t_n - s\tau$ we would have

$$\left(x(t_n - s\tau)\text{ based on observations up to }t_n\right) \approx \left(x_0{}^*_{n-s,\,n} = \left(p^*(s)\right)_n\right) \tag{13.3.26}$$

The double subscript $n - s,\, n$ in $x_0{}^*_{n-s,\,n}$ means that its validity instant is whatever numerical value is assigned to s, and that $x_0{}^*$ was derived from data whose most recent time stamp was t_n.

We also require estimates of the time-derivatives of $x(t)$. To obtain these we recall from (13.3.1) that t and s are related by

$$t = t_n - s\tau \tag{13.3.27}$$

which means that

$$dt = -\tau \, ds \tag{13.3.28}$$

from which

$$d/dt = (-1/\tau)\, d/ds \tag{13.3.29}$$

This then means that

$$d/dt \left(p^*(s)\right)_n = (-1/\tau)\, d/ds \left(p^*(s)\right)_n \tag{13.3.30}$$

and in general that

$$d^i/dt^i \left(p^*(s)\right)_n = (-1/\tau)^i \, d^i/ds^i \left(p^*(s)\right)_n \tag{13.3.31}$$

If we now let D mean d/dt, and assume for simplicity that $m = 3$, then we obtain the following vector of estimates based on observations up to t_n, in terms of the derivatives of $\left(p^*(s)\right)_n$:

$$\begin{pmatrix} x(t_n - s\tau) \\ Dx(t_n - s\tau) \\ D^2 x(t_n - s\tau) \\ D^3 (t_n - s\tau) \end{pmatrix} \approx \begin{pmatrix} x_0{}^* \\ x_1{}^* \\ x_2{}^* \\ x_3{}^* \end{pmatrix}_{n-s,n} = \begin{pmatrix} \left(p^*(s)\right)_n \\ (-1/\tau)\, d/ds \left(p^*(s)\right)_n \\ (-1/\tau)^2 \, d^2/ds^2 \left(p^*(s)\right)_n \\ (-1/\tau)^3 \, d^3/ds^3 \left(p^*(s)\right)_n \end{pmatrix} \tag{13.3.32}$$

Using (13.3.23) for $\left(p^*(s)\right)_n$ in (13.3.32) then gives us the following general expression for any degree m and any derivative i from 0 to m:

$$x_i{}^*{}_{n-s,n} = (-1/\tau)^i \sum_{j=0}^{m} \left(\sum_{k=0}^{\infty} \varphi_j(k, \theta)\theta^k y_{n-k} \right) d^i/ds^i \varphi_j(s, \theta) \tag{13.3.33}$$

We now rearrange the terms in (13.3.33) and multiply both sides by $\tau^i/i!$ obtaining

$$(\tau^i/i!)\, x_i{}^*{}_{n-s,n} = \sum_{j=0}^{m} (-1)^i/i! \, d^i/ds^i \varphi_j(s, \theta) \left(\sum_{k=0}^{\infty} \varphi_j(k, \theta)\, \theta^k y_{n-k} \right) \qquad 0 \le i \le m$$

(13.3.34)

The FMP filter does its internal computing using the normalized vector $\boldsymbol{Z}^*$ and we recall from (12.2.4) that the elements of $\boldsymbol{Z}^*$ and $\boldsymbol{X}^*$ are related by

$$z_i{}^* = (\tau^i/i!)\, x_i{}^* \quad 0 \le i \le m \tag{13.3.35}$$

This means that (13.3.34) can be written as

$$z_i{}^*{}_{n-s,n} = \sum_{j=0}^{m} (-1)^i/i!\, d^i/ds^i \varphi_j(s,\theta) \left(\sum_{k=0}^{\infty} \varphi_j(k,\theta)\, \theta^k y_{n-k} \right) \quad 0 \le i \le m \tag{13.3.36}$$

From now on we will focus on the normalized estimate vector $\boldsymbol{Z}^*$, and only when we are completely done will we denormalize.

Setting $s = 0$ in (13.3.36) gives the elements of the current-estimate vector $\boldsymbol{Z}^*_{n,n}$ as

$$z_i{}^*{}_{n,n} = \sum_{j=0}^{m} (-1)^i/i!\, d^i/ds^i \varphi_j(s,\theta)\Big|_{s=0} \left(\sum_{k=0}^{\infty} \varphi_j(k,\theta) \theta^k y_{n-k} \right) \quad 0 \le i \le m \tag{13.3.37}$$

and setting $s = -1$ we obtain elements of the 1-step prediction vector $\boldsymbol{Z}^*_{n+1}$ as

$$z_i{}^*{}_{n,n} = \sum_{j=0}^{m} (-1)^i/i\, d^i/ds^i \varphi_j(s,\theta)\Big|_{s=-1} \left(\sum_{k=0}^{\infty} \varphi_j(k,\theta)\, \theta^k y_{n-k} \right) \quad 0 \le i \le m \tag{13.3.38}$$

We will return to these three equations further down when we derive the expressions for the FMP covariance matrices.

13.3.8 The FMP recursive formulation

The true power of the FMP filters lies in their recursive structure. However, as with the EMP filters, the road from the starting point (13.3.37) or (13.3.38) to the recursive formulations in Appendices 13.2 and 13.4 is a long one and we will not follow it here. The reader who is interested in seeing how it was done is referred to Reference 6, starting on p. 504.

In addition to the recursive algorithms themselves, there are two other results of importance that are also derived. We touch on these only briefly, and the interested reader can find the details Reference 6.

13.3.8.1 The general expression for the FMP weights for any degree:

Define the matrix $F(0, \theta)$ whose i, jth element is (see Appendix 13.5)

$$[F(0, \theta)]_{i,j} \equiv (-1)^i/i!\, d^i/ds^i p_j(s, \theta)\Big|_{s=0} \qquad 0 \leq i, j \leq m \qquad (13.3.39)$$

Also define the $(m+1)$-vector $L(\theta)$ whose jth term is

$$[L(\theta)]_j \equiv \theta^j/c(j, \theta) \qquad 0 \leq i, j \leq m \qquad (13.3.40)$$

Then the general expressions for the weights $\alpha, \beta, \gamma, \ldots$ in the FMP algorithms (both 1-step-predictor and current-estimate) for any degree m is given by

$$W(\theta) = F(0, \theta)L(\theta) \qquad (13.3.41)$$

in which $W(\theta) = (\alpha, \beta, \gamma, \ldots)^T$ is called the *weight vector.*

Proof for this statement is provided in Problem 13.8.

13.3.8.2 Stability properties of the FMP filters

The FMP filters are recursive, and so it is necessary to show that they are stable, i.e. that any unwanted disturbances will die out. Starting on p. 507 of Reference 6 we prove that the natural modes for the mth degree algorithm are of the form

$$f(n) = p_m(n)\, \theta^n \qquad (13.3.42)$$

in which n is the cycle number of the filter and $p_m(n)$ is a polynomial of degree m.

Because $0 < \theta < 1$, it follows from (13.3.42) that the natural modes will all die out as n increases, and so the FMP filters are unconditionally stable.

In the language of linear control theory they are *critically damped.*

13.4 The FMP covariance matrices

13.4.1 The normalized FMP covariance matrices

At the beginning of Section 13.3 where we began the discussion of the FMP filters we made the statement that a number of significant problems arise when deriving the expressions for their covariance matrices. These problems come about because of the presence of the weight function θ^k and the summations from 0 to ∞, and did not arise for the EMP filters because there the weight function was 1 and the summations were from 0 to n.

Rather than discussing those problems here we provide only a brief summary and refer you instead to Reference 6 for full details.

Starting on p. 519 of that book it is shown that when the covariance matrix of the observations has the form

$$\boldsymbol{R}_n = \sigma_\nu^2 \boldsymbol{I} \tag{13.4.1}$$

in which $\boldsymbol{I}$ is an infinite-dimensional identity matrix, then the expressions for the normalized covariance matrices for the FMP filters can be obtained as follows.

For the FMP filters of degree m:

◊ Define the matrix $\boldsymbol{A}(\theta)$ by

$$[\boldsymbol{A}(\theta)]_{i,j} = \Big((i+j)!/(i!\,j!)\Big)\Big((1-\theta)/(1+\theta)^{i+j+1}\Big) \qquad 0 \le i, j \le m \tag{13.4.2}$$

◊ Define the matrix $\boldsymbol{F}(s)$ by

$$[\boldsymbol{F}(s,\theta)]_{i,j} \equiv (-1)^i/i!\; d^i/ds^i p_j(s,\theta) \qquad 0 \le i, j \le m \tag{13.4.3}$$

Then the covariance matrix of the normalized FMP estimate vector $\boldsymbol{Z}^*_{n-s,\,n}$ will be

$$\boldsymbol{S}^*(\boldsymbol{Z}^*_{n-s,\,n}) = \sigma_\nu^2\, \boldsymbol{F}(s,\theta)\boldsymbol{A}(\theta)\boldsymbol{F}(s,\theta)^T \tag{13.4.4}$$

For the *1*-step-predictors we use $s = -1$, obtaining

$$\boldsymbol{S}^*(\boldsymbol{Z}^*_{n+1,\,n}) = \sigma_\nu^2 \boldsymbol{F}(-1,\theta)\boldsymbol{A}(\theta)\boldsymbol{F}(-1,\theta)^T \tag{13.4.5}$$

and for the current-estimate filters we use $s = 0$, obtaining

$$\boldsymbol{S}^*(\boldsymbol{Z}^*_{n,\,n}) = \sigma_\nu^2 \boldsymbol{F}(0,\theta)\boldsymbol{A}(\theta)\boldsymbol{F}(0,\theta)^T \tag{13.4.6}$$

The matrices $\boldsymbol{A}(\theta)$ and $\boldsymbol{F}(s,\theta)$ are displayed in Appendices 13.4 and 13.5.

13.4.2 Denormalization

The covariance matrix for the denormalized FMP estimate vector $\boldsymbol{X}^*$ is obtained by use of the diagonal matrix $\boldsymbol{D}(\tau)$ defined in (12.2.11) as follows:

◊ For any validity instant:

$$\boldsymbol{S}^*(\boldsymbol{X}^*_{n-s,\,n}) = \boldsymbol{D}(\tau)\boldsymbol{S}^*(\boldsymbol{Z}^*_{n-s,\,n})\boldsymbol{D}(\tau) = \sigma_\nu^2 \boldsymbol{D}(\tau)\boldsymbol{F}(s,\theta)\boldsymbol{A}(\theta)\,\boldsymbol{F}(s,\theta)^T\boldsymbol{D}(\tau) \tag{13.4.7}$$

◊ For the *1*-step-predictor:

$$S^*(X^*_{n+1,n}) = D(\tau)S^*(Z^*_{n+1,n})D(\tau)$$
$$= \sigma_\nu^2 D(\tau)F(-1,\theta)A(\theta)F(-1,\theta)^T D(\tau) \tag{13.4.8}$$

◊ For the current-estimate filter:

$$S^*(X^*_{n,n}) = D(\tau)S^*(X^*_{n,n})D(\tau) = \sigma_\nu^2 D(\tau)F(0,\theta)A(\theta)F(0,\theta)^T D(\tau) \tag{13.4.9}$$

The algebraic expressions for the diagonal elements of the FMP *1*-step-predictor covariance matrix $S^*(X^*_{n+1,n})$ for degrees from *0* to *4* were displayed in Figure 12.21(a) and 12.21(b).

Those expressions and some of their plots in Figure 12.22 contain essential information for understanding the FMP filters and their properties. They were obtained by manual expansion of the diagonal elements of (13.4.8).

In the files

Miscellaneous\Polynomial_filters\1-step_predictors_up_to_8th_degree

and

Miscellaneous\Polynomial_filters\Current_ estimate_up_to_8th_degree

we display the algebraic expressions for the diagonal elements of the denormalized FMP covariance matrices up to 8th degree, obtained by the use of Symbolic Matlab.

For the entire *1*-step-predictor FMP covariance matrices or for any of the items related to the current-estimate covariance matrices we will have to content ourselves with computer implementations that give us numerical values at run-time.

In Project 13.3 we discuss how that is carried out.

In Project 13.5 we consider how you can obtain approximate expressions for the entire covariance matrices of the FMP filters in a very useful form. These approximations are quite accurate for $\theta = 0.99$ and become extremely accurate when θ increases beyond *0.99* and approaches *1*, and since θ is usually very close to *1*, the approximations are almost always highly accurate.

For $\theta \approx 1$ the approximations are valid for both the *1*-step-predictors and the current-estimate filters.

In ***Miscellaneous*** you will find the program ***Problem_13.3.exe*** which enables you to derive the approximate FMP covariance matrices up to 8th degree.

13.4.3 Validation of the expressions for the FMP covariance matrices

Project 12.7b demonstrates that the FMP filters are ECM consistent. By running that verification, you will observe how – time after time and under almost all possible circumstances – the error/sigma ratios and the Chi-squared values for the FMP filters always lie within their prescribed bounds.

Such performance can only take place if we have used the correct expressions for the covariance matrices when creating that program, and so verification that the FMP filters are ECM consistent also constitutes proof that the expressions for the FMP covariance matrices derived in this section are correct.

The following items are in the supplementary material

Problems for Chapter 13

Solutions to Problems for Chapter 13

Projects for Chapter 13

Appendix 13.1

The discrete Legendre orthogonal polynomials

$$p_j(s, n) = \sum_{v=0}^{j} (-1)^v \binom{j}{v} \binom{j+v}{v} \frac{s^{(v)}}{n^{(v)}} \qquad j = 0, 1, 2, \ldots, n$$

$$p_0(s, n) = 1$$

$$p_1(s, n) = 1 - 2s/n$$

$$p_2(s, n) = 1 - 6s/n + 6s(s-1)/n(n-1)$$

$$p_3(s, n) = 1 - 12s/n + 30s(s-1)/n(n-1) - 20s(s-1)(s-2)/n(n-1)(n-2)$$

$$p_4(s, n) = 1 - 20s/n + 90s(s-1)/n(n-1) - 140s(s-1)(s-2)/n(n-1)(n-2) + 70s(s-1)(s-2)(s-3)/n(n-1)(n-2)(n-3)$$

$$\left((c_j(n)\right)^2 = \frac{(n+j+1)^{(j+1)}}{(2j+1)\, n^{(j)}} \qquad j = 0, 1, 2, \ldots, n$$

$$\left(c_0(n)\right)^2 = n + 1$$

$$\left(c_1(n)\right)^2 = (n+2)(n+1)/3n$$

$$\left(c_2(n)\right)^2 = (n+3)(n+2)(n+1)/5n(n-1)$$

$$\left(c_3(n)\right)^2 = (n+4)(n+3)(n+2)(n+1)/7n(n-1)(n-2)$$

$$\left(c_4(n)\right)^2 = (n+5)(n+4)(n+3)(n+2)(n+1)/9n(n-1)(n-2)(n-3)$$

Appendix 13.2

The three Legendre *P* matrices

1. *The matrix* $[\boldsymbol{P}(s,n)]_{i,j} = 1/i!\ d^i/ds^i\ p_j(s,n) \qquad 0 \le i,j \le 4$

$p_{0,0} = 1$

$p_{0,1} = 1 - 2s/n$

$p_{0,2} = 1 - 6s/n + 6s^{(2)}/n^{(2)}$

$p_{0,3} = 1 - 12s/n + 30s^{(2)}/n^{(2)} - 20s^{(3)}/n^{(3)}$

$p_{0,4} = 1 - 20s/n + 90s^{(2)}/n^{(2)} - 140s^{(3)}/n^{(3)} + 70s^{(4)}/n^{(4)}$

$p_{1,0} = 0$

$p_{1,1} = -2/n$

$p_{1,2} = -6/n + 6(2s-1)/n^{(2)}$

$p_{1,3} = -12/n + 30(2s-1)/n^{(2)} - 20(3s^2 - 6s + 2)/n^{(3)}$

$$p_{1,4} = -20s/n + 90(2s-1)/n^{(2)} - 140(3s^2 - 6s + 2)/n^{(3)} + 70(4s^3 - 18s^2 + 22s - 6)/n^{(4)}$$

$p_{2,0} = 0$

$p_{2,1} = 0$

$p_{2,2} = 6/n^{(2)}$

$p_{2,3} = 30/n^{(2)} - 10(6s-6)/n^{(3)}$

$p_{2,4} = 90/n^{(2)} - 70(6s-6)/n^{(3)} + 35(12s^2 - 36s + 22)/n^{(4)}$

$p_{3,0} = 0$

$p_{3,1} = 0$

$p_{3,2} = 0$

$p_{3,3} = -20/n^{(3)}$

$p_{3,4} = -140/n^{(3)} + 35(8s-12)/n^{(4)}$

$p_{4,0} = 0 \quad p_{4,1} = 0 \quad p_{4,2} = 0 \quad p_{4,3} = 0 \quad p_{4,4} = 70/n^{(4)}$

2. The matrix $[\boldsymbol{P}(n+1, n)]_{i,j} = 1/i!\ d^i/ds^i\ p_j(s, n)\Big|_{s=n+1} \quad 0 \le i, j \le 4$

$$\begin{pmatrix} 1 & \frac{-(n+2)}{n} & \frac{(n+3)^{(2)}}{n^{(2)}} & \frac{-(n+4)^{(3)}}{n^{(3)}} & \frac{(n+5)^{(4)}}{n^{(4)}} \\ 0 & \frac{-2}{n} & \frac{6(n+2)}{n^{(2)}} & \frac{-2(6n^2+27n+32)}{n^{(3)}} & \frac{10(2n^3+15n^2+41n+38)}{n^{(4)}} \\ 0 & 0 & \frac{6}{n^{(2)}} & \frac{-30(n+2)}{n^{(3)}} & \frac{10(9n^2+39n+47)}{n^{(4)}} \\ 0 & 0 & 0 & \frac{-20}{n^{(3)}} & \frac{140(n+2)}{n^{(4)}} \\ 0 & 0 & 0 & 0 & \frac{70}{n^{(4)}} \end{pmatrix}$$

3. The matrix $[\mathbf{P}(n,n)]_{i,j} = 1/i!d^i/ds^i(p_j(s,n))\Big|_{s=n} \qquad 0 \le i,j \le 4$

$$
\begin{pmatrix}
1 & -1 & 1 & -1 & 1 \\
0 & \dfrac{-2}{n} & \dfrac{6}{n-1} & \dfrac{-2(6n^2-3n+2)}{n^{(3)}} & \dfrac{10(2n^2-3n+5)}{(n-1)^{(3)}} \\
0 & 0 & \dfrac{6}{n^{(2)}} & \dfrac{-30}{(n-1)^{(2)}} & \dfrac{5(18n^2-6n+10)}{n^{(4)}} \\
0 & 0 & 0 & \dfrac{-20}{n^{(3)}} & \dfrac{140}{(n-1)^{(3)}} \\
0 & 0 & 0 & 0 & \dfrac{70}{n^{(4)}}
\end{pmatrix}
$$

Appendix 13.3

The discrete Laguerre orthogonal polynomials

$$p_j(s,\theta) = \theta^j \sum_{v=0}^{j} (-1)^v b^v \binom{j}{v}\binom{s}{v} \qquad j = 0, 1, 2, \ldots \qquad b = \frac{1-\theta}{\theta}$$

$$p_0(s,\theta) = 1$$

$$p_1(s,\theta) = \theta(1-bs)$$

$$p_2(s,\theta) = \theta^2\left(1-2bs + b^2 s(s-1)/2!\right)$$

$$p_3(s,\theta) = \theta^3\left(1-3bs + 3b^2 s(s-1)/2! - b^3 s(s-1)(s-2)/3!\right)$$

$$p_4(s,\theta) = \theta^4\left(1-4bs + 6b^2 s(s-1)/2! - 4b^3 s(s-1)(s-2)/3! + b^4 s(s-1)(s-2)(s-3)/4!\right)$$

$$\left(c_j(\theta)\right)^2 = \theta^j/1-\theta \qquad j = 0, 1, 2, \ldots$$

$$\left(c_0(\theta)\right)^2 = 1/(1-\theta)$$

$$\left(c_1(\theta)\right)^2 = \theta/(1-\theta)$$

$$\left(c_2(\theta)\right)^2 = \theta^2/(1-\theta)$$

$$\left(c_3(\theta)\right)^2 = \theta^3/(1-\theta)$$

$$\left(c_4(\theta)\right)^2 = \theta^4/(1-\theta)$$

Appendix 13.4

The Laguerre $A(\theta)$ matrix

$$[\boldsymbol{A}(\theta)]_{i,j} = \left((i+j)!/i!j!\right)\left((1-\theta)/(1+\theta)^{i+j+1}\right) \qquad 0 \le i,j \le 4$$

$$\begin{pmatrix}
\dfrac{1-\theta}{1+\theta} & \dfrac{1-\theta}{(1+\theta)^2} & \dfrac{1-\theta}{(1+\theta)^3} & \dfrac{1-\theta}{(1+\theta)^4} & \dfrac{1-\theta}{(1+\theta)^5} \\[2ex]
\dfrac{1-\theta}{(1+\theta)^2} & \dfrac{2(1-\theta)}{(1+\theta)^3} & \dfrac{3(1-\theta)}{(1+\theta)^4} & \dfrac{4(1-\theta)}{(1+\theta)^5} & \dfrac{5(1-\theta)}{(1+\theta)^6} \\[2ex]
\dfrac{1-\theta}{(1+\theta)^3} & \dfrac{3(1-\theta)}{(1+\theta)^4} & \dfrac{6(1-\theta)}{(1+\theta)^5} & \dfrac{10(1-\theta)}{(1+\theta)^6} & \dfrac{15(1-\theta)}{(1+\theta)^7} \\[2ex]
\dfrac{1-\theta}{(1+\theta)^4} & \dfrac{4(1-\theta)}{(1+\theta)^5} & \dfrac{10(1-\theta)}{(1+\theta)^6} & \dfrac{20(1-\theta)}{(1+\theta)^7} & \dfrac{35(1-\theta)}{(1+\theta)^8} \\[2ex]
\dfrac{1-\theta}{(1+\theta)^5} & \dfrac{5(1-\theta)}{(1+\theta)^6} & \dfrac{15(1-\theta)}{(1+\theta)^7} & \dfrac{35(1-\theta)}{(1+\theta)^8} & \dfrac{70(1-\theta)}{(1+\theta)^9}
\end{pmatrix}$$

Appendix 13.5
The Laguerre $F(s, \theta)$ matrix

$$[\boldsymbol{F}(s,\theta)]_{i,j} = (-1)^i / i! \, d^i/ds^i \, p_j(s,\theta) \qquad 0 \le i,j \le 4$$

$$b = (1-\theta)/\theta$$

(See next page)

$$
\begin{pmatrix}
1 & \theta^2(1-bs) & \begin{matrix}\theta^2\big(1-2bs\\ +b^2(s^2-s)/2\big)\end{matrix} & \begin{matrix}-\theta^3\big(1-3bs+3b^2(s^2-s)/2\\ -b^3(s^3-3s^2+2s)/6\big)\end{matrix} & \begin{matrix}\theta^4\big(1-4bs+6b^2(s^2-s)/2\\ -4b^3(s^3-3s^2+2s)/6\\ +b^4(s^4-6s^3+11s^2-6s)/24\big)\end{matrix} \\
0 & \theta b & \theta^2\big(2b-b^2(2s-1)/2\big) & \begin{matrix}\theta^3\big(3b-3b^2(2s-1)/2\\ +b^3(3s^2-6s+2)/6\big)\end{matrix} & \begin{matrix}\theta^4\big(4b-6b^2(2s-1)/2\\ +4b^3(3s^2-6s+2)/6\\ -b^4(4s^3-18s^2+22s-6)/24\big)\end{matrix} \\
0 & 0 & {}^1/_2\,\theta^2 b^2 & {}^1/_2\,\theta^3\big(3b^2-b^3(s-1)\big) & \begin{matrix}{}^1/_2\,\theta^4\big(6b^2-4b^3(s-1)\\ +b^4(12s^2-36s+22)/24\big)\end{matrix} \\
0 & 0 & 0 & {}^1/_6\,\theta^3 b^3 & {}^1/_6\,\theta^4\big(4b^3-b^4(2s-3)/2\big) \\
0 & 0 & 0 & 0 & {}^1/_{24}\,\theta^4 b^4
\end{pmatrix}
$$

References

1 Kalman, R.E., 'A new approach to linear filtering and prediction problems', *Transactions of the ASME, Journal of Basic Engineering*, 1960, vol. 82D, pp. 33–45

2 Kalman, R.E., Bucy, R.S., 'New results in linear filtering and prediction theory', *Transactions of the ASME, Journal of Basic Engineering*, 1961, vol. 83D, pp. 95–108

3 Bar Shalom, Y., Xiao-Rong Li, X., Kirubarajan, T., *Estimation with Applications to Tracking and Navigation*, Wiley, 2001, ISBN 0-471-41655-X (See p. 243 for the extract referred to in our Chapters 11.)

4 Gauss, K.F., *Theory of the Combination of Observations Least Subject to Errors*, Translated by G.W. Stewart, SIAM, ISBN 0-89871-347-1

5 Gauss, K.F., *Theory of the Motion of the Heavenly Bodies Moving About the Sun in Conical Sections*, 1809, English Translation, Dover Publications, New York, NY, 1963

6 Morrison, N., *Introduction to Sequential Smoothing and Prediction*, McGraw-Hill, 1969. Free down-load of PDF available from either *http://opencontent.uct.ac.za/Engineering-and-the-Built-Environment* or *http://goo.gl/CM9cN*

7 Swerling, P., 'A proposed stagewise differential correction procedure for satellite tracking and prediction', Rand Corp, Santa Monica, CA, Report P-1292, January 8, 1958

8 Swerling, P., 'First order error propagation in a stagewise smoothing procedure for satellite observations', *Journal of Astronautical Sciences*, Autumn 1959, vol. 6, pp. 46–52

9 Swerling, P., 'Topics in generalized least-square signal estimation', *SIAM Journal*, September 1966, vol. 14–15, pp. 998–1031

10 Aitken, A.C., 'On least-squares and linear combinations of observations', *Proceedings of the Royal Society Edinburgh*, 1935, vol. A, no. 55, pp. 42–48

11 Hogg, R.V., Craig, A.T., *Introduction to Mathematical Statistics*, 5th edn, Prentice Hall, p. 372 *et seq*.

12 Stuart, A., Ord, J.K., Arnold, S., *Kendall's Advanced Theory of Statistics*, 6th edn, vol. 2A, Oxford University Press, New York, NY, p. 9 *et seq*.

13 Cox, D.R., Hinkley, D.V., *Theoretical Statistics*, Chapman & Hall, London, p. 254 *et seq*., ISBN 0-412-16160-5

14 Chapra, S., Canale, R., *Numerical Methods for Engineers*, 2nd edn, McGraw-Hill, pp. 596–609, ISBN 0-07-100412-2

15 Bellman, R., *Introduction to Matrix Analysis*, McGraw-Hill, 1960

16 Mirsky, L., *Introduction to Linear Algebra*, Oxford, 1961

17 Stigler, S.M., *The History of Statistics*, Belknap Press of Harvard Press, 1986, pp. 11–16

18 Roy, A.E., *Orbital Motion*, 2nd edn, Adam Hilger Ltd, Bristol, 1982, pp. 69–88, ISBN 0-85274-462-5

19 Levenberg-Marquard algorithm in Wikipedia. Available at http://en.wikipedia.org/wiki/Levenberg-Marquardt_algorithm

20 Rao, C.R., *Linear Statistical Inference and Its Applications*, 2nd edn, John Wiley and Sons, p. 188, Item (vii), ISBN 0-471-70823-2

21 Allen, H.J., Eggers, A.J., Jr., 'A study of the motion and aerodynamic heating of ballistic missiles entering the earth's atmosphere at high supersonic speeds', NACA Technical Report 1381, Forty-Fourth Annual Report of the NACA, 1958, Washington, DC

22 Brown, R.G., Hwang, P.Y.C., *Introduction to Random Signals and Applied Kalman Filtering*, Wiley, Hoboken NJ, ISBN 0-471-12839-2

23 Zarchan, P., Musoff, H., *Fundamentals of Kalman Filtering: A Practical Approach*, Progress in Astronautics and Aeronautics, vol. 190, American Institute of Aeronautic and Astronautics, ISBN 1-56347-455-7

24 Brookner, E., *Tracking and Kalman Filtering Made Easy*, John Wiley and Sons, 1998, ISBN 0-471-18407-1

25 Welch, G., Bishop, G., 'An introduction to the Kalman Filter', University of North Carolina. Available at http://www.cs.unc.edu/~tracker/media/pdf/SIGGRAPH2001_CoursePack_08.pdf (free download)

26 Young, G.A., Smith, R.L., *Essentials of Statistical Inference*, Cambridge, 2005, p. 123, ISBN 13-978-0-521-83971-6

27 Kolman, B., *Elementary Linear Algebra*, Collier Macmillan Pub. Co., New York, NY, ISBN 0-02-366080-5

28 Grossman, S.I., *Elementary Linear Algebra*, Wadsworth Inc., Belmont, CA, 1980, ISBN 0-534-00746-5

29 Autonomous systems in Wikipedia. Available at http://en.wikipedia.org/wiki/Autonomous_ system_(mathematics)

30 Blackman, S., *Multiple Target Tracking with Radar Applications*, Artech House, 1986, p. 26, ISBN 0-89006-179-3

31 Blackman, S., Popolis, R.B., *Design and Analysis of Modern Tracking Systems*, Artech House, 1999, p. 159, ISBN 1-58053-006-0

32 Orderud, F., 'Comparison of Kalman filter estimation approaches for state space models with nonlinear measurements', *Proceedings of Scandinavian Conference on Simulation and Modeling*, SIMS, 2005

33 Joseph, J., LaViola, J.R., 'A comparison of unscented and extended Kalman filtering for estimating quaternion motion', *Proceedings of the 2004 American Control Conference*, IEEE Press, June 2004, pp. 2190–95

34 Julier, S., Uhlmann, J., 'Unscented filtering and nonlinear estimation', *Proceedings of the IEEE*, March 2004

35 Julier, S., Uhlmann, J., Durrant-Whyte, H., 'A new approach for filtering non-linear systems', *Proceedings of the 1995 American Control Conference*, IEEE Press, June 1995

36 van der Merwe, R., 'Sigma-point Kalman filters for probabilistic inference in dynamic state space models', Workshop on Advances in Machine Learning, Montreal, June 2003

37 Wan, E., van der Merwe, R., 'The unscented Kalman filter for nonlinear estimation', *Proceedings of IEEE Symposium 2000* (As-SPCC), Lake Louise, Alberta, October 2000

38 Arulampalam, S., Maskell, S., Gordon, N., Clapp, T., 'A tutorial on particle filters for non-linear/non-Gaussian Bayesian tracking', *IEEE Transactions on Signal Processing*, February 2002, vol. 50, no. 2, pp. 174–88

39 Banani, S.A., Masnadi-Shirazi, M.A., 'A new version of unscented Kalman filter'. *World of Academic Science and Technology*, 2007, vol. 26. Available at http://www.waset.org/journals/waset/v26/v26-37.pdf

40 Sputnik-1 in Wikipedia. Available at http://en.wikipedia.org/wiki/Sputnik_1

41 Peter Swerling in Wikipedia. Available at http://en.wikipedia.org/wiki/Peter_Swerling

42 Blackman, R.B., 'Methods of orbit refinement', *Bell Systems Technical Journal*, May 1964, vol. 43, no. 3

43 Bellantoni, J.F., Dodge, K.W., 'A square root formulation of the Kalman-Schmidt filter', *AIAA Journal*, July 1967, vol. 5, no. 7, pp. 1309–14

44 Bierman, G.J., *Factorization Methods for Discrete Sequential Estimation*, Academic Press, New York, NY, 1977

45 Kaminski, P.G., Bryson, E., Jr., Schmidt., S.F., 'Discrete square root filtering: AZ survey of current techniques', *IEEE Transactions on Automatic Control*, December 1971, vol. AC-16, no. 6, pp. 727–36

46 Maybeck, P.S., *Stochastic Models, Estimation and Control*, Academic Press, New York, NY, 1979

47 Grewal, M.S., Andrews, A.P., *Kalman Filtering Theory and Practice*, Prentice-Hall, Englewood Cliffs, NJ, 1993

48 Kaplan, E.D. (ed.), *Understanding GPS: Principles and Applications*, Artech House, Boston, MA, p. 394, ISBN 0-89006-793-7

49 Rank (linear algebra), http://en.wikipedia.org/wiki/Rank_(linear_algebra)

50 Bar Shalom, Y., Fortmann, T.E., *Tracking and Data Association*, Academic Press, 1988

51 Cramér, H., *Mathematical Methods of Statistics*, 1st ed., Princeton University Press, 1946

52 Rao, C.R., 'Information and the accuracy attainable in the estimation of statistical parameters', *Bulletin of Calcutta Mathematical Society*, 1946, vol. 37, p. 8189

53 Kleeman, L., 'Understanding and applying Kalman filtering', Department of Electrical and Computer Systems Engineering, Monash University Clayton, Australia. Available at http://www.ecse.monash.edu.au/centres/irrc/LKPubs/Kalman.PDF

54 Chou, K.C., Willsky, A.S., Nikoukhah, R., 'Multiscale systems, Kalman filters and Riccati equations', *IEEE Transactions on Automatic Control*, March 1994, vol. 39, no. 3. Available at ssg.mit.edu/~willsky/publ_pdfs/P-2152_IEEE.pdf

55 Siouris, G.M., *Missile Guidance and Control Systems*, Springer, 2003, p. 241, ISBN 0-387-00726-1

56 Stone, L.D., Barlow, C.A., Corwin, T.L., *Bayesian Multiple Target Tracking*, Artech House, 1999, p. 72, ISBN 1-58053-024-9

57 Wertz, J.A. (ed.), *Spacecraft Attitude Determination and Control*, Kluwer Academic Publications, 2002, p. 463, ISBN 90-277-0959-9

58 *Ibid.*, 'State estimation: attitude determination methods', Chapter 13

59 Andrews, W., 'A square root formulation of the Kalman covariance equations', *Journal AIAA*, 1968, vol. 6, pp. 1165–66

60 Melkanoff, M.A., Sanada, T., 'Nuclear optical model calculations', *Methods in Computational Physics*, vol. 6, Academic Press, New York, NY, 1966

61 Wilde, D.J., *Optimum Seeking Methods*, Prentice-Hall, Englewood Cliffs, NJ, 1964

62 Marquardt, D.W., 'An algorithm for least-squares estimation of nonlinear parameters', *SIAM Journal*, 1963, vol. 11, p. 431

63 Bevington, P.R., *Data Reduction and Error Analysis for the Physical Sciences*, McGraw-Hill, New York, NY, 1969

64 Morrison, N., 'Smoothing and extrapolation of time series by means of discrete Laguerre polynomials', *SIAM Journal*, May 1967, vol. 15, no. 3, pp. 516–38

65 Morrison, N., 'Smoothing and extrapolation of continuous time series by means of Laguerre polynomials', *SIAM Journal*, November 1968, vol. 16, no. 6, pp. 1280–304

66 Bruckner, J., Ford, L.R., Jr., 'Recursive digital smoothing', Releasable Internal Memo, Defense Research Corp., Santa Barbara, CA, February 1964

67 Duffin, R.J., Schmidt, T.W., 'An extrapolator and scrutator', *Journal of Mathematical Analysis and Applications*, September 1960, vol. 1, no. 2, pp. 215–27

68 Duffin, R.J., Schmidt, T.W., 'A simple formula for prediction and automatic scrutation', American Rocket Society Flight Testing Conference, Daytona Beach, FL, March 1959

69 Helms, H.D., 'Maximally reliable exponential prediction equations for data-rate-limited tracking servomechanisms', *Bell Systems Technical Journal*, December 1965, pp. 2337–62

70 Golub, G.H., van Loan, C.F., *Matrix Computations*, 2nd edn, Johns Hopkins University Press, 1989

71 Long, N., Zuo R., Varshney, P., 'A sensor selection approach for target tracking in sensor networks with quantized measurements', IEEE International Conference on Acoustics, Speech and Signal Processing, (ICASSP), Las Vegas, Nevada, March 2008

72 Mathworks, 'Matrix exponential function — *expm()*'. Available at http://www.mathworks.com/access/helpdesk/help/techdoc/ref/expm.html

73 Sanders, M.A., 'Characteristic function of the central chi-square distribution'. Available at http://www.planetmathematics.com/CentralChiDistr.pdf

74 Abramowitz, M, Stegun, I.A. (eds.), *Handbook of Mathematical Functions with Formulas, Graphs, and Mathematical Tables*, Dover Publications, New York, NY, 1965, Chapter 26, ISBN 0-486-61272-4. Available at http://www.math.sfu.ca/~cbm/aands/page_940.htm

75 Johnson, N.L., Kotz, S., Balakrishnan, N., *Continuous Univariate Distributions*, 2nd edn, vol. 1, John Wiley and Sons, 1994, Chapter 18, ISBN 0-471-58495-9

76 Alexander, M., Graybill, F.A., Boes, D.C., *Introduction to the Theory of Statistics*, 3rd edn, McGraw-Hill, 1974, pp. 241–46, ISBN 0-07-042864-6

77 Box, G.E.P., Hunter, J.S., Hunter, W.G., *Statistics for Experimenters*, Wiley, p. 46 *et seq*., ISBN 0-47-09315-7

78 Wilson, E.B., Hilferty, M.M., 'The distribution of chi-square', *Proceedings of the National Academy of Sciences*, vol. 17, pp. 684–88

79 Chi-squared distribution in Wikipedia. Available at en.wikipedia.org/wiki/Chi-square_distribution

80 Distcalc.exe©. Available at http://www.mi.infn.it/~camera/lab-fisica/dispense/

81 Sivaraman, M.R., Sahay, R.R., 'Applications of Kalman filter in WAAS', Satcom and IT Applications, Area, Space Applications Centre, Ahmedabad, India. Available at http://www.gisdevelopment.net/technology/gps/techgp0033.htm

82 Horn, R., Johnson, C., *Matrix Analysis*, Cambridge University Press, 1985

83 Anderson, B., Moore, J., *Optimal Filtering*, Prentice-Hall, 1979

84 Gelb, A., *Applied Optimal Estimation*, MIT Press, 1974

85 Li, C., Wang, X., 'On convergence of the Gauss–Newton method for convex composite optimization', *Journal of Mathematical Programming,* January 2002, vol. 91, no. 2

86 Kalogerakis, N, Luus, R., 'Improvement of Gauss–Newton method for parameter estimation through the use of information index', Department of Chemical Engineering, University of Toronto. Available at http://pubs.acs.org/doi/abs/10.1021/i100012a015

87 Nesterov, Y., 'Modified Gauss–Newton scheme with worst-case guarantees for its global performance', Catholic University of Louvain – Center for Operations Research and Econometrics (CORE) December 2003, CORE Discussion Paper No. 2003/86. Available at http://papers.ssrn.com/sol3/papers.cfm?abstract_id=981420

88 Macconi, M., Morini, B., Porcelli, M., 'A Gauss–Newton method for solving bound-constrained underdetermined nonlinear systems', Universitá di Firenze, Italy. Available at http://www.informaworld.com/smpp/content~content=a909063185~db=all~order=page

89 Wapedia – Wiki, 'The Levenberg-Marquardt algorithm (LMA)'. Available at http://wapedia.mobi/en/ Levenberg%E2%80%93Marquardt_algorithm

90 Wapedia – Wiki, 'The method of steepest descent'. Available at http://wapedia.mobi/en/Method_of _steepest_descent

91 Swerling, P., Obituary, *SIAM News*, vol. 34, no. 3, April 2001. Available at http://www.siam.org/ news/news.php?id=526

92 Kalman, R.E., Biography. Available at http://www.answers.com/topic/rudolf-kalman

93 Bucy, R.S., Biography. Available at http://comsys.sharif.edu/it/profiles/2832.htm

94 Newton's Biography. Available at http://www.newton.ac.uk/newtlife.html

95 Aitken, A.C., Biography. http://en.wikipedia.org/wiki/Alexander_Aitken

96 Likelihood function. Available at http://en.wikipedia.org/wiki/Likelihood_function

97 Likelihood function. Available at http://www.statistics.com/resources/glossary/l/likehdfunc.php

98 Prosecutor's fallacy. Available at http://en.wikipedia.org/wiki/Prosecutor%27s_fallacy

99 Gauss. Available at http://en.wikipedia.org/wiki/Carl_Friedrich_Gauss

100 Bar Shalom, Y., Xiao-Rong Li, *Multitarget-Multisensor Tracking*, 1995, p. 38 *et seq.*, ISBN 0-9648312-0-1

101 Leon-Garcia, A., *Probability and Random Processes for Electrical Engineering*, 2nd edn, Addison Wesley, 1994, pp. 158–9, ISBN 0-201-50037-X

102 Todhunter, I., *A History of the Mathematical Theory of Probability, from the Time of Pascal to That of Laplace*, Macmillan and Co., 1865

103 Central limit theorem in Wikipedia. Available at http://en.wikipedia.org/wiki/Central_limit_theorem

104 Abraham de Moivre in Wikipedia. Available at http://en.wikipedia.org/wiki/Abraham_de_Moivre

105 Lane, D., History of the normal distribution. Available at http://cnx.org/content/m11164/latest/

106 Bertsekas, D.P., Castanon, D.A., 'The auction algorithm for the transportation problem', February 1989, LIDS-P-1850 (Google: Auction algorithm)

107 Bertsekas, D.P., 'Auction algorithms' (Google: Auction algorithm)

108 Bertsekas, D.P., 'A distributed algorithm for the assignment problem', Laboratory for Information and Decision Systems Working Paper, MIT Press, Cambridge, MA, 1979 (Google: Auction algorithm)

109 Bertsekas, D.P., *Linear Network Optimization: Algorithms and Codes*, MIT Press, Cambridge, MA, 1991

110 Bertsekas, D.P., 'Auction algorithms for network flow problems: a tutorial introduction', *Computational Optimization and Applications*, 1992, vol. 1, pp. 7–66 (Google: Auction algorithm)

111 Long, Z., Ruixin, N., Varshney, P., 'A sensor selection approach for target racking in sensor networks with quantized measurements', *ICASSP 2008*, March 2008

112 Reyneke, P., Morrison, N., Kourie, D., de Ridder, C., 'Smoothing irregular data using polynomial filters', *Proceedings of IEEE ELMAR (Electronics in Marine) Conference*, 15–17 September 2010, Zadar, Croatia

113 Kepler, J. Available at http://en.wikipedia.org/wiki/Johannes_Kepler

114 Lagrange's method of undetermined multipliers, Wikipedia. Available at http://en.wikipedia.org/w/index.php?title=Lagrange_multipliers&printable= yes

115 Morrison, N., Lord, R.T., 'The Gauss–Newton algorithm applied to track-while-scan radar', *Proceedings of IEEE Edinburgh 2007 Radar Conference*

116 Morrison, N., Lord, R.T., Inggs, M.R., 'Differential correction in passive aircraft tracking using Doppler and bearings', *Proceedings of IEEE Edinburgh 2007 Radar Conference*

117 Gauss, K.F., *Theory of the Combination of Observations Least Subject to Errors*, Translated by G.W. Stewart, SIAM, p. ix, ISBN 0-89871-347-1

118 Reyneke, P., Morrison, N., Gibbons, I., de Ridder, C., 'Finding the Weights for the *n*th-degree polynomial filters using symbolic matlab', Paper to be submitted to RadarCon 2011

119 Neuman, C.P., Schonbach, D.I., 'Discrete Legendre polynomials – a survey', *International Journal for Numerical Methods in Engineering*, June 2005, vol. 8, no. 4, pp. 743–70

120 den Brinker, A., Bastiaans, M., *Modern Signal Transformations*. Available at http://www.sps.ele.tue.nl/members/m.j.bastiaans/courses/index.htm

121 Aburdene, M.F., Dorband, J.E., 'On the computation of discrete Legendre polynomial coefficients', *Multidimensional Systems and Signal Processing*, 1993, no. 4, pp. 181–86

122 Larson, R.E., 'Analysis of the exponentially weighted polynomial filter', Tech. Memo 7, Contract DA-01-021-AMC-90006(Y), Stanford Research Institute, Menlo Park, CA, 1964

123 Varga, R.S., *Matrix Iterative Analysis*, Prentice-Hall, Englewood Cliffs, NJ, 1962, Theorem 1.4, p. 13

124 Ragazzini, J.R., Franklin, G.F., *Sampled-Data Control Systems*, McGraw-Hill, New York, NY, 1958

125 Lanczos, C., *Applied Analysis*, Prentice-Hall, Englewood Cliffs, NJ, 1956, p. 348

126 Ill-conditioned matrices. Available at http://en.wikipedia.org/wiki/Condition_number

127 Ill-conditioned matrices. Available at http://mathworld.wolfram.com/Ill-ConditionedMatrix.html

128 Ill-conditioned matrices. Available at http://www.padtinc.com/epubs/focus/common/focus.asp? I=39&P=article3.htm

129 Cheney, W., Kincaid, D., *Numerical Mathematics and Computing*, Brooks/Cole Publishing, Pacific Grove, CA, ISBN 978-0-495-11475-8

130 Trefethen, L.N., Bau, D., *Numerical Linear Algebra*, SIAM, 1997, ISBN 0-89871-361-7

131 Moszynski, K., 'On solving linear algebraic equations with an ill-conditioned matrix', *Applicationes Mathematicae*, 1995, vol. 22, no. 4, pp. 499–513

132 Spectral theorem. Available at http://en.wikipedia.org/wiki/Spectral_theorem

133 Strang, G., *Linear Algebra and Its Applications*, 3rd edn, Harcourt, Brace, Jovanovich, San Diego, CA, pp. 71–73, ISBN 0-15-551005-3

134 The non-central Chi-squared distribution, Wikipedia. Available at http://en.wikipedia.org/wiki/ Noncentral_chi-square_distribution

135 Generalised inverse. Available at http://en.wikipedia.org/wiki/Generalized_inverse

136 Hamming, R.W., *Numerical Methods for Scientists and Engineers*, McGraw-Hill, 1962, p. 18 *et seq.*

137 Hildebrand, F.B., *Introduction to Numerical Analysis*, McGraw-Hill, 1956

138 Gottlieb, M.J., 'Concerning some polynomials orthogonal on a finite or enumerable set of points', *American Mathematics Journal*, 1938, vol. 60, pp. 453–58

139 Rice, R.J., *Approximation of Functions*, vol. 1, Addison-Wesley, 1964, p. 120 *et seq*.

140 Davenport, W.B., Root, W.L., *Introduction to the Theory of Random Signals and Noise*, McGraw-Hill, 1958

141 Lindgren, B.W., *Statistical Theory*, Macmillan, 1962

142 Miller, K.S., *Multidimensional Gaussian Processes*, John Wiley, New York, NY, 1964

143 Plackett, R.L., 'A historical note on the method of least squares', *Biometrika*, 1949, vol. 36, pp. 458–60

144 Hastings, C., *Approximations for Digital Computers*, Princeton University Press, 1955

145 Joksch, H.C., 'Random errors of derivatives obtained from least-squares approximations of empirical functions', *SIAM Review*, January 1966, vol. 8, no. 1, p. 47 *et seq.*

146 Joksch, H.C., 'Reduction of the variance', *SIAM Review*, April 1966, vol. 8, no. 12, p. 211 *et seq.*

147 Wilkinson, *Rounding Errors in Algebraic Processes*, Prentice-Hall, Englewood Cliffs, NJ, 1963

148 Whittaker, E.T., Robinson, G., *The Calculus of Observations*, Blackie, London, 1924

149 Brouwer, D., Clemence, M., *Methods of Celestial Mechanics*, Academic Press, New York, NY, Chapters 8 and 9, 1961

150 Levine, N., 'Increasing the flexibility of recursive least squares data smoothing', *Bell Systems Technical Journal*, 1961, vol. 40, pp. 821–40

151 Claus, A.J., Blackman, R.B., Halline, E.G., Ridgeway, W.C., III, 'Orbit determination and prediction, and computer programs, (Telstar)', *Bell Systems Technical Journal*, July 1963, vol. 42, pp. 1357–82

152 Kalman filter. Available at http://en.wikipedia.org/wiki/ Kalman_filter

153 Pentzer, J., Armstrong, B., Bean, T., Anderson, M., Edwards, D., Schmehl, V., 'Preventing extended Kalman filter instabilities during two transponder long baseline navigation with real time fuzzy logic parameter adjustment', Center for Intelligent Systems Research, University of Idaho, Moscow, ID

154 Houtekamer, P.L., Mitchell, H.L., 'Data assimilation using an ensemble Kalman filter', Direction de la recherche en Météorologie, Atmospheric Environmental Service, Dorval, Quebec, Canada

155 Andreasen, M.M., 'Non-linear DSGE models, the central difference Kalman filter, and the mean shifted particle filter', 2008. Available at http://ftp.econ.au.dk/creates/rp/08/rp08_33.pdf

156 Masreliez, C.J., Martin, R.D., 'Robust Bayesian estimation for the linear model and robustifying the Kalman filter', *IEEE Transactions on Automatic Control*, 1977

157 The auction algorithm. Available at http://en.wikipedia.org/wiki/Auction_algorithm

158 Zhu, J., Zheng, N., Yuan, Z., Zhang, Q., Zhang, X., He, Y., Xian Jiaotong University, Xian, China, 'A SLAM algorithm based on the central difference Kalman filter', *Intelligent Vehicles Symposium*, 2009, IEEE

159 La Scala, B.F., Bitmead, R.R., James, M.R., 'Conditions for stability of the extended Kalman filters and their application to the frequency tracking problem', *Mathematics of Control, Signals and Systems*, vol. 8, no. 1, pp. 1–26

160 Anderson, B.D.O., Moore, J.B., *Optimal Filtering*, Prentice Hall, New York, NY, 1979

161 Hamilton, J., 'The Kalman filter', *Time Series Analysis*, Princeton University Press, Chapter 13, 1994

162 Bucy, R.S., Joseph, P.D., *Filtering for Stochastic Processes with Applications to Guidance*, John Wiley and Sons, 1968; 2nd edn, AMS Chelsea Publishing, 2005, ISBN 0-8218-3782-6

163 Jazwinski, A.H., *Stochastic Processes and Filtering Theory*, Academic Press, New York, NY, 1970, ISBN 0-12-381550-9

164 Simon, D., *Optimal State Estimation: Kalman, H∞ and Nonlinear Approaches*. Wiley Intersci- ence, 2006

165 Stengel, R.F., *Optimal Control and Estimation*. Dover Publications, 1994, ISBN 0-486-68200-5

166 Bozic, S.M., *Digital and Kalman Filtering*. Butterworth-Heinemann, 1994
167 Haykin, S., *Adaptive Filter Theory*, Prentice Hall, 2002
168 Moriya, N., *Primer to Kalman Filtering: A Physicist's Perspective*, Nova Science Publishers, Inc., New York, NY, 2011, ISBN 9781616683115
169 Chui, C.K., Chen, G., *Kalman Filtering with Real-Time Applications*, Springer Series in Information Sciences, 4th edn, Springer, New York, NY, 2009, ISBN 9783540878483
170 Costa, E.F., Astolfi, A., 'On the stability of the recursive Kalman filter for linear time invariant systems', *American Control Conference*, 2008, Seattle, WA, pp. 1286–91 (IEEE Xplore digital library)
171 Kalman filter instability. Available at http://www.nntpnews.info/threads/4628950-Kalman-Filter-Instability
172 Barnet, V., Lewis, T., *Outliers in Statistical Data*, 2nd edn, John Wiley and Sons, 1984, ISBN 0 471 90507 0
173 Kalman filter for dummies. Available at http://www.google.com/#hl=en&cp=17&gs_id=y&xhr=t&q=kalman+filter+for+dummies
174 Kalman filter, Wikipedia. Available at http: //en.wikipedia.org/wiki/Kalman_filter
175 Levenburg, K., 'A method for the solution of certain non-linear problems in least squares', *Quarterly of Applied Mathematics*, 1944, vol. 2, pp. 164–68
176 Hartley, H.O., 'Modification to the Newton-Raphson Method for the Fitting of Nonlinear Functions by Least Squares' *J. Chem. Phys*. 39, 1307 (1963)
177 Morrison, D.D., *Methods for nonlinear least squares and convergence proofs, Tracking Programs and Orbit Determination,* Proc. Jet Propulsion Lab Seminar, *Quarterly of Applied Mathematics*, 1944, vol. 2, pp. 164–68
178 Strang, G., *Linear Algebra and Its Applications,* 3rd edn, Harcourt, Brace, Jovanovich, San Diego, CA, pp. 153–62, ISBN 0-15-551005-3
179 Kuan, Chung-Ming. *Linear and Matrix Algebra,* Chapter 1, 2002. Available at http://www.math.upatras.gr/~vpiperig/Mul/Algebra.pdf
180 Graybill, F.A., Marsaglia, G., 'Idempotent matrices and quadratic forms in the general linear hypothesis', Oklahoma A. and M. College. Available from http://www.jstor.org/pss/2237227
181 Gauss–Markov Theorem. Available at http://en.wikipedia.org/wiki/Gauss–Markov_theorem
182 Plackett, R.L., 'Some theorems in least squares', *Biometrika*, 1950, vol. 37, pp. 149–57
183 The Homoscedastic Property. Available at http://en.wikipedia.org/wiki/Homoscedasticity
184 Flinn, C, 'Discussion of the Gauss–Markov Theorem', Available at http://www.nyu.edu/econ/user/flinnc/courses/0266/gauss-markov%20 theorem.pdf
185 Radar Tracker. Available at http://en.wikipedia.org/wiki/Radar_tracker
186 Orguner, U., 'Derivation of the IMM Filter'. Available at http://www.control.isy. liu.se/student/graduate/TargetTracking/IMMderivation.pdf

187 Farmer, M.E., Hsu, R-L., Jain, A.J., *Interacting Multiple Model (IMM) Kalman Filters for Robust High Speed Human Motion Tracking*, Department of Computer Science & Engineering, Michigan State University

188 Blom, H.A.P., Bar-Shalom, Y., 'The interacting multiple model algorithm for systems with Markovian switching coefficients', *IEEE Transactions on Automatic Control*, August 1988, vol. 33, no. 8, pp. 780–83

189 Herrero, J.G., Besada Portas, J.A., Molina López, J.M., de Miguel Vela, G., García, J.P., 'Interacting multiple model filter optimisation tool for air traffic control applications'. Available at http://isif.org/fusion/proceedings/fusion01CD/fusion/searchengine/pdf/TuB23.pdf

190 Simple harmonic motion, Wikipedia. Available at http://en.wikipedia.org/wiki/Simple_ harmonic_motion

191 Jacobi matrix, Wikipedia. Available at http://en.wikipedia.org/wiki/Jacobian_matrix_and_ determinant

192 Bell, J.W., 'Disambiguation of orthogonal projection underlying Kalman's filter'. Available at http://site.infowest.com/personal/j/jeffbell/Kalman Filter Disambiguation.pdf

193 Weighted least squares. Available at http://en.wikipedia.org/wiki/Least_squares#Weighted_least_squares

194 F-distribution, Wikipedia. Available at http://en.wikipedia.org/wiki/F-distribution

195 F-distribution, *NIST/SEMATECH e-Handbook of Statistical Methods*. Available at http://www.itl. nist.gov/div898/handbook/eda/section3/eda3665.htm

196 Bernstein polynomial. Available at http://en.wikipedia.org/wiki/Bernstein_polynomial

197 Least squares. Available at http://en.wikipedia.org/wiki/Least_squares

198 Linear regression. Available at http://en.wikipedia.org/wiki/Linear_regression

199 Booth, G.W., Peterson, T.L., *Nonlinear Estimation*, 1958, IBM Share Program, Pa. No. 687 WLNL1

200 Draper, N. R. and Smith, *Applied Regression Analysis,* 2nd edn, Wiley, New York, NY, 1981

201 Hartley, H.O., 'The Modified Gauss–Newton Method for Fitting Non linear Regression Functions by Least Squares', *Technometrics*, 1961, vol. 3, pp. 269–80

202 Reyneke, P.V., Morrison, N., Kourie, D.G., de Ridder, C., 'Polynomial Filtering: To any degree on irregularly sampled data', *Automatika*, 2012

Index

www.ingramcontent.com/pod-product-compliance
Lightning Source LLC
LaVergne TN
LVHW011857200926
848484LV00002B/464